S0-BIP-015

Under the editorship of

John E. Horrocks, The Ohio State University

Personality Development

From Infancy to Adulthood

Charles Wenar

Houghton Mifflin Company · Boston
New York · Atlanta · Geneva, Illinois · Dallas · Palo Alto

For Johnerik and Leif

Printed in the U.S.A.

Library of Congress Catalog Card Number: 76-133335

ISBN: 0-395-05527-X

Contents

Foreword

Although relatively short in terms of the entire life span, the first two decades of life are developmentally the most action-fraught of all the years given to man. During these years, a child constructs his psychological world as he builds the structure determining who and what he will be for the rest of his life. These are the years of becoming when a passionate, curious learner encounters experience and evolves a cognitive-affective system enabling him to explore, to judge, to make decisions, and to integrate all that he has experienced. Professor Wenar has taken these vital decades as the setting for the story he tells of the psychological development of the individual child through infancy, childhood, and adolescence. This is a book on the age-old question of the nature of man and of the ontogenetic origins and development of his behavior. It tells how an individual carves a self and a social personality out of the manifold of his personal experiences and interactions with others.

As a developmentalist Professor Wenar is interested in change over time and, most particularly, with those aspects of change of psychological significance. Continuously throughout the book he addresses himself to the what, the why, and the how of development. He deals with the what by describing the pace and incidents of development, and in answering the why he deals with the reasons for what happens. The how represents an exposition of the style or manner characterizing children's progression through their formative years. In answering these questions, the author starts with birth and carries the discussion through to maturity. Unlike most books covering an appreciable developmental span, Professor Wenar's text is organized in terms of topics such as values, play, or social feelings rather than in terms of the more artificial age levels. Such topical organization gives the reader a picture of process and a much better view of the progressive dynamics of development. In the final analysis, a child is a complex functioning personality rather than an example of a specific chronological age group. Individual differences and varying experiences from child to child place stringent limitations upon the value of describing the developmental process in terms of the confining limitations of chronological age. As a matter of fact, Professor Wenar's focus is upon the study of development in the individual child. As useful as the normative view may be, a child does not develop as a norm. He develops as a highly individual entity creating his own idiosyncracies as he pursues his essentially inductive approach to life, drawing inferences about himself and others from his own personal experiences. His story is his own story, not that of someone else. Although Professor Wenar does deal with normative implications, he never loses sight of an individual child who finds his own meanings in his environment and who, over time, builds himself first as a physical entity and then as a psychological entity, separate from his own specific environment, yet related to it.

Professor Wenar writes in the clinical tradition of concern for the individual, but the range of his work goes far beyond the clinical transaction to encompass all that the child is, does, and thinks. Scholarly without making a point of it, the author has selected widely from the pertinent literature and presented the reader those aspects best serving to illuminate the story of a child's coming of age. The literature, research, and theory on childhood and adolescence is immense. There is no way to include it in its entirety between the covers of one book. An author has to be highly selective but fair as he tries to seek those references for citation which will best illustrate the story he has to tell. A book on children should be about children and not clutter up the continuity of its presentation with a catalog of the extraneous inserted to enhance a scholarly reputation. Professor Wenar has steered well between the Scylla of unsupported opinion and the Charybdis of over-citation. Every study he cites helps his story and is skillfully interwoven so that it does not break the continuity of the tale. He presents two kinds of data in support of his discussion, clinical and objective, the first leading to psychological understanding and the second to the possibility of prediction and control. Freud, Adler, Piaget, and the learning theorists are frequently cited, and psychoanalytic interpretations are often made. However, Professor Wenar does not claim to identify the psychoanalytic tradition with all of clinical methodology. He tends to look to clinical studies for hypotheses and to objective studies for validating evidence. He writes, "We shall often present ideas from the clinical approach and then ask what conflicting evidence there is from objective studies." In general the slant is clinical, and the author explores pathology and deviant behavior to a greater extent than is true of most books in general development, but the material on aberrant behavior is well used to illuminate the course of normal development.

The concern of the book is with personality development in the American middle class, and contrasting development in different cultures and different socioeconomic classes is not considered. The author makes the specific point that what he says is not universal and applicable to all children wherever they may be. But, the focus of the book is on the child, and what is said is more generally applicable than the author is willing to admit. True, there are special cases and conditions, but the behavioral dynamics of the perceiving, processing, and responding human organism present a pattern that transcends culture, time, and place.

This is a developmentalist's book, kind and insightful. In it the author speaks informally, though scientifically, about children, as he would talk over lunch with his colleagues. That the author has known and worked with children as well as read about them comes through clearly. So many textbooks about children discuss them as though they were inhabitants of another culture to be examined and described impersonally, but not to be known as people. That he does not do this is the great virtue of Professor Wenar's book. He is talking about his own kind from his personal experiences with them. One gets the impression that he likes children and that he was once one himself. The discussion displays the personal warmth of the writer. Professor Wenar is concerned about children and transmits that warmth and concern to his readers who, in their turn, will have not only to work with children but also to understand them.

Who should read Professor Wenar's book? The answer is anyone interested in children. The discussion is scholarly and can stretch the imagination of the most advanced graduate student, but at the same time its warmth and fluid style will enable a parent or an undergraduate to read it with profit and pleasure. Certainly this book is not for the "happy few" who equate jargon and turgidity with scholarship. Next to having a child of one's own and observing his progression from infancy to maturity, this book will give a better understanding of the "becoming" of development than any other source. And even if one has a child of one's own, this book can offer a map to follow intelligently the sequence and unfolding of his experiences as he grows and realizes his potential. The present writer would think that every professional worker with children, no matter what his discipline or training, would not only keep it at hand but would turn to it again and again.

John E. Horrocks

Acknowledgments

My deepest gratitude goes to Dr. John E. Horrocks and Dr. George G. Thompson, in part for encouraging me to write this book, but primarily for being good colleagues. My gratitude to Dr. Ann M. Garner goes back in time and has many sources; the immediate occasion is her critical and constructive reading of many of the chapters which follow. Dr. Samuel H. Osipow was helpful in evaluating the discussion of work, and Dr. Jean Straub generously gave her time to reviewing the manuscript. Miss Margaret J. Connors and Mrs. Diane Faissler taught me that intolerance of ambiguity can be a highly prized virtue when it characterizes editors. Mrs. Clara E. Vaughn was an invaluable auxiliary ego in attending to the many details involved in preparing the manuscript. Mrs. R. Ervin Walther was impeccable in her ability both to type and to decipher.

And finally, I am grateful to Solveig for bringing her love to my labor.

1 Overview

What shall we discuss if we are to understand the development of personality? The question has as many answers as there are books on the subject. In this section we shall present our own.

The pivotal event of the first year of life is the development of the *bond of love* between infant and mother. The fruition of love comes rapidly and its expression is remarkably pure. It is the first intense human relationship, and there are those who claim that it sets the pattern for all future intimacy: if the mother-infant interaction is warm and gratifying, the child will approach other deep attachments with love and trust; if the mother-infant interaction is empty or frustrating or frightening, the child will be wary of closeness, becoming infuriated or terrified as he senses the danger of an intense emotional commitment to another person. We will take a critical look at this proposition, since the importance of mother love is being challenged by counter-claims that personality development can proceed as well without an intense attachment to a single caretaker.

The bond of love does more than set the pattern for intimacy; it becomes a critical variable in the socialization of the child. Socialization—i.e., the learning of acceptable modes and standards of behavior—often goes against the grain. The child has no natural inclination to do much of what his parents (and his culture) tell him to do; nor is socially proper behavior always intrinsically pleasurable—it is more likely to be neutral or negative, initially. The child comes to behave acceptably because he gains parental affection by doing so or jeopardizes affection if he fails. Thus, the bond of love lies at the core of parental discipline and contributes greatly to its effectiveness.

Having become attached to the mother in the first year of life, the toddler turns his back on her, literally and figuratively. He ventures out and explores the world in a spirit of self-reliant enterprise. He not only seizes the *initiative* but also willfully asserts his right to do what he wants. His fragile ego is imperious. Inevitably he is confronted with parental *No's* and *Don't's* and during the "terrible twos" his willfulness may be supplemented by a negativistic defiance of authority. More important, the lifelong process of weighing compliance against resistance begins. For their part, the parents are faced with the problem of preserving the toddler's initiative while making sure he does what they rightfully expect of him. There is no consensus among psychologists regarding how this problem should be solved, but it is instructive to examine their disagreements. Finally, we shall consider the conditions in which clinging demandingness, rebellious defiance, uncertainty, and self-doubt may develop instead of self-reliance.

Parental *No's* and *Don't's* set the stage for the development of *self-control*, since the child is increasingly expected to inhibit unacceptable behavior when the parent is not present. At one extreme the danger is that the child may become overly controlled because of an excessive fear of parental censure; at the other extreme, the danger is that he may be unable to control the expression of unacceptable impulses because of parental indulgence or laxness. We shall examine the effects of parental discipline on the kind of self-control the child develops, as well as the patterns of family life which support or undermine control. But self-control is more than conformity to parental demands; the child must learn to decide for himself what he will and will not do. *Judgment* develops as the child becomes increasingly capable of formulating his own stance and increasingly practiced in weighing alternatives and deciding a course of action. Ultimately, the young adult must become his own legislator, making new rules to serve as guides, retaining serviceable old mandates, and rejecting those which are inappropriate. Judging and legislating have been investigated most intensively under the rubric of moral judgment, which is one aspect of our next topic—moral development.

Parental behavior conveys values, explicitly or implicitly. Some are idiosyncratic to the parent or the family; others are the shared social mandates which we call moral values or morality. Values are internalized by the child and become part of his *conscience*. Because he adopts them before he can comprehend their real meaning, his early conscience is marked by rigidity, authoritarianism, and punitiveness. We shall explore the factors which enable the child to develop a conscience that will serve as a humane, flexible, rational guide to human behavior. Such a conscience involves the integration of thinking (understanding the meaning of moral values), feeling (experiencing guilt, shame, or self-esteem), and behaving (acting in a moral manner), each with its own developmental history. We shall be concerned with tracing the process of integration and with understanding the deviant kinds of consciences which result when development and integration do not follow their expected courses.

Thinking plays a special role in the achievement of self-control, since ideas perform the dual functions of delaying and guiding action. The more a child is capable of ideating, the less he is likely to be a victim of impulsive acts and the more potential he has for gearing his behavior to the parameters of reality. Understanding the nature of the physical world, the social world, and the self are tasks which occupy the child for at least the first decade of life. His initial efforts at understanding are both fantastic and reasonable—fantastic by adult standards, but reasonable in light of the information available to him and the thinking of which he is capable. Like early conscience, early ideating must be modified and transformed. Mature thinking can be achieved if all goes well with the child's overall biological and psychological development, but he is not fated to be rational. We shall pay particular attention to the powerful emotional forces which make the child cling to, or return to, the fantastic ideation of early childhood and to the distorted interpretations of reality which result.

If thinking plays a special role in the development of self-control, so do *impulses*. Not all of them are equally important in determining personality characteristics. Sex and aggression are crucial in deciding whether the child will be on good terms with his impulse life or will be terrified or overwhelmed by it. Anxiety generated by the socialization of sex and aggression can either facilitate the attainment of control or become a terrifying force which itself must be controlled, typically through mechanisms of defense. Although there is agreement about the importance of sex, aggression, and anxiety, each has been conceptualized in a variety of ways. Sex, for example, has been defined as the learning of appropriate roles; that is, learning the behaviors and feelings appropriate to being a boy or a girl in a given society. But sex has also been defined as an imperious drive for pleasure, Freud's concept of libido and psychosexual development being the prime example of sex-as-drive. As we shall see, each definition makes a special contribution to our understanding of the complex variable being examined.

Thus far, we have been concerned with love, initiative, and various aspects of self-control. Personality development includes more. *Peer relations* are as important in their own way as the relations between parent and child. Peers share the same world of fantasy, play, sports, and social events, and they participate in a special children's culture handed down from generation to generation independent of adults. Peer relations revolve around *esteem* rather than love, which is the central affect in the parent-child relationship, and peers can best foster the capacity of mutuality which is essential to mature social interactions. The intimacy of friendship and the group's relatively impersonal demands for conformity offer the child comfort and security, but they also jolt him into facing aspects of himself that he was not cognizant of or was trying to avoid. Finally, peer relations come to be the child's principal testing ground for participating in the adult world, which will be largely a world of contemporaries.

We have touched on the role of thinking in enabling the child to understand social and physical reality. *Play*, in its early manifestations, is free of socialization pressures and largely free of the constraints of physical reality. The child can go anywhere his ideas and feelings lead him. We shall trace the development of play from the infant's pleasure in sheer doing, through the preschooler's luxurious fantasy play, to the daydreams and the organized games of middle childhood. We shall see that play is the natural vehicle for the preschooler's expression of his fleeting interests and his deepest concerns. It also serves as one of the main bridges between the preschooler and his peers. As play becomes more organized and more governed by rules in middle childhood, fantasy and daydreams take over as expressions of personal feelings. Concomitantly, the very structure, organization, and social nature of games makes them an appropriate setting for studying the child's ability to accept discipline and mutuality while achieving a balance between impulsiveness and control.

As play is central through middle childhood, *work* is central from middle childhood into adulthood. The forerunner of work is initiative, that intrinsically rewarding, self-reliant exploration which is so evident in the toddler. But the toddler is free to explore as he wishes; for the school child, the environment dictates activities and levels of achievement, and free choice is replaced by required tasks. Initiative takes on an element of being "hard" or disagreeable and begins to approximate the adult notion of work. The adolescent is future oriented, and finding a proper vocation figures prominently among his concerns about his adult life. He must engage in a complicated process of assessing himself and evaluating the vocational possibilities afforded by adult society, and then he must try to harmonize the two. If he is successful, he re-experiences the sense of congruent expansiveness he had as a toddler. More is at stake than just finding a good job; the adolescent is seeking an identity in the adult world he is about to enter.

Heredity and the prenatal history of the neonate undoubtedly affect personality, but we have chosen to be highly selective in our discussion of their influence. We shall begin with a generally accepted statement: behavior is the result of heredity and environment. Then we shall present three studies which concretely illustrate certain of the empirical, methodological, and conceptual ramifications of that statement. We shall see how exceedingly complex the parent-child relationship becomes when hereditary factors are included; e.g., we shall learn that it is just as plausible to contend that good children make good parents as it is to contend that good parents make good children.

Then we shall turn to that prime product of heredity, the human body, whose external contours and inner workings have far-reaching effects on personality development. While the *visible body* (the body from the skin out) is relatively unimportant in the preschool years, it becomes increasingly important until a climax is reached in adolescence, at which time having an acceptable body is closely akin to having an acceptable personality. Our study of physically handicapped children will show that the visible body is a potent social stimulus; lack of physical intactness can elicit special attitudes from parents and peers which may affect significantly the child's evaluation of himself. The *inner body* (the body inside the skin) is as intimate as the visible body is public. It is the source of intense drives, such as hunger, and of intense affect, such as sexual excitement and anger. Being on good terms with the inner body is an important element in the child's overall sense of well-being. Feeling alienated, on the other hand, places a special psychological burden on him. The body then becomes an inescapable enemy diverting the child's attention and energy away from mastering developmental challenges. Thus, the visible body and the inner body furnish important frameworks for understanding the development of the self and of self-confidence. Learning to interpret the language of the child's body is an essential element in learning to understand his personality.

Authors not only select the topics they discuss; they also select the approach they take to understanding them. Our chosen topics could have been approached through a cross-cultural frame of reference in which the effects of different cultural patterns on personality development were examined. They could have been discussed in terms of socioeconomic classes within a given culture, such as the upper, middle, and lower classes of American society. Or the family itself could have been taken as the basic unit of study in order to determine the effects of various family patterns on the child's developing personality. All of these approaches are valid, but none of them is used here. Our focus is on the *individual child*. We shall try to understand the psychological world of the infant, the school-aged child, and the adolescent, and how his various experiences with parents and peers, with the physical world, and with his body enhance or impede his personality development.

Our general picture of the child can be inferred from the topics we have chosen to consider. He is passionate: the classical affects of love, hate, and fear are manifested with remarkable clarity by two years of age. He is curious: the magnitude of what he must learn about people, objects, and himself is more than matched by the innate drive to explore and to know. He is a learner: at times he proceeds slowly by trial-and-error, at times by sudden insight, at times by imitation. Through learning, parental standards of acceptable behavior become his own—that is, he becomes socialized. But the child does more than conform to his environment; he is appraising in the sense of evaluating situations, and he is calculating in the sense of weighing outcomes. He has his own vested interests—pleasures he wants to attain, activities he wants to pursue—which serve as a counterforce to conformity. He is a decision maker, acting crudely at first, but ultimately with great sophistication and subtlety. He continually strives to integrate his developing self with the changing requirements of family, peers, and society in general, and he learns to arbitrate the conflicts which develop within his own personality.

Passionate, exploring, learning, appraising, deciding, integrating—these are the basic characteristics of the child as we shall view him.

Questions Concerning Substantive Areas

In our discussion of the topics we have outlined, we shall continually raise the questions, *What*?, *Why*?, and *How*?

The question, *What*?, will be answered with descriptive data. Stated more fully, the question becomes, What is the developmental picture of a particular area, such as affection or play, from infancy to early adulthood? By describing what happens, we learn what behaviors must be accounted for, what facts must be explained. But description will also teach us something about development. As we shall see, development is rarely only a quantitative change, a simple matter of "more of the same." Physically, the infant is not a miniature adult who merely must increase in size. Psychologically, the infant is not equipped with minute amounts of adult understanding, morality, sociability, and so on, which need only to expand. Early manifestations can bear only a vague resemblance to adult forms, and the developmental process can include qualitative as well as quantitative changes. Sexual development, for instance, is not a matter of an infant becoming increasingly enamored of a peer of the opposite sex until he is ready to take a mate. Nor is moral development a matter of the toddler having a basic understanding of right and wrong which he applies to an ever-widening circle of experiences. Adult sexuality and morality are points in a series of changes and transformations. Moreover, there is no guarantee that every child will reach this point. Descriptive data will raise the question of why some children progress through all the stages to adulthood and others do not.

Descriptive data also tell us about the pace of development. Normal development is never a steady, regular progress toward maturity. Rather, advances are followed by regressions. To use the common image, development follows a pattern of two steps forward, one step back; periods of integration and rapid progress alternate with periods of instability and upset during which the child is apt to return temporarily to immature modes of behavior. In addition, the normal child will not make equal progress in all areas of development. More than likely, he will spurt ahead in some, lag in others; for instance, he may advance intellectually but reach a temporary impasse in his social relations. The image of the child who is well-rounded at every stage of development is an ideal which rarely corresponds to the reality of normal development.

The heart of our discussion will consist of answering the question, *Why*? Our goal here is to explain the descriptive data of development. *Why*? is difficult to answer because behavior, even when it is easy to describe, is often the result of a number of variables, each of which needs to be understood. Take the example of self-control; the phenomenon itself could not appear simpler. A mother repeatedly says *No*! and slaps the toddler's hand as he reaches for a vase. Gradually the toddler learns to say *No* to himself and avoids the vase even when the mother is not present. We say he has learned to control himself. But this achievement is contingent upon a variety of factors: the bond of affection between parent and child; the child's ability to understand and remember what the mother says and does; his ability to grasp antecedent-consequent relations so that he can connect punishment (consequence) with reaching for the vase (antecedent); his ability to use past experience to guide present behavior; the utilization of covert speech to control motor activity (so that the toddler can say *No* to himself); the complex consequences of reward and punishment; and the equally complex dynamics of imitation (as when the toddler slaps his own hand in imitation of the mother). If we could fully understand all the variables which enabled the toddler to avoid the vase, we would understand many of the major determinants of personality itself.

But there is a further complication in our search for explanations of developmental data; namely, the disagreement among psychologists concerning which variables are the necessary and sufficient ones to account for behavior. Freud, after observing sexual behavior, postulated a pleasure seeking instinct to account for his observations; social role theorists, after observing sexual behavior, postulated the learning of socially appropriate behavior and feelings. An explanation in terms of instinct is obviously different from an explanation in terms of social learning. We shall present different explanations on the grounds that each sheds some light on the behavior being discussed. In juxtaposing them, we shall try to determine whether the difference is merely one of terminology, and, if not, whether the explanations are complementary or contradictory. At times, psychologists say essentially the same thing but use different labels. Some explanations differ but are equally valid; e.g., the Freudian theorist and the social role theorist may be addressing themselves to different facets of sexuality. In other instances we shall be forced to conclude that different explanations are at loggerheads.

After presenting an explanation, we shall examine the empirical evidence which can be marshalled to support it. We shall also use empirical evidence to help decide the relative merits of contradictory explanations. At times we shall be able to make a reasonable choice among them; more often we shall be left with divergent viewpoints and inconclusive data. In general, we must be prepared to find that there is no definitive answer to many of the questions we raise concerning personality development.

One final caution: empirical evidence will be presented uncritically, but it should not be accepted uncritically. Too often we will have to rely on the impressions of a single experienced clinician or on a small number of studies which are far from definitive. Rarely shall we leave an issue with a feeling that it has been settled. More often, we shall be impressed with the amount of work yet to be done.

How? is a less pressing question, but one which will arise continually. It refers to the style or manner of behaving. The *What?* of behavior may be the same, while the *How?* differs. Two ten-year-olds may be equally skilled at sprinting, yet one approaches the race with high spirits and bravado while the other is abstracted and aloof. Children have different styles of thinking, of being affectionate, of coping, of controlling their feelings, which mark their behavior as distinctly their own. The question of *How?* is minor only in the sense that it has failed to engage the interest of a majority of investigators of personality development and therefore has generated a smaller body of research. It has considerable importance in understanding personality itself.

The Objective and Clinical Approaches

In all of our discussions, we shall utilize ideas and evidence from two investigative approaches. The first we shall label "objective"[1] and the second "clinical." The labels are idiosyncratic and are not drawn from the literature on scientific method. In fact, we shall not be concerned with the formal characteristics of scientific inquiry, since this topic is covered in books on general child development (see Baldwin, 1960, and Thompson, 1962). Instead, we should like to explore the characteristics of the two approaches which have been most fruitful in providing insights into personality. The approaches differ with regard to the problems they have explored, the populations they have studied, and the procedures they have used. Such differences are not systematic, so we cannot compare the approaches at every point. There are common features as well as contrasting ones, both of which have been determined as much by history as by principle. By labeling one approach "objective," we do not imply that

[1]"Experimental" might seem a better term to use, but many of the studies we shall cite are not experimental in the sense of holding all variables constant except one which is then systematically varied while its effects on behavior are observed. For example, consider a study which shows that punitive parents have defiant children, while reasoning parents have cooperative ones. The study is dealing with different kinds of parental discipline rather than systematically varying a given discipline; as a parent becomes less punitive, he does not necessarily become more reasoning.

the clinical approach is subjective and intuitive; we are dealing with differences, not antitheses. The approaches themselves are exemplified by two of the giants of modern psychology: Pavlov and Freud, and by the behavioristic-learning tradition and the psychoanalytic tradition.

The *objective approach* values precise, operationally defined terms and aptly designed, explicitly delineated tests of hypotheses.

The objective investigator is constantly raising the question, What exactly is meant by a given term? Clarity is particularly difficult to achieve in personality research because critical terms tend to be global and vague—adjustment, hostility, values, creativity, identification, body image—in fact, it would be difficult to find a term whose definition is generally agreed upon. Since the terms often are drawn from the language of the layman, they also have a deceptive aura of being well understood. *Everyone knows* what we mean by anxiety, or mothering, or sexual relations. But in reality this is not true. Perhaps everyone knows enough to converse adequately, but the meanings used are apt to be imprecise and shifting. When the same label has different meanings, the likelihood of pseudocommunication and pseudocontroversies runs high. Contradictions and confusions are revealed only as investigators challenge themselves and one another to be explicit. The challenge itself marks a significant advance in scientific thinking over the illusion of clarity.

The objective approach stresses the importance of operationalizing terms by means of observable, public behavior. The investigator states which behaviors will be regarded as indicators of the term he has defined, and such behaviors can be viewed by all members of the scientific community. Again, personality research presents a vexing problem. Conceptually rich terms run the risk of shrinkage when translated into observable behavior. Level of moral development, for example, can be operationally defined as a score achieved on a questionnaire concerning moral issues; however, this definition taps only intellectual understanding of right and wrong, and omits the elements of action (actually doing the right or wrong thing) and affect (feeling guilty after a transgression) which are also part of moral development. Since conceptual complexity militates against a simple translation, the objective investigator must be mindful of the limitations of his data. In actuality, even simple terms, such as weaning and toilet training, have been defined by different behavioral criteria, resulting in confusion in the research literature.

Having precisely defined and operationalized terms, the investigator is then required to state explicitly the relationship to be studied. What is the hypothesized connection between X and Y—between good mothering and preschool social adjustment, or between parental harmony and delinquency in adolescence? When their study is completed, objective investigators are particularly alert against the possibility of inferring a spurious relationship. Studies show that children reared in orphanages are intellectually dull. Can we conclude that orphanages have a dulling effect on children? Not necessarily. Infants sent there may have been intellectually inferior to the general population of infants, or the brighter ones may have been adopted more readily and left the orphanage earlier than the duller ones. Or again, a mother tells her eight-year-old boy, "If you keep gulping your food, you will have an ulcer by the time you are forty." When he is a man of forty, he has an ulcer. Was the mother right? We will never know. Gulping food might have been the cause, but so might the boy's unexpressed anger at his mother's nagging. The objective investigator develops the kind of critical intellect which seeks out contaminating factors in a study and challenges a given interpretation of results with plausible alternatives. His knowledge of research design serves as a powerful check on the tendency to draw unwarranted and wishful conclusions. Ideally he accepts a relationship only after all reasonable alternatives have been eliminated. Definitive research, like good drama, has a quality of inevitability.

The classical experiment, in which a single factor is systematically varied and its effects on behavior are observed while all other relevant factors remain constant, is held in high esteem by objective investigators. It is regarded as the design of choice for obtaining precise, definitive data. While much research on personality development does not utilize the experimental method in a laboratory setting, the basic features of control and systematic variation of a single factor are often retained. For example, in studying the effects of parental punitiveness on autonomy in children, two groups may be equated for age, intelligence, and socioeconomic status (i.e., these factors are held constant), while the autonomy of children of highly punitive parents is compared with the autonomy of children whose parents are low in punitiveness (this being the factor which is varied).

The objective approach insists on explicit delineation of methodology. The age, sex, and intelligence of the subjects, a detailed account of the instruments and apparatus involved, and a sequential description of what was done with the subjects, are a few of the specifications which are often given. This delineation of methodology is the most obvious feature of journals devoted to research. It is the standard procedure of objective investigation and, as such, it performs the crucial function of placing a study in the public domain of the scientific community. In essence, the investigator is saying, "I will describe what I did so clearly and in such detail that, if you wish, you can replicate my study and see for yourself." One reason he prefers standardized techniques and instruments is that they facilitate replication by others. Inspiration, insight, and ingenuity are in the investigator. They lie beyond rules and procedures. But the testing of the merit of the investigator's ideas must be open to the scientific community. Reliance upon his authority alone is not permitted.

The *clinical approach* will be discussed first in terms of Freud and those investigators who modeled their procedures after his. Followers of the *psychoanalytic tradition* (Adler, Sullivan, Anna Freud, Erikson, Jacobson) will be referred to frequently in our discussions of personality development, so it will be helpful to understand the procedures they utilized and the kinds of evidence they relied upon. However, we do not intend to identify "the clinical approach" with "the Freudian tradition," and, in the next section we will present the former in more general terms.

The psychoanalytically oriented clinical investigator has been indifferent to the laboratory experiment, which he regards as focusing on too narrow a range of behavior. Nor does he typically set out to manipulate a single variable while holding others constant. He prefers to grapple with a variety of factors which are in continuous interaction in the hope of coming closer to life as it is. He has gravitated toward the tradition of naturalistic observation. The behavior he observes is not completely natural, since his subjects are typically in psychotherapeutic sessions or in institutional settings, but it is closer to life than behavior in a laboratory. The empirical core of the psychoanalytic approach consists of accurate observation and recording of human behavior in these special therapeutic settings. Accuracy of observation is difficult to achieve and demands discipline and skill. At times, the observer is blinded by special taboos, anxieties, and prejudices; e.g., before Freud, sexual behavior in children was literally unthinkable, so people either did not see it or did not think about it if it were seen. More typically, the observer's insensitivity must be overcome by training, experience, and, most important, a desire to learn through observing; e.g., all Orientals look alike until one is motivated to understand them, or infants may seem to be doing nothing but crying and squirming until one is motivated to trace their emotional or cognitive development. Since child development has had a skimpy phase of naturalistic study, there is much still to be learned from looking (see Shakow, 1953).

But the psychoanalytic investigator is more than a sensitive observer. Like his objective counterpart, he conceptualizes, hypothesizes, tests, and revises with a similar concern for clarity of definition and definitiveness of relationships among variables. Freud himself provided the model. His concern with accuracy of conceptualization was evidenced in the major revisions he made regarding the nature of anxiety and hostility (see Chapter 6). His hypothesis testing was evidenced by his abandoning of his initial idea that hysterical symptoms were due to an actual sexual trauma in childhood when subsequent evidence indicated that a traumatic event could not have taken place. Until his death, Freud was expanding and amending his theory as observations of additional disturbed adults provided new data. Consequently there is no "Freudian theory" in the sense of a final accounting for the development of normal and pathological personality.

In spite of the overlap just described, the psychoanalytic tradition differs markedly from the objective approach in a number of ways. It differs most obviously in its concern with understanding deviant behavior. It arose from the study of emotional disturbance and has continued to focus upon deviancy. This point should be kept in mind, since the psychoanalyst's picture of normal development is largely an extrapolation from nonnormal populations. In addition, his data have not been as public as those of the objective investigator, because, not infrequently, they have been gathered in the context of psychotherapeutic procedures. Freud used psychoanalysis, for example, both to obtain information about the etiology of adult disturbances and to help his patients master their emotional problems. Patients have an ethical right to privacy, and clinicians are sensitive about exposing intimate thoughts and feelings to the general scientific community. As it turned out, the early psychotherapists were overly cautious. Patients frequently give their consent to having therapeutic sessions recorded, and with proper safeguards concerning identity, the data can be made available to qualified professionals. However, investigators are still concerned when a subject is a patient as well. In short, free access to data can never be assumed.

Psychoanalytic investigators have been as casual about methodology as objective investigators have been self-conscious about it. As we have seen, the objective investigator utilizes standardized techniques and instruments and explicitly delineates his procedures so that the reliability of his finding can be assessed by other researchers. Psychoanalytic research often employs psychotherapeutic techniques or intensive interviews, which furnish only the most general guides concerning procedure. Consequently, the probability of confounding data with the method of obtaining it runs high. To illustrate: Clinician A interviews mothers of asthmatic children and finds them possessive and hostile. Clinician B interviews another group of mothers of asthmatic children and fails to find possessiveness and hostility. Has B failed to confirm A's findings, or has he failed to conduct his interview in a comparable manner? We shall never know as long as we have no way of comparing the interviewing procedures themselves.

But psychoanalytic investigators are not necessarily concerned about the vagueness of their methodology and the paucity of standardized techniques—or maybe we should say they are willing to put up with them as necessary evils. They frequently are looking for intimate and disturbing ideas and feelings, just the kind of information a person would not or could not give readily. Standardization limits the investigator's freedom to modify a technique to fit the personality of the subject, to follow leads as they develop, to overcome a subject's tendency to conceal information or to say what is socially appropriate rather than what is personally meaningful. In psychoanalytic investigation, it is more important to individualize procedures for the possible gain in richer data than to standardize them at the risk of obtaining less personal information.

The psychoanalytic approach tends to be intensive. An analyst thinks nothing of seeing a child five times a week for a three-year period in order to understand the etiology and nature of his problem. The objective investigator typically is satisfied with a few contacts of an hour or so. In addition, the psychoanalytic approach readily addresses itself to the complexity of human behavior. After a series of intensive interviews, the investigator does not flinch from giving a detailed account of the subject's personality and the formative psychological events in his life. He is trained to evaluate a host of variables and their intricate patterns of interaction. His model is different from the experimental one of isolating a single variable at a time, studying it intensively, then learning how it interacts with another and yet another variable until complicated interrelations are ultimately revealed. The psychoanalytic investigator *starts* at a level of complexity which lies on the distant horizon for the objective investigator. But he pays a price for his scope. As complexity increases, the chances of arriving at definitive relationships decreases, particularly when there is no research design to serve as a check on confounding variables and spurious relationships. The psychoanalyst's data lend themselves to a variety of interpretations, and little is known as to why he arrives at one interpretation rather than another, why he chooses to emphasize one piece of information and minimize another, or connects one event and not another. His accounting for his data is often a matter of intuition and insight. Recall that intuition and insight have a place in objective research, and neither the objective investigator's explicit methodology nor his program of going from simple to complex phenomena can be substituted for them. Our point here is simply that the psychoanalytic approach places a greater reliance on the investigator's ability to have valid insights into his data, while exposing him to a wider array of potential sources of error in the data themselves.

We shall now describe the *clinical approach* in general and compare it with the objective approach. Both are concerned with understanding personality development. Both have rules for the conduct of an investigation, but both prize the ingenuity, the insight, and the occasional inspiration of the investigator. Both rely on the time-honored processes of observing, hypothesizing, and testing. Both continually scrutinize their terms, constructs, and theories in order to make them internally more consistent, more comprehensive, or more congruent with new data.

Frequently, the clinical and objective approaches represent different emphases rather than polarities. Keeping this in mind, one can make certain distinctions. The goals of the two approaches are different, at least as expressed in their official stances. The objective investigator is concerned with the prediction and control of behavior. He asks, What variables in what combination will produce the behavior I am interested in? If he is studying agression in children, for example, he would like to make a statement of this form: Given parents with characteristic X who reinforce behavior Y in their children while serving as a model for behavior Z, a large number of aggressive responses in the children is highly probable. The goal of the clinical approach is psychological understanding in the broad sense of the term. To take an illustration from our discussion of play (see Chapter 9): Redl became interested in children's games and set out to understand them in terms of psychological variables. He was particularly struck by the ways in which games excite the child while providing controls so that his excitement does not get out of hand. In addition to this excitement-control factor, he isolated other psychological variables, such as the amount of choice available to the individual, the power of one participant over others, the possibility of exchanging roles. In all of his analyses, his general goal was to arrive at an understanding of the psychology of playing a game. He had no interest in predicting how the variables he isolated would combine to produce one kind of behavior rather than another. The goals of the objective and the clinical investigator are not mutually exclusive, obviously; the objective investigator must understand in order to know what variables are important, and the clinician can eventually make statements which lead to prediction and control. However, a clinician may spend his professional life making only postdictions or reconstructions of data already obtained, while the typical objective investigator repeatedly challenges himself to predict an outcome.

A number of contrasts can be drawn between the sharply focused, precise, objective approach, and the broadgauged, general, clinical approach. In the objective approach, terms are precisely defined and refer to specific behaviors. In the clinical approach, terms tend to be general and comparatively imprecise, reflecting the investigator's interest in a broad range of behaviors. To illustrate: Erikson, the clinician, speaks of parental discipline as being "meaningless" to the child, a term which is global and covers a wide range of behaviors, feelings, and attitudes. Erikson feels no compunction to spell out exactly what is involved in "meaningless" discipline. Becker, in reviewing objective studies, deals with specific techniques of discipline, such as spanking and reasoning, and specific affects, such as warmth and hostility. Each study focuses on a limited aspect of the discipline and defines its variables explicitly (see Chapter 4).

In the objective approach, the classical experiment tends to be the design of choice and, when it is not suitable, as many of its features are preserved as possible. The clinical investigator has more options when it comes to the actual conduct of his investigation than does his objective counterpart. Some accept the psychoanalytic model and are unconcerned with criticism from outside this school. Others are well informed about the objective approach and try to meet its requirements. They may decide not to—for example, they may forego the higher reliability of a standard questionnaire for the flexibility of a less reliable, unstructured interview—but they know the price they are paying for the advantages they gain. To put the matter another way, the question in objective research is: How well did the investigator obey the rules in conducting an objective study? In clinical research another question must be asked first: To what extent did the investigator accept the rules for conducting an objective study?

A few additional contrasts can be mentioned, although they are less important. The objective investigator is explicitly concerned with the generality of his findings and therefore pays particular attention to the issue of sampling the population he is studying. The clinical investigator tends to be casual about sampling and adds new cases unsystematically rather than with representativeness in mind. (See Chapter 10, p. 392, for an example of the problem of generalizing from a study of a few cases.) The objective investigator tends to use a large number of subjects, to spend a limited time with each, and to be relatively unconcerned about the individual child except to secure his cooperation and protect him from distress. He also tends to study normal children and normal development. The clinical investigator is more attracted to the intensive study of a limited number of children, and he has a special interest in psychopathology and deviant development.

Having described the objective and clinical approaches, we would like to make three comments. First, nothing we have said should be taken to imply that one approach is "more scientific" than another. We are not concerned with a formal discussion of the nature of science; we only want to describe the investigative traditions which have significantly advanced our understanding of personality. Second, there are a number of investigators who fit neither pattern. Piaget is the foremost example, since he uses a method of gathering data which has some of the formal properties of the objective approach but which also has the freedom to improvise of the clinical interview. Finally, our concern with traditions does not imply that each approach will be used in the future as it has been in the past. Current clinical investigators utilize data which are available to the scientific community; they are explicit in describing procedures, are cognizant of research design, and employ sophisticated strategies. Concomitantly, advances in research design and in the technology of data analysis have enabled the objective investigator to deal with complex segments of behavior and to reduce the shrinkage involved in translating concepts into behavior.

In drawing on the two traditions in our subsequent discussions, we often shall present ideas from the clinical approach and then ask what validating evidence there is from objective studies. This procedure is more the result of historical developments than a matter of principle. The clinical approach, with its individualized, intensive, flexible, complex procedures, has produced a majority of the seminal ideas concerning personality development. It has been the "variable-discovering" approach. The objective approach has been strong on methodology and procedures but has contributed less to understanding the substantive issues which will concern us. In a general way, then, it makes sense to look to the clinician for hypotheses and to the objective investigator for validating evidence.

The Limitations on Our Discussions

Selectivity inevitably limits. Our choice of topics necessarily excludes some topics which psychologists regard as important; our focus on the individual child precludes other foci. We will not discuss personality development in different societies or in different classes within a society. We will be concerned primarily with middle class American children. The choice was difficult because the study of different cultures and different classes provides an invaluable perspective on personality. It serves as a check on the tendency to regard the familiar as basic to human nature. The important consequence of our decision is that nothing we say can be regarded as universal and applicable to all children. Nor will our discussion contain any clues regarding what is universal and what is specific to the American middle class. The issue is too knotty and too important to be treated casually.

To compensate for this loss of scope, we shall explore psychopathology and deviant development more extensively than is commonly done. Psychopathological conditions highlight certain aspects of personality development which might be overlooked or underestimated if one studies only normal children. We shall not be concerned with psychopathology in its own right; rather, we shall use it to enrich our understanding of personality development in all children.

Other limitations have less serious consequences. We shall be selective in our choice of theories and research in answering the question, *Why*? Often there will be more to a theory than we shall present and more explanations and research concerning a given topic. To compensate, we shall continually refer to more comprehensive accounts of theories, more exhaustive presentations of various explanations, and more thorough reviews of the research literature.[2]

[2]Freud, Adler, Piaget, and learning theories will frequently appear in our discussion. Sullivan and the gestalt tradition as exemplified by Werner and Lewin will also be utilized selectively. For a general background, see Baldwin (1967). For more extensive presentations of the psychoanalytic tradition, see Munroe (1955) and Brenner (1955). Piaget's theory is discussed in detail in Flavell (1963).

Nomenclature for Age Groupings. There is no standard nomenclature for the various age groupings between infancy and adulthood. We have adopted the following designations:

Neonate—Newborn, including the first two weeks of life.
Infant—First year of life, exclusive of the neonate period.
Toddler—1 to 2½ years.
Preschooler—2½ to 6 years.
Middle child—6 to 11 years.
Preadolescent—11 to 13 years.
Adolescent—13 to 20 years.
Young adult—20 to 22 years.

Our discussions generally will end with the adolescent period although, in certain instances, we shall briefly explore aspects of the young adult's personality. When we use the term "adult" or "mature," then, we will mean behavior appropriate to the twenty- to twenty-two-year old. Unless otherwise specified, maturity will not have the usual meaning of the fully developed personality. The young adult is still inexperienced as a mate, a worker, a parent, a citizen. Too much of his development lies ahead for us to treat him as mature or to extrapolate from his present status to his fully developed one.[3]

[3]Much of the illustrative and anecdotal material in the text is taken from the author's unpublished research and his clinical experience.

2 Attachment and Affection

We shall start with love, for love is as important at six months as it is at sixteen years or at sixty. In infancy the bond of love is between infant and caretaker, typically the mother. We shall be concerned with the origins of this bond and we shall see that its formation is only partially understood. In exploring the origins of love, we shall also arrive at a better understanding of the nature of mothering itself. Next, we shall trace the developmental course of love; as early as the second year of life, the basic delights and distresses of love can be seen clearly, while subsequent periods bring their special mixture of gratification, jealousy, and anger. Then, we shall consider the consequences of being unloved or of being loved ambivalently. We shall raise the questions of whether love is, as some authorities claim, a matter of life or death, psychological health or psychosis, and whether caretaking should be done by a single individual rather than dispersed among many.

Normal Development

The Formation of Attachment

Before beginning our descriptive account of the formation of attachment, we must make a general point about the neonate. The neonate has no knowledge of the world of people or the world of things; he even lacks knowledge of himself. He does not know who his mother is; in fact, he does not know what people are or how people differ from a rattle or, for that matter, from a stomachache. His experience consists of sensations which vary in quality, intensity, and duration. Out of this primitive stuff he will slowly evolve an understanding of the animate and inanimate worlds, and will grasp the nature and distinctiveness of each. He will also come to see himself as a separate and special entity. How the infant learns about himself and his social and physical environment will be discussed in Chapter 6. What concerns us here is the newborn's cognizance of and emotional investment in his caretakers. The important point to remember is that the neonate does not respond to people—he responds to stimuli. In fact, psychologists are trying to unravel the mystery of which stimuli emanating from people the neonate is capable of perceiving and how he comes to grasp the general idea of "human beings" and the specific idea of "mother." To understand the formation of attachment, therefore, we must understand both the evolution of the idea of "mother" and its ability to arouse that special constellation of affects known as love.

A second general point will help set the stage for our account of attachment. The human offspring has a longer period of helpless infancy than any other species. He must be cared for by more mature human beings for many months. At the barest minimum, the caretaker must see that the infant is fed; more ideally, conscientious and loving adults are constantly responding to him in ways which maximize his comfort and pleasure. Thus, attachment develops in the context of the caretaking situation. We shall subsequently examine in detail how the nature of this caretaking context can affect both the strength and the quality of attachment.

Earliest Responses to Humans. The development of the infant's attachment to his mother or primary caretaker follows a typical and fairly predictable course. In his fleeting moments of attention to the world around him, the neonate is no more and no less interested in his caretaker than he is in any other external object which comes his way. However, as early as two weeks of age, there is evidence that he prefers the human voice over other sounds and that by four weeks he prefers the mother's voice over other human voices (Wolff, 1963). In the second month, the eyes of another person come to be a source of special delight. At this time there is still no evidence that he clearly distinguishes his mother from other human beings or that he distinguishes human beings from inanimate objects; an oval pillow with two artificial eyes swinging back and forth elicits a smile as effectively as a nodding human face (Spitz and Wolf, 1946). Rather, what seems to happen is that those fragmentary human stimuli to which the infant is capable of responding begin slowly to form a pattern. To illustrate: the sight of a nodding face plus the sound of a human voice come to elicit a more intense positive response than either stimulus alone. We can infer that the infant has evolved the concept of "face-speaking," or, more accurately, "animated oval with two dark, round objects (face) goes with rythmic, expressive, auditory sensations (speaking)." The pattern at this point is neither so detailed nor so unique that it serves to distinguish a human being from other face-like, voice-like, animated stimuli in the environment. Suppose you were presented with the pattern 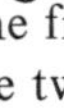 and asked, "What is it?" It could suggest a person or a monkey or a rabbit. The figure does not contain enough information for a definitive choice. Similarly, the two-month-old has grasped certain characteristics of human beings but not a sufficient number to differentiate them from similar animate objects.

The Indiscriminate Social Smile. Between the third and fourth months, the precursors of a human relationship reach a cognitive and emotional climax. The infant smiles a distinctly social smile. In form it differs from the shadowy, insubstantial smile of the neonate by being an expression of intense and sustained positive affect. It is social because now it is predominantly a response to human beings. The pattern of stimuli which signifies a human—and particularly a human face—is sufficiently cohesive and complex to be distinguished from other patterns the infant encounters. Equally important, this pattern of stimuli is a source of special delight. The idea "people" and the affect "pleasure" are fused. (For an excellent review of the data and the theories on smiling, see Freedman, 1964.)

During the stage of the social smile, the infant is indiscriminate and hedonistic. He may have his favorite people, lighting up unaccountably when a father or brother or aunt approaches; but, generally speaking, he responds to anyone who hits upon that combination of grimacing and gooing and bouncing and tickling which is his special delight. And adults prove malleable pupils, freely setting aside mature behavior in order to engage in any antic necessary to please him. We learn two things from these antics. First, they show how strongly adults are motivated to elicit a positive response from the infant and how pleased they are with their success. The simple fact that delighting the infant is highly rewarding to the adult is one of the strongest guarantees that an attachment will be formed. The interaction is mutually satisfying; it is the exact opposite of the vicious circle. Secondly, the antics show which deviations from normal adult behavior serve as the most effective stimuli for eliciting the response of delight. "Kitchy-Coo" suggests that the high-pitched voice is more effective than a conversational tone and that nonsense syllables are as effective as meaningful words. Turning to the research literature, we find that Wolff (1963) presents evidence that, as early as the second week, a feminine voice or a high-pitched masculine voice is more effective than any other sound in eliciting a smile. "Ah-boo!" suggests that the face approaching at a moderately rapid rate is more effective in eliciting a smile than a face at a set distance. Again, research by Kistiakovskaia (1965) shows that, around the second month of life, the infant frequently smiles as he watches a red ball come close to and then recede from his eyes. Thus, the fond adult provides the researcher with leads concerning the infant's positive responsiveness to stimulation. (In our subsequent discussions, this stage of indiscriminate smiling will be termed social responsiveness.)

Attachment, Separation Anguish, and Stranger Anxiety. Between the sixth and ninth months there is a fascinating and incompletely understood development. The infant ceases his indiscriminate responsiveness and becomes highly selective. He shows a decided preference for his mother (the primary caretaker). His greatest delight is in response to her stimulation and his greatest comfort is in her presence. In addition, two negative affects appear which further attest to the exclusiveness of the attachment. The first is separation anxiety, which occurs whenever the mother leaves him. The reaction ranges from mild fretting to loud, bitter crying. Although commonly called "anxiety," the label is misleading if one regards anxiety as akin to fear. (For a definition of anxiety, see Chapter 7.) The infant does not seem to be frightened. His response would be better described as anguish—that painful blend of protest and despair which wells up when a crucial source of pleasure disappears and one is helpless to bring it back. (We shall use "separation anxiety" when citing psychologists who employ this term, and "separation anguish" when presenting our own ideas. The behavior referred to is the same.)

The infant also develops stranger anxiety, which is a fear of unfamiliar people. His indiscriminate openness to pleasure, regardless of the source, is gone. The grandmother whose jostling and grimacing produced a smile a few months earlier now elicits wariness, crying, and a frightened clinging to Mother. Note that great anguish and great fear are parts of the normal development of attachment. In an unvarnished form, the one-year-old learns that openness to love brings with it a vulnerability to hurt and distress. At his own level and in his own terms, he is experiencing both the pleasures and the price of being totally committed to loving another human being. It is also no exaggeration to state that if the loss of love were not painful—if, for example, the infant were unconcerned and went about his own business—the development of personality would be radically altered.

Attachment as a Sequence of Phases. Ainsworth (1962), in her study of twenty-three Uganda infants, presents a detailed empirical account of the formation of attachment, which both reviews and enriches our picture. She distinguishes five phases: (1) The Undiscriminating Phase. The neonate is unresponsive to social stimulation except for specific postural adjustments when being held or reflex adjustments to the feeding situation. However, later in this phase, social responsiveness develops, epitomized by the indiscriminate smile. (2) The Phase of Differential Responsiveness. The baby now begins to discriminate between the mother and other people. The behavioral signs are differential crying, smiling, vocalization, and postural adjustment; e.g., the infant cries when held by someone other than the mother but stops when returned to the mother, or when crying he can be comforted more readily by the mother. (3) The Phase of Differential Responsiveness at a Distance. Here the discriminations of the previous phase can be effectively made through distance receptors, especially vision; e.g., around twenty to twenty-four weeks, the infant cries when his mother leaves the room and greets her return when seeing her approach from a distance. (4) The Phase of Active Initiative. The infant now takes the initiative in establishing, sustaining, and renewing contact and interaction with the mother; e.g., he follows her when she leaves the room, lifts his arms or claps his hands in greeting her, or clings to her. During this period (beginning roughly between twenty and forty weeks of age), the infant is somewhat reserved with strangers but not anxious. He selectively seeks interaction with other familiar figures, some infants being highly specific in their preferences. The infant also shows initiative in exploring the physical environment, although he uses the mother as a "secure base" from which to explore. (5) The Phase of Stranger Anxiety. The infant is now distressed by strangers and typically clings to the mother, although another person might serve as a "haven of safety" in the mother's absence. In addition to delineating these phases of attachment, Ainsworth claims that the process of forming an attachment is orderly, with the phases always following the above sequence; e.g., stranger anxiety would never precede differential responsiveness at a distance.

Ainsworth's study makes an important point. Attachment is a process extending over a period of months. There is no one point in time before which the infant was not attached and after which he is. The question, When is the infant attached? may have different answers depending on the criterion used. If separation anguish is the criterion, he is attached in Phase 3; if stranger anxiety is used, then he is not attached until Phase 5. An investigator may be interested in studying one phase exclusively, or he may choose to use a single behavior as evidence of attachment, but such single criteria cannot be regarded as epitomizing the sequence as a whole.

Viewing attachment as a sequence of phases also helps us to see each phase in the perspective of the entire process. Such perspective is particularly important in the case of separation anguish, which is frequently used as a criterion of attachment, and which is regarded in the clinical literature as an important variable in the etiology of psychopathological behavior. One might conclude from this emphasis that the infant has a continual and even desperate need for closeness to the mother. Note, however, that the phase characterized by separation anguish is followed by the phase of initiative in which the infant is not only active in seeking contact, but also begins to leave the mother in order to explore the physical environment. This upsurge of initiative colors all aspects of the infant's behavior and is a strong counterforce to chronic dependency (see Chapter 3). Next, the phase of initiative suggests that separation per se may not cause anguish but that an element of helplessness must also be present. The infant is distressed when he can do nothing to prevent the mother's leaving and is unable to follow her. When he has mastered his motor apparatus sufficiently to crawl after her, his anguish subsides. When we discuss play in Chapter 9, we shall return to the theme of helplessness or inability to control a situation as one antecedent of distress.

Factors Involved in the Formation of Attachment

Although the formation of attachment can be described with reasonable accuracy, the factors responsible for bringing about the observed changes are not well understood. Until recently, the formation of attachment was most frequently explained in terms of either classical conditioning or psychoanalytic theory. We shall postpone our discussion of the latter until Chapter 7 (see pp. 235-236) and concentrate on the former.

The Classical Conditioning Model. Recall that in classical conditioning a neutral stimulus, called the Conditioned Stimulus, is repeatedly paired with an Unconditioned Stimulus which regularly elicits an Unconditioned Response. Eventually the Conditioned Stimulus comes to elicit the response which is then called the Conditioned Response. A well-known example of this process is Pavlov's experiment in which he sounded a bell every time he produced salivation in a dog by giving it food. After repeated presentations of the bell and food together, the sound of the bell alone came to elicit the salivation response.

Unsentimental as the idea may be, is not classical conditioning the perfect paradigm for analyzing the development of mother love? Let us look at the situation in detail. The neonate is completely helpless and subject to frequent and intense physiological distress. The sensitive mother is especially adept at relieving his distress. She feeds him, she removes cold, wet diapers, she adjusts clothing so he will be warm enough in winter, cool enough in summer. In addition, she knows how to delight the infant by physiological stimulation such as tickling him, rocking him rhythmically, or bouncing him on her lap. In terms of our classical conditioning model, the mother is a complex stimulus who is often present when bodily distress is diminished and bodily pleasure is heightened. Through repetition of this association, the mother comes to elicit the responses of physiological relaxation and pleasure. Finally, merely being in her presence produces in the infant a feeling of well-being and delight.

Limitations of the Classical Conditioning Model. The explanation of attachment in terms of classical conditioning is certainly plausible. However, its acceptance in the past was also due, in part, to the fact that there was very little empirical evidence concerning early attachment. In effect, the explanation was never put to a test. This situation is rapidly changing, and the more we learn about the early months of life, the more we suspect that a classical conditioning model is, at best, incomplete. The model does not account for the appearance of separation anguish and stranger anxiety, nor does it explain why the bottle, which should act as a conditioned stimulus during feeding, fails to become permanently associated with relaxation. To offer a fretful child an empty plastic bottle at bedtime would be cold comfort indeed.

The most serious challenge to the classical conditioning model, however, comes from an empirical study of attachment done by Schaffer and Emerson (1964). They found that attachment was not related to the amount of physical caretaking, such as feeding, bathing, and comforting by means of cuddling or rocking. Nor was attachment related to the amount of positive physiological stimulation the infant received from the mother; the mother who kept her distance and stimulated her child through talking to him or providing him with toys fostered as intense an attachment as the mother who engaged in a good deal of physical contact. In addition, the infant could become attached to family members who had been totally uninvolved with his physical care. Such data strongly suggest that the infant can love a mother who is minimally involved in relieving physiological distress and who infrequently stimulates pleasurable body sensations by direct physical contact.

Stimulus Bias Favoring Attachment. One might argue that the classical conditioning model is not at fault; the fault lies in an overemphasis on bodily satisfactions and positive sensations, with a relative neglect of the pleasures derived from the distance receptors of vision and hearing. However, this defense does not settle the issue. In classical conditioning, the Conditioned Stimulus is initially neutral in regard to the Unconditioned Response. Yet, Walters and Parke (1965) present compelling evidence that a human being, far from being neutral, is a stimulus of special attractiveness and interest to the infant.They do not claim that the infant is automatically drawn to people as such; as we have seen, the infant initially has only a vague and incomplete concept of what "people" are. Rather, they argue that (1) infants are strongly attracted to special kinds of environmental stimuli, and (2) a human provides a richer array of such attractive stimuli than do most other objects in the environment. To illustrate their thesis, they cite current research on vision. Until comparatively recently, the neonate was regarded as being functionally blind. We now know he can see, although his range of clear vision is limited to objects eight or nine inches away (Haynes, White, and Held, 1965). (Incidentally, this finding means that the mother's face is in the area of clearest vision when the infant is being nursed. Infants have been observed to interrupt sucking at the breast in order to gaze at the mother's face in a pensive manner. This alternation of feeding and gazing provides anecdotal evidence suggesting that the need for visual stimulation can be strong enough to vie with the need for food.)

However, the neonate is not equally interested in all visual stimuli. On the contrary, he shows definite preferences for certain kinds. Complex stimuli are preferred to simple ones (Fantz, 1958), while mobility also increases the attractiveness of a stimulus (Salzen, 1963). The human face, as a face, probably means nothing to the neonate, but it has just those qualities of complexity and mobility which cause him to attend with a special intensity. Walters and Parke maintain that this purely perceptual interest is as basic to forming an attachment as the physical satisfaction of being fed. Although research data are meager, it is conceivable that the characteristics of the human voice may likewise prove especially attractive to the infant. Rheingold (1961) succinctly expresses the same view as Walters and Parke when she writes, "The human being is but another complex of stimuli, but because he is living, he is more interesting..." (p. 168).

Response Bias Favoring Attachment. While a number of investigators are exploring the role of environmental stimuli in the formation of attachment, others are concerned with the responses involved. Like Walters and Parke, Bowlby (1958) maintains that relief from physiological distress in general and sucking in particular do not provide a sufficient basis for explaining the origins of mother love. He accounts for attachment in terms of an updated instinct theory. Each species has responses which are innate and characteristic. These responses are called species-specific behaviors. Bowlby does not claim that such instinctual behaviors inevitably appear regardless of environmental or physiological factors, since extreme conditions in either realm can affect them. However, he does state that their adaptive appropriateness during evolution makes their appearance highly probable. Bowlby also does not dichotomize behavior into learned and unlearned responses; instead, he postulates a continuum with learning playing a minor role in determining a response at one extreme, and a major role in determining a response at the other. Species-specific behavior falls in the former end of the continuum. It is readily elicited and/or rapidly learned under ordinary conditions.

In man, five species-specific responses underlie attachment, according to Bowlby: sucking, visual and locomotor following, clinging, crying, and smiling. Each follows an independent course of development in earliest infancy, but all become integrated into the complex pattern of behavior called attachment. Ainsworth (1964) adds vocalization and reaching for an object as unlearned behavioral components of attachment, and the list is still regarded as incomplete. Remember that these components do not guarantee that attachment will develop; however, they are behaviors which form the basis of the developmental stages of attachment, such as those delineated by Ainsworth, and they make attachment a highly probable outcome under normal caretaking conditions. Let us take an example. An essential feature of attachment is that mother and infant are brought into close physical proximity. In fact, Bowlby claims that the evolutionary function of attachment was to protect the infant from predators, rather than to insure that he be fed. In his first few weeks, the infant can be observed to respond to anyone in his vicinity by orienting toward the person, tracking him with his eyes, reaching, and grasping. As much as his primitive response repertoire permits, he is maintaining contact. His innate crying response, while not directed toward anyone, serves as a signal to the caretaker and brings her to him; at a later date, his smile may have the effect of prolonging the caretaker's attention to him. Thus, many of the infant's innate responses conspire to make physical proximity with a caretaker a highly likely outcome. Attachment is further developed and elaborated in the context of this proximity; e.g., the infant begins to discriminate the mother and to prefer her over other caretakers, and later, when locomotion permits, he follows her when she leaves. Note that attachment is not inevitable; physiological deficiencies in the infant and/or extreme indifference on the part of the caretaker can prevent it from following its normal course. Barring such unusual circumstances, attachment is more likely than any other response, such as indifference to or fear of human beings. (For a more elaborate discussion of Bowlby's theory, see Ainsworth, 1969.)

The Pleasure of Recognition. Piaget (1952) calls attention to another factor which may be present in the early stages of forming an attachment. He maintains that, in the first few months of life, recognition of the familiar is accompanied by feelings of pleasure. Recall that everything is new for the neonate, and he must slowly piece together an understanding of the world around him. The process of developing an image of his mother takes time, and only gradually do facial features and body contours and voice and smell and dress become integrated into a distinctive unit. However, once the image of his mother has been partially formed, the actual sight of the mother and the recognition of her is an occasion for intense pleasure. Behaviorally, the infant's pleasure in recognition is expressed by the social smile. Let us make an analogy. Suppose a person were suddenly forced to live in an alien culture in which the inhabitants, the language, and the customs were totally new. What is more, he would have to figure out the culture on his own and adapt to it as best he could. After a few weeks of struggling to cope with a strange and baffling world, he suddenly sees another American. His reaction is an upsurge of joy because he can at last recognize something which has an established meaning. Our example suggests that, in adult life, recognition causes intense pleasure only under special conditions, since our world has long since become a familiar one. However for the infant, who is still in the process of understanding a largely unknown environment, the appearance of a person whom he recognizes can produce a special reaction of pleasure.

Conclusions. What can we conclude from such diverse accounts of factors responsible for the formation of attachment? That all the facts do not fit comfortably into the classical conditioning model or into any other single model? Obviously. That the field is in a state of ferment? Obviously. However, we leave the obvious only to enter the realm of uncertainty, partly because our data are too meager to support one explanation over another, and partly because theorists have only begun to discuss their differences.

Our own conclusion is that all the factors discussed might well be involved in the formation of attachment. Evolving the concept of "mother" is a cognitive achievement which itself entails a number of variables such as attention, perception, memory, and assimilation of information. This cognitive evolution is not coldly intellectual, since recognition of the mother, according to Piaget, is a source of special pleasure to the infant. (We shall treat the development of the concept of mother in more detail in Chapter 6.) There seems to be a "stimulus bias" favoring the development of attachment, in that the infant is especially interested in the kinds of complex, animated stimuli characteristic of human beings. There may be a "response bias" favoring the development of attachment, since its component behaviors are highly likely to appear in the caretaking situation. It is also possible that the order of their appearance follows a predetermined sequence. Classical conditioning may well be involved in certain facets of attachment, particularly those contingent upon the reduction of bodily distress and the production of pleasurable body sensations. To be somewhat teleological—the formation of an attachment is too crucial to personality development to depend upon a single factor, so that, for example, it would be jeopardized if a mother did not nurse the infant warmly and sensitively, or if she did not frequently talk to or play with him. We prefer to think that the formation of attachment has a wide margin for error, so that if some avenues are blocked, others can be utilized.

Let us make the same point about multiple determinants but this time focus on the emotional components of attachment. Experts disagree about the factors involved in producing the infant's affective reactions of pleasure, anguish, and anxiety. Champions of the classical conditioning model and psychoanalytic theorists alike claim that the passionate nature of attachment is due to the life-and-death importance of relieving the infant's physiological distress, particularly his hunger. They claim that visual and auditory stimulation from the mother may account for the infant's interest in her but not for the depth and strength of his emotional reactions to her. At the other extreme, Kistiakovskaia (1965) argues that visual and auditory stimuli are the principle sources of positive affect in infancy; physiological processes alone produce only the negative affect of distress or the neutral affect of comfort. Our conclusion is that such polarization of factors does not do justice to the complexity of the infant's emotional involvement with the mother. Attachment encompasses both comfort and pleasurable excitement: the mother is a source of security and her presence serves as a signal that distress will be relieved; at the same time she is an interesting and delightful stimulus because she is human, and because she stimulates the infant's body so as to arouse pleasurable sensations. Love is neither security nor excitement, it is both; the mother's caretaking caters now to one component, now to the other. The task facing psychologists is to delineate definitively the multiple sources of the multiple affects which comprise the infant's love.[1]

Having brought the caretaking mother into the discussion, we will reserve our final comment for her. We have been discussing the role played by physiological relief and pleasurable stimulation in the infant's attachment; now let us look at caretaking from the mother's viewpoint and speculate concerning her sources of gratification and anxiety. Our impression is that more of her positive and negative emotional reactions stem from physiological caretaking than from visual and auditory stimulation. Mothers who successfully breast-feed find the experience extremely gratifying. Regardless of technique, however, all mothers must feed their infants and are responsible for their general physiological well-being. A mother takes pride in the good eater and the generally healthy infant; she is concerned about the poor eater and the generally fretful, distressed one. Finally, the cultural stereotype is one of mother-as-feeder and mother-as-comforter. Thus, a host of forces conspire to make the infant's physiological well-being the source of the mother's greatest emotional investment. We can think of no comparable forces concerning visual and auditory stimulation. While an alert, responsive, inquisitive infant is undoubtedly a source of pride to the mother, the cultural image is not one of mother-as-stimulator and the responsibility for stimulation is not inescapable. If our speculation is correct, we would infer that physiological caretaking plays a more important role than sensory stimulation—not in the formation of attachment but in the balance of positive and negative affects in the mother and consequently in the infant as attachment develops.

The Nature of Mothering. The interest in understanding the process of attachment, together with the interest in the effects of maternal deprivation (which we shall discuss in a later section), have generated an inquiry into the nature of mothering itself. "Mothering" is one of those terms "everyone knows the meaning of," until he must conceptualize and operationalize it (see Chapter 1). Actually, there is no generally agreed upon definition; rather, different psychologists have emphasized different facets of the complex caretaking which mothers perform. We shall summarize the factors which have been mentioned frequently by clinical and objective investigators. Certain investigators prefer the terms "primary caretaker" and "caretaking" to "mother" and "mothering," since the biological mother is not necessarily the individual who cares for the infant. We shall use the terms interchangeably, since our only concern is with the individual who functions as caretaker, whether she is the biological mother or not.

The emphasis in classical conditioning and psychoanalytic theory traditionally has been on the mother-as-comforter who protects the infant from extreme physiological

[1]Attachment includes more variables than we have presented. Initiative (see Chapter 3) and constitutional factors (see Chapter 10) are involved. For a review and evaluation of theories of attachment, see Ainsworth (1969).

distress, especially the distress of hunger. These theorists also point to the pleasurable body sensations the mother provides the infant through her caretaking. In psychoanalytic theory, the mother's affect, particularly her warmth, is crucial; the essence of mothering lies in the feelings mobilized by caretaking.

Recently psychologists have emphasized the mother-as-stimulus, who is a fascinating, delightful, visual and auditory stimulus in her own right. The mother is also a mediator of visual and auditory stimuli, bringing the infant in contact with a variety of interesting sights and sounds, such as rattles and rings and mobiles. The mother's sensitivity and timing are important elements in her caretaking. Infants are highly individualistic in their needs and preferences for gratification; the alert mother knows how to decode the messages her infant sends and what particular interventions are most gratifying. She knows when he needs immediate attention and when he can tolerate delay; in fact, the adeptness of her timing may be more important than the sheer quantity of attention she gives. Another variable is the mother's stability over time. Just by being the same day after day she becomes a source of security and of fulfilled expectancies in the infant's kaleidoscopic world.

Mothering has also been conceptualized in terms of the mother-infant dyad, with an emphasis on the reciprocal nature of the interaction. While formerly the mother was regarded as the active agent and the infant the passive recipient, the current picture is of two human beings actively accommodating to one another in the context of the caretaking situation. The mother modifies her caretaking on the basis of its effect upon the infant, and the infant learns to adjust to the mother's special caretaking techniques. The vigorous, voracious infant makes different demands on the mother than the placid, moderate one; just as the mother who feeds by propping the bottle makes different demands on the infant than the mother who breast-feeds. What matters in caretaking are the harmonies and disharmonies engendered by the interaction of the participants. Psychoanalysts tend to stress the emotional exchange rather than the behavioral interaction. Benedek (1949) uses the term mother-child symbiosis to describe the emotional interdependence and mutual gratification of mother and infant; e.g., the contentedly nursing infant is enjoying maximal pleasure while at the same time providing maximal pleasure to his loving mother.

Now let us refer back to our discussion of attachment. Both attachment and mothering turn out to be complex patterns of behavior. The question, How does mothering affect attachment? is therefore too general to be meaningful. Instead, we must ask; How do the various components of the complex pattern labeled "mothering" affect the various components of the complex pattern labeled "attachment"? Yarrow's (1963) research on early maternal care provides one of the best models of this more sophisticated inquiry. Unfortunately for our purposes, he is concerned with general personality characteristics of the infant, but we can readily extrapolate from his design to one dealing with attachment.

Yarrow has four maternal variables: (1) need-gratification and tension-reduction (e.g., how quickly and adeptly the mother responds to the infant's expressions of needs); (2) stimulus-learning conditions (e.g., the amount and appropriateness of maternal stimulation of vocal communication and social responsiveness); (3) affective interchange (e.g. the mother's warmth, sensitivity, and individualization of the infant), and (4) consistency. The infant characteristics he studied were intelligence, handling stress, exploratory behavior, social initiative, autonomy, and adaptability. Using this design, Yarrow was able to determine which maternal variables were correlated with which infant characteristics, as well as those characteristics which were independent of the maternal behavior he studied; e.g., maternal stimulation was highly correlated with intelligence in the six-month-old infant; need-gratification and affectional interchange were correlated with the ability to handle stress; none of the maternal variables correlated with autonomy. Thus we learn that the mother's stimulation serves to advance the infant's intellectual development; her warmth and comforting make him emotionally secure, but her caretaking may be unrelated to his autonomy.

Applying Yarrow's model to attachment, we can ask which of his maternal variables will be related to the component behaviors which comprise attachment, such as preference for the mother in times of physiological distress, separation anguish, using the mother as a secure base for exploration, stranger anxiety, assuming initiative in maintaining contact with the mother. We can also anticipate our discussion of maternal deprivation by pointing out that our question there becomes, If the infant is deprived of specific components of the complex pattern of mothering, what components of the complex pattern of attachment are affected? The question is essentially the same whether one is interested in normal or deviant development.

Consequences of Attachment

Love and Trust. In psychoanalytic theory, the bond of affection between mother and infant often is considered the cornerstone of normal personality development. The argument runs as follows: Initial experiences are particularly potent models for subsequent experiences. Attachment to the mother is the infant's first significant relation to a human being and it tends to be the prototype for future human relationships. It is also a special relationship in that it is both intimate and highly charged emotionally. It develops in the context of caretaking and at those times when the infant is both extremely distressed and completely helpless. Closeness to the mother can take on a life-or-death quality—her availability and sensitivity as a caretaker make the difference between comfort and prolonged periods of helpless distress. Thus, psychoanalists write about attachment affecting "deep" and "basic" attitudes, the kinds of feelings an individual has about other people "when the chips are down," in contrast to casual or superficial social contacts. If the mother is loving, the infant comes to believe that others are loving and that they will be understanding and comforting when necessary. The child and adult will therefore feel free to establish intimate human relations. We are not necessarily talking about sexual intimacy, but about any relationship in which one can reveal his intimate thoughts and feelings without fear and know that his cry for help will be answered. In sum, as the infant develops a loving, trusting relationship to the mother, he will be inclined to approach future close relations with openness and trust (Erikson, 1950).

Psychologists also claim that sensitive caretaking is an essential element in producing general emotional security in the infant and child. The mother, in protecting the infant from prolonged periods of distress and in establishing positive expectations from closeness with a human being, is preparing the infant to take future stress in stride and to find pleasure in relationships with other people. Two objective studies support and elaborate the clinical observations. Yarrow, as we have seen, found that a close, satisfying relation with the mother was correlated with the ability to cope effectively with frustration and stress in six-month-old infants. In these early months, at least, exposure to distress does not toughen the infant but makes him more vulnerable. Murphy (1962) found that a gratifying feeding experience in infancy was correlated with a number of healthy traits in the preschool child; a sense of self-worth, a positive attitude toward others, resistance to confusion under stress, and a sound perception of reality. Both studies suggest that the well-cared-for infant has a positive orientation toward others and toward himself and an ability to take life's problems in stride.

Prerequisite for Socialization. Attachment has a practical consequence. It is a prerequisite for most future socialization. The reward of love and the threat of withdrawing love are basic techniques for redirecting the infant from asocial behavior to socially acceptable behavior. In a variety of ways parents convey the message, "If you do X, then you will receive my affection and approval, but if you do Y, you will put my affection and approval in jeopardy." Even in physical punishment the real threat to the child often lies in the parent's angry rejection rather than in the physical pain.

> Every token of love from the more powerful adult, then, has the same effect as the supply of milk had on the infant. The small child loses self-esteem when he loses love and attains it when he regains love. That is what makes children educable. They need supplies of affection so badly that they are ready to renounce other satisfactions if rewards of affection are promised or if withdrawal of affection is threatened. The promise of necessary narcissistic supplies of affection under the condition of obedience and the threat of withdrawal of these supplies if the conditions are not fulfilled are the weapons of authority (Fenichel, 1945, p. 41).

Socialization frequently goes against the grain. The growing child does not inherently want to share his toys or wear clean clothes, let alone obey the Ten Commandments. The standards of parents and of society often run counter to natural impulses and natural pleasures. Why should the child ever abandon his asocial, primitive ways? The answer is that something more compelling is at stake, namely, the pleasure of being loved. Love is also important in socialization in that it counterbalances and counteracts the resentment engendered by the barrage of *No, Don't, Wrong, Bad,* which the child inevitably faces.

Multiple Attachments, Jealousies, and Rivalries

Multiple Attachments and Intensity of Attachment. Let us return to the developmental picture of affection and concentrate on the span between, roughly, eighteen months and five years of age. Attachment involves a narrowing down, a concentration of affection on a single adult, typically the mother, although the evidence is not incontrovertible. Ainsworth (1963) claims that the mother is always the first object of the child's affection, while Schaffer and Emerson (1964), using a different criterion, found that infants may become attached first to the father or to a number of individuals rather than to the mother. Interestingly, the narrowing down inherent in attachment is soon followed by a selective broadening; e.g., it is not unusual to find and eighteen-month-old child having both an intense love for one individual and circle of favorites—not only parents but a relative or caretaker or even the teenager next door. Schaffer and Emerson have some evidence that the infant chooses these favorites not on the basis of the time they spend with him but on the basis of their ability to interact intensely and pleasurably with him. That is why a vivacious teenage neighbor may be preferred to a dour aunt who lives in the same house and has daily contact with the infant. These investigators also found that infants who relate intensely relate extensively as well; that is, they form attachments to a number of people. It is not true that a strong attachment to the mother somehow drains the infant so that he has no affection left for others; on the contrary, the infant with a strong tie to his mother is just the one who has other favorites. In like manner, the intensely stimulating mother does not bind the infant to herself exclusively, except if she actively prevents other people from caring for and interacting with him.

Sibling Rivalry. As the toddler becomes more cognizant of others, especially of members of his family, new difficulties develop in the form of jealousies and rivalries. Like separation anguish and stranger anxiety, such difficulties are expected concomitants of attachment and are not considered deviant behavior. From age three to five, the child's strongest feelings are still concentrated on the mother and, to a lesser degree, the father; if parents shift their affection or attention away from the child, he experiences anew the anguish of loss of love. Even though the child has made remarkable strides in expanding his awareness of and interest in other people, he is still possessive and selfish rather than generous and sharing when it comes to love. In addition, his time perspective is limited; a shift of affection which is only temporary from the adult's point of view may have an aura of permanence to the child. Because the child's stake in being loved is still so high, and because his social feelings and temporal sense are still weak, the best of parents have difficulty convincing him that there is enough love to go around. The child is more apt to be uncompromising. As he sees it, the love and attention given to others are taken away from him. He reacts to this perceived deprivation with jealousy and an angry desire to eliminate his rival.

The most obvious expression of jealousy is sibling rivalry. This rivalry does not always show itself the moment a little brother or sister arrives. The older child, caught up in the general excitement surrounding the new baby, dazzled by the presents, and slightly bewildered as to what has really happened, might regard the sibling as a special new plaything. The naive parent might even boast about how much the older child loves his brother or sister. But then as the weeks go by, it dawns on the child that the party is over and the baby has not gone home. The shift of attention and affection away from him was not a temporary but a permanent rearrangement. Both at home and outside the home, the baby continues to steal the show, and the older sibling smarts under his decline in status. To make matters worse, the infant is permitted just those behaviors the older child recently was forced to relinquish under the pressure of socialization—the soiling, the messing, the demandingness, the total irresponsibility. Parental praise for being a "big boy" may have a somewhat hollow ring because the temptations to babyish actions are so strong and the pleasures of socialized behavior so uncertain. The final indignity comes when the rage he expresses toward the baby is punished and only serves to alienate him further from his parents. At first the child's resentment takes direct, crudely destructive forms—hitting, throwing objects at the baby, pushing him down. As these are punished, the child develops devious techniques: he becomes adept at "accidentally" hurting the baby or feigning solicitousness after making him cry or luring him to do things for which he, the baby, will be punished, or claiming with righteous indigation, "But he hit me first." By three or four years of age, subtlety and inventiveness are remarkable.[2]

Jealousy of Parents. There is also a more serious and difficult-to-manage jealousy which is directed toward a parent rather than siblings. As we have seen, the child between two and five years of age is selfish rather than sharing with affection, and any apparent withdrawal of love produces anguish and anger. Ironically, even the love of the mother for the father can be perceived as a threat. As the child comes to learn that the mother has a special kind of sharing and intimacy which belong to the father alone, he may feel excluded and deprived of the affection which he feels should be exclusively his. The two-year-old who used to run exuberantly to his father shouting, "Daddy, Daddy, Daddy!" now, at five, begins to say to his mother, "Wouldn't it be nice if Daddy didn't come home so that we could be together all the time?"

Siblings are fair targets for hostility because they are unwanted intrusions, but anger toward a parent is a different matter. Even if the father is the less preferred parent, he is still loved, and the child's feeling toward him is a mixture of anger and affection. The general term used to describe the presence of both positive and negative feelings in relation to another person is ambivalence. The preschooler's ambivalence is particularly difficult because his anger jeopardizes the parent's love and, consequently, the primary source of his own security.

On their part, parents have their own preferences, values, and negative feelings which contribute to the child's ambivalence toward them. They may resent the investment of time and energy the child requires and the curtailing of social, personal,

[2]For an account which captures the complexity of the reaction to the birth of a sibling, see Anonymous (1949). A somewhat more objective analysis of the factors contributing to rivalry can be found in Sewell (1930). For a detailed account of how sibling rivalry can affect doll play and how some of its tensions can be relieved through doll play, see Levy (1936).

and sexual freedom; they may have their own uneasiness about the child's dependence upon them for love and may encourage him to be independent; they may have a preference for a child of the opposite sex or for a more docile or more intelligent or more physically attractive child. To take a simple example: a boy's ambivalence toward a father who is particularly proud of him will be different from his ambivalence toward a father who is critical of him. In the first instance, feelings of closeness may predominate with occasional angry flareups; in the second instance, the boy may primarily feel anger, perhaps mixed with a longing to be acceptable to the father.[3]

In sum, attachment to the caretaker is elaborated into the family triangle in the preschool years and further complicated by sibling rivalry. Clearly, affection is more complex now, but the child himself is more complex also. He is less a creature of impulse; he has both a richer repertoire of behaviors with which to express and manage his feelings and a greater self-awareness which enables him to see his role in shaping the events going on around him. He can express his anger and jealousy in a variety of direct and indirect forms, and he can sense how his behavior might jeopardize the parental love which is essential to his security. Jealousy and rivalry themselves are more complex affects than separation anguish; their very existence implies that the child now has a network of meaningful human relations instead of a relationship to the mother only.

Up to this point we have used the child's relations with siblings and parents to illustrate the jealousies and rivalries in the two- to five-year period. However, each set of relations is important in its own right. We shall return to the family triangle when we discuss Freud's controversial concept of the Oedipus complex (see Chapter 7), and sibling relations will be covered more fully in the discussion of social relations (see Chapter 8).

Affection in Middle Childhood

Gesell's Concept of Equilibrium-Disequilibrium. Relatively little is known concerning the normal course of affection from around six years to adolescence. The richest source of information is Gesell and Ilg's (1946) study of middle class children. The authors interpret their data in terms of periods of psychological equilibrium in which the child functions smoothly, followed by periods of psychological disequilibrium, during which he is at odds with himself or his environment or both. Age six is a time of general disequilibrium and highly charged ambivalence toward the mother. The child craves affection but is temperamental, touchy, given to tantrums and rebellion. "I love you," may be followed in minutes by, "I wish you were dead." Age seven is a calm, inward time. The child is companionable, sympathetic, and anxious to please, but he tends to be at odds with himself. He may not accept affection, although he

[3]There are a number of studies of parental preferences and attitudes; the following list is a sampling of the literature. Parental preference for a boy or a girl is examined in Freedman, Freedman, and Whelpton (1960). Tasch (1955) studies the joys and problems the parents perceive in relation to their children and also how one parent's idea of his role coincides with or differs from the other parent's idea. For a general discussion of factors influencing parental behavior, see Brim (1958).

can give it. Eight is again stormy. The child demands all of the mother's attention, and is exacting, rude, and fresh. At nine the child is ideal, from the parents' point of view. He is busy with his own life and makes few demands on them. He can become angry, but he is eager to please and is both proud and affectionate.

What Gesell and Ilg describe is a spiraling kind of developmental pattern; certain themes recur again and again but with quantitative and qualitative changes. Thus, the raw anger of early attachment is changed into rudeness and impudence, while physical expression of affection is refined into companionability and pride. The themes are increasingly varied but not new.

Peer Relations. Between the ages of six and twelve, the child's increasing involvement in peer relationships and group activities dramatically changes the course of affection. As we shall see (Chapter 8), a variety of peer relationships can flourish as early as the second year, but if the family is intact and attachments are strong, these are peripheral. When the child goes to school and is exposed to other children daily, his membership in a group takes on a continuity and an import which it lacked previously. A new dimension is introduced into his life. In the home the child had to be loveworthy; now, with his contemporaries, he must be esteemworthy (White, 1964). New sources of prestige and status, new social techniques, new interests and emotional commitments emerge and form a complex pattern of social forces which compete with those in the home. The family ceases to occupy center stage in his life.

Adolescent Emancipation and Adult Affection

Adolescent Emancipation. In adolescence, major revolutions take place: a biological revolution brought on by sexual maturation (see Chapter 10); an interpersonal revolution involving a shift away from the family and toward the peer group (see Chapters 3 and 8); and an identity revolution involving a struggle to find a fulfilling adult role (see Chapter 9). Affection is caught up in these revolutions. Parental expressions of affection become a source of embarrassment to the adolescent, and the very idea of once having been an infant seems intolerable. The standard outcry, "I'm not a baby any more," carries an implicit addendum, "And I am sorry I ever was one."

Parental affection is not only inimical to the adolescent's urgent need to grow up, but it may also be a temptation for him to regress. The typical adolescent is not equal to the task of making a rapid transition, and the discrepancy between goals and capabilities is likely to lead to feelings of impotence and defeat. In the past, it was at just this juncture that the sensitive parent intervened with comfort and diversions. The adolescent may once again long to cry for help, but the unacceptability of the longing may motivate a denial of the need and a renewed effort to assert his independence.

The adolescent's revolt against parents in order to gain independence is often referred to as a struggle for emancipation (see Horrocks, 1969, and Ausubel, 1954). Emancipation cannot be regarded as a prerequisite for psychological maturity. In many societies the family is presided over by a patriarch or a matriarch, and family ties remain primary even after the children are grown and married. There is no reason to believe that such arrangements necessarily stunt the growth of personality. However, emancipation of the adolescent is a central issue in our society (see Chapter 3), and we shall explore it more fully in our discussion of initiative.

Affection in Adulthood. Emancipation is not the ultimate or necessary fate of attachment. Rather, emancipation is often followed by rapprochement at some stage in adulthood. An adult who vigorously protests any showing of parental affection impresses us as being inappropriately adolescent. The ability to feel and express love for parents, without guilt or embarrassment over being close to them, marks the return of an old theme in more mature variation.

Psychoanalysts have been more daring than other theorists in relating caretaking in infancy to adult characteristics. To them, one outcome of consistent, sensitive caretaking is an openness to involvement in deep human relationships and a trust that people will help when help is needed. From this openness and trust grow a general valuing of one's life and a positive anticipation of the future. The psychoanalysts do not mean that the well-cared-for infant will inevitably be a manifestly happy adult, but they would expect him to have a basic belief in his own worth and in the worthwhileness of his life, regardless of the balance of pleasures and pains.

An adult's need to care for others and to be cared for by them also derives from earliest childhood, according to the psychoanalysts. These needs should not overshadow all others in adulthood; if they do, the adult will be regarded as infantilizing others on the one hand, or as being babyish on the other. But the caretaking theme weaves in and out of adult life, and attitudes toward it are tested at critical points: in times of helplessness, such as in severe illness, or during periods of bafflement and defeat, or when the adult, as a parent, is faced with the demandingness and the life-and-death dependence of his own infant. At these points the rightness of protective love and the trustworthiness of care again become manifest.

Deviant Development

Psychopathology and Lack of Affection

There is a voluminous literature concerning distortions of the normal process of forming attachments. In fact, much more has been written about abnormal development than about normal development. The psychoanalysts, in particular, have been concerned with the ways in which faulty mothering affects personality development. According to their theory, the most severely pathological conditions in adulthood—schizophrenia and manic-depressive psychosis—can be traced to the absence of a healthy attachment in earliest infancy. Since the psychoanalysts have the most explicit hypotheses relating infantile experiences to subsequent disturbances, we will present a summary of their theory. First, however, we must describe the kinds of pathological behavior they are attempting to explain.[4]

Schizophrenic individuals are those whom the layman will call "insane" or "crazy" because they talk and act and feel in ways which do not make sense. A schizophrenic may stop drinking tap water out of a conviction that the Communists are poisoning the reservoirs with lethal rays. He may use words in an idiosyncratic manner and his progression of ideas may be illogical; he might say, for example, "The mysteries of incestion are quite premature, sir." He may be chronically indifferent to his own fate and the fate of others, or, if he is a manic-depressive, he will alternate between wild activity and stuporous inactivity.

Although such behavior appears senseless to the layman, as behavior it must be considered lawful, and clinical psychologists have been particularly intrigued with the meaning of irrationality. One hypothesis is that schizophrenia represents a retreat from reality and a return to primitive modes of thinking and feeling. "Primitive" in this instance means characteristic of earliest childhood. The toddler and preschooler have only a vague and patchy conception of the world, and the line between fantasy and reality is blurred. In a roughly analogous way, the man who is convinced that his water is being poisoned is like the three-year-old who believes literally that there is a wolf under his bed. The illogical thinking of the schizophrenic is roughly akin to that of a three-year-old who asks earnestly, "Are we going in the springtime tomorrow?" (The issue of primitive thinking has been oversimplified at this point to provide a general orientation. The separation of fantasy from reality and the change from illogical to logical thinking will be discussed in Chapter 6.)

What has all this to do with mothering? In psychoanalytic theory the connecting link is the assumption that the infant does not learn automatically to distinguish fantasy from reality, nor to think and speak logically. He becomes interested in his human environment only as something that offers him pleasure and comfort. That "something" typically is the mother and her sensitive caretaking. As she becomes loved and valued, the world, especially the world of people, becomes important. It is the bond of love which motivates the infant to learn more about the mother, and subsequently, about people in general. Equally important, love engenders the desire to communicate, and in order to communicate, the infant must replace primitive thought with understandable sequences of ideas.

If love is weak, the bridge between the primitive world and reality will be weak. There will be no deep interest in people which will lead the child to expand his understanding of the human environment, and there will be no compelling desire to communicate which will lead to increasingly clear and logical thinking. The psychoanalytic theory also stresses the importance of timing. The early months of life are uniquely suited to the formation of a healthy attachment. Once those critical months have passed, a deep attachment is extremely difficult or impossible to form. The individual has a lifelong tendency to retreat into the world of primitive fantasy. He will not

[4] For a detailed discussion of schizophrenia and of the manic and depressive disorders, see White (1964).

necessarily become schizophrenic, since his subsequent development might be favorable; but the psychoanalysts would claim that he will always be vulnerable and inclined to withdraw from people instead of participating fully in interpersonal relations.

Psychoanalysts claim that other kinds of personality damage may stem from faulty mothering. If good mothering engenders feelings of trust and optimism, its opposite engenders suspicion and dejection. The damage can be severe. Mistrust may lead primitive thinking in the direction of paranoid delusions: the individual may be convinced that he is the victim of destructive forces (e.g., his drinking water is being poisoned), or he may defend himself against imagined destruction by delusions of grandeur (e.g., believing that he is the Son of God). Dejection may take the form of a psychotic depression: the victim, weighed down by guilt and despair, may sit tearful and inert, monotonously rehearsing real and imagined transgressions (e.g., "I broke my mother's heart–I made her suffer when she needed me most–I ought to die"). Routine activities, such as eating or answering a simple inquiry, are done with great effort and heavy sighs.

The effects of faulty mothering do not have to be so devastating. Instead, they may leave their mark on general attitudes and overall mood. They may produce underlying suspiciousness which might not be immediately apparent. The person may function well socially but react to offers of love and deep friendship with mistrust, asking implicitly, "What are they after?" He is chronically on guard because closeness is equated with the infantile state of being helplessly in the power of a hostile caretaker. In his general outlook on life and in his mood he tends to be pessimistic and bitter.

Up to now, we have been describing adult psychopathologies. Bizarre behavior can also be found in childhood schizophrenics. However, one pathological condition, of special interest because it occurs so early in life and is so dramatic in its behaviorial manifestations, is called early infantile autism (Kanner, 1943)[5]

Autism has three characteristics. First, there is extreme aloneness; autism represents the furthest imaginable retreat from contact with humanity. The three- or four-year-old can be almost totally oblivious to the people around him. When an adult tries to look him in the eye, he will look through the adult as if he did not exist. Or, when he needs a door opened, he will take hold of an adult's hand and bring the hand in contact with the doorknob as if it were an inanimate tool rather than part of a human being. Second, he has a pathological need for sameness. He may spend hours repeating one activity, such as running water into a basin and letting it out. He reacts to even minor changes in the environment with uncontrolled rage and panic. As with an infant, these affective outbursts are not directed to any one person or thing. They are purely emotional explosions. Finally, the autistic child either is mute or, if he speaks, does not communicate. For instance, one child, after repeatedly flushing the toilet said, "The hamburgers are in the refrigerator." Autistic children are sometimes mistakenly called deaf or dumb or mentally retarded; however, they are capable of amazing feats of memory and may have advanced mechanical skills. In some instances, the mothers report that the children were different from birth; instead of being

[5]See also, Kanner (1949) and Kanner and Eisenberg (1955). For a general review, see Rutter (1968).

comforted by cuddling, they were indifferent to caretaking or reacted as if it were noxious. Kanner's description of the parents of autistic children is especially relevant to our concern with attachment. He found them to be above average in intelligence but cold, formal, overintellectualized, humorless, and socially isolated. Psychoanalysts would conclude that an infant could not become attached to such a mother, and that, with love absent, no significant interpersonal, social, or intellectual development is possible.[6]

Maternal Deprivation

The psychoanalytic theory of the etiology of severe adult disturbances was primarily reconstructed from the psychoanalytic treatment of adults. It stimulated clinicians to examine directly the effects of extreme deviations from optimal caretaking on the personality development of the child. A host of clinical studies appeared to confirm the psychoanalytic hypothesis that a faulty mother-infant relationship would result in pathological development. Subsequently, more rigorous studies introduced a number of qualifications into the hypothesis. We shall now examine the initial evidence and then turn to the better controlled research which followed. Much of the investigatory activity centered around the concept of maternal deprivation.

Maternal deprivation includes a variety of phenomena representing significant quantitative departures from optimal maternal care. The infant-child does not receive sufficient mothering at critical stages in his development. In one cluster of studies, infants were inadequately mothered from the earliest weeks of life. The term privation might be used to indicate that the infants or children never had a close relationship with a mother or a mother substitute. Many of the privation studies involved infants reared in institutional settings. In another cluster of studies, infants or children were at different points in forming an attachment when it was disrupted by separation. Typically, the children in these studies were hospitalized or placed in foster or adoptive homes. There is some literature on multiple mothering. These studies assumed that a single mother is essential to the fruition of affection, while dispersing mothering activities among many individuals stunts the growth of affection. Settings for these studies included hospitals, societies in which a number of women care for the infant, and homes with working mothers. (For a review and clarification of the research on maternal deprivation, see Yarrow 1961, 1964.)

Initial Evidence. The clinician's emphasis on the importance of mothering has been buttressed by an unusually large number of studies. A few of these will be described briefly.[7] Spitz's (1946) investigation of early and extreme privation is one of the

[6]The etiology of autism is still a mystery. While some investigators stress the mother's role, it is equally possible that the infant is constitutionally deviant and responds to the mother's love with distress. We shall discuss this point more fully in a subsequent section. Numerous other explanations have been offered (see Rutter, 1968).

[7]For a comprehensive review and evaluation, see Bowlby (1952). Also, for a reappraisal of the situation, see World Health Organization Public Health Papers, 1962, No. 14.

most dramatic. He describes a foundling home in which certain infants, for some unknown reason, were wasting away and dying. He labeled their condition marasmus. The infants had no known physical disease, their diet was satisfactory, and their surroundings were sanitary. However, they were cared for in a strictly routine, impersonal manner. Between caretaking periods, they were not only isolated from all human contact but also were prevented from visually exploring their surroundings. Spitz recommended that the infants be "mothered," i.e., that they be given the tender love, attention, and handling a normal infant might have. When his suggestions were implemented, all the infants began to thrive. A second example of the effects of early and severe privation comes from Bakwin's (1949) chilling report that, as late as 1915, 100 percent of infants under two years of age who had to be hospitalized for an extended stay died. This was the era of the cult of a germ-free, sterile environment; the higher the mortality rate, the more the infants were isolated from other human beings and deprived of the very contact that was so important to their well-being. These studies suggest that mothering is not a sentimental luxury; rather, it is necessary for the infant's growth, and in certain instances it may be essential to his survival.

There have been studies of institutionalized infants in which caretaking was more adequate than in the institution Spitz described, but still significantly less than an infant usually receives in a home setting. Such studies frequently describe the child's relative indifference to adults at an age when most children in home settings have formed close attachments. The infant or child may derive some brief, limited pleasure from being with an attentive adult, but there is neither intense nor sustained involvement. He is strikingly passive; he neither seeks help from adults when distressed nor anticipates pleasure from being with them. There is no spontaneous playfulness, no distress at separation. The infant does not discriminate among faces; to him, "all adults look alike" just as, to the disinterested adult, "all Chinese look alike." Finally, the infant rarely imitates adults. Instead of a vividly experienced and behaviorially complex pattern of relatedness, there is blandness and apathy.[8]

A less frequently described reaction to deprivation has been labeled affect hunger. Here the child is excessively demanding of adults, is easily disappointed, and readily shifts to another adult, only to repeat the pattern of demandingness and disappointment. Yarrow (1961) speculates that the blandness described in the preceding paragraph may result when the infant has never had a deep relationship with an adult, while affect hunger might represent a desperate reaction to the loss of a relationship. Maternal deprivation not only affects attachment, it also impedes growth in other areas. Language, in particular, is retarded. There is some evidence that this apparent retardation might be due, in part, to the child's social apathy. Provence and Lipton (1962), for instance, note that, while consonant and vowel sounds emerged on schedule, they were used only minimally because of the infants' lack of interest in communicating with adults. Those investigators also report apathy with regard to exploring physical objects. This decrease in curiosity might well be the forerunner of the decline in intelligence that is found in older children. It might also be related to a

[8]See A. Freud and Burlingham (1944) and Provence and Lipton (1962) for two comprehensive reports concerning how infants fare under different amounts of mothering.

tendency to be literal and to deal only with the here-and-now, rather than to enjoy the challenge of solving complex problems and planning for the future. Motor development, such as sitting up, crawling, and walking, is least affected by maternal deprivation perhaps because it is more strongly influenced by maturational than by social and motivational factors.

A final group of studies is concerned with the infant's reaction to separation from the mother. Here, the age of the infant seems to be a crucial factor. Schaffer and Callander (1959) found that infants under seven months of age showed little evidence of disturbance when they were hospitalized. They were responsive to the new adults who paid attention to them and they accepted the change in routine. Recall that, in this age range, the infant has not formed a strong tie to an individual and is still in the stage of general social responsiveness. However, Schaffer and Callander noted that the infants, on returning home, did show some disturbance; e.g., they had periods of extreme preoccupation during which they stared at their surroundings with a blank expression. When infants over seven months of age were hospitalized, the outcome was quite different. Typically their initial reaction was one of great distress, marked by clinging and bitter protest as the mother started to leave, strenuous crying, motor restlessness, and active reaching out to people, perhaps in an attempt to bring the mother back or to find a substitute for her. In the next stage, adults were actively rejected and the infant lapsed into a state of apathy, lack of interest in people, and motor inactivity. Bowlby (1953) labels this reaction "mourning" and likens it to depression in adults. Spitz and Wolf (1946) call it anaclitic depression and emphasize the loss of the love which is essential to the infant's psychological well-being. The final state was marked by denial or detachment from deep involvement. The child showed no apparent recognition of the mother and used any adult in a superficial and manipulative manner. Naive hospital personnel might think he was "adjusting nicely" and miss the significant shift from deep involvement with the mother to indifference, superficiality, and self-centeredness.

The Question of Reversibility of Damage. An important issue in all studies of maternal deprivation concerns the permanence or reversibility of damage. There are studies supporting the psychoanalytic contention that damage is either irreversible or extremely difficult to rectify. Provence and Lipton followed their group of infants after they had been placed in foster homes. The children were now between two and five years of age. They observed "dramatic gains," and to the casual observer the children appeared no different from those reared from birth in their own homes. However, closer study revealed that their relationships were marked by an indiscriminate friendliness, the substitution of one adult for another, and failure to turn to adults for comfort and help; their self-control and ability to wait were weak, and their thinking was still limited to the here-and-now. Goldfarb (1943a, 1943b, 1945) found similar results when he compared two groups of adolescents; those in one group had spent the first three years of life in an institution before going to a foster home; and those in the other group had been in foster homes from birth. The adolescents in the institutionalized group were lower in intelligence, had more language and speech difficulties, had weaker self-control (as evidenced in temper tantrums, lying, stealing, and hitting other children), were emotionally cold, and were capable of only superficial relations with others.

These studies should be regarded as suggestive rather than definitive. Of necessity the investigators could not exercise control over the home environments of the children, so we do not know whether the children did, in fact, receive the attention and affection a child usually has in a family setting. Well-intentioned adults can begin neglecting a child who is unresponsive, and it is possible that the adoptive parents gave less because the institutionalized child was less rewarding than the average child. But there is a more important criticism. It may be that, after being institutionalized, the child will need remedial measures—special attention, affection, and understanding—in order to reverse the damage. Is it not asking too much of a damaged child that he be resilient enough to overcome all the consequences of neglect just because he becomes part of a family group? Before the question of reversibility can be answered, therefore, we must know (1) whether the damage can be remedied in a setting where the child is, in fact, treated like other children, or (2) whether the damage can be remedied in a setting where the family makes special accommodations to the child in order to meet his special needs. Only if the child fails to form an attachment under the latter condition can we speak of irreversible damage with more assurance. At present, psychologists are far from knowing the answer to this crucial question.

Re-evaluation. The voluminous literature on maternal deprivation constituted an impressive testimonial to mother love and documented the damaging consequences of being inadequately loved, of being separated from a loving mother, or of having multiple mothers (i.e., multiple caretakers). During the 1950s, clinicians stressed the importance of a "one-to-one relationship" between mother and child; "tender loving care," "motherliness," and breast-feeding had their heyday. Neglect was bad, but separation and institutionalization were also bad; even working mothers were suspect. Perhaps in reaction to an excessive emphasis on the early mother-infant relationship, the evidence was re-examined critically, contradictory evidence was unearthed, and new studies were done. As a result of this ferment, psychologists are now re-evaluating the role of early attachment in personality development. We shall discuss certain criticisms of the initial evidence and the new evidence which was added to it.

Many of the clinical studies were marred by a number of weaknesses: biased sampling, lack of a control group, inadequately defined terms, and the confounding of variables. Not all of the studies suffered from these methodological defects, but the sheer number of apparently confirming cases is appreciably diminished when only acceptably designed studies are considered. We shall examine each shortcoming briefly.

The clinician is concerned primarily with the disturbed child (see Chapter 1). When he observes a group of infants, he naturally focuses on the deviant ones, since these interest him most. He is therefore vulnerable to the pitfalls of biased sampling. One is that he generalizes to a total population on the basis of its most disturbed members; e.g., that he will characterize all institutionalized children in terms of the most damaged ones. As a result, the effects of maternal deprivation tend to be documented by descriptions of its most devastated victims. Clinicians can also fail to consider the possibility that institutionalized infants may be a biased sample of infants in general; e.g., their mothers might be inadequately nourished during pregnancy or may have a higher incidence of diseases which damage the fetus, and there is a greater likelihood that the infant was unwanted, illegitimate, or that his mother was judged unfit to care for him. In order to generalize concerning the effects of maternal deprivation on all infants, the population studied must not be initially inferior and consequently more readily damaged by inadequate care. The frequent lack of control groups arises from the fact that clinicians typically study only disturbed individuals and rarely have contact with those who are functioning adequately. Without a normal control group, it is difficult to evaluate their findings relating adult disturbances to maternal deprivation. If a clinician reports, "Seventy percent of the schizophrenic adults I treated were neglected by their mothers as infants," we cannot really know how impressive the relationship is until we know what percent of adequately functioning adults were also neglected. If, to continue the hypothetical case, 40 percent of normal adults were neglected, we should still regard maternal deprivation as a contributing factor to schizophrenia, but the probability of neglect eventuating in severe psychopathology would be considerably decreased.

A number of the clinical studies defined their population of infants in terms of settings, such as "institutionalized" or "hospitalized" infants. Such a definition is useless, since it bypasses the essential step of delineating the environment in terms of psychologically relevant variables. The same criticism applies to populations defined as "reared in the home." Mothers, as well as institutions, vary widely in their caretaking, and maternal deprivation can be found in the home as well as in the hospital or orphanage (see Robertson, 1962). Maternal care is the variable being studied and settings are useful only as they provide variations in maternal care. Thus the investigator is called upon to delineate and to quantify caretaking activity. Rheingold's (1961a) research serves as a model. She used a time-sampling technique to determine the frequency of caretaking acts, including looking at, talking to, playing with, holding, feeding, and showing affection to the infant, and then statistically determined the difference between the home and institutional settings.

The most telling criticism concerns the confounding of factors in many of the studies. As we noted in Chapter 1, it is difficult to pinpoint relationships in clinical investigation because so many variables are involved. When a child goes to the hospital not only is he separated from his mother, but he is also subjected to many other stresses—the stress of being ill, of living in unfamiliar surroundings, of strange and painful medical procedures. Similarly, an institutionalized child often is deprived not only of mothering, but also of objects and toys to play with, a variety of sounds and sights to stimulate him, opportunities for peer interaction. Or again, many severely damaged adolescents and adults have a background which includes not only institutionalization but also repeated subsequent traumas, such as numerous, unsuccessful foster home placements. Thus, separation and institutionalization can be only two of many significant factors in the infant's or child's life, and it is impossible to disentangle the effects of maternal deprivation from the effects of other noxious influences.

Results of Better-Controlled Studies. The critics of the maternal deprivation literature have not merely carped; they have obtained data from better-controlled studies. They find that, as stresses and deprivations other than those associated with maternal deprivation are eliminated, dramatically deviant behavior disappears. The child is not unaffected, but neither is he severely disturbed or permanently damaged.

In regard to separation uncomplicated by other sources of stress, Heinicke (1956) found that children between fifteen and thirty months of age had no traumatic reaction when sent to good day care centers and to residential nurseries, although the latter experience was more stressful than the former. Institutions in which maternal deprivation is not accompanied by general environmental bleakness do not produce personality disturbances and severe developmental retardation. This conclusion was reached after studying home management houses which serve as training centers for students in home economics (Gardner, Hawkes, and Burchinal, 1961) and hospital settings (Du Pan and Roth, 1955), both of which provided warm, stimulating environments. In a like manner, increasing the amount of stimulation an institutionalized child receives results in an increase in his social responsiveness (Rheingold, 1961) and an advance in his measured developmental age (Sayegh and Dennis, 1965), while the course of intellectual and social deterioration can be reversed by placing children in more stimulating institutions (Skeels and Dye, 1939). Evidence concerning effects of multiple mothering or, at least, of having a primary caretaker in addition to the mother comes from Rabin's study (1958) of children reared from infancy in the Israeli kibbutzim. These children were cared for in nurseries by a substitute mother, although they also maintained close relationships with their own mothers. Rabin found no intellectual retardation and, if anything, the boys had better self-control and greater maturity than boys reared at home.

The relation between maternal deprivation and the development of severely psychopathological conditions was not confirmed by studies in which deprivation was uncomplicated by a host of other noxious influences. Bowlby and others (1956) studied sixty children between six and fourteen years of age who were hospitalized during their first four years for treatment of tuberculosis. The children showed tendencies toward withdrawal or aggressiveness, but there was no evidence of intellectual damage or disturbance in their ability to relate to peers. Only a small minority developed serious personality disturbances, although there was a significantly higher incidence of maladjustment than was found in a control group. As found in studies of separation, the children were not severely damaged although they were negatively affected by the experience. Maas (1963) found no intellectual damage or severe personality pathology in adults who had been placed in stimulating nurseries as infants during World War II. However, there was a suggestion that the experience was more damaging to infants who had been placed there when less than a year old than it was to older infants.

Comment on the Criterion Problem. In the studies cited above investigators used different criteria to assess the effects of maternal deprivation on personality development. Not all of the criteria may be equally valid. A psychoanalyst might not be convinced by much of the evidence on how well children adjust without benefit of mothering by a single caretaker, if the criteria of adjustment consists of behavior in a group, teacher's ratings, and other evidence of socially acceptable behavior. In his work with adults he is accustomed to seeing the "successful man" who suddenly tries to commit suicide and the "ideally married couple" who becomes violently antagonistic. An adult individual whose life is publicly acceptable can live in a private hell of anxiety and despair or he can be bedeviled by perverse thoughts and fantasies. A child may be such an adult in the making. Therefore the psychoanalyst is concerned with the highly personal ideas and feelings a child reveals when he spends many hours with an interested adult whom he trusts (see Chapter 1). The crucial question about mothering must be answered in terms of damage to the child's inner life, for here is found either the resilience or the vulnerability which will affect his future.

Even if the psychoanalyst's emphasis on inner feelings and thoughts is rejected as parochial, his warning against judging adjustment by a single sample of public behavior is generally valid. As we have seen, the infant under seven months of age who is hospitalized seems to be unaffected by separation until he returns home, and hospitalized older children can be regarded as making a good adjustment when they have, in fact, become detached from the mother. Parents report having qualms about going away on a trip for the first time and then being surprised on their return that the child seems not to have missed them; yet the same child might subsequently have sleep disturbances, fits of jealousy or possessiveness, stomachaches, and other signs of emotional upset. Even a two-year-old can protect himself from the pangs of separation by feigned indifference. Before concluding that separation has had no effect, therefore, one must evaluate the child's behavior in many situations and over a period of time.

Integration of Evidence. Let us see what generalizations can be made when all the studies of maternal deprivation are taken into account. It appears that impersonal gratification of physiological needs, without additional care and stimulation, has devastating effects on the infant's development and, in certain cases, even on his life. The adverse effects which occur following separation from the mother depend on the age of the child, the stress involved in separation, the length of the separation period, the availability of substitute gratifications and stimulation during separation, and the amount of subsequent trauma in the child's life. Thus, separation can be either a temporary stress or a trauma blocking healthy growth. If the infant is separated while forming and consolidating a close attachment, if separation is accompanied by additional stresses (such as illness) and followed by a prolonged stay in an impersonal, unstimulating environment (such as a bleak institution), and if the child has a subsequent history of traumatic experiences (such as repeated, unsuccessful placements in foster homes), he is likely to show severe emotional and intellectual damage later. However, the damage is less as efforts are made to ameliorate any or all of these conditions: to separate the infant before attachment develops, to reduce the stress present at the time of separation, to make the new environment warm and stimulating, to return the child to his mother or to find an adequate mother substitute. If damage occurs, it is likely to affect the ability to form a deep attachment, language, and abstract thinking; motor responses are less affected. Studies suggest that the damaged infant has difficulty forming attachments subsequently, but the evidence is meager and inconclusive.

The Issue of Multiple Mothers. Whether the child emerges unscathed from a group setting with multiple mothers remains an open question. He clearly is not severely damaged but, to date, the best institutional care is still inferior to the best home care. However, other countries such as Israel (Rabin, 1965) and Russia are using institutions in ways which are more imaginative and more congruent with the culture than is true of America (Ainsworth, 1962). The Israeli experiment with rearing infants in kibbutzim has been mentioned. Russia has a more extreme program in which infants are reared almost completely in institutions (see Bronfenbrenner, 1970, and Chauncey, 1969). They receive the best of care and stimulation, but extensive individual attention is avoided on the basis that it would lead to egotism and selfishness rather than to a concern for others. Group participation is introduced at an early age, and what we would call individuality is systematically discouraged. Even in our own country, day care centers are supplementing the home, and the child is spending an increasing amount of time with a group of peers under the care of strangers. Since this experience is occurring earlier and earlier in the child's life, the effects of multiple mothers is of practical as well as theoretical interest.

In light of the meager data, our discussion will be largely speculative. Our first question is: Can maternal care be as effectively administered by several persons as by a single caretaker? Our impression is that it would be difficult if not impossible for a number of caretakers to duplicate the behavior of a single mother. It is doubtful whether they could know the infant as intimately and thus time their ministrations as sensitively as a single mother can, or whether they can match her total emotional commitment. At best, they might be more varied in their care and stimulation and less fatigued by their responsibilities.

But there is a more perplexing question: Should an institution try to duplicate the caretaking of a single mother? Mead (1962) claims that the very idea of care by the biological mother is culturally determined rather than God-given. From her observations of other cultures, she reports that the child who is reared by a number of good mother figures develops a rich personality, is trusting of others, and is less vulnerable to separation anguish. On the other hand, Freud and Burlingham (1944) believe that a deep attachment to a single mothering person is essential to emotional growth and is well worth the price the child pays in increased anguish, anxiety, and jealousy. To develop fully, they maintain, the child must be able to feel deeply and learn to master the difficult consequences of such feelings; bypassing an intense emotional commitment may produce a more docile and socially pleasing child but one whose emotional range is limited.

Our own speculation is that differences in the number of caretakers will affect the way a child relates to other individuals. If he is deeply loved by his mother, he will develop the capacity to love a single person deeply, but he will experience the anguish, ambivalence, and guilt which often are part of such a relationship. If he is loved well by a number of caretakers, he will develop the capacity to enjoy closeness with a variety of people. He may not be capable of a total emotional commitment to a single individual and may not experience the full depth and intensity of love, but he will be exempted from the negative consequences of a close relation. Different backgrounds may produce different kinds of attachments, such as concentrated and intense or varied and less intense; they do not produce a child who is inferior in other aspects of personality (e.g., self-control, conscience, social relations, intelligence) or one who is more vulnerable to psychopathology. The choice of whether to foster a single deep attachment or varied attachments becomes a matter of values–what one regards as intrinsically important for the child to experience in life; it is not a matter of jeopardizing normal development.

Negative Closeness

Clinical Descriptions. In its purest form maternal deprivation is a significant quantitative deviation from optimal care. The clinical literature abounds with descriptions of mothers who do not deprive their infants but whose care is contaminated by attitudes inimical to the formation of a healthy attachment. There is the overindulgent mother who is excessively solicitous in her care, the narcissistic mother who uses the infant as a showpiece, the ambivalent mother whose love is mixed with anger and anxiety, the inconsistent mother who shifts back and forth unpredictably between positive and negative feelings, the cold mother who is distant, intellectual, and perfectionistic, and so on.

Clinical studies indicate that each pattern of maternal behavior can have diverse roots. A mother might be overindulgent for a number of reasons, for example. She may feel intensely fulfilled when caring for the infant and unconsciously try to prolong his state of helpless dependence. She may have a strong but unacceptable desire to be rid of the infant, which must be masked by excessive concern. A mother can have her own separation anguish, her own fears of eventually being left alone which force her to bind the child tightly to her. A mother may irrationally equate the infant's self-assertiveness with hostility and anxiously overindulge him in the hope of placating an imaginary monster of destruction. In like manner, ambivalence in mothers can have many origins. The helplessness of the infant, instead of stimulating protectiveness, may terrify the mother, and the responsibility of caretaking may take on a life-or-death quality. Helplessness may aggravate the mother's own unfulfilled needs to be loved and protected, and she may spitefully give the care she longs to receive herself. A mother may be deeply disturbed and disgusted by the infant's "animal-like" behavior, or she may resent the infant as an intrusion in her life or regard him as a symbol of the end of her youth and freedom. All of these negative forces, often not recognized or only dimly grasped, affect the mother's caretaking. She may be insensitive, or out of phase in her timing, or tense or abrupt or rough in her handling. Consequently, the infant experiences intimacy as a highly charged mixture of gratification and pain; he is cared for and distressed at the same time.

The clinical literature is fascinating to read and has served to delineate a number of deviant patterns of maternal care. However, each investigator has been satisfied to report only his own findings, and there have been few systematic attempts to review and conceptualize all of the studies. Nor is the literature in this area undergoing the kind of critical examination which has revitalized the study of maternal deprivation. Because of the paucity of well-designed studies and objective data, it is difficult to decide whether the clinicians are describing dramatic oddities or have hit upon generally valid insights into the etiology of deviant personality development.

Since the literature lacks an organization of its own, we shall present it under an arbitrary choice of headings. First we shall discuss studies relating negative closeness to the development of schizophrenia in adults. The research strategy here involves studying a group of schizophrenic adults and trying to verify the hypothesized negative maternal attitudes in infancy. Then we shall turn to studies of the effects of negative closeness on the infant and child. We shall review Spitz's attempt to relate specific maternal behaviors to specific pathogenic conditions in infancy. Then we shall turn to objective studies which have utilized specific feeding practices and general patterns of maternal behavior as indices of negative maternal attitudes.

The "Schizophrenogenic" Mother. As we have seen, psychoanalytic theory postulates that faulty mothering in infancy hinders the formation of attachment and that this in turn sets the stage for the development of schizophrenia. Psychoanalytic theory does not claim there is a single cause for schizophrenia; constitutional factors and trauma at other developmental stages can also be crucial factors. However the emphasis on infancy stimulated a large number of clinical studies which apparently confirmed the hypothesized relationship between negative closeness in infancy and adult schizophrenia. Reichard and Tillman (1950), after reviewing seventy-nine histories of adult schizophrenics described two predominant patterns of maternal behavior. Most frequently they found a covertly rejecting mother who is manifestly overprotective. In these cases, the mother's need to infantilize overrides all of the child's needs to grow. She is omnipresent, literally or symbolically, since she is concerned with and involved in every detail of her child's life. She may indulge him excessively out of her need to bind him tightly to herself, or she may be exacting in her demands for conformity. These maternal behaviors place the child in the intolerable situation of being the center of the mother's universe, while at the same time he senses her underlying rejection. Another less frequent pattern is overt rejection. The mother is cold, sarcastic, destroys confidence by nagging, makes excessive demands for neatness and politeness, pushes the child to fulfill her own social ambitions, praises him in public but punishes him later. Reichard and Tillman also found, as have other investigators, that the mother frequently is dominant and the father passive, especially in the case of male schizophrenics. Other clinical investigators describe similar patterns of maternal behavior (see Friedlander, 1945; Yerbury and Newell, 1943-1944; Gerard and Siegel,

1950). Recall, also, that Kanner (1949) found parents of autistic children to be cold and mechanical in their behavior. However, these two patterns of compensatory overprotection and rejection do not exhaust the list of maternal characteristics. Jackson and others (1958) add Machiavellian manipulation and helplessness, while Rank (1949) describes an "as if" mother who tries to mask her basic emotional emptiness by living up to an image of how the "good wife and mother" should behave. Finally, Hill (1955) claims that the mother is not characterized by a single pattern; rather, her basically disordered personality makes her a baffling mixture of a number of traits.

These efforts to relate patterns of negative closeness to schizophrenia reached an apotheosis of sorts when the term "schizophrenogenic" mother was coined. As in the case of maternal deprivation, the evidence for schizophrenogenic mothers has not gone unchallenged. Data to the contrary are not as extensive, but there is the same tendency for better-controlled research to modify the original clinical observations. All the studies cited above were done either with schizophrenics alone or with a group of schizophrenics and a control group of normal subjects. When McKeown (1950-1951) added to these a group of neurotic children, he found that mothers of schizophrenics and neurotics were different from normals but not different from one another. The most challenging bits of evidence come from two studies cited by Bell (1958) which indicate that mothers of physically damaged children have attitudes similar to mothers of emotionally damaged children. Bell suggests that any deviant child tends to elicit unusual maternal reactions, such as overconcern and intrusiveness. His thesis that maternal behavior may be a reaction to the behavior of the schizophrenic child rather than being the cause of such behavior has been championed by Escalona (1948) and by Peck, Rabinovitch, and Cramer (1949). The argument is that schizophrenic children (and autistic ones as well) are constitutionally different. Their development is uneven and erratic; precocity exists side by side with slowness, and developmental advances unaccountably give way to immature patterns. In their interpersonal relations they are puzzling and unpredictable; at times they are extremely remote, at times hypersensitive and irritable. The concerned mother is chronically baffled, frustrated, enraged, and defeated in her efforts to comprehend and care for the child. Her overprotection or isolation are outgrowths of a process set in motion by the child himself. In Chapter 10 we shall present further evidence that children can engender excessive anxiety and guilt in parents because their temperament makes them difficult to rear.

In sum: the study of schizophrenic children and adults is a hazardous research strategy for attempting to understand the effects of negative closeness in infancy. Since schizophrenia may have many origins—heredity, organic brain dysfunction, peculiar patterns of family interaction, socioeconomic factors—the specific influence of the mother is difficult to isolate. Even if confounding variables were not present in the clinical studies, the concept of schizophrenogenic mother is open to question. There is no definitive evidence that the observed maternal attitudes are unique to mothers of schizophrenic children, since they are found in mothers of other deviant children as well; and there is no definitive evidence that the observed maternal attitudes are causative, since they may be reactions to a constitutionally deviant child. While negative closeness may play a significant role in the development of certain cases of schizophrenia, the conditions in which it is the principal causative factor have not been determined.

Pathogenic Conditions in Infancy. Spitz (1951) is one of the few clinicians to postulate specific relations between maternal attitudes and pathology in infancy. As an example, he hypothesizes that colic (excessive crying) in infants between three weeks and three months of age is produced by "primary anxious overpermissiveness." In this pattern, the overconcerned mother continually responds to distress by feeding the infant, thus aggravating his colic and producing further crying, which she again tries to relieve by even more strenuous efforts to feed him. Infantile neurodermatitis, he suggests, is produced by "hostility in the garb of anxiety." Although she is unconsciously hostile, the mother consciously regards the infant as fragile and fears the damage she might do to him. She is reluctant to handle him, and the infant is deprived of cutaneous stimulation. Excessive rocking is due to the mother who vacillates frequently between pampering and hostility, thereby preventing the infant from forming a gratifying attachment and forcing him to seek pleasure from the bodily sensations of rocking. Spitz presents eight of these specific relationships between maternal attitudes and infant disturbances.

Spitz's psychoanalytic explanations are too complex to be presented here. However, his approach to understanding the consequences of negative closeness is sounder than the study of schizophrenic adults. Instead of trying to reconstruct the infancy of a group of disturbed adults, he deals with the mother-infant relationship as it happens. Since he bypasses the errors involved in recalling events which happened in infancy and early childhood, the potential gain in control is appreciable. Unfortunately, his interesting hypotheses have produced little objective research.

Objective Studies and the Criterion Dilemma. We turn now to objective studies relevant to the effects of negative closeness on attachment and personality. One group of studies is concerned with the effects of specific feeding practices—breast versus bottle; scheduled versus demand; early versus late weaning—on subsequent personality development. The rationale for this research derives from Freud's theory that attachment is formed in the context of the feeding situation. Bottle feeding, scheduled feeding and early weaning are considered to be negative maternal behavior. Caldwell (1964) admirably reviews the data and finds no convincing evidence of a relation between feeding practices in infancy and either child or adult personality. However, a majority of the studies can be criticized for oversimplifying both psychoanalytic theory and the feeding situation. Relatively few studies have been concerned with the mother's attitude. This neglect is difficult to understand, since psychoanalytic theory stresses affect (for example, warmth, gratification, pleasure) as the key to attachment rather than feeding techniques and schedule. As Brody (1956) has shown, holding the infant and breast-feeding him do not necessarily require or imply sensitivity and emotional rapport. Interestingly, one study which did take both the mother's specific behavior and her attitudes into account came up with a number of positive findings. Heinstein (1963) related infant feeding practices and maternal warmth to problem behavior, such as disturbing dreams, finicky eating, and nailbiting, between six and twelve years of age. The most pronounced maladjustment occurred in boys who were nursed for long periods by cold mothers; boys nursed for long periods by warm mothers were relatively free of behavioral problems. Breast feeding per se was not related to problem behavior in the boys; girls, on the other hand, had fewer problems if breast-fed by a warm mother or if bottle-fed by an impersonal mother. The author speculates about the damage to personality when intimacy and coldness or oversolicitousness and hostility are intermingled.

A second cluster of studies is concerned with the effects of maternal attitudes or general patterns of maternal behavior, such as rejection, control, and hostility, upon personality. (The research literature includes both terms, "maternal attitude" and "maternal behavior." The distinction between the two is not important to our discussion.[9]) There is little doubt about the potency of such variables. For instance, Sears, Maccoby, and Levin (1957) found maternal coldness to be associated with feeding problems, persistent bed-wetting, high aggression, emotional upset during toilet training, and slow conscience development. Behrens (1954) found that the adjustment of preschool children was related not to the mother's specific child-rearing techniques but to her total personality, including her general "child-centeredness" along with her self-control, her affect, her range of interests, her interpersonal relations, and her fulfillment in being a mother.

Levy's (1943) research further illustrates the fruitfulness of using maternal attitudes as criteria of negative closeness. In his classical investigation twenty cases of maternal overprotection were studied. The mean age of the children was ten years, and all but one were boys. In contrast to the "schizophrenogenic" mothers, these women were not trying to compensate for underlying feelings of hostility and rejection; instead, they evidenced "exaggerated maternal love." Their behavior toward the child was characterized by excessive contact, such as constant companionship of mother and child, prolonged nursing care, excessive fondling, or sleeping with the child long after infancy; infantilization, such as continuing to feed, dress, and bathe the child far beyond the age at which the child usually begins to do these things for himself; prevention of social maturity and independence, such as allowing an eight-year-old child to play only in the mother's sight, excusing the child from all chores, or blocking the formation of friendships. One group of mothers in the study were extremely dominating and actively controlled every aspect of the child's life; mothers in another group were extremely indulgent and placed few if any restrictions on the children's behavior, no matter how aggressive or selfish. The dominated children were found to be unusually docile and submissive; they were clean and neat, diligent and polite, but they were timid and fearful in school and were regarded as sissies by their peers. The indulged children were tyrants; at home they reacted to frustrations with tantrums, kicking, and throwing food on the floor; they were impudent, demanding,

[9]See Schaefer (1959) for an attempt to order various attitudes into a simple but comprehensive schema. Ausubel (1958) discusses the effects of maternal attitudes on child development.

and disrespectful; with peers they were bossy and uncooperative, and frequently they were friendless. In spite of their dramatically deviant behavior, only four of these children were found to be severely disturbed in a follow-up study, and several were doing very well. This finding suggests that exaggerated maternal love does not necessarily have the noxious consequences which are attributed to "schizophrenogenic" mothering in which maternal love is only a disguise for underlying hostility.

Garner and Wenar's (1959) research on psychosomatic disorders focused on the mother-infant relationship. The investigators hypothesized that faulty caretaking has both physiological and interpersonal consequences. Because the mother is inept at comforting the infant, he must endure long periods of physiological distress. He develops a pattern of reacting to stress with erratic and extreme physiological activity which may be the forerunner of psychosomatic disorders, such as ulcers, asthma, and ulcerative colitis. At the interpersonal level, intimacy is not pleasurable but is contaminated with a host of negative feelings, such as anger and mistrust. The authors specifically hypothesized that the mother-child relation underlying psychosomatic disorders is a close but mutually frustrating one, both parties being irresistibly attracted to a highly charged but ungratifying interaction. The hypothesis was confirmed: they found that the mothers of children with psychosomatic disorders were ambitious and controlling women who had high positive anticipation during pregnancy but derived little pleasure from actual caretaking activities. Because of their strong emotional investment, they were unable to achieve any psychological distance from the child but became either subtly manipulative or relentlessly dominating. The children, for their part, associated intimacy with lack of gratification and were generally more emotionally distant and psychologically disturbed than were children in control groups, one of neurotic children and one of children with nonpsychosomatic physical illness.

One fundamental difficulty with all these studies is that maternal attitudes and patterns of maternal behavior are too general to enable us to pinpoint the specific consequences of negative closeness in infancy. Love or hostility, for instance, characterizes the way a mother handles many interactions during a given period and is highly stable over time (Schaefer and Bayley, 1963). The mother who nurses her infant irritably might also undermine his autonomy, be resentful of his attachment to the father, criticize his friends, belittle his achievements in school, and so on. It is impossible to disentangle the specific effects of hostility on attachment and the specific effect of attachment on subsequent personality development. No criticism of attitude studies is intended here; we are merely pointing out the problem encountered in using attitudes as criteria in our inquiry into the effects of negative closeness.

In summary, we have reached an impasse in our search for an operational definition of negative closeness which would illuminate its effect on attachment and on personality development. Studies using specific feeding techniques as a criterion have shown little relation to personality variables. Mothering must be more broadly conceptualized, and a wider spectrum of maternal behaviors must be tapped. Attitude studies have been fruitful in relating maternal behavior during infancy to the child's personality development; but attitudes, especially the basic ones of love and hostility, tend to be pervasive and stable over time, so that the specific effects of negative closeness during the formation of attachment cannot be isolated.

We shall suggest one possible solution to the dilemma of operationalizing the concept of negative closeness and explore its implications concerning personality development. Suppose we could find a group of mothers who, like some fathers in our culture, have genuine difficulties in caring for infants but genuinely enjoy the child once he is past the stage of helpless dependency. During the first year of the child's life, the mother would show just those negative attitudes which the clinicians describe so well—she is anxious or irritable or mechanical or intellectual, and so on. She greets the child's walking and talking—those developmental milestones signaling the end of infantile helplessness—with relief. But, more than that, the mother now starts enjoying her child and handles capably the problems of autonomy, social adjustment, and intellectual development. However, she reverts to her negative attitudes whenever the child becomes helplessly dependent, such as during a severe illness, or whenever he acts in a "babyish" manner. If we could locate families with these well-defined relationships, how would the children develop and what kinds of adults would they become? They probably would not be schizophrenic or severely depressed. Such devastation, like that attributed to maternal deprivation, is probably due to noxious influences which are pervasive and chronic. Personality development does not seem to be like a row of dominoes which collapses once the first member has been knocked over. It seems more reasonable to speculate that the child would develop along lines similar to those of his mother—he might function quite well except in times of physical or psychological helplessness or when he himself has to care for a helpless infant. His vulnerability might come to light in specific situations and/or in general personality traits—he might be an "impossible" patient when he is physically ill, or a "demon" when given a birthday present, or perhaps he "would rather die" than ask directions when lost on a car trip; he might make an absolute virtue of activity, self-sufficiency, and leadership, while having to contend with a psychosomatically caused ulcer. In general, one would expect that the child would, as an adult, wall off any experience which served as a reminder of the distress of being cared for, while managing many other areas of adult life effectively.

Whether our solution proves practicable remains to be seen, since the problems involved in implementing it are prodigious. However, it does serve the didactic function of clarifying the issue concerning the effects of negative closeness in infancy on attachment and personality development. As such, it can be used as a guide in evaluating future clinical observations and objective studies.

Conclusion

We have been particularly concerned with answering two questions: How does the complex behavior of caretaking in infancy affect the formation of the complex behavior of attachment? and, How does attachment in infancy affect the subsequent course of affection and of personality in general? The answer to these questions is much less clear today than it was twenty years ago. At that time it seemed that the formation of attachment was mediated primarily by the relief of physiological distress, although pleasurable body sensations provided by the caretaker also played a role. The classical conditioning model seemed adequate to account for the learning of attachment. From clinical studies it seemed clear that the well-loved infant became the emotionally secure child and adult. Deviations from the ideal of warm, sensitive caretaking by a single mother were apt to produce emotional insecurity or, in the case of extreme deprivation or negative closeness, severe psychopathology.

The current situation does not permit so neat a summary. The concept of mother as physiological comfortor and stimulator must be supplemented by the concept of mother as a source of visual and auditory stimulation, since distance receptors have been found to play a major role in the development of attachment. At the very least, the classical conditioning model must be supplemented by a cognitive factor which enables the infant to grasp the idea of "mother," by innate biases which make the infant respond with particular interest and delight to stimuli emanating from human beings, and by instinctual forces which make the appearance of attachment responses highly probable.

While extreme deprivation of maternal care seems to have a devastating effect on infant development, deviations from a one-to-one relationship do not necessarily impede normal growth. In regard to separation, only temporary distress can be expected when the infant is well cared for before and after separation; extreme disturbances and permanent damage are most likely to occur when other noxious influences precede and/or follow the event. Negative closeness, as part of a pattern of maternal behavior extending beyond infancy, may be one of many determinants of severe psychopathology but it is not a unique correlate of severe psychopathology (since it characterizes mothers of moderately disturbed and physically deviant children), nor can a correlation be regarded as causative of (rather than reactive to) severely deviant development. Negative closeness, when confined to caretaking in infancy, may produce a subsequent vulnerability to intense distress in the area of caring for others and of being cared for by them.

Turning to the normal end of the caretaking continuum, we find that adequate mothering still is one prerequisite for normal personality development in infancy, but the role of the mother herself is not clear. A one-to-one relationship can no longer be considered the only form of optimal caretaking; we must at least entertain the possibility that good maternal care dispensed by a number of caretakers may produce different kinds of normal attachments rather than jeopardize the development of affection or personality growth in general.

While the clinical literature of the 1950s seems to have exaggerated the significance of a one-to-one relationship, it served to call attention to the importance of attachment and affection in personality development. There is nothing in our present reevaluation which is intended to question the centrality of the bond of love. It is significant in its own right and it plays a prominent part in the development of many other aspects of personality which we shall discuss.

3 Initiative, Willfulness, and Negativism

As love is central to the first year of life, initiative is central to the second year. In an upsurge of independence, the toddler literally and figuratively stands on his own two feet and turns his back on his mother. He avidly explores the tangible world of objects and the subtle world of human relations. As he does so, he begins to sense his capabilities and his power, to develop an embryonic notion of himself as an agent, a doer, a causer. No sooner has this happened than his inability to moderate and modulate propels him into a stage of marked willfulness and negativism. The fragile "I" becomes imperious. The toddler also runs headlong into the curbs and restrictions of his socializing parents. For the first time he is confronted with *No* and *Don't*—those pivotal words which will reverberate throughout his life. More and more, he finds he must choose between doing what he wants to do and doing what he is told. Thus, the universal conflict between freedom and conformity becomes part of the toddler's life, while the temptation to disobedience confronts him with a dilemma as old as original sin.

In this chapter we shall first describe initiative and willfulness as they appear in the toddler and trace their roots in infancy. We then shall examine the nature of the socialization of initiative and explore the issue of permissiveness versus restriction. We shall round out our picture of the toddler by discussing negativism and the diverse attempts to explain it. In the middle childhood period we shall be particularly interested in discovering the conditions for preserving the toddler's initiative, with its valuable characteristics of expansiveness, congruence, and confidence. We shall explore the concept of emancipation in adolescence and raise the question of whether the turmoil ascribed to the adolescent is actually typical of him. Finally, we shall examine the conditions under which self-reliant enterprise can be replaced by clinging demandingness or rebellious defiance, by uncertainty and self-doubt, and possibly by a profound retreat from involvement in the world of people and the world of objects.

Normal Development

Description of Initiative and Willfulness in the Toddler

In the first year of life a host of forces combine to bind the infant to his caretaker: the infant's helplessness, his vulnerability to physiological distress and concomitant need for comfort, and the pleasures he experiences through the ministrations of his caretaker; the mother's deep fulfillment in caretaking, her delight in the infant's smile and growing attachment, and society's valuing of human life and mother love. In the second year of life, another set of forces conspires to send the infant venturing out on his own: maturation of his motor apparatus enables him to move about and to manipulate objects with dexterity; his relative freedom from physiological distress both increases the time available for exploration and decreases the need for watchful parental care; while parents, for their part, respond to the developmental milestone of walking with enthusiasm and praise.

And venture out he does. The toddler is all over the place and into everything. His rapidly maturing body opens up a new world of motor activities, and he becomes absorbed in cruising, walking, climbing, jumping, trotting, sitting, and running. Some children use locomotion instrumentally—they walk because they have somewhere to go or they scale a sofa or chair "because it is there." For other children, locomotion for its own sake holds a special delight—they prance for the sheer pleasure of prancing, they crow with delight as they crawl upstairs, they jump off a step with glee. In addition, common objects in the environment hold endless fascination—pots and pans, chairs, knobs, paper bags, light switches, washcloths. A mother remarked, "I bought him a toy, but he played with the box." This everyday world is fascinating because it is totally new. The infant cannot have known previously that a drawer exists, that it can be pulled out and pushed in, that objects can be placed in it, that these objects disappear when the drawer is pushed and reappear when the drawer is pulled. The nature of everyday objects, as well as their functional properties and varied uses, must be discovered. Such exploration and discovery absorbs much of the toddler's time and energy.

Certain features of the toddler's venturing out are noteworthy. First are the spontaneous, congruent qualities. The toddler explores continuously and wholeheartedly. He explores because exploration is intrinsically interesting and because it is the most absorbing activity in his life at the time. The next feature is the toddler's remarkable capacity for sustained attention, especially in contrast to the ephemeral nature of attention in the neonate. Even by adult standards he has a well-developed ability to "tune out" external and internal distractions so as to concentrate on the many objects which interest him.

We call the toddler's venturing out initiative, having in mind the dictionary definition "self-reliant enterprise." His enterprising nature is obvious, and he literally can rely on himself to do things which previously his parents had to do for him. But we mean something more by self-reliance. We assume that, as the toddler repeatedly experiences his capabilities in managing his body and in exploring objects, he begins to develop a concept of himself as an agent of effective action; through doing he grasps the idea of himself as doer. With repeated mastery he also begins to trust his ability to succeed in what he undertakes; he begins to develop self-confidence.

We have been discussing initiative in terms of locomotion and exploration of the physical environment. The toddler's initiative in relation to the human environment is equally significant for his personality development but, ironically, it has received relatively little attention from psychologists. We might say that the toddler is continually trying to understand and manipulate human beings as well as physical objects, but such an equation is not helpful; since people and objects differ in fundamental respects, we cannot extrapolate from one to the other. We must learn what aspects of human behavior the toddler is capable of understanding, how accurate or distorted his understanding is, how he evolves techniques for manipulating others, what needs such manipulations satisfy, and the effects of successful and unsuccessful manipulation on his self-reliance. Since we know so little about any of these topics, we must content ourselves with observational evidence that the toddler can be socially perceptive and adept at eliciting responses from others. Note the following account of an eighteen-month-old boy who was being toilet trained:

Ronnie is in the living room and his mother is nursing his baby sister. Mother is absorbed in nursing. Ronnie tries to get her attention by climbing in and out of a nearby chair. Mother ignores. He goes to the stairs and starts jumping flatfooted off the bottom step, laughing excitedly. Mother ignores. He gets a stick and starts walking around banging it on the floor, on the playpen rail, on the stairs, becoming more excited and noisy. Mother ignores. Suddenly, he stops dead still. He looks at his mother and whispers, "Potty." She immediately puts the baby down and rushes him to the bathroom. (A Freudian theorist would say that it is no accident that we refer to the toilet as "the throne.")

Or, again, consider the following behavior in a boy who was observed at periodic intervals between twelve and eighteen months. At twelve months he was fascinated by a gold box which was one of his mother's prized possessions. His early attempts to explore it were promptly and severely punished. Because he was a lively and resilient toddler, he persisted in his efforts in spite of loud shouts and hard slaps. Gradually, he began to shift his interest from the box to the mother. Now he would stand in front of the box and slowly bring his hand closer and closer, meanwhile glancing at his mother out of the corner of his eye. He was testing to find out at what point she would strike. By eighteen months, he had completely lost interest in the box per se and ignored it when the mother was not in the room. However, as soon as she entered, he would reach toward it with a provocative smile. Almost invariably, the mother would fly into a rage. This toddler had experimented with producing a reaction in his mother just as he had experimented with the workings of physical objects, and he succeeded in gaining complete control over her. Infuriating her became a goal in itself, one which was more important than the pleasure of exploration or the pain of the inevitable punishment.

In these examples we see toddlers manipulating the mother in order to satisfy their need for attention, affection, and power. We can speculate that initiative is compounded of the inquisitive toddler's cognizance of how people behave plus his successful evolving of techniques necessary for eliciting the behavior which will satisfy his needs.

Initiative is characterized not only by the toddler's venturing out but also may be marked by an insistence on doing things himself. Let us watch the development of the "battle of the spoon" between a one-year-old and his mother. Previously this little boy had been willing to let his mother feed him; now, with no prompting from her, he seizes the spoon and tries to feed himself. Because of his inexperience and poor coordination, sometimes he hits his mouth, sometimes his nose or his hair or his ear. He also has not learned to make the subtle adjustments of the wrist necessary to keep the spoon level; since arm, wrist, and spoon form a rigid unit, the spoon tips as it comes close to his face, and cereal spills on bib, diapers, and chair. The fastidious mother, seeing her child festooned with oatmeal, tries to take the spoon, but her son holds on for dear life. Finally, he is overcome by brute force. He then clamps his mouth tightly shut and refuses to be fed, flails about wildly, and knocks the bowl of cereal to the floor. The war is on.

Or, note this observation of an eighteen-month-old child. Bruce is sitting by the door to the patio trying to fit an old-fashioned key into a keyhole. He has the idea of putting it into the keyhole and taking it out but lacks the skill and the understanding necessary to accomplish what he wants to do. The idea of reversing his arm movement in order to remove the key is particularly baffling. As he continues to fail, he becomes increasingly upset and angry. Finally, his mother, who has been watching with interest in the background, approaches, talks to him a soothing tone of voice, and gently tries to guide his hand. He turns on her furiously, hitting her on the face and shoulder. She retreats for a minute, and then tries a second time to help, and again he strikes out blindly at her. She retreats and leaves him fretting but still determined to remove the key from the hole. Her offers of help were rejected as infuriating intrusions.

We shall call the toddler's insistence on doing things himself *willfulness*. The label is apt because it captures the qualities of self-assertiveness and of arbitrariness which are so characteristic. The toddler is willful in spite of the inefficiency of his behavior and the exasperation of his parents; doing something himself takes precedence over efficiency and social approval. Like sustained attention, willfulness grows with remarkable rapidity and reaches something of a climax in the "terrible two's." The two-year-old can be exacting in regard to the most trivial details of routine care. He does not want to sit in this chair but in that one; he does not want corn flakes but raisin bran; he does not want his cereal in a small bowl but in a large one; he does not want his milk in a cup but in a tumbler; and he screams with rage if his mother puts sugar on his cereal before he tells her to. He acts as though arbitrariness were an inalienable right and is petulant when his demands are denied.

Concepts Related to Initiative. Before turning to the precursors of initiative, we would like to examine briefly other concepts which are relevant to ours but not equivalent to it. Heathers (1955) distinguishes instrumental and emotional independence. In the former, the child initiates his own actions and copes with difficulties without asking for help. In our terminology, he shows self-reliant enterprise. Heathers speculates that the more frustrations the child encounters and the more he expects to be helped, the more he will tend to seek help; the more the child expects he can reach his goal unaided and the more reassurance he receives while performing an activity, the less likely he is to seek help. Heathers defines self-confidence as the ability to master feared situations and is concerned primarily with techniques for coping with hurt without turning to others for aid. This emphasis on the mastery of fear and/or rejection tolerance is not particularly relevant to our concept of self-reliant expansiveness.

Heathers's article shows how two similar concepts, such as independence and initiative, are mixtures of identical and disparate components. Each conceptualization is equally valid but generates different lines of investigation. The same is true of the relations between initiative and achievement. The latter has been conceptualized (Crandall, Katkovsky, and Preston, 1960) as the child's attempt to gain approval or avoid disapproval for his performance in situations where standards of excellence are applicable. Such standards can be reflective, when the child looks to others, or autonomous, when he has his own criteria. The latter condition is similar to our concept of self-reliance. However, the study of achievement focuses on what is called attainment value or the importance the child attaches to the approval he gains and the disapproval he avoids. It is also concerned with the complexities of the standards the child uses in judging his competence. Again such issues are not our primary concern.[1]

In general, there is a good deal of overlap between our term initiative and what others have labeled independence, competence, mastery, or achievement. The conceptualizations tend to complement rather than contradict one another. It might even be possible to have an omnibus definition which would embrace them all. Such a definition, however, would run the risk of riding off in all directions at once. Our more limited term serves the purpose of providing a point of reference in a sprawling territory.

Precursors of Initiative and Willfulness in Infancy

Initiative, willfulness, and negativism (which we shall discuss subsequently) is as striking a behavioral constellation as smiling, attachment, and separation anguish (see Chapter 2), and it is equally significant for personality development. Unfortunately, psychologists have paid comparatively little attention to exploring its origins,[2] e.g., there is nothing comparable to Ainsworth's delineation of the stages of attachment. Lacking large-scale empirical investigations and extensive theorizing, we are forced to rely on fragmentary data and tentative conceptual schemes. We shall present four precursors of initiative and willfulness which are present in infancy: exploratory behavior, the concept of the self as a causal agent first in relation to physical objects and next in relation to people, and, finally, willfulness in self-feeding.

Exploratory Behavior. A number of investigators have been struck by the early appearance of exploratory behavior, by its persistence, and by the fact that it is

[1]For a summary of empirical findings, see Crandall (1963).

[2]For a vivid description of early initiative, see Murphy (1962).

intrinsically rewarding. Stott (1961) observed his son between the ages of one month and eighteen months and recorded a number of behaviors which were unrelated either to organic need or to social interaction. The activities were marked by concentration, perseverance, gratification with results or annoyance at failure. Typically, they were neither initiated by the parents nor sustained by parental approval; in fact, certain activities persisted in the face of parental annoyance or concern. In the first six months of life, the infant's behaviors were simple; e.g., he studied his hand movements (at 78 days) and repeatedly blew "raspberry" noises (at 164 days), in exploring the workings of his body. In other instances, he was fascinated with his ability to produce changes in the environment, such as his ability to make a noise by banging a napkin ring on the high chair tray. Stott was also impressed by his son's preference for newness even at the expense of efficiency in meeting a need; for instance, he tried new ways of feeding himself even when he got less food as a result.

Rheingold (1961b), on the basis of more extensive and systematic research, discusses her observations of the three-month-old infant. She stresses the infant's activity. He is alert and attentive. He not only responds to environmental stimuli but also actively searches for environmental stimulation. His eyes, in particular, move freely and often, focusing on one object after another. When he is picked up, his first activity is to look around and take in his surroundings visually. Rheingold describes the typical sequence of behavior:

> Intent and fixed visual regard is accompanied by a cessation of physical activity. The infant is showing the orienting reflex of Pavlov, as if asking, "What is it?" On occasion, however, there follows the response I have already enumerated, the brightening of the face, the smiles, the marked increase in total bodily activity, and the vocalizations, a sequence which is often repeated again and again.
>
> The simple enumeration of the components of the response, however, omits what is so striking, but so obvious as to be ignored, that the infant is displaying what seem to be the emotions of delight and glee...
>
> This pattern of behavior in the infant is all the more remarkable and worthy of study because of its maturity relative to his other patterns of behavior. It is already in the form it will keep throughout his life. Fetal as he still is in many ways, especially in comparison with other mammals, in this pattern of behavior he is precocious beyond his age (p. 167).

Thus, Rheingold observes in the three-month-old infant the beginnings of characteristics which lie at the heart of initiative—active engagement and the capacity for sustained attention.

Exploratory Behavior, Curiosity, and Theories of Motivation. We should like to present some theoretical accounts of exploratory behavior and curiosity, its ally.[3] In order to do so, we must first review certain features of motivational theory in general. In the past, motives were traditionally divided into physiological or primary drives, which are directly related to organic needs, such as hunger, thirst, pain, and sex, and learned or secondary drives, which result from the organism's transactions with his environment, such as fear and the desire for wealth or prestige. Primary drives were considered basic in two ways: they are essential to survival and they serve as the base for the acquisition of secondary drives. The classical conditioning model of attachment (see Chapter 2) is an example of this view of motivation; attachment (a secondary drive) develops only because the mother is constantly associated with the gratification of hunger, thirst, and other primary needs. A number of psychologists have studied the nature of organic needs, since these were considered to be prime movers. The study of secondary drives tended to be considered less important.

Exploratory behavior or curiosity was an embarrassment to this neat bifurcation of motives. It was not organic like hunger, since it did not originate in any physiological imbalance and did not seem essential to survival, but it also was not secondary, since no one could give a satisfactory account of how it could be learned. Even more serious, curiosity ran counter to the time-honored homeostatic model of motivation. According to this model, physiological imbalance, or tension, produces persistent activity culminating in a consummatory response which returns the organism to a state of quiescence. Hunger is the best example of the homeostatic model–the organism progresses from satiation to tension to food-seeking to eating to satiation. Exploratory behavior does not conform to this pattern. While the tension of primary drives is noxious to varying degrees, the tension of exploration is positive–the organism is interested, stimulated, even excited. Since tension is not noxious, quiescence is not a sought-after goal; on the contrary, it is an inimical state of inaction or boredom which must be avoided. Being well fed is a comfortable state; having nothing to do produces restlessness and dissatisfaction. Thus, exploratory behavior reverses the affects associated with both tension and quiescence.

To contrast primary needs and curiosity further: Primary needs are intense, peremptory, and satisfied by relatively specific consummatory responses. Curiosity, or the need to explore, is more moderate but more pervasive; it can be diminished or overruled by stronger drives, but it readily reappears when such diversions are removed. Curiosity has no specific consummatory response and it is not tied to tissue needs but to the novelty or complexity or suprisingness of environmental stimuli.

Because exploration and curiosity were an embarrassment to popular theories, they have often been neglected in the past. Renewed interest in observing infant behavior has been a potent factor in forcing psychologists to come to grips with them. The data clearly show that a need to explore is as basic to the human infant as hunger and thirst. It does not depend upon the primary drives for its existence, nor is it

[3]For a review of the evidence for and explanations of curiosity and exploratory behavior, see Maddi (1961).

derived from social approval. It is primary in the sense of being intrinsic to the human organism, without being organic in the traditional sense of being essential to physical survival. Since traditional theory was unable to accomodate such a motive, efforts are being made to formulate new concepts which will fit the data.[4] Woodworth (1958) writes about a primary motive to deal with the environment, to take in the environment through the senses and manage it effectively through action. In a similar vein, White (1960) writes of an effectance motive:

> My proposal is that activity, manipulation, and exploration, which are all pretty much of a piece in the infant, be considered as aspects of competence and that ... we assume one general motivational principle lies behind them. The word I have suggested for this motive is effectance, because its most characteristic feature is seen in the production of effects upon the environment (p. 102-103).

Mittelman (1954) favors an independent motor urge which impels the child to locomote and manipulate and which is intrinsically pleasurable. Hendrick (1942) proposes a broader "instinct to master" or an inborn urge to do and to learn how to do. Hunt (1960) attempts to integrate the data on drive reduction and exploration by means of the concept of an optimum level of activation: when an organism is below this level (as in the case of a monotonous environment or a boring task) variation in stimulation raises the level and is reinforcing; when the organism is above the optimum level (as when physiological needs or anxiety dominate) a decrease in stimulation is reinforcing.

Evaluating the merits and limitations of these different conceptualizations can be left to the student of human motivation. The important point for our discussion is that the homeostatic model can no longer be considered comprehensive; we must have a model which can account for the need to reduce tension at certain times and to increase it at others. In Chapter 2 we saw an example of this dual need when we discussed mother-as-comforter and mother-as-(visual and auditory) stimulus. The infant needs to be relaxed and calm, particularly when primary drives are intense and he is in a state of physiological distress; but he also needs visual and auditory stimulation, particularly when he is in a state of alert wakefulness and ready to explore his surroundings.

The Self as Causal Agent: The World of Objects. In the process of exploring the physical world, the infant begins to evolve the concept of himself as the agent of change. While the adult observer clearly perceives that the infant is making things happen—that he is pulling his blanket up or shaking a rattle or bringing a bead to his mouth—the infant himself has no such understanding. Adults are so used to thinking in terms of personal causation that they rarely realize the idea must be learned. When they write, it is obvious that the movement of their hand is the essential event causing the movement of the pen across the page. For the neonate, there are no such "obvious"

[4]The definitive article on the ferment concerning motivational models is White (1959).

connections between his actions and environmental changes. Remember that he experiences only a succession of sensations differing in quality, intensity, and duration (see Chapter 2). The movement of an arm, an ache in the stomach, the sound of a voice, are undifferentiated elements in a flowing stream. The infant only gradually learns to organize his experiences in terms of stimuli which belong to his body and stimuli which come from outside–to distinguish "me" from "not me." (For a more detailed discussion of the development of this distinction, see Chapter 6.) However, the distinction must be at least dimly grasped before the infant can learn that the patterns of sensations which comprise "me" affect the patterns of sensations which come from "out there."

Piaget (1954) traces the steps by which the infant learns that he is a causal agent. During the first few months of life, the infant has, at best, made only a vague connection between sensations representing actions on his part and consequent sensations representing changes in the external world. By about the middle of the first year, however, a definite connection has been established. While lying in the crib, for example, the infant might begin to kick vigorously and, by chance, notice that a mobile attached to the crib bounces jauntily in the air. The sight delights him, and he proceeds to kick repeatedly and watch the mobile bounce. However, his understanding of the nature of the external world and of his place in it is still quite primitive. In his experience, the association is between the two temporally contiguous events, kicking and the delightful spectacle he sees. He does not grasp the true sequence–that his kicking shook the crib, the crib shook the rod holding the mobile, and the vibration of the rod caused the mobile to bounce. In essence, the infant says, "All I need to do is act and the world will change to suit my wishes." Subsequently, he may be observed kicking vigorously whenever he wants an exciting spectacle to last, even though the spectacle might be an electric fan moving from side to side, or a pocket watch dangling in front of him, or any happening in the physical world which may be totally independent of his actions and wishes. This faulty understanding, in which the infant believes his actions can effect changes in objects which, in reality, are independent of them, is called magical thinking.

The infant is approximately a year old before he achieves a more realistic notion of causality. The clarification occurs as the failure of magical gestures forces him to revise his ideas of their effectiveness and as his exploration of the environment enables him better to understand the nature of the physical world. One important lesson he learns is that temporal succession is not the essence of causality in the physical world, but that spatial relations may be more important; e.g., when a rolling ball knocks over a tower of blocks, the temporal sequence of events is not as important as the change in the spatial relations between ball and tower. To influence an object, merely doing something is not sufficient; one must do something *to* the object. (We shall discuss causality more fully in Chapter 6.)

The Self as Causal Agent: The World of People. Unfortunately, we know more about the infant's attempts to master the physical environment than about his attempts to understand and influence his social environment. We do know that the neonate fails to differentiate people from physical objects and that he gradually makes the distinction as he develops. People are more variable in their behavior than objects and thus have a higher interest value and a greater potential for satisfying the infant's need for novel stimulation. People alone can bring him comfort in time of physiological distress and thereby become cues for relief. And people are the only environmental objects which are both motivated to give him pleasure and adept at finding ways to delight him. Thus, people become more highly saturated with interest, comfort, and delight than objects, and his involvement with them is more complex and emotionally charged. By contrast, the one-year-old's exploration of the physical world typically is without affect; he concentrates intensely but signs of pleasure or distress are infrequent.

Assuming that people come to be more attractive objects, by what process does the toddler learn that he can influence their behavior? How does he learn that if he does X, another person will respond with Y? We do not know, but we do have some interesting leads. The prerequisite for social manipulation may be the attentive parent's consistent response to the infant's behavior in the first few months of life. The baby cries and squirms, looks and grasps, pouts and turns his head. Initially, these behaviors are solely manifestations of nonsocial needs; the infant is not using them even for communication, much less for control of his environment. The mother must "read" his messages and minister to the baby accordingly. Only as she does so can the infant establish a connection between his behavior and the mother's and subsequently take the significant step of utilizing his repertoire of responses to capture her attention (see Latif, 1934). Sander (1962, 1964) cites observational evidence demonstrating that between the fifth and ninth months the infant has taken this step of initiating social contact with the mother and of stimulating her to respond. Whereas, in the past, it was the mother who took the initiative in eliciting pleasurable responses from the infant, it is now the infant who attempts to set the playful exchange into motion. He may, for instance, reproduce some of the joyful excitement which has characterized their exchanges as an invitation for the mother to join in the fun. Thus, responses which were once reactions to the mother now are utilized to elicit reactions from the mother. Between nine and fifteen months the infant not only attempts to manipulate the mother but also becomes quite demanding of her. In fact, Sander reports that some mothers resent the constant time and attention which the infant requires at this stage. It may be that the infant is exploring the effects of his behavior in the interpersonal world as avidly as he explores the effects of his behavior in the physical world. When speech becomes more intriguing than locomotion, at the end of the second year, the child has a renewed interest in influencing the adult world through this new means of communication.

Instead of having a realistic grasp of his ability to control others, however, the infant-toddler probably has a highly exaggerated notion of his power. Freudian theory has been quite explicit in postulating such an inflated sense of power. As the infant experiences distress followed by the appearance of the nurturant mother in the first months of life, and as he builds up an expectation that she will appear when he is in need, he establishes a primitive connection between his expectation and her appearance. Since he has no concept of the mother as an independent individual with a volition of her own, he believes that it is his need and his wish and his cry alone which cause her to appear. In short, the infant regards himself as *omnipotent*. In the second year of life, when he begins to speak, the omnipotence of words is added to the omnipotence of wishes and actions. He discovers that when he produces a sound such as *mama* by chance, his mother reacts with delight; certain other sounds also will inevitably capture her attention and even make her do his bidding. Speaking becomes a more potent and more versatile means of controlling parental behavior than crying and the limited motor repertoire of infancy.

The Freudian concept of infantile omnipotence is similar to Piaget's concept of magical thinking which we just presented. Piaget addresses himself to the infant's struggle to grasp the nature of causality in the physical world while Freud is concerned with the struggle to grasp causality in the interpersonal world. Piaget uses the example of an erroneous concept of power over objects, Freud, of erroneous power over people. According to either view, the infant has not yet realized that a person or an object has an existence and a nature wholly apart from his own needs and wishes and actions. To put the issue another way, it would be remarkable if an infant did *not* regard himself as omnipotent! Let us look at what realistic thinking would entail. When his mother appears after he cries, the infant would have to think, "I cried; mother was busy in another room but she heard me; she could either have continued to work or come to me, and she chose to come." Not even an infant Einstein could reconstruct such a sequence, any more than Piaget's infant could reconstruct the steps between his kicking and the dancing mobile. The infant accepts his immediate experience at face value and makes causal connections between experiences which are associated (see Chapter 6). The result is magical or omnipotent thinking.

Willfulness in Self-Feeding. Willfulness, which can be so characteristic of the toddler, can also be present in the infant. Spock (1963) observes that the breast-fed infant is easily weaned to a cup around six or seven months of age. Feeding even becomes something of a social hour—the infant seems to prefer to smile and coo at the mother rather than to suck at her breast. In contrast, infants who have been bottle-fed, and particularly those who have been allowed to hold the bottle, show great resistance to taking milk from a cup. Interestingly, they will drink other liquids, such as juice or water, from a cup, but milk must be in the bottle. As these infants become increasingly attached to the bottle at seven, eight, and nine months, they become increasingly resistant to attempts to make them drink milk from the cup. "The clamp their mouths shut when the cup is offered and knock it away. A few of them permit the mother to pour the milk into their open mouths but, grinning, let it all run onto their bibs (p. 361)." Just as the toddler will later "stand on his own two feet," the bottle-fed six-month-old infant takes feeding "into his own hands." The battle of the spoon may have its origins in the battle of the bottle.

Spock accounts for his observations partly in terms of the infant's need to outgrow his dependent relationship with his mother. This explanation of behavior in terms of a corresponding need—the infant becomes more independent because of a need to be independent—is unsatisfactory. However, it would be difficult to find a satisfactory explanation of willfulness, or, indeed, to find a psychologist who has addressed himself to accounting for its origins in infancy.

Restriction of the Toddler's Initiative

Parental reactions to the toddler's venturing out vary widely. Some mothers may be relieved that infancy is over. They may have found caretaking wearing or upsetting and derived little pleasure from what they regarded as the infant's demanding, obscure, alien behavior. They see the toddler as having become a human being, whom they can talk to or read to, who is more comprehensible and companionable, and whose independence gives them more time for themselves. Other mothers, who found caretaking intensely gratifying, may be reluctant to close the door on this phase and may tolerate the toddler's autonomy while secretly cherishing bedtimes or illnesses or other occasions which revive his need for tender care. Still other mothers equate the growth of the child's initiative with their own loss of control over him; to them, walking means "getting into trouble," and they brace themselves for the new role of disciplinarian.

Techniques for dealing with the toddler also vary widely among parents. Some mothers like to maximize the toddler's freedom to do what he wants. They "baby-proof" the house by removing all physically harmful and highly prized objects, and they intervene only on those occasions when he is in physical danger. These mothers believe that nothing is more important than the child's learning on his own. Other mothers value expertness over discovery and begin directing the toddler's curiosity into channels they regard as important. They make a point of stimulating him verbally, labeling objects for him, reading to him, and reciting nursery rhymes; they buy educational toys, explain the nature and workings of household objects, and participate in his play in order to guide him to the next level of complexity. They can make activities pleasurable and intriguing for the child, but he grows through their mediation, not on his own. A third kind of mother may begin a deliberate program of restricting the toddler's freedom. She may purposely leave out objects which must not be handled, and use them as opportunities to teach the child, gently but firmly, not to touch. These mothers do not want the toddler to believe that every wish must be granted and that everything is his for the taking. He will be part of a society in which he is one among many, where the wishes as well as the property of others must be respected; his distress in the face of immediate frustrations will be more than compensated by the rewards of socially responsible behavior.

These three techniques and their rationales represent only a sampling of parental behavior. Most parents use a variety of techniques to meet different situations, as dictated by reason, temperament, and mood. However, all parents, sooner or later, and with varying degrees of persistence and emotional involvement, begin to curb or redirect the toddler's autonomous behavior. *No* and *Don't* appear on the scene. In place of unrestricted expansiveness, a conflict now exists between the toddler's desire to do what he wants and the parent's demand that he inhibit or redirect his behavior. If the child is docile and the parent benign, the conflict may be minimal; but if a lively and determined toddler runs headlong into an insistent and demanding parent, the battle of the bottle and the battle of the spoon may give way to what mothers aptly label the "battle of the wills." The field of battle may be the toilet (as Freud emphasized), or a wall socket the toddler wants to explore with his finger, or a gas jet on the stove he wants to turn on. Regardless of the setting, the willful toddler is capable of resisting and evading parental prohibitions as well as exasperating and exhausting the parent.

We shall devote our next chapter to a detailed examination of the processes involved in the toddler's learning to inhibit his behavior. At this point we shall concentrate on only one aspect of the confrontation between parent and child—namely, the controversy concerning restriction of initiative.

The Controversy Concerning Restricting Initiative. In our society, affection is good; mother love is sacred. The value of initiative is ambiguous. One cultural tradition prizes individualism, autonomy, self-assertiveness. For a young adult to be subservient to the state, to an employer, or even to his own family is un-American. He should be independent at least, a hero at best. On the other hand, there is an equally strong tradition of conformity. The average man, the solid citizen, the all-American boy are respected public symbols, as is the "normal," well-adjusted, well-rounded child who conforms to approved stereotypes of social behavior. Thus, society presents an ambiguous mandate to the parent, who is left to decide for himself at what point liberty becomes license, independence becomes irresponsibility, and compliance becomes automatism. Both the parent of the well-behaved child and the parent of the self-expressive child will find friends and foes among his neighbors.

We find a similar divergence among theorists on the issue of management of initiative. Some theories are predicated on what might be called the growth principle—the essence of man's nature is to expand and develop, to set new tasks and master them, to venture out and assimilate an ever-widening range of experience. In these theories, growth typically is regarded as supremely valuable. Parents are assigned a supportive or facilitating role. Adler, reacting against the philosophy of "breaking the child's will," championed a child-centered approach to socialization, in which the parent accomodates himself to the child's natural bent. Rogers (1959) takes an even more extreme stand, maintaining that the child is the best judge of what will enhance and what will impede his growth. The parent may present alternative behaviors and may make the child mindful of unpleasant consequences of certain choices, but he should never preempt the child's natural urge to try out a new experience and judge its value for himself. Such growth-oriented theories echo Rousseau's doctrine that the natural man is good and that society corrupts.

The growth principle is alien to classical Freudian psychology, in form and in spirit. The emphasis there is on the irrationality and the excesses of infancy and early childhood and on the infant-toddler's weak self-control which is easily overwhelmed by possessiveness, jealousy, and rage. While the toddler's newfound independence should be encouraged as much as possible and conformity should be amply rewarded with love, the parent must function as the decision maker. The sympathetic but authoritative parent protects the toddler from his own grandiose impulses and disrupting passions. The firm parent also serves as a model of mature control.

It is no accident that the champions of unfettered initiative emphasize the toddler's constructive expansiveness, while the champions of authoritative control emphasize the intensity of his passions. The two theories convey different images of human nature. While each has an internally consistent stance concerning the socialization of initiative, their positions differ because they are based on different premises. The inconsistencies in American middle class ideology and in parental attitudes toward initiative have their counterparts in personality theories.

Negativism in the Toddler

The industrious one-year-old becomes the terrible two-year-old. He is imperious and temperamental. Deny him a simple request and he throws his toys around or falls on the floor in a fit of temper; ask him to do something which does not fit his interest at the moment and he stubbornly or violently resists. Some parents achieve a temporary victory by asking the child to do the very thing they want him *not* to do, but such maneuvering soon taxes their ingenuity. Most of them take comfort in the fact that negativism in toddlers is so common that it is considered normal and that it will pass in time. In spite of its prevalence, negativism has failed to capture the interest of present-day developmental psychologists. While theoretical speculation abounds, objective data and critical evaluations are meager. A search for objective studies, for instance, quickly takes one back forty years or so (Reynolds, 1928).

Theories of Negativism. In theoretical accounts, the term negativism usually includes the child's determined effort to ignore a parental request, reacting by saying *No* or by doing the opposite of what is requested.

Gesell and Ilg (1943) have presented the most benign, constructive interpretation of negativism. They view it as a natural consequence of the toddler's inexperience in making choices. The two-year-old vacillates between two opposing actions because he is not used to making decisions and does not know where he wants to go. By the time he is three, he is conforming and has himself well in hand. Murphy (1947) interprets negativism in the context of learning theory. He cites experimental evidence indicating that when two incompatible responses compete, the dominant response not only wins out but also is strengthened indirectly by the weaker one. Thus, the child's determination to do what he wants is strengthened by the parent's demands that he do something else. Except in cases where the child has no strong investment in continuing an activity or where the parents make heroic efforts to disrupt and redirect his activities, demands to "*Do this*" only reinforce the child's determination to go his own way. Paradoxically, continual parental pressure builds up in the child a habit of ignoring which the layman terms stubbornness.

Other theories deal with negativism in the context of general personality development. In the classical Freudian theory, negativism is derived from the power struggle inherent in toilet training. The toddler, who revels in his newfound control over the muscles of his body and his concomitant ability to act independently, is keenly aware of the pleasures to be gained from retaining and evacuating his feces. Being forced to go to the toilet at a specified time and place threatens to deprive him of his sense of self-determination and to interfere with his pleasure in defecating at will. If the mother adequately rewards him with love, and if she helps him to be proud of his ability to control his bowel movement, then he is compensated for his loss of autonomy and pleasure. But if she is harsh, demanding, or manipulative, her efforts at toilet training become a noxious intrusion, and the toddler reacts with anger and a stubborn determination not to yield. Withholding feces is the physical embodiment of the psychological attitude of willful uncooperativeness, and *No* becomes an angry defiance of parental authority (see Chapter 7).

Ausubel (1950) stresses the dramatic change in status which occurs when the infant becomes a toddler. The infant is the center of the universe, his wish is the parent's command, and he is omnipotent. Although locomotion (walking) enhances his sense of omnipotence, it ushers in a change in parental attitudes. Parents now demand self-control, obedience, and conformity. Instead of being the center of the family, he becomes just another member. This change in status causes a crisis in the toddler's life. He is both afraid and reluctant to relinquish his illusion of power. Negativism is his desperate, last-ditch attempt to bend his parents to his will. He fails, as he must, because the illusion is no longer tenable. Parents who indulge the toddler and cater to his tantrums do him a disservice by postponing an inevitable step in his psychological growth.

Levy (1955) sees negativism as "autonomy raised to the nth degree." The toddler is reacting against infantile dependency and surging ahead toward self-determination. He resists all obstacles which bar the way. Sometimes his resistance is passive and takes the form of stubbornness which protects him from excessive parental demands for cleanliness, self-control, and intellectual achievement. Sometimes the resistance is active opposition which both protects self-determination and serves to retaliate against a strict and unloving mother. Negativism plays a particularly important role in preventing suggestible and compliant toddlers from becoming overly submissive. It also aids in the growth of self-control, since it keeps the locus of control within the toddler rather than making him dependent upon external directives.

Confronted with such varying explanations, we must decide which ones are complementary and which are incompatible. Negativism, like affection, may well have many roots and may serve a variety of functions. The toddler's inexperience in making choices, the tendency for weak responses to reinforce strong but incompatible ones, and the threat socialization poses to the infant's omnipotence and self-determination, all may be involved. However, there are also differences among the speculations, and these are difficult to reconcile. The classical Freudian emphasis on toilet training is not found in any of the other theories and, in fact, seems irrelevant to them. Regarding negativism as a maneuver which insures the growth of self-determination is incompatible with regarding it as a futile attempt to cling to infantile omnipotence.

The paucity of objective evidence on negativism does not help us to choose among these interpretations. The Freudian emphasis on the toilet training situation has not fared well (see Caldwell, 1964), but none of the other theories has been put to the test. Only one study (Frederiksen, 1942) tests the idea, common to many theories, that negativism is positively related to the amount of frustration produced by parental interference. The results were generally in the predicted direction but not statistically significant. There is also some evidence that obstinate parents have obstinate children (Caldwell, 1964). In an extensive normative study, Macfarlane and others (1954) found that negativism in boys was correlated with personality traits indicating an angry dependence in an emotionally vulnerable and irritable child. In girls, negativism was associated with tempers and attention-demanding but was less related to dependence. Finally, negativism showed no significant relation to health, intelligence, or any of the rated characteristics in the mother, such as nagging or a tendency to criticize.

All we can glean from these meager findings is that negativism seems to be part of a general pattern of anger in the child and that it may be enhanced by the parent's serving as a model of stubbornness or by a close but distressing parent-child relationship. The data might be interpreted as supporting Levy's observation that negativism is necessary in compliant and suggestible boys, but they throw little light on the important issues raised by the various theories. As we noted in Chapter 1, it is impossible to choose among divergent explanations when evidence is limited.

Legitimate and Arbitrary Negativism. Both Freud and Levy imply that negativism can be a legitimate protest against being coerced, pressured, and unloved. Levy goes so far as to claim that, while irritating to parents, negativism may serve a self-protective and even a growth-protective function for the toddler. Studies of preschoolers' behavior while taking intelligence tests offers further evidence that negativism can be a legitimate reaction (Murphy, Murphy, and Newcomb, 1937). A good deal of legitimate resistance appears when a child is required to do a task which is beyond his abilities. In another form of realistic opposition, the child is more interested in the task at hand than in the one the adult wants him to perform; as anyone would, he protests an intrusion. When the child is capable of doing a task, he may become negativistic if the activity involved calls attention to himself; e.g., when he must name parts of his body or stand on one foot. Negativism in these instances may arise out of the preschooler's acute self-consciousness. As we have seen, the dramatic increase in his ability to manage his body and objects and to express himself in words is accompanied by a clearer concept of himself as an independent entity. For some reason this emergence of self-awareness is not always accompanied by feelings of pleasure or triumph; instead, the child may be embarrassed or shy or he may forceably resist being the center of attention.

Keeping Freud, Levy, and the objective studies in mind, let us try to distinguish between legitimate and arbitrary resistance. Legitimate negativism allows the child to protest adult intrusions into his developmentally appropriate activity. He wants to venture out on his own, to follow his own interests, to set his own pace. Adult behaviors which pressure, divert, or embarrass him are frustrating and he reacts vigorously to ward them off. Even when the parent is acting in the child's own best interests, the toddler's protest is legitimate if the parent does, in effect, intrude upon his initiative. No matter how sympathetically one views negativism, however, one cannot regard all of its manifestations as legitimate. There is a "*no* for the sake of *no*" defiance, which is all out of proportion to the parental intrusion. It is as if any communication is an affront and the toddler must defend himself, even when his initiative has hardly been called into question. In addition, negativism can be used for the ulterior motive of manipulating and controlling the parent by exasperating him. Here again, the child's defense of his initiative is no longer the issue; instead, negativism is the child's special technique for expressing anger and achieving power for reasons which have nothing to do with protecting self-reliant venturing out. In general, the guidelines for deciding the appropriateness of negativisms must be based upon the child's developmental status in regard to initiative and on the conditions necessary for insuring its continued development.

Initiative from the Preschool Period to Adolescence

We shall begin this section by sketching some major areas of development of initiative from the preschool period to adolescence. However, our primary interest will not be with descriptive data but with the concept of the self-as-agent and the feelings of self-confidence which lie at the heart of initiative. We shall comment briefly on the self-defining aspects of initiative in the preschool period and then examine the concept of locus of control, which is pertinent to our interest in self-reliance. Our primary concern will be in discovering the factors in the child and in the parent-child relationship which insure the continued development of initiative from the toddler period to adolescence.

Areas of Development. In order to set the stage for our subsequent discussions, we shall give a brief and admittedly incomplete account of some major areas of expansion and their developmental trends. Fascination with physical objects, so pronounced in the toddler, continues throughout childhood. What is it? and, How does it work?, first asked of ordinary household objects, expands to cover the objects in the world and in outer space. The toddler also discovers that an object can be used as a tool—a stick can be used to rake in a crayon which is out of reach or a chair can be climbed in order to look out an otherwise inaccessible window. In early childhood, actual tools begin to come into their own, giving the child added power and precision—the pounding bench is replaced by hammer and nails, the blunt scissors by the sewing kit; ahead lie tinkering with a jalopy, learning to type, and, perhaps ultimately, programing a computer. The body also continues to be used for experimentation and discovery. Walking is only the beginning of a series of activities which includes running, jumping, hopping, kicking, and, subsequently, the highly developed skills required in athletics and dancing. The preschooler delights in words. He learns their meaning, experiments with expressive nonsense sounds, deliberately misuses labels, discovers rhymes, and enjoys the sheer sound of words apart from their meaning (Werner, 1957). The ten-word vocabulary of the eighteen-month-old expands to the eighty thousand word vocabulary of the high school freshman. Close on the heels of verbal expansion comes formal schooling, beginning with the basic skills of reading, writing, and arithmetic, and gradually encompassing the cultural heritage of society. Social relations with peers have their own developmental trend, from fleeting curiosity in infancy to the transient relationships of the first grader, the spontaneous and cohesive groups of the fifth grader, and the adolescent's strong identification with his chosen gang.

The Self-As-Agent: Self-Definition, Pride, and Locus of Control. When we defined initiative we stated that we were not only interested in the toddler's manifest behavior but also in the evolution of the concept of himself as an agent of change and in the self-confidence which accompanies his venturing out. Murphy (1962) sees self-reliant enterprise as the beginning of a sense of identity. Through doing, the child begins to discover who he is, what he is like. "I can do" often precedes "I am"; for instance, "I can jump" comes before "I am three." Just as the young woman learns what kind of mother she is through actually caring for an infant, just as the bridegroom learns what kind of husband he is by meeting the specific challenges of marriage, just as the freshman learns what kind of student he is through managing the complexities of college life, so the preschool child begins to form an image of himself in the process of coping with the tasks appropriate to his stage of development. Murphy also sees in the preschooler's mastery evidence of a healthy pride, a spontaneous sense of his self-worth. His positive self-regard is quite different from vanity or conceit. "I can do what I want to do"; "I enjoy doing what I do"; "I am proud of my successes"—these are invaluable feelings which can sustain the child as well as the adult through many difficulties encountered during the course of development. Narcissism, or self-love, can be as healthy as a love relation with another person.

Rotter's concept of locus of control (Rotter 1966; also see Lefcourt, 1966) is also pertinent to the concept of self-as-agent. Rotter argues that rewards and punishments do not act in simple, mechanical ways to strengthen or weaken response patterns. The important variable is whether or not the person perceives a causal relation between what he does and the ensuing reward or punishment. If the person perceives an event as contingent upon either his own behavior or some characteristic of himself, he believes in internal control of reward and punishment. If the person perceives rewards and punishments as due to some factor other than himself, such as luck or chance or fate, he believes in external control. These perceptions are crucial in determining whether rewards will strengthen or punishment will weaken behavior patterns; e.g., if a ten-year-old is given a dime for shining his shoes, he is apt to continue shining them in order to make money, but if he finds a dime on a street corner, he will not continue to return in the hope of finding more dimes. Note that Rotter is dealing with self-perceptions, not with feelings of self-confidence or with the sense of identity and pride Murphy delineated. Locus of control concerns the individual's perception of personal responsibility for events.

Although most of the research stimulated by Rotter concerns adult behavior, there are a number of interesting objective studies of children. Bialer (1961) constructed a Locus of Control scale, consisting of questions such as these: "Do you really believe a kid can be whatever he wants to be?" "When nice things happen to you, is it only good luck?" The scale measured the extent to which a person believed that achievement of a goal is due to ability and that failure is due to personal shortcomings. In a group of children between six and fourteen years of age, Bialer found a progressive increase in the belief that one's behavior influences outcome. However, the increase was more closely related to mental age than to chronological age. Battle and Rotter (1963) found that middle class sixth and eighth graders had stronger beliefs in internal control than did lower class children. Lower class Negroes believed more strongly in external controls than did middle class whites or Negroes.

Crandall, Katkovsky, and Crandall (1965) conducted the most extensive series of studies with children but addressed themselves exclusively to intellectual-academic situations. They devised an Intellectual Achievement Responsibility Questionnaire (IAR) which they administered to children between the third and twelfth grades. Their most relevant finding for our purposes was that the IAR was correlated highly with the amount of time boys chose to spend on intellectual activities during free play and with the intensity of their striving in such activities. This finding agrees with studies of adults–the individual who feels responsible for his success or failure shows initiative in seeking out and assimilating information that is available in the environment. A related finding comes from a national survey of grammar schools which was concerned with equality of educational opportunity (Coleman et al., 1966). The survey reported that a pupil's belief in his own control of his destiny was more important to achievement than any of the "school" factors measured, such as facilities, teachers, and curriculum. Except for Orientals, racial minority pupils have less conviction than white pupils of their ability to affect their environments; when they are convinced of their effectiveness, their achievement is higher than that of white pupils who lack such conviction. The feeling that education does not matter, that it will make no difference, is the most significant deterrent to academic achievement.

Crandall, Katkovsky, and Crandall make an important point about the development of ideas concerning locus of control. While there is evidence of a generalized expectancy for controlling reinforcements in adults (Rotter, 1966), there is no reason to assume such generality in children. The generality of attitude is a developmental issue to be charted and understood in its own right. To illustrate this thesis, the authors present data from the IAR suggesting that assuming responsibility for academic success is different from assuming responsibility for academic failure, the latter being more stable. They speculate that punishment might have a greater impact than reward upon external-internal responsibility beliefs and produce a more durable effect. There is also evidence suggesting that young children learn personal responsibility for success and for failure separately, while a more generalized attitude of responsibility begins to emerge only in the older children.

The Conditions for Preserving Initiative

The initiative of the toddler is marked by an eagerness in venturing out, a congruence between what he does and what he wants to do, and self-confidence. All these qualities are worth preserving as he develops. How can this be done?

White's (1960) is the first hypothesis we will consider and also the simplest. A person's competence (which corresponds roughly to our term initiative) depends upon his history of being effective or ineffective in the many developmental ventures he undertakes. As each new venture succeeds, his self-respect and the respect he receives from others are increased; as he fails in doing what he should be able to do, he feels ashamed and suffers from being shamed by others. Since few individuals succeed in all their undertakings, people build up different pictures of their competence in various areas.

White's hypothesis not only has the appeal of simplicity, but it also has the empirical support of a number of studies. Silber and others (1961) have characterized competent adolescents as reaching out for new experiences, being active and purposeful, enjoying the challenges of problem solving, and taking initiative and responsibility for planning their futures. More to the point, they draw upon reservoirs of past successes to reinforce their confidence in themselves. In a similar study of healthy adolescents, Offer and Sabshin (1966) list among their findings the mastery of previous developmental tasks without serious setbacks. Sears (1940) tested the hypothesis that the goals children set for themselves, i.e., their levels of aspiration, are determined by their histories of success or failure on the particular undertakings. She found that successful children, defined both in terms of objective evidence—achievement in reading, arithmetic, and school grades—and in terms of subjective feelings of success during their entire school experience, characteristically set their goals slightly higher than their past achievements. Translated into our terminology, this means they realistically challenged themselves to master the next higher step. In contrast, a group who had known primarily academic failure set goals which were unrealistically high or slightly below their previous achievements. Sears interprets the behavior of the failure group as an attempt to gain social recognition by at least saying they will do better or as an attempt to bolster their self-esteem by exceeding their low estimate.

Also relevant is Keister's (1943) study of a group of three-to six-year-olds who were particularly immature in the face of failure—they retreated or repeatedly asked for help, rationalized, or had emotional outbursts, such as crying and yelling. She gave these children special training in mastery, beginning with simple tasks which could be done easily and gradually increasing the difficulty of the tasks. The children were never helped directly, but they were encouraged to persist and their successes were praised. These graduated experiences of successful mastery enabled the children to eliminate their previously immature behavior and to become persistent and self-reliant in the face of difficult problems. (Also see Jack, 1934.)

In contrast to this confirmatory evidence stands Macfarlane's (1964) sobering finding with a population of normal subjects who were systematically evaluated at intervals from birth to thirty years of age. Although certain competent children became competent adults, there were also many outstanding adults with histories of expulsion from school, academic failure in spite of high intelligence, and tendencies to be withdrawn and isolated. On the other hand, a group of children who had progressed satisfactorily during childhood turned out to be brittle, restless, puzzled adults. Macfarlane's data are not retrospective, hence are less susceptible to bias than those of Silber and his co-workers and Offer and Sabshin. Her findings can be supplemented by innumerable clinical cases of adults with long histories of successful undertakings who are tortured by feelings of inadequacy and despair.

Thus, White's hypothesis has been confirmed only within limits. While it may be generally true that "nothing succeeds like success," it is not inevitably true. Something must be missing from the simple formula. White (1964) himself suggests one possibility, namely, the evaluation of the individual's competences by those he loves, admires, and respects, by his peers, and by society in general. A high level of competence might contribute little to self-esteem if it is not positively valued by significant individuals, or, conversely, a low level of competence which is instrumental in gaining love or respect might inflate self-esteem unduly.

The evaluations of people the child loves or prizes highly—our second clue to the preservation of initiative—is basic to the Rogerian theory of personality. Rogers (1959) stresses the importance of experiencing directly and fully all facets of development. Moreover, the self should always be the ultimate judge of the goodness or badness of an experience. Rogers postulates both an innate impulse to grow, which he calls an actualizing principle and an organismic valuing process, which enables the infant or child to evaluate each new experience in terms of whether it enhances or impedes psychological growth. Both are inherent in human makeup, and together they work toward the full realization of the child's potential as an individual. What, then, prevents him from developing fully? A more powerful force than self-actualization, called the need for positive regard, which includes the need for warmth, sympathy, acceptance, and respect. The child will do anything, Rogers says, even sacrifice his organismic valuing process, to satisfy his need for positive regard. For example, if a parent demands that a child be "good" in order to receive warmth and acceptance, the child will begin to value experiences in terms of the parental image of goodness rather than in terms of his own reaction. Instead of being free to find out for himself how it would feel to play hooky or put a frog in his sister's bed or steal a pencil from Woolworth's, he prejudges these experiences as "bad" and condemns them. The more a child must live up to a parental label—"he's a real scrapper"; "she's a doll"; "he's a brain"—the wider the gap between his actual experience and his conscious interpretation of the experience. Thus, the "good boy" cannot acknowledge feelings of anger or defiance as part of himself, even though the feelings continue to exist outside awareness and continue to be potentially disruptive.

How can this discrepancy between direct experience and self-image be avoided? Rogers introduces the concept of unconditional positive regard. To promote healthy development, the parent must completely accept, value, and prize the child as a person. This does not mean he must approve or condone everything the child does. The parent may not like the child's behavior, but he must be able to understand the way the child feels; the parent may convey his dislike or even forbid the behavior, but he must do so without demanding that the child disown his honest reactions. Acceptance also implies a sense of perspective on the parent's part; for instance, he reacts to the demanding willfulness of the two-year-old not as a personal affront but as a part of the growth process itself. Nothing in development is so extreme as to give the parent a reason to say, in essence, "If you do this (or, if you feel this way), then I can no longer value you." Only when the child is given such unconditional positive regard can he continue to have access to his basic feelings and preserve the self as the unbiased evaluator of his experiences.

While Rogers's general personality theory and his client-centered therapy have had a significant impact, his developmental theory has not been fully worked out and has generated little empirical research. A study by Ausubel (1954), although not in the Rogerian framework, is relevant. He found that fourth and fifth graders who perceived themselves as extrinsically valued asked for more assistance in doing things they were capable of doing on their own than did other children. Extrinsic value in this context is defined as the child's being accepted not because of what he is but because of something the parent values. In Rogerian terms, the dependent child does not receive unconditional positive regard.

We are not interested in Rogers's theory per se but in his answer to the question of preserving self-reliance. Rogers stands as a provocative champion for the child-centered approach. We do not have to accept his entire theory in order to learn something from it. He tells us, in essence, that the child will value himself as he is valued by his parents (and other significant individuals) and that he will have confidence in himself to the extent that his parents have confidence in him. We do not believe Rogers has the entire answer to the question, but his message is valid and represents an advance over the simple formulation that "nothing succeeds like success."

Our third set of leads concerning conditions for preserving initiative has been generated by Rotter's concept of locus of control. Parental values, especially as these reflect class or ethnic differences, are significant. Strodtbeck (1958) found that Jewish families, in contrast with Italian families, believe more strongly in an orderly world which is amenable to rational mastery. Consequently, the individual can more readily control his destiny, especially if he plans well. The Jewish families are also more approving of an individual's leaving home to make his way in the world, and they prefer individual to collective credit for work done. In short, they stress individual initiative. Rotter (1966) speculates that the middle class values regarding responsibility are taught to the child directly; however, he believes the child may also be influenced significantly by his perception of the realistic chances for success. The lower class child, then, would believe in external control of his destiny because he has very little chance of obtaining the rewards offered by the culture. (Here Rotter echoes White's simple formula concerning experiencing success.)

These findings and speculations sound eminently reasonable. The child who is told the world is orderly and that he can succeed if he tries, whose parents serve as models of this philosophy, and who sees many opportunities to experience success might well perceive himself as having control over his fate. The child who is told that life is capricious and governed by forces beyond his control, who is discouraged from becoming independent, and who sees little likelihood of success no matter how hard he tries, is apt to feel little responsibility for his fate.

We have two reservations about applying these speculations concerning familial, class, and ethnic values to initiative. An analogy from Adlerian psychology serves to introduce one reservation. Adler noted that a child may seem to be well adjusted when his environment protects him from stress, but he may develop behavior problems if placed in a less protective setting. Adler then distinguished adjusting to one particular environment from a general ability to adjust to varying environments and to the ever-changing demands produced by psychological growth. In a like manner, we can distinguish a child whose belief in internal control is dependent upon continual environmental reinforcement from the child whose belief persists in non-supportive settings. The distinction–conceptually an important one–may be difficult to make in a group of children who have not been put to the test of coping with unfavorable settings. While we may not know what makes the difference between a perception of responsibility which merely reflects and can only survive in a special environment and a deeply ingrained self-perception which persists under a variety of conditions, favorable and unfavorable alike, only the latter meets our requirements for self-reliance. A second reservation concerns the concept of locus of control itself. There is a difference between a self-perception and a feeling about the self. "I am responsible" differs from "I am confident." Internal locus of control can be accompanied by a variety of feelings, from assurance to uncertainty to despair. Locus of control does not deal directly with confidence and pride, which are essential affective components of self-reliance.

Our final set of leads concerning conditions for preserving initiative comes from two objective studies of parent-child interactions. Consequently, the leads are more specific and detailed than those we have presented so far. Both studies involved very bright preschoolers who had professional, well-educated parents. Baumrind (1967) studied a group of preschoolers who were self-reliant, self-controlled, curious, and content. Behavioral criteria for self-reliance included ease of separation from parents, a matter-of-fact rather than a dependent manner with the teacher, pleasure in mastering new tasks, leadership, and interest in making decisions and choices. Baumrind found that the mothers of these children were consistent, loving, conscientious, and secure; they respected the independent decisions of their children but would hold firmly to their own point of view once they took a stand; they accompanied direction with reason and communicated clearly; they demanded that the child perform up to his intellectual, social, and emotional abilities but were supportive of the child. The home atmosphere was sometimes highly charged with feeling and conflict but not disorderly and discordant. In general, then, these mothers were highly nurturant, highly controlling, highly demanding, and communicated clearly. They reinforced age-appropriate behavior, and, because of their warmth and involvement, their rewards and punishments had high potency. In their explicit communication and encouragement of verbal give-and-take, they maintained control without provoking rebellion or intimidating the child into passive acceptance. They served as models for activity, initiative, and openness in human relations.

Baldwin (1949) found similar results. He described an active "democratic," home environment in which the parents explained decisions, considered the child's wishes, and let him have a voice in policy making. Parents in these homes also gave the child responsibility for his own actions, expected mature behavior of him, placed a high premium on exploration and intellectual attainment, were warm, and imposed moderate controls. The children were active, curious, and participated fully in preschool activities. They were original, creative, socially outgoing, friendly, hostile, and dominating. In their bossing of others, they were usually successful. (Baumrind's children did not have the domineering, aggressive qualities found in some of Baldwin's preschoolers.)

Both of these studies strongly suggest that warm, dynamic, demanding, communicative parents have children who are high on initiative and self-reliance. The picture is significantly different from the warm, accepting, nonintrusive parents idealized by Adler and Rogers. Rather, they are pacesetters and models for pacesetting behavior. As long as they are loving, let the child have his say, and avoid overwhelming him with demands, they seem to stimulate vigorous, self-assured expansiveness.

Summary and Comment. We would like to summarize and comment on what we have learned concerning the conditions for preserving initiative. A history of successful mastery is important to insuring initiative, so the environment should provide the child with ample opportunities to succeed, both in his own eyes and in the eyes of others whom he loves or prizes. The parent who believes that an individual can control his destiny through planning, who rewards independent effort, and who serves as a model of enterprising behavior, enhances the child's feeling that he is responsible for his successes and failures rather than feeling that he is a victim of forces beyond his control. Rogers champions an extreme child-centeredness in which the parent values and supports the child's desire to test out experiences for himself, while Baumrind's and Baldwin's research suggest that parents who are standard setters and pacemakers, who are warm but somewhat strict and authoritative, have children high on initiative. The two parental images are incompatible, but there is no reason to assume that only one kind of parental behavior enhances initiative.

Our first comment concerns the relevance of the literature to our particular conceptualization of initiative. Remember that we are interested both in the manifest behavior of venturing out and in the general feelings of congruence ("I am doing what I want to do") and self-confidence ("I can take care of myself") which accompany it. Internal locus of control is relevant to self-reliance in that the child regards himself as responsible for events rather than believing that what happens is beyond his control. However, it seems to be dependent upon social settings which reinforce individual responsibility, and offers no explicit guides to producing a general self-perception which would persist in uncongrenial settings as well. The relation of locus of control to feelings of congruence and confidence is also uncertain. Baumrind's and and Baldwin's research tap attitudes of self-reliance and self-confidence, but the population they studied was a highly select one; we do not know how these children fared in meeting subsequent developmental challenges, or whether initiative in other populations (say, children of non-professional, lower middle class parents) would flourish under different parental treatment. White's and Rogers's concepts are relevant to self-reliance and self-confidence, but White's has not been fully supported by objective data, while objective evidence for Rogers's position is meager. Our search for ways of preserving the kind of self-generating, wholehearted, pervasive initiative characteristic of the toddler has been only partially successful.

Our final comment concerns a peculiar imbalance in the developmental account of initiative. In our discussion of initiative in the toddler and its roots in infancy, parents were pictured primarily as reacting to but not responsible for the child's behavior. In our present search for factors which sustain initiative, parents were assigned the major responsibility while characteristics of the child were ignored. The reversal in emphasis is not complete: e.g., in the earlier developmental stages the mother stimulates exploration and, through her sensitive caretaking, fosters in the infant the idea that he can influence others; and, in the later developmental stages, the child's self-confidence is reinforced by experiences of success apart from parental reactions. Yet, it is only a slight exaggeration to say that the picture of initiative is "parent-less" in the first three years, and "child-less" subsequently. Such discontinuity does not reflect reality. Parents undoubtedly play an important part in the early stages of initiative, while the effectiveness of parental acceptance, of class-determined values, of pacesetting, democratic treatment, is influenced by the strength of the child's need to explore, the vigor of his motor behavior, and his determination to pursue his own interests and to control his parents (see Chapter 10). A satisfactory answer to our question concerning the preservation of initiative, therefore, will require an integration of what we learned about the infant and toddler with the leads we have uncovered concerning parental behavior.

Initiative in Adolescence

An Example of Initiative. Initiative in adolescence deserves special consideration. Let us start with a concrete example of self-reliant enterprise. Silber and others (1961) provide a detailed picture of fifteen competent high school students in the process of deciding which college to attend. These students never doubted that they would go to college, since education was highly valued by their parents and reinforced by their peers. In making the decision, they were both self-reliant and enterprising. They actively gathered information about various colleges, reading catalogues and visiting campuses. Then each engaged in a process of trying to match his self-image to the image of a college. In their self-assessments, some students said they desired new experience, a more liberal intellectual outlook, a chance to be in daily contact with the opposite sex; others said they wanted to strengthen present values in a setting in which they could immerse themselves in familiar and rewarding pursuits without the distractions of a variety of courses or a demanding social life.

The adolescents were mindful of parental preferences and values when making their decisions. They could understand their parents' emotional investment in having them attend a particular college, major in a certain subject, or achieve intellectual recognition. They wanted to live up to parental ideals while maintaining their own independence. The parent-child relation can be described as collective bargaining based on mutual respect. Their final decisions represented attempts to satisfy both parental ideals and their own notions of where they should go. For example, a boy who was not a top student but whose father wanted him to make Phi Beta Kappa chose a college whose student body was not composed of the brightest, most intellectually aggressive students. For their part, the parents discussed their preferences openly and unequivocally. However, they also gave their children a large measure of freedom. When a child's decision ran counter to their wishes, the parents were disappointed, but they did not threaten the child with loss of love or respect.

It is instructive to juxtapose this picture of the eighteen-year-old with that of the eighteen-month-old. Certain general features are similar: the venturing out and, more important, the sense of congruence which comes from doing what they are most interested in doing; the ability to take matters into their own hands and to stand on their own two feet; the awareness of the disharmony between pursuing their own interests and accommodating to the directives of parents. Yet, the two situations are also vastly different. There is the difference in cognitive complexity: the eighteen-month-old trapped in the present, unable to anticipate five minutes ahead of time; the eighteen-year-old amassing, assimilating, ordering, and comparing quantities of information in order to reach a decision which lies months in the future. Although the eighteen-month-old has a notion of himself as an active agent, he has none of the self-evaluative capacity which enables the eighteen-year-old to take stock of himself in terms of his needs, goals, and values. The composition of the need to explore is different. The eighteen-month-old's need is pure and simple—intrinsically he wants to explore; the eighteen-year-old's need is now a mixture of intrinsic interests and the values he has acquired from the significant persons in his life, such as parents and peers. The confrontation with the parents is significantly different, since the parent is no longer an incomprehensible intrusion but a distinctive individual who has a point of view and reasons for this point of view, as well as feelings of his own. "Collective bargaining based on mutual respect" implies the ability to give in without fear of being taken over, the ability to compromise while still being in charge of one's destiny—all of which are beyond the toddler. In subsequent discussions we shall deal extensively with the topics of cognitive complexity (see Chapter 6), the assimilation of values (see Chapter 5), the ability to see another person's point of view (see Chapter 5), and the willingness to compromise (see Chapter 8).

Adolescent Turmoil. Studies of initiative in healthy adolescents are rare because the notion of a healthy adolescence in middle class America is itself rare. Like the "terrible two's," adolescence is typically pictured as a period in which turbulence is to be expected. Instead of initiative, willfulness and negativism abound. While turmoil is not inevitable, there are a number of factors in our society which make for such difficulties. The adolescent's ambiguous status is of special relevance to initiative; he is neither child nor adult but is caught in contradictory demands that he be both grown up and obedient.

The accepted American image of the young adult is that of the independent individual, free to choose his mate and his friends and to find his special vocational niche. The young married couple frequently moves into an apartment or a house of their own, expressing in concrete form their desire to be separated from their families. Even the acceptable psychological term for becoming independent of parents is "emancipation," which suggests that continued closeness to them is a kind of slavery. That elders are a reservoir of wisdom and close family ties a source of comfort and security are ideas which run counter to our culture. (The ideas are clearly culture-bound, however. Other societies hold the opposite view.) The adolescent feels, for the first time, that society's image of independence is within his grasp. Physically and sexually, he is no longer a child but a young adult. His intellectual capabilities also are approaching maturity. He is socially sensitive, can form meaningful friendships, and can participate in group activities. He has the capacity to work and to devote himself to an ideal.

What baffles the adolescent is the fact that while society offers the lure of independence, it does not give him a clear mandate to be on his own. On the contrary, there is a prevailing feeling that he lacks the experience, judgment, and maturity necessary for assuming adult responsibilities. Until recently, the minimum age for voting was twenty-one, and states vary in regard to the legal minimum age for driving a car, owning property, and purchasing liquor. The increasing importance of higher education serves to prolong financial dependence on parents and to hinder vocational independence. In spite of the sexual revolution, the older generation still look askance at free sexual experimentation and does not condone teen-age marriages.

Parents also tend to be inconsistent in their expectations. On the one hand, they want the adolescent to be more responsible, to start earning his own money, to date, and to be popular. Yet, they want to continue being the adolescent's main guide to proper conduct and to make major decision about friends and dating as well as minor decisions about clothes and grooming. They feel responsible for the adolescent's education and for guiding his plans for the future. Equally important, parents have their deeper, more personal uncertainties. Their status is as much in flux as the adolescent's. They must face the prospect of relinquishing the child who has been a prime source of gratification to them and of relinquishing the responsibility and authority which epitomized their parental function as socializers. If the adolescent must face the problems of growing up and no longer needing his parents, parents must face the problem of growing old and no longer being needed. While the plight of the adolescent has been analyzed by psychologists, sociologists, artists, and observers of the American scene in general, the readjustments and turmoil of the parents have received scant attention.

In addition to encountering inconsistent expectancies from society and parents, the adolescent has his own ambivalence concerning independence. Although he may desire autonomy, he cannot become adult overnight. Finding his niche in a complex society and in the intricacies of sexual relationships inevitably takes time, and the challenges he faces are momentous and baffling. In his uncertainty he is tempted to fall back on the security of parental guidance. If parents gratify this need his pride is hurt, and he rejects them with a defiant, "I am not a child any more"; if his parents fail him, he covers his anxiety with an angry, "You never understand me." Subsequently, he may feel guilty about his outbursts and become loving and tractable until the next cycle of needfulness occurs. The adolescent, like his parents, alternates in his feelings of warmth and resentment. Unlike the parent-child relationship described by Silber and others, disharmonies can strike the sensitive nerve of love and respect. When this happens, the battle of the wills becomes the battle of the generations; parents view the adolescent as wild, disrespectful, amoral, while the adolescent willfully seizes upon the latest fad in dress or music or social behavior to proclaim his independence. He may also cover his uncertainty by bragging, or he may assume a sophisticated pose of knowing everything in order to conceal his frightening ignorance of what lies ahead.

How can the adolescent transition be made less stressful? Typical recommendations include the gradual withdrawal of parental protectiveness, encouragement of greater responsibility, acceptance of the adolescent, and avoidance of premature pressures for independence. Healthy peer relations and emotional attachments outside the family serve to fill the adolescent's need for affection, respect, and guidance. Opportunities to participate in vocationally or socially useful activities promote realistic images of adult life and develop skills in planning and making independent choices.

These recommendations may be well taken. Remember, however, that our special concern is with the preservation of initiative, not with the production of a socially acceptable model of adult independence. The two are not synonymous. The adult who conforms to a prescribed social role may or may not be self-reliant. He may be using conformity as a means of working out a fulfilling personal destiny, but he may also be seeking the safety of social conformity out of a feeling of powerlessness to manage his life. In like manner, the offbeat adult may be floundering helplessly or he may be assured, decisive, and have a strong sense of direction. We do not know the relevance of the above recommendations to the preservation of initiative as distinct from the production of a socially acceptable example of an independent adult.[5]

The Problem of Depicting the Average Adolescent. Up to this point we have been discussing adolescence as a period of turmoil. Have we been describing the average American adolescent? Douvan and her associates (1966) claim we have not. We have been dealing with only a small segment of the population, whose dramatic status has caught the fancy of psychologists. The modal adolescent is quite different. Adolescent girls, in particular, show little rebelliousness. In middle childhood they have been docile, sociable, and loving. In adolescence their transition is from being the recipient

[5]For two summaries of the topic of emancipation in adolescence, see Blos (1941) and Mussen, Conger, and Kagan (1969).

of love to being the giver of love—as a girlfriend, as a wife, and as a mother. They generally accept parental control as justified and plan to model themselves on their parents. Again, the shift is from recipient to agent. Boys are more in conflict. Their lives revolve around assertion, work, and proving their worth in a competitive world. Since they feel they must "become their own man," they chafe against parental advice.

Douvan minimizes the battle of the generations by pointing to the fact that it is fought in the arena of fads not values. Length of hair, style of clothes, and kind of music are typical foci. The modal adolescent does not challenge basic values—that he is loved and respected by his parents, or that love and respect are essential. Nor does he typically become criminal, promiscuous, or disreputable; instead he rebels within a middle class framework of provocation. Douvan argues that there is a tacit understanding between generations that central values will not be violated and that both parties implicitly agree to wage a pseudowar so that neither will be basically threatened.

Douvan's research is a valuable antidote for the tendency to see the typical adolescent as delinquent or alienated. Whether she is correct in her interpretation of the battle of the generations is another matter. If intense affect is generated over small differences, does it follow that the generation gap is small rather than large? One could argue that the emotional involvement must be high indeed when matters as trivial as fads become the center of heated arguments. And if this is a pseudobattle, it is difficult to understand why parents, in the mature years of life, become so trapped by it. Surely the well-informed, middle class parent knows that adolescence is the time for fads; surely he remembers his own fads; surely he realizes that the majority of "crazy kids" turn out to be the respected citizens and parents of tomorrow. Yet, this perspective is not enough to prevent the feeling of personal affront.

Douvan's suggestion that there may be less sound and fury in this battle of generations than meets the eye and ear, will have to be tested by further investigation. For the present, Campbell's (1964) more conservative evaluation is more acceptable. Like Douvan, he maintains that the differences between adult and peer group values are usually differences of degree, not of kind. He also pictures the period as one of lessening emotional attachment to the family rather than of active rejection. Yet, he also underscores the divided loyalties, the difficult conflicts and compromises which beset the adolescent. For example, an intelligent girl may hide her ability in spite of parental pressure, in order to be acceptable to other girls and attractive to boys. Thus Campbell includes both continuity and conflict in his picture of typical adolescent development, without stressing one at the expense of the other.

Deviant Development

Distortions of Initiative

Clinicians are constantly confronted with deviations from the normal pattern of development. Failures in initiative particularly concerned Adler, who prized it highly and delineated the kind of parental indulgence and parental neglect which undermined it. The clinician must also deal with the crippling indecision and doubt which may hamstring self-reliant enterprise. Freud addressed himself to the issue of the etiology of indecision, while Lewin offered a classical account of the nature of indecision itself. There is also some speculation that the collapse of initiative is one of the factors producing severe psychopathology in children. Clinical research on excessive willfulness and negativism is comparatively meager, but is interesting in its own right and for the light it sheds on the distinction between the mature and the immature rebel. (Recall the issue of legitimate and arbitrary negativism in the toddler.) As always, our concern will not be with a detailed explanation of deviations but with utilizing them as aids in understanding general development.

Adler versus Freud on Socialization. We shall approach our examination of deviant development indirectly by returning to a theme introduced in the section on the restriction of the toddler's initiative. As we have seen, personality theories differ fundamentally in regard to the existence of a growth principle and the kind of socialization which enhances growth. In the Freudian theory, the process of civilizing the child always is in conflict with his basic, primitive nature. The control the toddler feels over his muscles, his body in general, and his anal sphincter in particular, is highly charged with omnipotence. He wants to do what he wants when he wants. A demand to delay or inhibit or redirect his activities is a challenge to his power. He relinquishes his autonomy only because of the rewards of love he receives, typically from the mother. Toilet training epitomizes the socializing bargain: the toddler foregoes the pleasure of total control over the timing of his bowel movements for the pleasure of receiving the mother's love.

But being socialized always remains a bargain, no matter how successful. Deep within every man there is the infantile urge to break loose and assert his power: I will do whatever I wish whenever I wish. A Freudian theorist would say it is not by chance that we have described so many battles—of the bottle, of the spoon, of wills, of generations. We naturally fight intrusions. And the battles which once were external continue to be fought internally. The tension between being social and being selfish has no final resolution.

Adler, on the other hand, would deny the validity of the Freudian position. There is no natural antagonism between child and society; on the contrary, man's basic nature is social. One of Adler's prime theoretical constructs is termed social feelings, which is an innate, positive responsiveness to the social environment. Social feelings are present from the beginning, the mother and the nursing infant providing a prime example of cooperation and mutual adaptation. In addition, the infant is especially attuned to human beings and derives his most intense pleasures from them. However, the fulfillment of his social potential is not automatic or maturationally determined; it must take place in the proper environment. Social feelings must be cultivated. Unlike the good "Freudian" mother, who caters to the infant's need to avoid physiological distress and maximize physiological pleasure, the good "Adlerian" mother gives the child his first experience with a trustworthy human relationship. She also facilitates interest in and attachments to other members of the family. The father, instead of being a disciplinarian, should also foster the spirit of companionship, sharing, and participation. In such an environment, the infant or child will naturally expand his social interests. More important, he will feel at home with his fellow men.This does not mean that he will be happy or well adjusted in any stereotyped sense; it means that he will be able to participate fully in the experience of being human.

Excessive selfishness does not represent the outcropping of a basic narcissism; it is a sign that the cultivation of social feelings is going wrong. Perhaps the mother is using the toddler to compensate for love which she is not receiving from her husband or to satisfy her vain striving to have a showpiece; perhaps she is unloving, thereby inflating the child's natural feelings of helplessness and inferiority. Under conditions such as these, the child loses touch with his human heritage. He may become clinging and demanding out of an exaggerated sense of his own importance, or he may become guarded and suspicious, striking out angrily at an unloving, untrustworthy world. The stage is set for the development of neurosis, delinquency, or psychosis.[6]

Thus, a Freudian and an Adlerian would interpret the same behavior differently because of their different pictures of human nature. Who is right? The issue must remain among the imponderables. We have chosen Adler as the more fitting clinician to introduce us to deviant behavior, since "self-reliant enterprise" exactly captures the Adlerian spirit of healthy psychological functioning; moreoever, Adler spent much of his professional life trying to understand the conditions which produce neurotic feelings of discouragement, helplessness, and inferiority.

Adler on the Undermining of Initiative. To appreciate fully Adler's position we must discuss some general features of his theory of personality development. Adler maintains that inferiority underlies all personality and that to be human is to be inferior. Infancy is the best example, since the infant is helpless and powerless to manage any aspect of living, including his own survival. Inferiority, therefore, is a term to conceptualize ineffectuality in mastering life's challenges. It is not necessarily a feeling of inferiority, since it can be accompanied by a range of affective states, such as anxiety, guilt, distress, and depression. Fortunately, inferiority is countered by an equally strong and innate urge to master. The very helplessness of the infant spurs him on to learn about his physical environment, to seek love, security, and companionship from adults and peers, to achieve maximum fulfillment and self-esteem. What we have called the growth principle Adler came to call a striving for perfection. If all goes well in development, particularly if the child's potential for social feelings are encouraged, he is able to overcome his inferiority and evolve a life pattern marked by social involvement and cooperativeness, courage, and a realistic evaluation of himself and his relations with others.

However, under certain conditions (which we shall soon discuss), the child's inferiority is exaggerated and becomes burdensome or exceedingly painful. As a result, the compensatory striving for perfection becomes a selfish drive for power. Coopera-

[6]For a resume of Adler's position and an interesting comparison of Adler and Freud, see H. L. Ansbacher and R. R. Ansbacher (1956). For a bridge between Adler's clinical approach and an objective orientation, see Rotter (1962).

tiveness is replaced by competition; the child is concerned with his own aggrandizement and security. Adler's clinical descriptions are fascinating to read because of his insight into the myriad disguises which the drive for power can assume—not only the ruthless robber baron and the chest-thumping he-man, but also the pathetic wallflower who manages to stand out as being strikingly unattractive, or the silent sufferer who somehow lets everyone know he is in pain, or the chronic invalid who uses sickness to play on the guilt of family members—all, in their own way, are living selfishly, are demanding special consideration or concessions, or are making excuses for not facing life's problems constructively. Even conscious feelings of inferiority may be used as an excuse not to act reasonably and courageously and may represent a means of gaining power by appealing to the sympathies and solicitousness of others.

An example may be helpful. A student psychologist talked of his reactions to giving an eight-year-old girl an intelligence test:

> She looked so pathetic; she was pale and skinny and her dress was poor and old-fashioned. She sat kind of "crouched down" in the chair like she wanted to get away from me or drop through the floor or something. She spoke so low I could hardly understand her. I felt like giving her an ice cream cone instead of a Binet. I did everything possible to be friendly and make the test interesting but she didn't change at all. At the end of two hours I still had to ask her, "What did you say?" I guess something is wrong with me, because I found myself getting angrier as the test progressed. I wanted to take her and just shake her. That was awful.

An Adlerian would say that the psychologist's reaction may have been appropriate. The little girl might have been well practiced in the art of being pathetic in order to receive special attention, to irritate, and to reduce adults to a state of helpless, guilty exasperation.

Using this background, let us discuss two ways in which parents subvert the child's urge to master, according to Adler. The first, more insidious way is by pampering. The pampered child comes to regard his wishes as commands and expects that everything will be given him on request. His parents are his slaves. He lacks initiative because he is prevented from doing things on his own and discovering his capabilities and limitations. He has little tolerance for delay and little ability to take other people's feelings into consideration. He is a failure with his peers because they refuse to cater to him. He is unprepared for work because he has never been required to make an effort in order to accomplish a task. He loves only himself and evaluates others as they fulfill or frustrate his demands. As an adult he may try to hide behind a veneer of good will or personal charm, but his selfish demandingness quickly comes to the surface. Since most people will not tolerate his need to be given to without giving in return, he can become peevish, bitter, and vengeful. Adler was particularly unsympathetic toward these spoiled tyrants and regarded them as a very dangerous element in the community.

The neglected child, on the other hand, is unloved or rejected and therefore fails to develop the feeling that other human beings can be loving and trustworthy. He is discouraged, guarded, or suspicious, and sensitive to criticism. His principle technique for covering feelings of helplessness and discouragement is to attack. He is aggressively assertive, hard, and envious, and cannot bear to see others happy. To the ordinary demands of social living he angrily reacts with, "What's in it for me?" He may grow up to be a petty dictator or "the enemy of mankind," although he may also disguise his destructive competitiveness with energy and skill and a show of concern for the public welfare.

In both the pampered and neglected child, one sees a disregard or hatred of others which, in Adler's view, prevents them from developing a healthy sense of self-reliance. The individual may well have an illusion of control over his destiny or power over others, but he is invariably vulnerable to inferiority and its crippling effects.

Adler's basic position concerning the mastery of inferiority through striving for perfection lies beyond objective validation. However, there is evidence that pampering can be inimical to initiative. Levy's (1943) classical study of maternal overprotection of twenty such children between the ages of five and sixteen years has already been discussed in another context (see Chapter 2). Our concern here is with the effects on the children of excessive maternal indulgence and the prevention of social maturity through active attempts to undermine self-reliance, particularly in relation to peers and school (i.e., through discouraging the development of friendships). The indulged, overprotected children were disobedient, impudent, excessively demanding, tyrannical, and given to tantrums.

> An eight-year-old ordered his mother around until she was exhausted obeying his commands. He struck her when angry, spit on her when given something he disliked, threw food on the floor when it was not to his taste. He shot a toy pistol close to her face and, although she disapproved, continued doing so until she wept. (p. 162)

All the children were friendless, being bossy, selfish, cocky, and dominating. However, eight of the eleven overindulged children presented no special problems in school, perhaps because even their overly permissive mothers were strict in regard to schoolwork, perhaps because of a basic flexibility in the children which made them responsive to the demands and the gratifications of scholastic achievement. In spite of the children's dramatically deviant behavior, and in spite of Adler's ominous predictions concerning pampering, only four of the children were seriously maladjusted in a follow-up study several years later, and a number of them were doing quite well.

Levy explains his findings in terms of a relationship "rich in love and poor in discipline," which allows the aggressive, demanding aspects of the child's personality to luxuriate. Yet, Adler's emphasis on the cultivation of social feelings seems more apt than Levy's emphasis on discipline. These children were not only undisciplined in their social relations, but they also lacked respect and consideration for others as well as an ability to see themselves as only one human being in a complex human society. Empathy and self-perspective are surely more than a matter of discipline.

In terms of our definition of initiative, the overprotected children clearly failed to venture out in their social relations (even though they did achieve scholastically). In their feelings for others, and in their social techniques, they remained at an early stage of development (see Chapter 8). But what about their self-reliance? Since Levy did not deal specifically with this variable, we can only speculate. The indulged children seemed to have a grandiose idea of their power. Like infants, they continued to rule their mothers, while their attitude of "Let mother do it" prevented them from acquiring serviceable techniques for managing their own affairs. If the tyranny should ever fail to work, they would be reduced to a state of helplessness. Thus, an unrealistic notion of one's control over others is a hair's breadth away from rage or discouragement. This is just Adler's point. The rarefied atmosphere of an overprotective relation ordinarily is not tolerated outside the home; the pampered child is being deprived of experiences which will enable him to meet life realistically and flexibly, with a sense of being able to deal with whatever it has to offer.

McCord, McCord, and Verden (1962) provide data relevant to Adler's concept of the neglected child. In studying nine- to thirteen-year-old boys, they singled out a group who were particularly dependent, in the sense of seeking adult approval, being followers, and using familial rather than peer values as the basis for behavior. These boys openly expressed feelings of worthlessness and inferiority, had grave doubts about their abilities, and tended to escape from problem situations rather than face them. Their family backgrounds were characterized by (1) a lack of emotional security, indicated by frequent and open conflict between parents, low mutual regard between the parents, either extreme domination or withdrawal on the part of the mother, and a relatively high incidence of neurosis, psychosis, incest, and illegitimacy; and (2) parental rejection, in the form of open disparagement, regarding the boy as a burden, wishing he had never been born, and labeling him an "oddball" or a "failure." These boys illustrate what Adler would regard as the discouraging effects of neglect rather than the compensatory defiance and anger. When we discuss discipline (see Chapter 4), however, we shall see that a combination of parental neglect and brutality is apt to produce in the child the angry defiance of society described by Adler.

In general, Adler's concepts of the pampered and neglected child have a compelling face validity, and his observations concerning the undermining of initiative have been reasonably well documented by objective studies.

Uncertainty and Indecision. It is unreasonable to expect a child to master all stages of development confidently. Normal growth is marked by phases of eager expansion and baffled retreat. The ideal of the well-rounded child rarely exists in reality; instead, different areas of functioning are managed with different degrees of self-reliance. It is only when the child is burdened by a chronic sense of uncertainty in a major developmental area that his behavior becomes extraordinary. For example, an eight-year-old girl had an excessive fear of failure in school. She would go over an assignment again and again at home, but no degree of mastery could reassure her that she would do well. In class, her nagging uncertainty would frequently cause her to recite poorly or even forget the overrehearsed passage. In spite of her good intellect, she was trapped in a vicious cycle and became incapable of functioning in school.

Indecision plays a relatively minor role in Adler's theory. Its purpose is to prevent change in a child who lacks the courage and the social feeling to adapt flexibly to life's demands. The indecisive child may try to maintain an illusion of power by fantasying how superior he would be if only he were not hamstrung. Incidentally, Adler was especially sensitive to the self-defeating aspects of the "if only" attitude: "If only I had had a college education...," "If only I had been born white...," "If only my mother had not been so selfish...." Regardless of whether the individual is right or wrong, he is in danger of wasting his life by nurturing past grievances. His fiction of superiority prevents him from discovering his true worth through involvement in his present environment.

Indecision was more carefully studied by Freud, who was intrigued with the exquisite balancing of pros and cons by which an individual immobilizes himself. Such doubting has its roots in toilet training. (We have already presented briefly the Freudian interpretation of toilet training. For a more detailed account of his theory, see Chapter 7.) For the toddler, stimulation of the anus through defecation produces highly pleasurable sensations, and control of the anal sphincter epitomizes his autonomy. In toilet training, conforming to maternal demands runs counter to the pleasure and power of defecating at will. If the mother is coercive, the conflict is intensified. Sacrificing total control in order to please her does not seem like an equitable bargain. In addition, the inevitable anger at having to relinquish control becomes a rage over being coerced. The toddler is thus caught between opposing forces: Should I give in to Mother's demands or maintain my autonomy? Should I be "good" and do what Mother wants or "bad" and obey my impulses? Do I love or hate? Am I loved or hated by my mother? It is this uncertainty about himself and about his central emotional attachment which sets the stage for subsequent doubting. The content of the doubt is unlimited because it depends on the particular life experiences of the child, but the underlying theme of love versus hate, compliance versus resistance, is the same. In our example of the girl who could never feel confident that she knew her lesson, performing in class and conforming to the teacher's demands had become her special variation on the theme of "doing her duty"—having a bowel movement to please her mother and thereby proving she was a good girl.

Indecision need not be approached in the context of either Adlerian or Freudian personality theory. Lewin (1931), perhaps the most brilliant representative of the gestalt tradition as applied to personality, was not so much concerned with tracing antecedents as with understanding the psychological situation of the doubting individual, child or adult. The adjective "psychological" must be stressed. Behavior is never a response to the purely objective features of the environment according to Lewin; rather, it is determined by the way a person perceives the environment and the particular meanings it has for him. The same object can have a multitude of meanings—a hammer can be a toy for a child and a tool for his father; a man can be a frightening enigma to his son and a brilliant thinker to his colleagues. To understand a person, therefore, we must see the world through his eyes.

One of the striking features of a psychological environment is that people and objects and events are rarely neutral. Typically they have varying degrees of attractiveness or repulsiveness. In Lewin's terminology, they possess positive or negative valences. We tend to approach positive valences and to avoid negative ones. Life would be simple indeed if each person, object, or event had but a single valence and if we could deal with them one at a time. We could then do what we liked and avoid what we disliked. However, we frequently find ourselves involved in conflicting tendencies to approach and to avoid.

Lewin delineated four types of conflict (for a summary of research, see Miller, 1944): In an *approach-approach conflict*, the individual must decide between two equally attractive alternatives. Having once decided, the chosen activity becomes increasingly attractive the nearer the individual comes to the goal. This happy state, in which a response, once started, increases in strength or decreases the strength of a competing response, is called unstable equilibrium. Indecision is minimal.

In an *avoidance-avoidance conflict*, the individual must decide between two equally unattractive alternatives. A six-year-old who is afraid of the dark must choose between his fear and the shaming of his parents who want him to go to bed alone. His psychological situation is highly fraught with conflict and results in much indecision. As he starts upstairs, his fear of the dark increases; as he returns to his parents, his fear of ridicule increases. The individual is in a state of stable equilibrium if a response, once started, produces effects which decrease its strength or increase the strength of competing responses. In an avoidance-avoidance conflict, the individual frequently attempts to escape from the situation altogether. Thus, the six-year-old may try a number of evasive tactics, such as suddenly becoming interested in a television program or saying he is hungry. Fantasy can also offer an avenue of escape, as the individual tries to relieve the pressures of his dilemma by retreating into an imaginary world.

In an *approach-avoidance conflict*, the individual has strong tendencies to approach and to avoid the same goal. A child wants to go swimming, but the water is uncomfortably cold; an adolescent wants to date an attractive girl, but he is afraid she will turn him down. The individual vacillates back and forth. Incidentally, this analysis of the approach-avoidance conflict is congruent with Freud's notion that people are unconsciously attracted to things they consciously fear. The old maid who looks under her bed every night may be consciously afraid of finding a man hiding there, while secretly wishing she would find one. A persistent and self-initiated exposure to, rather than avoidance of, a feared situation suggests that a strong, often unacknowledged attraction also exists.

In a *double approach-avoidance conflict*, the individual must choose between two alternatives, each of which has positive and negative valences. In this case, the nearer he comes to one goal the less attractive it seems, while the more attractive the alternative becomes. Indecision is heightened if the achievement of one goal entails giving up an equally desirable goal. A girl who is invited to two equally attractive parties which are being given at the same time is in such a conflict, since she does not want to miss either and cannot attend both. The individual might remain forever suspended by indecision if other factors did not enter—some re-evaluation of the situation, some rationalization, or external pressure to make a decision.

Lewin was not interested in antecedents of conflict or in the sequence of their appearance in childhood, so his conceptualizations shed little light on early development. However, his analysis of types of conflict has become a classic and a discussion of indecision would seem incomplete without it.

Loss of Initiative and Severe Psychopathology. There have been a few attempts to relate loss of initiative to the kind of severe psychopathology which is manifested in a profound retreat from participation in human relations. White (1965) gives one of the most explicit accounts of the process. In normal behavior, action is "focalized, intended, and effortful," and the individual can observe the effects of his actions in bringing about changes. The infant learns that when he cries, mother appears; when he waves an object, it makes a rattling sound. A myriad of such experiences builds up a general feeling of efficacy, of having an effect on people and objects. It is just this feeling which is weakened to a pathological degree in the schizophrenic. He feels that nothing he does makes a difference, and, in his frustrated rage and despair, he withdraws into a bizarre world of his own creation.

This loss of interpersonal efficacy can be produced in many ways, according to White. The mother who cares for the child in a cold, highly routinized manner, timing her attentions by the clock rather than by her child's expressions of need, robs him of the experience of effecting change. Or, a parent may confuse the child by being flightly, inconsistent, and unpredictable. In addition, there may be a constitutional weakness in a child who becomes schizophrenic. Instead of being able to focus his attention and act decisively, he may be highly distractable and have difficulty keeping track of the enormous complexity of normal human behavior. He may also be hypersensitive, so that the inevitable distresses of normal development register as exceedingly painful. Out of his confusion and pain grow a profound sense of helplessness and an angry, baffled retreat.[7]

The etiology of severe emotional disturbances in childhood is too tangled and controversial an issue to pursue in detail. However, it is interesting to observe that Bettelheim (1967), a member of the psychoanalytic school, has proposed a theory consonant with the concepts of White and Lois Murphy. As we have seen (Chapter 2), psychoanalysts traditionally have stressed inadequate maternal care and affection during infancy as producing a vulnerability to psychotic disturbances. Regarding this position as oversimplified, Bettelheim maintains that the infant comes alive psychologically only as his actions have an effect on the environment and he feels that he is capable of doing for himself. The self is both created and defined through effective activity. If the infant or toddler fails to achieve this sense of effectiveness (either because of the caretaker's indifference or insensitive response or because of some intrinsic factor), he comes to view the world as overwhelmingly frustrating. He abandons his strivings to do things on his own, he becomes oblivious to other human beings, no matter how benign and loving, he is mute or speaks a private, incomprehensible language, and he endlessly engages in some simple activity, such as rocking or twirling a shoestring. In short, he becomes a negative image of initiative.

[7]Objective support for White's position is still meager, especially in the case of young children. For a study of young adults, see Coelho, Silber, and Hamburg (1962).

Excessive Willfulness and Negativism

Willfulness and negativism have failed to generate the same interest as has initiative among psychologists concerned with deviant development. There are some noteworthy clinical descriptions and theories and a few objective studies but a dearth of full-dress treatments.

The Anal Character. Freud listed stubbornness as one of the three character traits resulting from unsuccessful mastery of toilet training. True to his theory that psychological functions parallel bodily functions, Freud traced the origins of stubbornness to the infant's withholding of feces in the face of coercive attempts to train him. Out of this determination not to give in in terms of defecation grows a psychological disposition to resist compliance and a mulish belief in the correctness of his ideas and actions. While the stubborn child is less socially disruptive than his overtly defiant counterpart, he has the same bedrock determination to resist authority; he merely uses passive rather than active techniques.

The other character traits resulting from unsuccessful toilet training are excessive orderliness and parsimony, both of which are closely related to stubbornness. In fact, Freud combined the triad under the general label *anal character*. From the toddler's determination to keep his feces to himself grows a general attitude of ungivingness, a literal and an emotional miserliness, an overvaluing of objects and an undervaluing of human relationships. The excessive anger mobilized by coercive toilet training also leads the child to exaggerate the dangers of giving vent to his destructive impulses. He protects himself by being overly controlled, typically becoming excessively neat and orderly. Even the normal trial and error and give and take of living make him uneasy. His life must be perfectly ordered: "a place for everything and everything in its place" becomes his creed.

The striving for perfect order and the stubbornness of the anal character may seem incompatible with the indecision which, as we have seen, also results from faulty toilet training. However, Freud's point is that stubbornness is a caricature of self-reliance just as perfectionism in trivial matters is a caricature of self-control. Both are protective maneuvers. Beneath the surface the individual is profoundly uncertain concerning his basic feelings of loving or hating others and distrusts his ability to act freely and use his judgment flexibly.

Freud's picture of the anal character has been validated only partially by objective studies (see Caldwell's 1964 review of research). The traits of obstinacy, orderliness, and parsimony do seem to cluster together, especially in girls, but they seem to be related more to the mother's general need for order and her fear of sexuality than to toilet training per se. Coerciveness in toilet training, instead of epitomizing the struggle for control, seems to be only one of many manifestations of the general problem of socializing initiative—and not a critical one at that.

The Negatively Organized Child. On the basis of clinical experience, Ross (1964) depicts a "negatively organized child" whose personality is pervaded by opposition to the parents. In some cases, the opposition is overt; in others, the child goes through the motions of carrying out parental wishes but accomplishes nothing. Ross, along with a number of other clinicians, lists three conditions for the perpetuation of the battle of wills. The first is maternal restrictiveness, especially if the toddler is energetic and vigorous. Next, the mother may pressure the toddler into independence, demanding that he do things beyond his capacity or expecting him to conform too quickly. Finally, the mother may be strict and cold in her demands, having little patience with deviations and offering inadequate compensation for compliance.

The conflict over autonomy and compliance becomes the axis of subsequent parent-child relations. It colors the child's relations with other adults, especially teachers, and his attitude toward the normal demands for conformity. School work falls prey to the struggle, and the child often underachieves or becomes a discipline problem. He may be touchy and guarded with his peers, or he may attract children with a similar need to rebel. Parental anger over his selection of "undesirable" friends only reinforces his alliance with them. In adolescence, the defiant attitude of "Nobody is going to push me around" is apt to get him into trouble with the law, and he may have problems with authority throughout his life.

Ross cites the case of a fifteen-year-old boy who was brought in for psychotherapy because he had threatened to shoot his girlfriend's father who had forbidden her to see him. The boy's mother was ambitious, energetic, and intense. His father was also a striving individual who was determined to give his children the best education, even though he could barely afford the expense. No issue in the home was too large or too small to escape the battle of wills–from the boy's academic failure in spite of superior intelligence to his refusal to wear a tie. His parents' nagging, exhortations, pleading, and scolding were relentless, and he spent as little time as possible at home. He excelled in athletics and worked hard on a friend's farm where he could do as he pleased. But to most of life's demands he responded with a defiant, "I won't."

Comment on the Mature and Immature Rebel. We have described the normal fluctuations of willfulness and negativism, and we have seen how they can, in deviant cases, blind an individual to his own best interests, block his growth, and set him against society. Yet, Adler noted that the neglected child can, as an adult, carry off his hatred of man with great vigor, subtlety, and capability. His rebelliousness may be disguised as righteousness or a passionate, persuasive defense of the underdog. Thus, Adler was aware of the difficulty in distinguishing the self-centered rebel from the mature nonconformist. But he also offers the best guide to making the distinction. The immature rebel is using society as the arena for expressing his personal resentment over being coerced. With him, as with the toddler, the issue at stake is secondary to his defiant stance: "Nobody is going to tell *me* what to do!" As with the toddler, his fragile self must be defended even when the challenge is minimal or nonexistent, while aggravating those in authority becomes an end in itself. We can further speculate that such a rebel is dangerous because, if he is successful, he will primarily be concerned with preserving his own power and eliminating those who threaten him. "Destroy or be destroyed" is the language he understands, and issues are constantly confounded by his hypersensitivity. Such individuals are literally self-centered and power-centered. By contrast, the mature rebel has social feelings. He has been able to transcend self-centeredness so he can attend to the issues at hand, and he uses power instrumentally. His primary concern is with combating social injustices, and rebellion is merely his choice of technique. Norman Thomas (1966), the epitome of mature protest, summed up the matter: "...not dissent for its own sake, but only as necessary to advance the truth to which one is devoted."

Conclusion and Critique

Our examination of deviancy has enriched our understanding of initiative, especially as it is affected by the parent-child relationship. Pampering prevents the child from venturing out and from learning self-reliance through independent mastery, while it engenders a false sense of superiority. Coercion and pressure upset the balance between demands and love in the socialization situation and produce exaggerated resistance or defiance for its own sake. Neglect and disparagement lead to feelings of worthlessness or an angry vindictiveness. Combining certain of these findings with those from our discussions of normal development, we arrive at a continuum of parental behavior ranging from overindulgence through child-centeredness and authoritativeness (the "pacesetting" parent) to coerciveness and authoritarianism ("breaking the child's will"). At one extreme the child can do no wrong; at the other, he can do nothing right. Child-centeredness and authoritativeness seem to enhance initiative; the other parental behaviors impede it.

In spite of all that can be learned from the literature, the nature and developmental course of initiative are incompletely understood. One misses the invigorating interplay between the clinical and objective approaches, between theory and naturalistic observation, which we met in discussing attachment. We are not certain that all the variables affecting initiative have been isolated, nor are we certain how highly abstract concepts (e.g., the growth principle) and concepts derived from studying deviant children will fare when put to the acid test of accounting for the behavior of normal children. Even if we accept what we have learned at face value, we find that some basic questions are not answered satisfactorily. Champions of an intrinsic urge to grow and master give only a programmatic account of how this urge is related to constitutional and maturational factors on the one hand, and to learning and experience on the other. The concept itself tends to be so general (all living beings have it) or so specific (each child has his unique way of growing), that it is impossible to predict how a given kind of environment or a given pattern of parental behavior would differentially affect the child's growth. There is also a discrepancy between the advocates of child-centeredness and the research on authoritative parents. Our science is as uncertain as our society on the issue of how much freedom and how much control maximizes self-reliant enterprise. And, finally, we are left with the baffling issues of depth and time. What makes the difference between the child whose initiative is nothing but a reflection of the environment which teaches, rewards, and serves as a model for such behavior, and the child who is self-reliant and enterprising regardless of, or even in spite of, his environment? And what accounts for those disturbing reversals described by Macfarlane in which initiative in childhood and adulthood were negatively correlated?

Our final impression is that certain important factors involved in initiative have been delineated, some obvious, others subtle; but we do not understand initiative in detail or in depth, and there are baffling data which challenge everything we now know. Yet, this kind of taking stock of what we know, what we disagree on, and what we have yet to learn can serve as a basis for future efforts at understanding.

4 Self-Control and Judgment

The delight in people which will culminate in attachment (see Chapter 2) and the exploration of the environment which will culminate in initiative (see Chapter 3) are clearly evident in the three-month-old infant. To tell this infant "Wait a minute" when he cries or to scold him for having a bowel movement makes no sense. *Wait* and *Don't* become meaningful only in the second year of life. In essence, the parent begins to tell the infant, "Control yourself." This message will become more frequent and more insistent until patterns of self-control have become part of his personality. While specific acts of self-control may be simple, the variables involved are complex. Some hark back to our discussions of affection and initiative; others anticipate the discussions of cognition, values, and conscience.

However, we are not only interested in understanding how the child learns to conform to parental directives; we are equally concerned with the process by which he becomes able to judge what he will and will not allow himself to do. The socializing situation is a complex one for both parent and child. The parent must promote self-control while nurturing the child's capacity to decide for himself. The child, while first an executor of parental mandates, ultimately must be his own legislator. He must become cognizant of the conflict between his desires and parental prohibitions, and of the fact that he must decide between them. In the previous chapter we were concerned with the self-as-agent; here we will be concerned with the self-as-judge. While initiative can be seen clearly in the two-year-old, judgment does not appear until middle childhood, perhaps not until adulthood, depending upon how strict the criterion of wise decision making is. Because judgment is of late fruition, we shall use the term to cover the precursors and the development process, both of which will concern us more than the final ability to act with judgment.

Normal Development of Self-Control

Changes in Self-Control from Infancy to Adulthood

Self-control undergoes quantitative and qualitative changes from birth to early adulthood. We shall present a descriptive overview of both kinds of development. Self-control is weakest during the first few years of life, when reactions are intense. There is nothing as totally distressed as a crying infant, nothing as totally enraged as an angry one, nothing as ecstatic as a happy one. It takes time to harness such vitality. Fortunately, there is a complementary relation between control and affect–the strengthening of control is accompanied by a diversification in the objects of affection. Diversification is an aid to modulation. The exclusive attachment to the primary caretaker is supplemented by involvements with other adults and peers until the preschooler has several significant but moderate emotional relationships. In short, the child is increasingly capable of managing feelings which are becoming increasingly manageable.

The period between six and nine years of age is the high point of self-control. By this time the crises of early childhood are over. Separation anguish, intense family jealousy and rivalry, accommodation to peer groups, and the transition from home to school have been weathered. While the demands of home, school, and peer group continue to change, there is no dramatic new adjustment to be made. Mechanisms of self-control have been consolidated, and strong feelings can be countered with equally strong techniques for managing them. The child handles himself with a poise and assurance which foreshadows his adult status.

The calm of middle childhood is shattered by the physiological changes and explosive growth of puberty. The period between the twelfth and fourteenth years is marked by turmoil. The preadolescent is uncertain and confused, touchy, negativistic and solitary, aggressive, deliberately provocative, and resistive to authority. The adolescent period is not so dramatically unstable. However, the adolescent still has a host of unmastered problems: managing strong sexual urges, establishing his sexual identity, emancipating himself from his family, and finding a fulfilling adult role. The urgent need to become adult alongside uncertainty and conflict about how to achieve adult status, makes the period stressful. Stable self-control is achieved only as the adolescent finds acceptable solutions to the problems of love, work, and companionship.

These changes are accompanied by a gradual shift from direct expression of affect in the preschooler to more internalized reactions around ten or eleven years of age. Temper tantrums and destructiveness at three give way to moodiness and nightmares at eleven; attack is replaced by reserve and shyness. Instead of being acted out, strong feelings are inwardly disturbing. Even the poorly controlled ten-year-old or adolescent does not have the toddler's directness and abandon. His behavior may be unreasonable, but he is not incapable of reasoning and reflection.[1]

[1] For a documentation of these changes, see Macfarlane, Allen, and Honzik (1954).

A study of children between five and seventeen years of age by Sanford and others (1943) helps fill out the picture of change. In general, children's behavior becomes more orderly, organized, and precise with time. They are better behaved and more persistent. They make more efforts to overcome weaknesses and to keep trying after failure. They seek aid, protection, and sympathy less frequently and assume more personal responsibility for their failures. They become less aggressive and less possessive; they have fewer fights, and the tendencies to blame and ridicule, to be greedy and possessive, or to snatch and steal, also wane.

Sanford and his co-workers found that middle childhood is a high point of conformity. Middle children are practical, realistic, and concerned with the here-and-now. Disruptive feelings, such as anger, jealousy, and possessiveness, are successfully held in check. Incidentally, Gesell and Ilg (1949) also note the maturity of the nine- to ten-year-old. They describe him as industrious, businesslike, and self-motivated. Having himself and his skills well in hand, he functions smoothly and takes difficulties in stride. Moreover, his insights into his own behavior and that of others have a maturity they formerly lacked. Puberty marks a turning inward. The child is more preoccupied with his problems and his inner feelings. He tends to be sensitive, to personalize, and to be touchy and unreasonable. On the positive side, he is given to idealism and searching for basic answers to life's problems.

Factors Involved in Self-Control

An Example and an Overview. We have met self-control before (see Chapters 1 and 3). An eighteen-month-old girl becomes intrigued with a china figurine and reaches for it. The mother rushes over and says *No* and slaps her hand. The child reaches again, and again is prohibited. She withdraws. The following day she approaches the figurine with interest, but as she begins to reach for it she suddenly stops. She may shake her head or even slap her own hand. Then she withdraws. Subsequently, she no longer reaches for the figurine. In this miniscule area of behavior, she has achieved self-control.

Arriving at an understanding of this simple phenomenon will take us into many areas. In order to see the total picture, it will be instructive to list the variables involved in self-control which were proposed by Anna Freud (1965), a representative of the clinical approach, and by Justin Aronfreed (1968), a social learning theorist who represents the objective approach. The terminology and emphasis of these theorists differ, but the similarity in their work is impressive. Both agree that self-control depends upon a combination of emotional and cognitive factors: if the bond of love is tenuous, or if the child cannot comprehend parental demands, the achievement of self-control will be significantly crippled. Anna Freud writes about the importance of the child's "libidinal ties" and Aronfreed about a "positive attachment to an agent of nurturance." Behind this difference in terminology lies an idea we have already met: namely, that the bond of affection makes the infant "educable" (see Chapter 2) or capable of being socialized. Both authors stress the role of cognitive factors. Anna Freud states that the child must be able to remember parental messages in order to use them to guide his future behavior. He also must be able to grasp cause and effect relations so that he can connect his acts with their consequences ("Mother got angry because I touched the figurine"). Finally, the child must be able to integrate piecemeal experiences into guiding principles of behavior ("I should not touch anything in the living room," and later, "I should be careful with anything that belongs to somebody else"). Aronfreed agrees on the importance of memory, grasping causal relations, and integration, although he writes in terms of "representing and storing social experiences," using intentions and anticipations to govern actions, and evaluation.

Both Anna Freud and Aronfeed agree as to the importance of speech in achieving self-control. When the child learns to speak, he can begin to talk to himself and tell himself *No* and *Don't*. Such self-instructions help him to delay actions. As words become the principal vehicles of thought, they come to be important aids in integrating isolated experiences. Finally, both authorities assign imitation an important role in the development of self-control.

Aronfreed gives a detailed account of how reward and punishment affect self-control, while Anna Freud makes only passing reference to the child's ability to avoid adverse consequences arising from clashes with the environment. However, Aronfreed's analysis complements rather than contradicts her list of variables. On her part Anna Freud mentions two other aids in the achievement of self-control. One is the ability to wait or to postpone gratification. The other is the ability to do something else when thwarted, or the ability to accept substitute goals. We shall now examine a number of these variables in detail in order to understand how they facilitate the development of self-control.

Reward and Punishment. Let us review the sequence of events in the case of the little girl. First she sees the figurine, and it arouses her interest. Because she can now control her body, she can implement her interest by walking toward the object and reaching out for it. Then the mother intervenes, speaks in an angry voice, and slaps the little girl's hand. In Aronfreed's terminology, the girl's overt behavior has led to "aversive consequences." The immediate consequence is pain, and pain holds the key to the events which follow.

Pain is a response which also is a distinctive, discriminable, noxious stimulus. Because it is subjectively distressing, it stands out clearly in the stream of experience. The little girl knows that the slap hurts. In addition, pain is a response which can become conditioned; or, more accurately, certain components of the complex pain response, such as the "churning" of the stomach or the rapid heartbeat, can come to be elicited by previously neutral stimuli. In our example, such stimuli would be the sight of the figurine and the interest in it, the intention of examining it, and the motor behavior involved in approaching it. Typically, the stimuli present at the time of the pain response become more strongly conditioned. As the sequence of behavior is repeated, the antecedent stimuli become progressively more effective in producing the conditioned response. Initially, for example, the little girl may approach and begin to reach for the figurine before drawing her hand back; later, the mere sight of the figurine at a distance, or the intention of approaching it, may elicit a response of anxiety.

Note that we use the word anxiety. In learning theory, anxiety is conceptualized as the conditioned components of the pain response. It is not the total experience that is conditioned; as we have seen, the child does not re-experience the sharp slap on her hand, but certain elements in the response to the slap. In essence, anxiety is anticipatory; it presages the pain to come if the behavior sequence is continued. Like pain, anxiety is both a distinctive and a noxious stimulus and, because it is noxious, it motivates the organism to behave in such a manner as to diminish or eliminate it (Miller 1951, 1969). Psychoanalysts have a similar concept which they call "signal anxiety." After experiencing extreme distress, the individual comes to anticipate what lies ahead. This anticipation is a small fraction of the imminent trauma and serves to warn the individual, Danger Ahead! With the warning, he can take whatever steps are necessary to avoid re-exposing himself to the trauma. (For a more detailed discussion of anxiety, see Chapter 7.) According to both theories, anxiety motivates the child to find ways to diminish it and reinforces the behavior which brings relief from it. Avoidance is the simplest maneuver. The child no longer performs the forbidden act.

It will be sufficient for our present purposes to identify only two painful stimuli, namely, physical punishment and withdrawal of love. In the latter, the parent does not literally say, "If you do that again, I will not love you any more." However, his behavior and his manner clearly convey displeasure, and the child feels he is unworthy of love. As we saw in Chapter 2, withdrawal of love is painful to the child, once the bond of affection has been established.

Before we leave our prototype of self-control, we should add that socialization rarely remains at such a simple level. A four-year-old who has been punished for roughly pushing his two-year-old sister may stop briefly; but later he may adopt the trick of pushing her down and then tenderly consoling her as soon as he hears his mother approaching. This show of affection may deflect punishment, at least temporarily. In general, the parent's determination to socialize the child is matched by the child's determination to satisfy his own needs in one way or another, and self-control develops in the context of the interplay between the two forces.[2]

So far, we have talked only in terms of aversive stimuli and anxiety. Parents aid self-control by providing positive stimuli—they reward as well as punish. The mother who sees her eighteen-month-old turn away from the tempting figurine may smile or give other evidence of being pleased. The child's avoidance of the figurine now both diminishes fear and increases pleasure. Conscientious parents are concerned lest they hem in their children with a constant barrage of *No's*, and so, either naturally or by design, they reward acceptable behavior with affection and praise.

However, reward has a practical limitation with young children. In order to be rewarded, the behavior must first occur. As Aronfreed notes, socially acceptable behavior may appear very infrequently in young children. A three-year-old may not

[2]For a delightful account of children's constant efforts to outwit the socializing parents, see Loevinger (1959). Her thesis is that it is impossible for parents to win the battle of socialization, but it is imperative that they keep trying!

only be jealous of his infant sister, he also may realistically find her uninteresting. The probability of his striking out at her is high; the probability of his playing with her is low. The most nonpunitive parent is left, then, with little behavior to reinforce. Even lavish encouragement of fleeting moments of positive interest are not sufficient to compete with the strength of the older sibling's resentment.

Imitation. In rewarding and punishing the child, the parent conveys more meaning than he intends to, since he is also serving as a model of either controlled or uncontrolled behavior. Children are adept imitators. A doctor reported the following example: "I was watching my (verbally precocious) two-and-a-half-year-old last night. He went to the phone and, as usual, went through all the forms of dialing a number and listening. All of a sudden he turned to me and said angrily, 'Be quiet. I'm talking to a patient.'" The father was startled. The boy had picked up not only the content but the exact expressive quality of irritation in a remark made by the father. Moreover, the behavior had come "out of the blue." Certainly there had been no attempts to teach it by reward and punishment, and there was no evidence that the boy had said anything like this before.

Imitation has been amply demonstrated in objective research. In one study (Bandura, Ross, and Ross, 1963), groups of kindergarten children were exposed to an adult model, a film of an adult model, or a film of a cartoon character engaged in aggressive behavior toward a large Bobo doll. A control group was not exposed to any of these aggressive models. Some of the model's behavior was novel, such as sitting on the doll and punching it in the nose, or pummeling it on the head with a mallet, or tossing it in the air and kicking it. Subsequently, the children were mildly frustrated—they were interrupted in their play with some attractive toys and put in a room with aggressive toys (a gun, a mallet), nonaggressive toys (a tea set, crayons), and the Bobo doll. The frustration was introduced to heighten the tendency to be aggressive. Both generally aggressive behavior and behavior aping the model's novel means of expression were scored. All three experimental groups showed significantly more imitative aggression than the control group, and the results suggested that the filmed adult model was the most effective in producing aggression. The sex of the child was an important variable—the boys displayed more aggression than the girls. The sex of the model was important—the aggressive male model was more effective in eliciting aggression than the aggressive female.

Let us analyze the essential features of imitation. Obviously, there must be a model. There also must be a positive relation between model and child sufficiently strong for the child to observe the model's behavior with interest. The child subsequently reproduces the model's behavior with fidelity to the content and the expressive aspects (what the model does and the way he does it). The behavior appears fullblown, with no evidence of overt rehearsal. Finally, the behavior constitutes an addition to the child's repertoire of responses.

Bandura (1965), utilizing social learning theory, enriches our analysis. He defines imitation as the acquisition of a new response or a modification of an existing response on the basis of observing the behavior of others and the rewarding consequences of such behavior without overtly rehearsing the modeled response. Other names for imitation are observational learning or vicarious learning. Bandura argues persuasively that theorists have tended to ignore imitation for the past two decades, preferring to explore the two other principles of learning–reinforcement (reward or punishment) and contiguity. In both of these processes, an existing response is strengthened, sometimes rapidly, sometimes slowly. In imitation, a new response appears fullblown after it has been merely observed in another, and no environmental rewards are required for its occurrence. Imitation thus is regarded as the mechanism par excellence for expanding the child's behavior repertoire.

Bandura lists three effects of observing a model. First, the occurrence of a readily available response can be facilitated; e.g., if one person stops on the street and looks at the sky, soon a crowd of people will stop and look at the sky. Second, the occurrence of a previously inhibited response can be facilitated; e.g., if an inhibited child sees other children acting aggressively, there is an increased probability that he will also begin to act more aggressively. The third and most important effect, however, is the acquisition of new behavior on the basis of observation. The ability to do something new after having observed it enables learning to progress by great leaps. Bandura states that the acquired behavior need not be totally new to be called imitation, since the components may already exist in the child's repertoire, but at least the patterning of responses and the context must be novel.

It is important to keep Bandura's analysis in mind, since we are going to limit the meaning of imitation to the acquisition of a new response. A teacher leaves the room; some unruly boys begin to throw wads of paper at one another; one boy refrains for a while but finally joins in. Some psychologists might label the boy's behavior imitation, but we will not, since it is unlikely that throwing paper wads is a new response learned on the basis of observing others. More than likely, throwing paper wads had been more strongly inhibited in this boy, and the sight of the other boys changed the balance between avoidance and expression.

There are many different explanations of imitation. Piaget (1951) views imitation as a special instance of the two general processes involved in learning–accommodation and assimilation (see Chapter 6). Accommodation is the taking-in phase of learning, the registering of information from the real world. This taking-in is never a passive process; rather, the organism must actively and continually adapt itself to the environment. Piaget draws an analogy to infant feeding. Milk does not flow passively into the infant's mouth; instead, he must adapt (accommodate) his body, his head, his mouth, and his sucking reflex to the source of nourishment in order to take it in. Assimilation is the integration of new information into an appropriate body of existing information so that it becomes meaningful. Assimilation of new material in turn enriches the body of acquired and digested information.

Let us illustrate the point. The reader has just been introduced to the concepts of accommodation and assimilation. If the concepts are to register, he must attend with interest. If they are new and he knows little about cognition in general, he may be able to recite the preceding paragraph but feel that he does not really understand it. He has then accommodated to the information–he has gone through the processes necessary to take it in and remember it—but he has not assimilated it. The information is still "undigested" since it has not become part of a larger body of accumulated understanding. At this point he may read about cognitive development in children or talk to an expert about it. After much thought and effort, suddenly something clicks and he says, "Oh, *that's* what Piaget means by assimilation and accommodation." He has then assimilated the concepts. His now enriched understanding will enable him to accommodate to and assimilate even more complex ideas concerning cognition.

Ideally, the two processes are in balance. Accommodation basically keeps us in touch with the world around us; assimilation enables us to organize more and more information and so gives us an ever-increasing ability to understand the nature of the world. But the processes can be out of balance, especially in childhood. Accommodation can outweigh assimilation. This is exactly what happens in imitation. Just as an adult can memorize without understanding, the child can observe and act out behavior which he only vaguely comprehends. The doctor's son mentioned earlier had undoubtedly assimilated the idea of "Keep quiet"; but the idea "I'm talking to a patient" probably was only vaguely grasped.[3]

Piaget adds a new dimension to our discussion of imitation. While Bandura stresses the sudden appearance of a novel response, Piaget reminds us that this response can have various degrees of congruence with the child's general level of comprehension. In the Bandura, Ross, and Ross study (1963), the children clearly grasped the nature of aggressive behavior and were merely adding a novel mode of expression. By contrast, the doctor's son talking on the phone still had to learn the meaning of the phrase he was repeating. While imitation permits great leaps forward, the child—especially the preschooler—can overleap himself. He still must go through the process of comprehending (assimilating) what he is doing, even though his overt behavior may seem to demonstrate learning.

Learning theories account for imitation in terms of general principles of learning. Mowrer (1950) regards attachment as the basic prerequisite for imitation. Before the infant can imitate, Mowrer says, he must first be fond of someone, typically his caretaker. As the caretaker provides the infant with many pleasures, other, contiguous aspects of her behavior, such as her voice, her verbal expressions, and her mannerisms, also become sources of pleasure through classical conditioning (see Chapter 2). In addition, the infant is continually monitoring stimuli from himself, such as the sound of his vocalizations or the feel of his body as he performs various activities. Eventually, more or less by chance, he performs an activity which resembles that of the caretaker. Since responses tend to generalize to similar stimuli, he reacts with pleasure to his own behavior, just as he did to the caretaker's. The more closely his behavior resembles that of the caretaker's, the stronger his pleasure will be. Thus, the infant finds, by trial and error, ways of behaving that are close approximations to the caretaker's. The most obvious example of Mowrer's explanation is speech. A verbal expression of the mother's becomes highly pleasurable to the infant. Then, in

[3]For a more detailed exposition of early cognitive development in general and imitation in particular, see Flavell (1963).

the course of vocalizing, he hears himself make a sound which resembles the mother's. His own sound now becomes rewarding. The more nearly he approximates the mother's exact expression, the greater his pleasure will be.

Aronfreed, like Mowrer, includes motivational variables in his explanation. For imitation to occur, the model must be a source of nurturance and affection, and the imitative behavior must serve the basic functions of increasing pleasure and decreasing distress. However, Aronfreed believes that the trial-and-error period which Mowrer postulates is too long to account for the sudden emergence of imitative behavior. He also questions the role of stimulus generalization. The principle of stimulus generalization states that a learned response to a given stimulus is likely to occur to other, similar stimuli; e.g., a toddler frightened by a dog may react with fear to the sight of other animals that resemble a dog. Typically, generalization is to similar stimuli in the same sense modality; in our example, the child is frightened of animals which look like dogs. Aronfreed maintains that imitation does not conform to the model, since different sense modalities can be involved; e.g., the child copies with his body the model's behavior which he has seen with his eyes. The visual stimuli must be translated into kinesthetic sensations, and the leap from the visually perceived body of the model to the kinesthetically perceived body of the observing child must be more than a simple matter of stimulus generalization. Aronfreed argues that there must be a complex inner representation of the model, which he calls a cognitive template. How rewarding imitation will be is a function of the match between the template and the imitative behavior. Although Aronfreed does not elaborate the idea of a cognitive template, it suggests Piaget's concept of a meaningfully organized or assimilated area of information, which he calls a schema (see Chapter 6).

Bandura's (1965) explanation of imitation involves a three-stage process. First, the child must attend to the model, since mere exposure, although necessary, is not a sufficient condition for imitation. Motivational factors which enhance attention (e.g., interest in the model), and qualities which increase the attention-getting value of the model (e.g., its vividness or novelty) will facilitate the initial stage. The second and third stages are central processes. In the second stage an inner representation or an image of the observed behavior is formed. The phenomenon of image making is a common one, as we all mentally represent friends, entertainers, political figures, and others that we have seen. Verbal labels enhance the distinctiveness and durability of the image. Thus, if the child says to himself, "He's hitting," after observing someone in this aggressive behavior, he has another inner cue with which to link the model's behavior with his representation of it. Bandura maintains that the child may engage in covert rehearsal, especially if the model's behavior is so complex that he cannot assimilate all the incoming information. In the third and final stage, the representation serves as a cue to elicit responses corresponding to the ones observed. Responses already richly elaborated in the child's repertoire are more easily elicited than responses which have occurred infrequently; for example, young children imitate motor behavior more readily than verbal behavior. In sum, Bandura claims that imitation is not a passive mirroring of the behavior of others. Attention, representation, rehearsal, and the nature of the behavior repertoire are all involved.

Looking back over the four theories of imitation, one is struck by the differences among them. Piaget has little to say about motivation, especially in regard to affection. Mowrer and Aronfreed place affection and pleasure seeking at the heart of the imitative process, while Bandura assigns motivation the minor role of enhancing attention to the model. The relations between Piaget's "schema," Aronfreed's "cognitive template," and Bandura's "image" have not been clarified, and Mowrer lacks a cognitive factor altogether. Piaget's central concept of an imbalance between accommodation and assimilation has no counterpart in other explanations. We can hope that the renewed interest in imitation will eventually produce a more definitive theoretical accounting. (For a review of objective studies of imitation, see Flanders, 1968.)

Words. Inhibition induced by punishment is well documented in animal behavior. The anticipatory warning involves the "churning of the stomach" or the "fluttering of the heart" or "the tightening of the muscles" or whatever physiological concomitants of pain can become conditioned to preceding stimuli. With humans, however, a critical and unique variable is added. The toddler's conditioned anticipation of pain begins to include a verbal response. Along with shaking his head or slapping his hand, he begins to say *No*, sometimes aloud but usually to himself. By imitating the verbal prohibition of the parent, he has taken a giant step toward becoming his own regulator.

Verbal symbols or words have two important characteristics. First, they are distinctive stimuli. In the language of learning theory, they have cue value. Remember that to a toddler a word is much more a concrete thing with intriguing auditory and kinesthetic properties of its own, and speaking per se is a more challenging activity for him than it is to the habituated adult. When an adult tries to learn a foreign language, he may again be reminded that the sounds of words can convey agreeable or disagreeable impressions independent of their meaning; and he will re-experience the difficulty of matching the sounds he makes with those of his teacher. The toddler is intensively involved in just such a matching process, and he can be observed rehearsing pronunciation as he rehearses stair climbing. In addition to being distinctive stimuli, words have a special magic which objects and motor skills lack—they bring highly rewarding responses from parents. While other exploratory activities may go unnoticed, the toddler usually has an audience when he speaks.

When the toddler adds *No* to his repertoire of warnings, he has added not just another response but a kind of response which can expand his control until it becomes a complex and subtle system of rules and regulations. Words will enable him to categorize objects and behaviors, so that a prohibition need not be limited to a specific act and a single object. Let us take an example. An imaginative three-year-old is using his hot dog as a dive bomber at supper. His father says, "Don't use your hot dog as a dive bomber. Food is to eat, not to play with." The first part of the prohibition is object and behavior specific; but, in the second part, the father is making a distinction in terms of categories: there are certain objects called "food" toward which one must inhibit certain kinds of responses called "play." Food now takes on a new and distinctive meaning. It becomes a "not-play" object. Notice how the simple labels "food" and "play" facilitate the learning process. If the father had only said *No* every time his son began playing with food, eventually the boy would have gotten the idea, but only after a number of experiences and a good deal of trial and error.

As the child's comprehension develops, words play an increasingly important role in socialization. Conditional statements can supplement simple prohibitions; e.g., "Don't ever start a fight, but if someone hits you first it's alright to hit him back—unless it's a girl." Approved and disapproved acts become part of general guides to behavior. "These dolls are yours and you can do anything you want with them. But these dolls belong to your sister and she can do anything she wants with them. So you can't play with her doll unless she wants you to, but she can't play with yours either." Here approved and disapproved behavior are embedded in the elaborate structure of property rights, and the simple *No* of the toddler has become a set of instructions. For these instructions to be effective, the child must be able to comprehend them intellectually, and their impellent value must be high enough to override competing tendencies to behave otherwise. As we shall see when we discuss the development of conscience (see Chapter 5), neither the cognitive nor the motivational side of obeying rules is a simple matter. The rules so clearly stated by parents suffer many a change as the child struggles to assimilate them, while his ability to give himself good advice is no guarantee that he will be able to follow it.

Self-Control as the Ability to Delay Gratification

The Importance of Trust. *Wait* is different from *No*. It does not require the toddler to inhibit an activity entirely, only that he tolerate the tension of postponed gratification. We discussed delayed gratification when we described (in Chapter 2) the trusting relationship which develops between the infant and his caretaker. The neonate cannot tolerate any delay. In his experience mild hunger leads to intense hunger which leads to distress. Whatever anticipations he develops warn him that, if things are bad, they will only get worse. However, if feeding is sensitively timed to his needs and accompanied by affectionate care, hunger will begin to be associated with the pleasures of gratification. Eventually he develops a firm expectation that the mother will come when he is in need. In short, the infant learns trust. Concomitantly, the anticipation of pleasure comes to outweigh the anticipation of increasing distress and enables the infant to tolerate the discomforts of hunger. The anticipation spreads in time: the sight of the mother holding the bottle will calm the infant and, subsequently, the sight of the mother herself will quiet him. Schematically, the development is as follows: consistent gratification leads to a strong expectation of gratification or trust, which, in turn, enables the child to tolerate the tensions of delay.

Trust is equally important to the ability to postpone gratification in middle childhood. Here expectancy of reward rather than the period of delay per se is the determining factor in the decision to postpone gratification. Mahrer (1956), using seven- to nine-year-olds, found that if the promise of delayed reward was honored (specifically, if the child was always given the balloon he was promised the next day), the child preferred the delayed reward; if the promise was not honored (if the experimenter said next day that he did not have the balloon), the child preferred immediate rewards. The ability to tolerate delay, at least when the reward is in the hands of another human being, becomes a matter of trust. In a like manner, the untrustworthy adult, who makes promises which he does not keep, fosters in the child the attitude that he should take what he can get.

The Role of Thinking. Both the mentalistic Freud and the objective behaviorists postulate a reciprocal relation between action and thinking: the inhibition of motor activity is essential to the development of thought, and thinking serves to delay motor activity. Freud's theory is more elaborate. The neonate, he maintains, has no ability to delay motor reaction; on the contrary, there is a direct connection between stimulation and motor response. To continue with our example from the previous section, the stimulus of hunger produces the response of motor restlessness; the more intense the hunger, the more vigorous the motor reactivity. As we have seen, however, the well-cared-for infant begins to anticipate fragments of the gratifying experience. Perhaps he has some vague image of the mother's breast or face, perhaps he begins to suck as he did during nursing, but at least some representation of the gratifying experience is present when hunger arises. Freud labeled these anticipatory representations hallucinations, borrowing the term from psychopathological adult behavior. The same primitive connection between strong need and hallucinated gratification can also be observed in normal adults under severe stress: the man lost in the desert hallucinates water, the castaway, food. In infantile hallucinations a mental event now intervenes between need and motor action. This is the significant developmental step. Instead of the sequence need-motor activity, we now have the sequence need-mental activity. The mental image of gratification cannot satisfy the need, but it temporarily holds the motor activity in check. The direct connection between stimulus and response, which is the hallmark of primitive behavior, no longer exists.

Hallucinatory activity, however, is the most primitive form of ideation. It is magical in that the need directly conjures up an image of gratification–the desire itself produces the thing desired. Obviously, hallucinations are not adapted to reality. As the infant learns about the real world, the fleeting, magical image of gratification is replaced by ideation which is more sustained and more attuned to reality. The infant begins to think. For Freud, thinking is essentially experimental action at a symbolic level. Instead of actually doing something the infant rehearses an action symbolically to see whether it will lead to pleasure or pain. To illustrate: an angry three-year-old wants to hurt his younger sibling. His immediate, primitive idea is of striking the baby. However, he is mature enough to rehearse the action and its consequences symbolically. We will reconstruct his thoughts as follows: "If I hit the baby, Mother will punish me and that will be painful. If I run around and accidentally bump into the baby, I will still hurt him, but Mother may not punish me. Therefore, an accidental bump is better than a direct hit." The primitive sequence, need-ideated goal, which epitomized hallucinatory gratification, has become need-ideated goal-ideated action to achieve goal. Note that thinking still serves the primary purpose of need satisfaction in Freudian theory, but now it is concerned in part with actions designed to obtain that goal. As mental activity becomes more elaborate and stable, it is increasingly effective in serving the dual functions of delaying motor behavior and guiding it realistically.

The mentalistic nature of Freud's formulation has made it unpopular with objective investigators. However, some of them do accept the concept of thinking as internalized action. The infant, with his vast capacity for learning, does not remain a creature of immediate motor reaction; he comes to represent actions in an abbreviated manner. Words are particularly suited to such representation, and talking becomes the symbolic counterpart of acting. When the toddler says, "I played on the slide," he compresses fifteen minutes of intense physical exertion into a sentence which can be spoken in seconds. Subsequently, representation becomes covert, so that the child "talks to himself." This talking to oneself, or the covert manipulation of verbal symbols, is regarded as equivalent to thinking by some theorists, or as an important component of thinking by others. Some objective psychologists would also agree with Freud that thinking is experimental action at a symbolic level. In talking to himself, the toddler can symbolically engage in certain behavior and anticipate its consequences. (For a discussion of theories concerning the relation between speech and thinking, see Kohlberg, Yaeger, and Hjertholm, 1968.)

Luria (1961) has conducted one of the most extensive developmental studies of the ability of speech to inhibit and guide motor behavior.[4] Assuming that thinking involves covert speech, we can view Luria's research as an investigation of the overt counterpart of thought. Luria first studied the process by which a child learns to subordinate his actions to the adult's verbal instructions—that is, how he learns to obey the adult—and, next, how the child acquires the ability to subordinate his actions to his own speech—that is, how he learns to engage in voluntary behavior. Both acquisitions are difficult because of the potency of motor activity. During his first two years of life, at least, the motor impulse is so strong that verbal instructions cannot serve a braking function. In his first study, Luria presented each child between eighteen months and two years of age with a small balloon and instructed him to squeeze it. At this age, verbal instructions from the adult are sufficiently potent to initiate activity. However, once he begins to squeeze, the toddler tends to continue squeezing. Instructions to stop not only are ineffectual, but they may even lead him to squeeze harder. Thus, verbal instructions can initiate but cannot stop behavior.

It is interesting to note that Freud came to a similar conclusion from a completely different theoretical orientation. He said that there is no such thing as *No* in the unconscious. In infancy, which is the period of the most primitive functioning, feelings and fantasies are acted out immediately. The ability to inhibit one's own behavior is a later development.

Verbal instructions also cannot regulate behavior in a child of eighteen months. If you tell him, "When you see a light go on, press the balloon," he cannot grasp

[4]Luria's findings are also applicable to the preceeding section concerning the inhibition of behavior. However, his studies are easier to follow if presented as a unit.

each of the two phrases of the instruction and coordinate them so as to comply appropriately. Rather, when he hears the first part of the instruction, "When you see a light go on," he starts looking around for the light; when he hears the second phrase, he starts pressing the balloon. Again, the impulse to immediate action dominates. The child must be three to four years old before verbal instructions from an adult are sufficiently potent to regulate motor behavior. Even then, the child has some trouble inhibiting action. If he is told, "Press the balloon when the green light goes on, but don't press it when the red light goes on," he is more proficient in doing the former than the latter.

What about the three- to four-year-old's ability to regulate his own behavior by giving himself instructions? Here the difference between the potency of verbal facilitation and verbal inhibition is striking. If the child says "Press" every time he should press the balloon, his behavior is more proficient than if he did not give himself the instruction. Self-instruction facilitiates behavior. However, if the child says, "Don't press" every time he should inhibit a response, he is much less proficient in holding back than if he does not instruct himself at all. Self-instruction impedes appropriate behavior.

Luria maintains that the "impellent" action of speech is still more potent at this age than its meaning. His point was demonstrated in the final experiment of this series. The child was instructed, "Press the balloon twice when a light goes on." If he said "Go! Go!" each time he pressed it, his motor behavior was more proficient than if he said nothing. But if he said, "I shall press twice," his motor behavior was disrupted. In fact, he typically responded with one long press. The meaning of the instruction is no match for the impellent action of speech. Luria claims that the child must be four-and-a-half to five-and-a-half years old before speech has developed sufficiently to allow the meanings of words to serve as effective guides to behavior. While Luria's developmental timetable is significantly slower than Freud's, remember that he is dealing with thinking in terms of verbal symbols or words, while Freud was dealing with any kind of symbolic representation, such as an image or a gesture, as well as a word. Words are comparative latecomers on the symbolic scene.

Thus far we have been concerned with the relation between thinking and motor activity. Both Freudian and objective psychologists extrapolate from the ability to delay a motor response to the ability to delay gratification. Recall our example of how thinking resulted in the three-year-old's postponing his goal of hurting his younger sibling. Both schools of psychology would predict a positive relation between thoughtlessness and the need for immediate gratification, while planning ahead makes possible the postponement of rewards (see Singer, 1955).

The relation of delay of gratification to ideation has been explored in a series of objective studies. Typically, the child was given a choice between having a small reward immediately or a larger reward at some later time; e.g., for eighth graders, it might have been a choice between $1.00 or one hit record now, or $1.50 or three hit records in three weeks. Mischel and Metzner (1962) studied a group of children between five and twelve years of age and found that preference for the delayed reward increased with age and with intelligence. Further studies showed that delinquent thirteen-year-olds preferred immediate reward to delayed reward and that socially mature children preferred delayed reward (Mischel, 1961). Also, children who had high achievement need—those who evaluated their work against a high standard of excellence—had a stronger preference for delayed rewards. Thus, the ability to delay gratification is associated with intelligence, planning, social responsibility, and ability to find work rewarding. Preference for immediate rewards is associated with thoughtlessness, irresponsible or delinquent behavior, and a tendency to live for the pleasure of the moment.

The Effect of Success and Choice. Mischel (1966) extended the experimental procedure just described by having the child work for the larger, delayed reward rather than merely wait for it. He found, as predicted, that willingness to work for the reward depended upon expectancy of success or failure. Children who had already been successful on a certain task chose to work on another similar task for a larger reward more often than did children who had failed on the first task. Just as one's willingness to wait for a reward from another person depends upon that person's trustworthiness, one's willingness to work for a reward depends upon his confidence that he will be successful.

Mischel also investigated the variable of choice; that is, whether the child was given a choice as to delay or had the delay imposed upon him by conditions over which he had no control. When delay was voluntary, willingness to wait was a linear function of the perceived probability of obtaining the reward. By contrast, when delay was imposed and when the child believed he had no chance of obtaining the reward, the reward took on significantly higher value. The child longed for the goal which was impossible to obtain rather than realistically abandoning such a goal. In general, then, when a child had a choice, he realistically balanced delay against chances of attaining the goal; deprived of this choice, he unrealistically overvalued the goal he could not obtain.

While the research results are far from conclusive, they suggest that the capacity to delay gratification depends on factors in addition to the trustworthiness of others and the increasing complexity and potency of thinking. Delay may also involve the "self" variables which are aspects of initiative (see Chapter 2)—the self-confidence which comes from successful venturing out and the perception of being in control of one's fate rather than being a victim of external forces. Further research may reveal that these characteristics of the child's concept of himself are as important in determining his capacity to delay gratification as are any others.

Substitute Gratification as an Aid to Self-Control

Freud's Theory of Substitute Gratification. The final step in the development of self-control is the ability to find substitute activities for ones which are forbidden. The concept of substitute gratification harks back to Freud. Let us return to his conceptualization of the neonate period. The infant functions according to the pleasure principle, which means that he strives to maximize pleasure and minimize pain or unpleasant tensions. However, the self-centeredness of the pleasure principle makes it unsuited to the realities of social living, and the socializing parent inevitably begins to thwart it. He may wean the infant and thereby diminish the pleasure of sucking; or he may prevent the toddler from giving vent to his primitive rage and require him to sustain the tension of unexpressed anger. If the frustrations of socialization are introduced gradually and the demands for control are appropriate, the infant will be able to relinquish the pleasure principle in tolerable stages. He is also aided by the fact that he can find substitute gratifications instead of having to give up pleasures entirely or totally inhibit the expression of negative feelings. The infant who can no longer suck on the bottle can suck on a blanket or his thumb or a piece of candy; the toddler who is punished for hitting his brother can knock over a tower of blocks. Substitute satisfactions protect the child from becoming unduly resentful of or inhibited by the constraints of socialization and furnish a socially acceptable outlet for otherwise disruptive anger and anxiety. By providing some degree of pleasure and release, they keep negative affects at a manageable level of intensity and facilitate the achievement of self-control.

But Freud adds an important proviso. In the case of intensely pleasurable experiences which have an erotic component, such as sucking or masturbation, no substitute is fully satisfactory. There always remains a certain amount of frustration over not having experienced pleasure fully. The more the child—and especially the adolescent and adult, in the case of sexual pursuits—has to deny himself direct erotic gratification, and the more completely he must rely on substitutes, the more this frustration builds up, until he becomes generally restless and chronically dissatisfied.

Lewin on the Substitute Value of Tasks. As frequently happens, Freud's observations and conceptualizations have been accepted by some psychologists who are not sympathetic to his overall theory, especially his emphasis on erotic gratification (see Chapter 7). Lewin (1935) reinterpreted the concept of substitute gratification in light of his own theoretical framework and stimulated a series of objective studies designed to illuminate it (Escalona, 1943; Lewin, 1954). Typically, the procedure was as follows: A child is presented with a number of simple tasks, such as making objects out of clay or numbering sheets of paper. He is not allowed to complete them, but is given substitute tasks to complete instead. A substitute task has low substitute value if the child subsequently chooses to return to the incompleted task and resume working on it. If the child is no longer interested in the initial task after completing the substitute task, then the substitute task has high substitute value. We assume, then, that the original need has been satisfied by an alternate activity.

This procedure is based on earlier research which showed that there is a strong tendency to complete an interrupted task. Beginning a task sets up a need, or what Lewin calls a tension, which persists until the goal of completion is achieved. If the task is interrupted, the tension remains and influences subsequent behavior. The individual may continue to have the unfinished business on his mind and want to return to it. The tendency to resume or to remember incompleted tasks more than completed ones is called the Zeigarnik effect, after a student of Lewin's who investigated the phenomenon.

The research literature reveals a number of factors which contribute to substitute value. The more similar the new task is to the original one, the higher the substitute value will be. (Logically, an identical activity should have the highest substitute value because it is almost equivalent to a continuation of the interrupted task.) A child who is interrupted while drawing a dog on red paper and diverted to drawing a dog on green paper will complete the drawing on red paper less frequently than if he is diverted to putting together a jigsaw puzzle. The difficulty of the new task is also important. The substitute value of a similar but easier task is lower than that of a similar but more difficult task. If, for example, the child is interrupted while making a clay dog and is asked to make a clay snake, he is more likely to return to the dog subsequently than if he is interrupted to make a clay horse. This finding suggests that the child not only wants to do a specific task but also holds himself to a certain standard of achievement. In Lewinian terms, he has a certain level of aspiration, and a substitute activity is not satisfactory unless it presents a comparable challenge. Finally, the stage of the child's cognitive development is important, since it affects his ability to see similarities among tasks. Between the ages of seven and ten, children become increasingly able to accept substitutes because they are better able to grasp the notion of similarity. When a younger child is interrupted in building a house of blocks, he will not be satisfied with building another house. He is still at the stage where a given object is a unique thing rather than a member of a class of objects. The house he started to build is special in all the world, and nothing else can take its place. Only when he is able to think in terms of classes of objects, to subsume individual houses under the general category "house," does one house become as good as another.

Lewin uses these results to re-emphasize a major theoretical point: behavior is determined not by the objective situation but by the way an individual perceives the situation. The same objective setting may have different meanings to different people, and their reactions can be understood only if one knows their idiosyncratic interpretation. Thus, the same play slide may represent high adventure to one three-year-old and a terrifying threat to another. To evaluate substitute activities, one must understand what goals the child himself has when he undertakes a task. The instruction, "Make an animal out of clay," may be interpreted by one child as a command to make a specific animal, such as a dog; it may be interpreted by another child to mean any animal. For the first child, the task of making an animal other than a dog will be viewed as a different task; hence, its substitute value will be lower than it will be for the second child, for whom one animal serves to fulfill the instruction as well as another. The general finding that similarity increases substitute value is not as simple as it seemed at first. As we now know, two tasks may appear similar to an adult but different to a child if he is too young to understand that they belong to the same class; or two objectively different tasks may appear similar if the child has a very general goal in mind. In Lewin's terms, the tasks must be dynamically related—that is, they must be perceived as related by the child—before one can satisfy the tension set up by another.

Note that Lewin deals exclusively with delay—a goal is temporarily unavailable. He does not deal with punishment. One can speculate that, if a task had been punished rather than interrupted, both the tendency to resume it and the attraction to similar tasks would be weakened. The child might accept substitutes but have to content himself with less similar ones, for now he has two concerns: the new activity must be as similar as possible but it also must be as safe as possible.

Lewin on the Substitute Value of Play. Lewin was also intrigued with the substitute value of play and fantasy. Again, he took his lead from Freud, who regarded fantasy, daydreams, and night dreams as forms of substitute gratification. The phenomenon is familiar: the outwitted social climber lies awake fantasizing a dozen clever remarks she could have made; the timid Walter Mitty dreams of heroic exploits (see Chapter 9). In early childhood the line between reality and fantasy is not as firmly established as it is in adulthood, and fantasy is constantly weaving in and out of the child's activities; a bowl of cream of wheat may become a panorama of mountains and milk lakes, clumps of bacon may become bears. Does this mean that, in children three to six years of age, fantasy can serve as a substitute for reality?

Lewin varied the research procedures described above to answer this question. Instead of presenting new tasks, the experimenters introduced different toys or "pretend" objects into the ongoing sequence of activity in order to determine whether they would be accepted or rejected as substitutes. The results showed that the children can indeed separate fantasy from reality. If they start cutting with real scissors, they will not be content to take toy scissors and just pretend; nor is a piece of cardboard in the shape of candy an acceptable substitute for the chocolate they have begun eating. However, if the task itself is one of play, then any number of substitutes are readily accepted. At a tea party, cardboard squares are adequate substitutes for doll plates, or a stuffed sock with facial features can serve as the guest of honor.

Lewin accounts for the difference between realistic action and play in terms of the fluidity of the situation. The practical demands of reality, by their very nature, put constraints upon behavior. When the young child learns the proper uses of objects, he must turn his back on a host of equally fascinating alternative uses; knives are to cut with; crayons are to color with. In play, constraints are once again relaxed. Events can happen according to the child's wishes, and objects can regain their versatility of meaning—a knife can be an airplane, a crayon can be a piece of toast. Under the relaxed conditions of play, therefore, objects can more readily be perceived as similar and substituted for one another.

Finally, Lewin found that the effectiveness of a substitute depends upon the intensity of the need. When the need is weak, substitutes are more readily accepted. A child who is busily cutting with a pair of scissors will not accept toy scissors; however, after he has finished cutting, he will accept them more readily. Strong unsatisfied needs have the effect of increasing spontaneous substitute activity while making substitutes less gratifying. This seems paradoxical, but it corresponds to clinical observations. A man who is helplessly enraged at his boss because he did not receive a promotion may be preoccupied with fantasies of gettting revenge on the boss and may become irritable and hostile toward his family, while still remaining tightly in the grip of his anger. The rage which drives him to seek substitute outlets cannot be diminished by such substitutes.

Lewin ends his discussion with the following summary: "Substitute value is greater the more the substitute action corresponds, not to a new goal, but to another way of reaching the original goal" (p. 192). This is Freud's conclusion exactly, except that the implication of the conclusion is more daring and controversial in the context of Freudian theory. Unlike Lewin, Freud maintained that a human being has a limited number of goals. One prime goal is to obtain maximal erotic gratification. The goal itself is fixed and universal; the great variability among personalities is due to the versatility of human beings in learning socially acceptable means of achieving the goal. The infant sucks the breast; the adult puffs a cigarette. The goal of pleasurable sucking can be the same. Such a lineage seems reasonable. The four-year-old boy peers with excited curiosity at his sister's genitalia; the forty-year-old scientist peers with excited curiosity into his microscope. Is the goal the same? Such a lineage seems absurd. Yet Freud would claim that, in given instances, the goals may be the same, but social pressures have forced the substitution of scientific curiosity for direct sexual curiosity. We shall investigate Freud's motivational theory, with its emphasis on erotic gratification, in Chapter 7. At this point we shall only underscore the agreement between Freud and Lewin concerning the increasing effectiveness of substitute activities as they approximate the achievement of the original goal.

Initiative and Self-Control

In our discussion of factors involved in self-control, we have often referred to attachment and affection. Rarely have we referred to initiative. Is this because the two processes are independent, at least as independent as any processes can be in a living organism? On the face of it they seem to be. In exploring the world, the toddler is doing what he wants to do; in controlling himself, he often does what he would not choose to do. This is an important difference. Yet there are some striking parallels between initiative and control which make one suspect that the former might at least enhance the latter. Let us look at some of these parallels.

In our discussion of delay, we learned of the antagonism between ideation and motor activity. Whether one accepts the Freudian concept of hallucination or the learning theory concept of an anticipatory goal response, it is fairly clear that a representation of satisfaction serves to inhibit motor discharge. Evidence of an even earlier antagonism between action and attention can be observed in the neonate (Wolff, 1965; Wolff and White, 1965). The basic process of attending to the external environment involves checking immediate motor impulses. From the beginning, then, alert wakefulness necessitates motor quiescence. This state of alert wakefulness is most conducive to the neonate's and infant's exploration of the environment which, in turn, is a precursor of initiative. We can speculate that the capacity for attending may facilitate both the infant's exploratory behavior and his subsequent ability to utilize symbolic representations in order to brake motor discharge. Conversely, immediate motor reactivity may be the common antagonist of attention-exploration, and symbolization-self-control.

While engaged in exploratory activity, the infant experiences both frustration and delay in gratification. Instead of *No* and *Wait*, he is faced with the frustrations which arise from lack of skill and the delay which occurs when his goal exceeds his instrumental ability to achieve it. For example, a one-year-old may want to place one block on top of another before his motor coordination allows him to be successful; a two-year-old's desire to communicate by speech can exceed his ability to speak. Thus, the *No* and *Wait* of the socializing parent do not confront the child with a totally new experience. We can speculate that the infant and toddler's ability to project task goals further into the future and tolerate the delay inherent in achieving them may facilitate his ability to wait for rewards administered by the parent; e.g., a two-year-old's ability to persist in trying to build a tower of five blocks may help him tolerate the situation in which he wants a cookie and his busy mother says, "Wait a minute."

Finally, we have speculated that the self-confidence engendered by successful exploration of the environment, and the perception of the self as being in control of events—both of which are important facets of initiative—may play a significant part in tolerating the delay of socially administered reinforcements (see p. 107).

Environmental Influences on Self-Control

We have been dealing with self-control as an inner regulator which operates independently of the environment. Certainly a degree of independence is essential. The child who gives in to every temptation from without, like one who succumbs to every impulse from within, lacks self-control. However, even normal self-control fluctuates, and environmental factors are constantly at work either supporting or undermining it.

Modeling. The potency of situational factors is derived from the child's well-known tendency to pattern his behavior after the behavior of others—"monkey see, monkey do," as parents put it. A birthday party for three-year-olds ends in shambles, not only because the children become overly excited and tired, but because they also serve as models of uncontrolled behavior for one another. Self-control is difficult for a three-year-old when his one-year-old brother demonstrates just those babyish behaviors he has recently learned to inhibit. The tendency to act as one observes others acting is often called imitation; however, since we have reserved that term for learning a new response, we shall use the term modeling for the elicitation of existing responses.

There is objective evidence that a model can significantly increase or decrease the child's control of his behavior (Bandura and Walters, 1963). Observing an aggressive adult will significantly lower the child's subsequent resistance to acting aggressively. There is a sex difference: boys' aggression is increased by exposure both to an aggressive model and to a highly expressive but nonaggressive one; girls are relatively unaffected by aggressive models, but nonaggressive models produce a reduction in their aggressive behavior. The authors account for these differences in terms of the relative dominance of aggression in boys, which makes aggressive responses easier to disinhibit;

girls are relatively nonaggressive, so their tendency to control is more readily strengthened. However, if boys or girls observe an aggressive model being punished, their resistance to behaving aggressively is strengthened.

Family Themes. Research on family interaction contains revealing descriptions of the ways in which characteristic kinds and levels of self-control are constantly being reinforced. Hess and Handel (1959) made an intensive study of five families by means of observation, interview, and personality tests. They found that families have characteristic "themes" just as an individual has a characteristic personality. A family theme is an implicit and explicit notion as to "who we are" and "what we are all about." Although the themes are exhibited in a variety of ways, we will concentrate on the "intensity of experience."

For the Lansons, the family theme is quiet harmony and modesty. The very house itself is quiet and well cared for; the walls and carpets are in pale, neutral shades; the furniture has plastic covers. The members speak softly, and even the nine- and twelve-year-old daughters play quietly together. However, there is no sense of conscious restraint; rather, the family is leading the kind of life it values and finds most fulfilling. The members talk freely and openly. Behavior is evaluated in terms of whether it promotes or disrupts the equilibrium and harmony of the family. Cooperation is the prime value; the usual acts of selfishness are frowned upon, but so is anything which gives an individual private satisfaction. Each must work for the overall good of the family unit. Discipline is fair and firm; democracy is stressed. The family is not joyless, but it would never occur to any of them to seek pleasure for its own sake; even occasional surprises are suspect. In such a well-modulated setting, anger poses the greatest threat. However, the parents are in such basic accord that quarreling is almost unknown. Although they punish aggression, they never take sides and champion one girl over the other. The girls have completely accepted parental values; whatever anger they feel is expressed outside the family, and even this is moderate.

The Newbolds, by contrast, value activity, self-assertiveness, and achievement. (They very much suggest Baumrind's "authoritative" parents described in Chapter 3.) They are doers and natural leaders. Mastery, rationality, personal responsibility, self-discipline, and productivity are their values, as are zest and enthusiasm for seeking challenges. The father is strong, authoritative, and fair. He does not punish often, but his punishment is final. Both parents suppress conflict immediately and effectively, so that strong antagonism and anger do not often develop. In contrast to the democracy of the Lanson family, the Newbold parents have unquestioned power and authority. Childhood is valued only as preparation for adult life. The children can gain status primarily through being responsible and energetic within the family. High standards are set, but achievement is rewarded by parental pride, freedom, and privileges outside the home. By acting "like a Newbold" the child can achieve status with adults. The older adolescent son completely embodies family values. He has energy, initiative, leadership, and an assured expectation of success. Unaccountably, the younger son has a tendency to be reserved and somewhat isolated. He wants to be a farmer rather than a businessman and prefers the outdoors to the city. His deviation makes the parents uneasy. He confronts them with the paradox of championing individuality while tacitly wanting him to be "our kind" of adult.

Although the Lansons and the Newbolds are in many respects opposites of one another's way of life, they share a belief in the importance of self-control, and each family provides the children with a coherent set of values and a meaningful structuring of experience which are so important to its development. Parents in general are constantly defining for their children both the kind and the extent of self-control they expect, while the fabric of family life itself reinforces compliance and checks deviations.

The Concept of Styles of Control. Other families are more casual and emotionally free-wheeling than the Lansons and Newbolds. Feelings come quickly to the surface and are readily expressed. There is no effort to avoid anger and open confrontation. The need for involvement and intensity is high. One mother said, "If it got quiet around here, I'd go crazy"; and another, "If I didn't yell at the kids, they'd think I didn't care about them." Fighting and quarrels alternate with affection and tenderness. (Incidentally, both the Lansons and the Newbolds lacked tenderness.) It would never occur to such families that anger is disruptive or incompatible with love and respect. Even a parent's explosive response to being defied can be accompanied by pride in the child's spunkiness. Standards of proper behavior are championed and achievement is valued, but there is no feeling that a proper and useful life requires emotional abstinence.

Such families encourage emotional expressiveness in their children. But what can we say about self-control? Customarily, we think of a continuum of self-control, with inhibition at one extreme, flexibility in the middle, and impulsiveness at the other extreme. Should we put the Newbolds in the middle, with the Lansons at one extreme and the emotional families at the other? This does not seem justified. The Lanson girls are not inhibited in the sense of being psychologically hemmed in and struggling against impulses to defy convention. Their moderation provides a congenial structure which carries a deep conviction of its intrinsic rightness. Similarly, emotional reactivity is not equivalent to impulsiveness. Expressive children have no fear of losing control of their feelings; they do not seek excitement for excitement's sake; they have no potentially antisocial violence. If they allow themselves deviations from the norms of propriety, they never defy the norms themselves. We are not suggesting that a continuum of control is invalid; we shall use it in succeeding chapters. However, we are saying that the middle range of good control is not exhausted by a single stereotype. To regard one pattern of behavior as the model of good control and all deviations as tending in the direction of poor adjustment is to oversimplify and thus distort the situation. Instead, we must think in terms of diverse styles of control, each having its assets and liabilities, but each equally defensible in terms of promoting the growth of the child's personality.[5]

[5] For an extensive account of styles of coping and their development, see Murphy (1962). Adler introduced the concept of "life styles"; it was central to his theory, and it is relevant to our discussion. Since Murphy addresses herself specifically to children, her account is more pertinent.

Parental Discipline

Thus far we have taken a cavalier attitude toward parental discipline. *No, Don't, Good boy, Bad boy* is a compact formula for a richly varied realm of parental behavior. The problems inherent in disciplining children have concerned clinicians and objective investigators alike. We will begin our discussion with certain broad issues and then examine specific techniques of discipline and their consequences.

A Clinical Model. Clinicians have wrestled with the problem of how to discipline the child so as to maximize psychological growth and avoid both stultifying conformity and irresponsible impulsiveness. In reaction to Victorian strictness, they championed permissiveness and the image of parent-as-pal. As the dangers of this approach became apparent, the pendulum began to swing back (Bronfenbrenner, 1958), and both authority and superior status were once again granted to parents. While administering discipline, both parents, but especially the mother, must be loving. Without love, discipline either is ineffectual or becomes a bitterly resented imposition. On the other hand, an excess of love binds the child too closely to the parent and makes him feel needlessly guilty over normal anger and rebelliousness. Authority is also needed, especially from the father. The child must be protected from acting on his immature judgment and is reassured when parents help him to control impulses he cannot manage alone. At the same time, authority should not be so cold and severe that it frightens and intimidates the child into subservience.

There is often a child-centeredness theme in the clinical approach to discipline. This theme was championed by Adler, who strongly opposed the then accepted philosophy of breaking the child's will. On the contrary, he maintained that the child's will is just what must be preserved—his initiative, his freedom to experience and experiment (see Chapter 3). Adler also stressed the need to preserve the child's individuality. Not only do children's needs change as they develop, but each child also has his special tempo, interests, assets, and vulnerabilities. His uniqueness is his most valuable asset. Being attuned to individual needs requires flexibility in discipline, and, for Adler, flexibility was one of the primary signs of maturity.

Meeting the needs of the individual child may seem like a vague concept; yet even the most rigorously objective investigator must face the problem of individual differences in endowment and previous experience. For example, there are no universal rewards which can be used with all children at all ages. A middle class child is reinforced by being told he is *Right*; a lower class child is relatively unimpressed with being right but responds to being told that his performance was *Good*; some severely retarded and emotionally disturbed children are unresponsive to verbal approval and need tangible rewards, such as candy or toys. The sex of the adult and the age of the child affect the potency of rewards. A child may find it painful to be ignored, if the researcher has previously praised him generously for his success, but being ignored may be a welcome relief if the child has been told repeatedly that he has been doing poorly. To be meaningful and effective, rewards and punishments should be child-defined, not experimenter-defined (Thompson and Hunnicutt, 1944). Thus, the clinical concept of child-centeredness has its counterpart in objective research.

Meaningful Discipline. Erikson (1950) adds an interesting dimension to child-centeredness. He maintains that children become disturbed not because they have been punished but because they have been punished "meaninglessly." He does not elaborate on this point, but it is worth pondering.

Let us consider "meaningful" discipline. Discipline should be intellectually comprehensible. Particularly for very young children, comprehensibility is enhanced by explicitness. Telling a child, "You are bad because you did X," is more comprehensible than merely saying, "You are bad." "Bad" is a general label which effectively communicates parental disapproval but may leave the child uncertain as to what is being disapproved. Remember that discipline is something new to the toddler. Observe him encountering his first *No's* and *Dont's*. Often he does not know what to make of his mother's behavior. He may at first act as if this were a new game, or he may be stymied by her strange actions. Once he understands that his mother means business, he must struggle to figure out, business about what? To him an activity is either interesting or not interesting. He sees no reason why his mother should periodically descend with her prohibition. If the mother is explicit, the toddler at least will know what behaviors result in disapproval. To be comprehensible, discipline must also be appropriate to the child's level of development. To tell a child that it is wrong to steal before he has a clear idea of possessions, or to tell him that it is wrong to lie before he has separated truth from fantasy, will only bewilder him.

Meaningful discipline is for the child's own good (see Erikson, 1950). Its purpose is to help him manage the rage, jealousy, possessiveness, and greed which are personally disruptive and jeopardize his relations with others. It also prepares the child to live in the community of adults. Whatever fulfillment he will eventually find will be within a social context. In upholding the values and requirements of social living, the parent is preparing the child for full participation in adult life.

Meaningful discipline can also be understood by examining instances of meaningless discipline. A vague and uncertain parent is baffling to the child, who becomes uncertain himself as to what he can and cannot do. A toddler can be observed returning again and again to a vaguely forbidden activity, not to provoke his parent and not out of willfulness, but in an effort to clarify the prohibition. Discipline that is capricious and unpredictable is not meaningful. An alert youngster learns that an "irrevocable" *No* will give way to an indifferent shrug if he bides his time. There are highly volatile parents whose children learn to respond to their mood of the moment rather than to the content of their discipline. Avoiding parental anger and capitalizing on periods of good humor are more important than learning guides to behavior.

Discipline can be meaningless if the child's behavior touches off poorly mastered impulses in the parent. We are referring here not to normal anger and irritations, which most parents experience, but to intense overreactions—the mother who is so disgusted by her toddler's messy eating that she must leave him alone at mealtime, or the father who is so outraged by his son's disobedience that he wants to beat him to within an inch of his life. Children sense the inappropriateness of such parental reactions and are frightened by their intensity. Discipline becomes too grave a matter for the child, with too much of his security at stake. Finally, discipline is meaningless when it is used as a battleground for parental conflicts. Fathers and mothers are expected to disagree to a certain extent, since men are generally more absolute and objective in their reasoning about discipline than women (see Stolz, 1966). But discipline can become a vehicle for expressing deeper marital disharmonies. A father who is dissatisfied with his wife as a sexual partner may begin depreciating her and subverting her efforts to control the children; or a mother may try to make her child a model of good behavior to prove she is superior to her husband. Parental discipline in these instances is no longer motivated by a concern for the child's good, and the child is trapped in the distressing situation of having divided loyalties. It is this kind of contamination of discipline with irrelevant motives, conflicts, and feelings which, Erikson implies, may produce emotional disturbances in a child. Whether the parents spank him or not is unimportant.

Factors Influencing the Effectiveness of Punishment. We shall now examine some objective evidence concerning discipline. Aronfreed (1968) has conducted a series of studies which throw light on the effectiveness of prohibitions or punishment. He was primarily interested in discovering which variables in the socialization situation produce the most stable self-control in the sense of insuring that the child will avoid a prohibited behavior.

Typically, he confronted an eight- to ten-year-old with a pair of toys, one attractive and one unattractive. Using procedures to be described later, he modified the child's behavior so that he would choose the less attractive toy. Then the child was left alone. The effectiveness of the experimental procedure in producing self-control was evaluated according to how well it inhibited the child's desire to explore the more attractive toy.

The first series of experiments involved the timing of punishment. One group of children was told *No* and was deprived of candy which was to be theirs, just as they reached for the attractive toy; another group was punished immediately after they picked up the toy, and a third group six seconds after they had picked up the toy; a fourth group was punished after they had described the toy in response to a question. The earlier the punishment, the more strongly the children resisted the temptation to handle the attractive toy when left alone. Aronfreed's interpretation of these results is that anxiety—which is strongest at the moment when punishment occurs—suppresses behavior more effectively if it becomes attached to cues at the beginning of an activity rather than at the end (see p. 96). Punishing a child after a transgression may inhibit the final action, but will leave the initial phases of interest, intention, and approach relatively unaffected.

Aronfreed observes that parents are rarely hovering around waiting for a transgression to begin so that they can punish it efficiently. The most important facilitator of punishment is the parent's verbal behavior and cognitive structuring. Therefore, the procedure was modified so that six seconds after the child had chosen the attractive toy, he was both punished and told the toy was appropriate only for older children. This cognitive structuring greatly facilitated subsequent resistance to temptation, as compared with the condition in which the children were punished with no explanation. (Recall our discussion of words as aids to categorizing objects and behaviors, p. 102.) Again, the timing was important. If the cognitive structuring was delayed ten to twelve seconds after the punishment, its effectiveness was weakened, even though it still was more effective than no explanation at all.

In a final series, the children were punished not for their behavior but for their intentions. Under one condition they were informed that they were being punished for wanting to tell the experimenter about the toy, six seconds after they had been punished for choosing it but before they had had a chance to tell about it. The combination of cognitive structuring and delay was extremely effective, and these children suppressed transgressions as completely as the ones who were punished immediately upon reaching for the attractive toy. Once again, timing was important. For the children who were punished after the intended behavior of telling about the object had occurred, the cognitive structuring concerning intention lost its special potency.

Aronfreed's research program is a good example of an objective investigator teasing out significant variables and the intricate relations among them (see Chapter 1). The question, "Is punishment effective in producing self-control?" becomes too general to answer. Instead, we begin to learn the conditions under which punishment can and cannot be effective.

Love Oriented and Power-Assertive Discipline. Is it what the parent does or the way he does it which most affects the child? Is praise good and spanking bad, or is the technique of negligible value as long as the parent "really loves" the child? In order to answer these questions, we shall examine objective studies of the effects of different techniques and affects on the child's personality (see Becker, 1964).

The first studies concern the consequences of love oriented and power-assertive forms of discipline. Love oriented discipline includes two techniques: on the positive side, the parent uses praise and reasoning; on the negative side, he threatens to withdraw love by isolating the child or showing that he is disappointed in or hurt by or ashamed of him. In essence, the parent says, "If you do what I want, I will love you; if you do not do what I want, I will suspend my love." To answer the question we are posing, these positive and negative techniques should be utilized with both positive and negative affects. In most of the studies, however, love oriented parents have also been emotionally warm. This correlation robs us of the chance to determine whether it is their technique, their affect, or a combination of the two which is influencing the child's personality.

Love oriented discipline by warm parents tends to be correlated with a socially responsible and cooperative child. He is reflective and concerned about his own role in incurring disapproval. When he does something wrong, he tends to think in terms of his own responsibility, to feel guilty, and to confess. He also tends to be nonaggressive in his social relations.

We have only fragmentary information concerning cold, love oriented parents. Wenar, Handlon, and Garner (1962) have described a kind of mother who praises and reasons but who is emotionally cold. She knows all the right things to say and has the technique of child-centeredness down to perfection, but is unloving. Such a mother can have a devastating effect on her child. Because she is cold, the child feels deprived and resentful, but because she is perfect, he has no apparent reason for reproaching her. As a result, the child is in a chronic state of unrelieved rage and, in certain instances, may become severely disturbed emotionally.

Power-assertive techniques rely on physical punishment, commands, yelling, and physical threats. The parent adopts the authoritarian stance: "Do what I tell you because I tell you to do it." Reason is minimized, and physical pain or the threat of physical pain provides the affective charge. Power-assertive parents tend to be hostile; hence, technique and affect are again confounded.

Power-assertive techniques used by hostile parents tend to be correlated with aggressive and uncooperative behavior in the children. The child resists authority and uses power-assertive tactics with his peers. He is also externally oriented: when he does something wrong, he tends to fear punishment and to blame others rather than blaming himself. The age of the child is a significant factor in determining his behavior. The reactions described so far are typical of young children. Around middle childhood and early adolescence (twelve years or so), the child inhibits overt expressions of aggression. His anger comes out in a punitive attitude toward wrongdoing; e.g., wrongdoers should be punished rather than understood or helped. He also turns his aggression back on himself; and becomes angry with himself for his own transgressions.

Our discussion of the factors involved in self-control helps to explain the different effects of these contrasting disciplines. Love oriented techniques play up the child's pleasure in receiving parental love and anxiety over jeopardizing it. The warm, reasoning parent both serves as a model of rational control and provides the child with the meaningful explanations so important to the formation of an evaluative approach to conduct. Aronfreed points out another facet of love oriented discipline. Physical punishment is behavior-specific; when it is effective, the child learns only to avoid the punished act. The effects of withdrawal of love are more complex; the child who is distressed at having done something wrong now wants to do something right in order to regain love. He has an opportunity to think and to choose restitutive measures, which aid him in developing an inward, reflective attitude toward punishment.

Power-assertive techniques rely on the anxiety-inducing properties of punishment. The interplay of anger and fear is relatively unelaborated by reason or understanding, and the child's primary concern is to avoid pain, detection, and blame. But why is the child not merely fearful? Why is he also aggressive? Partly because the angry parent serves as a model for the very aggressive behavior he is trying to eliminate in the child. There is also some evidence that the power-assertive parent, although he does not tolerate the child's hostility toward himself, either condones or encourages his expression of anger toward peers or outsiders.

We have by no means exhausted the kinds of parental authority or the effects of warmth, hostility, and reason on the child's personality. We shall return to this topic when we take up the development of conscience in Chapter 5.

Restrictive versus Permissive Discipline. Another set of objective studies has been concerned less with techniques of discipline than with the degree of control the parent exerts over the child's behavior. Some parents impose many demands and restrictions and insist on compliance; others impose few demands and enforce them only weakly. The former parents will be called restrictive, the latter will be called permissive, lax, or indifferent as the case requires. Fortunately, these parental behaviors occur independent of affect; hence, we can study the effects of warmth and hostility with both degrees of control.

Warm, restrictive discipline tends to be correlated with a compliant, dependent child. He is polite, neat, obedient, and submissive, the kind of child who never causes the parents, teachers, or other authorities any trouble. There is very little aggression but also very little friendliness or creativity. The child with a hostile, restrictive parent is surrounded by angry directives and prohibitions. Although he may occasionally quarrel with his peers, he is typically shy and withdrawn. More important, he takes his angry feeling out on himself and experiences inner turmoil and conflict. Neurotic symptoms of worry, anxiety, unhappiness, crying, and even accident proneness and suicidal tendencies characterize children of this group.

Warm, permissive discipline tends to be correlated with an independent, friendly, creative, assertive child. The child's assertiveness can take the form of hostility and bossiness, but, perhaps because of his vitality, friendliness, and constructiveness, he is socially successful and enjoys high status among peers. The combination of permissive, lax, or indifferent discipline with hostility toward the child is the best breeding ground for aggressive children. Antisocial behavior in children flourishes when one or both parents is inconsistent, since a meaningless, unpredictable alternation of neglect and harsh punishment is the pattern found to be most likely to produce delinquency. The child receives no love to make self-control worthwhile, and the parent's capriciousness sensitizes him only to situational signs of punishment rather than providing general guides to correct behavior. The punitive parent both foments rebellion in the child and serves as a model of irresponsible, hostile action.

Age and Sex as Significant Variables. Objective investigators have studied the changes in the nature and effects of parental discipline over time. We have already noted the differential results of power assertion on young children and on adolescents. While maternal warmth and hostility tend to be stable over time, the degree to which the mother controls the child or gives him autonomy tends to change (Bayley, 1964). Kagan and Moss (1962) studied the results of maternal restrictiveness on children from age three to adulthood. Both the age and the sex of the child turned out to be important variables. In general, they found that early restrictiveness has greater inhibiting effects and produces more conforming, dependent behavior than does later restrictiveness. Restricting an older child unduly is more likely to produce resentment and be regarded as unfair. The older child is more capable of retaliating with aggression, although he is rarely uncontrolled and eventually conforms. Boys are more likely to oppose a restrictive mother and be successful than girls are. Thus, girls who were restricted early in life tend to remain dependent on parents in later years while boys do not.

The Kagan and Moss study demonstrates the importance of the sex of the disciplining parent in relation to the sex of the child. Studies show that in general the mother is regarded as more loving and the father as stricter and more fear-arousing. The mother is seen, especially by girls, as using more psychological control, and the father is seen, especially by boys, as using more punishment. Boys also feel they are punished more than girls are. The same-sex parent is regarded as being less benevolent and more frustrating, particularly by older children.

Bronfenbrenner (1961) explored the interactions between sex of parent and sex of child in one of the most sophisticated studies of discipline. We will summarize only that part of the research which was concerned with responsibility in adolescents. As expected, extremes of authority, rejection, or neglect are associated with lower responsibility in boys and girls. Aside from extremes, there is an optimal level of authority for producing responsibility, and most interestingly, this level is higher for boys than for girls. For boys, lack of discipline and power-assertion by the father lowers responsibility, while for girls discipline and power assertion have debilitating effects. There is also an optimal level of emotional support. In general, girls experience more warmth and less punishment than boys. Such a love oriented approach is aimed at producing the accepted feminine behaviors of obedience and inward control. The danger is that girls will receive too much affection, especially from the father, so that they become overly sensitive to correction and control. While affection is important for boys, it cannot be so high as to produce effeminate sensitivity. Boys are effectively disciplined by power-assertive techniques and are encouraged to be assertive and aggressive. The danger is that they will receive too little emotional support and become rebellious.

Another interesting finding is that the father holds the key to responsible behavior. If the father is absent (due to war, death, divorce, etc.), the boy often does not receive the discipline which is so important to responsible behavior. However, the father's absence is likely to increase responsibility in the girl. When she is not subjected to paternal affection, she does not have the hypersensitivity to correction and control which is so detrimental to responsibility. Bronfenbrenner concludes that, in regard to responsibility, boys thrive in a patriarchal setting and girls in a matriarchal one.

Evaluation of the Objective Studies. Looking back over the objective studies, we see that they have yielded many insights into the effects of parental discipline on personality. Undoubtedly, they could be expanded even further. The taxonomy of discipline could be elaborated beyond the simple dichotomies of love oriented versus power-assertive discipline and permissive versus restrictive discipline. Becker (1964) has suggested just such an extension. The dimensions warmth-hostility, permissiveness-restrictiveness, and involvement-detachment can be combined to derive a variety of parental behaviors. For example, indulgence can be conceptualized as a combination of permissiveness, warmth, and emotional involvement; overprotectiveness as a combination of restrictiveness, warmth, and emotional involvement; and neglect as a combination of permissiveness, detached hostility, and rejection.[6] The influence of each kind of parental discipline can then be investigated.

The studies also illustrate a point made in Chapter 1. Objective investigators prefer to begin with specific, well-defined variables, examine them one at a time, and then study the interactions among them. At this point significant variables in parental discipline have been uncovered and studied in isolation. Conclusions from most of the studies should be prefaced by "all other things being equal...." Now we are prepared to face the fact that all other things never are equal. Love oriented discipline, for example, is affected by a host of variables in the child (see Chapter 10), the parent, the family, the peer group, and the society. Instead of holding such variables constant, it is time to explore their interaction. Bronfenbrenner's study can serve as a model, since discipline was studied in the context of age, sex, family constellation and social roles. Now that objective investigators have the research designs, the statistical procedures, and the technology to do multivariable studies, there is every reason to anticipate an increasing number of fruitful, multidimensional explorations in the future.

Comments: The Use of Authority; The Ideal Discipline. At a speculative level, we can extrapolate from the objective studies of discipline to the use of authority in general. Among adults in authority, there are the "good" leaders who use variations of love oriented techniques by being warm, understanding, and reasonable. Then there are the "strict" leaders who use variations on power assertion by being demanding, critical, aloof and perfectionistic. The "good" leader controls his followers by his inspiring example and by the guilt they feel if they fail him. The "strict" leader motivates through a combination of anger and respect–he manipulates anger to spur his followers into exerting maximum effort, and they are rewarded by the knowledge that they have been stretched to the limit and have met the test. Each use of authority has its risks. The "good" leader runs the risk of being regarded as soft or weak if he fails to function effectively, so that his followers' guilt turns into pity or indifference. The "strict" leader runs the risk of becoming unreasonably demanding, so that the goal is not worth the frustration involved in reaching it and constructive efforts turn into rebellion.

Bosses, teachers, and coaches can be fitted into these two patterns, and, surprisingly, so can orchestra conductors and army sergeants. Moreover, in psychotherapy with children, the therapist must decide whether he will be loving or strict. Some children punish themselves unduly with guilt over normal resentment, rebelliousness, or hostility; the therapist offers them sympathy, warmth, and understanding so they can learn to forgive themselves for being human. Other children are impulsive and irresponsible, possibly because of indulgent or irresponsible parents; for them, the

[6]For another conceptual scheme limited to maternal behavior, see Schaefer (1959).

therapist, although he is always sympathetic, has definite standards of what is and is not permitted and holds the child responsible for his behavior. The child comes to require of himself the control that the therapist has required of him. The therapeutic approach which is exactly right for one kind of child is exactly wrong for the other. To be strict with a child who is continually punishing himself or to be accepting and indulgent with a child who has weak self-control would not be helpful. In clinical practice, as in life, it is not always clear at first whether a given child needs love or strictness, but the decision to offer one or the other is crucial to helping him.

We will conclude by returning to the clinician's concern that discipline enhance psychological growth. Clearly, extremes of neglect, rejection, brutality, and inconsistency should be avoided. Hostile control also has generally negative consequences. However, other disciplines are less clear-cut in their effects on growth. Love oriented discipline produces a responsible child but one who suffers more pangs of conscience than his thick-skinned peers. In boys, sensitivity may be a handicap in meeting the demands for aggressive, assertive behavior which are so potent from middle childhood to maturity. It might seem that the creative, lively social leader produced by warm, permissive discipline is superior to the docile, conforming product of warm restrictiveness. Yet, as we have seen, the Lanson girls found conformity a congenial and fulfilling role, as many other children might.

Whether discipline is aimed at producing a responsible or obedient or assertive or scrappy child is probably more a matter of values than of healthy psychological growth. There are a variety of defensible disciplines just as there are a variety of legitimate forms of self-control (p. 114). The conscientious parent can do no more than recognize what his values are and think through the assets and liabilities they involve for the child in the family setting, with peers, in the classroom, and in his adult life. If there is an absolute answer to the question of what discipline is best, psychologists have yet to discover it.

Deviant Development of Self-Control

Deviations are usually conceptualized in terms of self-control being too weak or too strong. In the first case the child is hemmed in by conformity and can allow himself little freedom; in the second case he is irresponsible and impulsive. We will use this model in our discussion of sex and aggression (see Chapter 7). For our present purposes, it will be more fruitful to discuss two deviations which hinder the development of self-control itself.

The Effects of an Unpredictable Family Environment

We have seen that a predictable environment is important for developing the child's capacity to postpone gratification. In the middle class, predictability is taken for granted. Of course the house contains dining areas and sleeping areas; of course breakfast is followed by lunch and lunch by supper and supper by bedtime; of course there is an inner circle of familiar people and wider circles of less familiar people; of course objects have their proper places and events have continuity, so that a model plane left on the bureau will be there an hour later, and the meal on the stove will be cooked and served. How could life be otherwise?

In reality, the family environment is not naturally ordered. Order is introduced, and in the middle class it is so strongly adhered to that it is scarcely noticed. Disorganization is equally possible. Minuchin and others (1967) in their study of disadvantaged families, describe the impermanence and unpredictability of the home environment:

> For example, beds shared by two or more children can be turned over to a different child or to a semipermanent visitor while the original occupants are crowded into a section of another bed... Meals have no set time, order, or place. A mother who prepares four individual and different dinners one day, according to the wishes of the children, will prepare nothing another day... Care for the young child is divided among many figures. Mothers, aunts, grandmother, as well as older siblings care for the young child. Sometimes they shower him with stimulation, and at other times he is left alone for long periods while he wanders through the house unattended... Parents' responses to children's behavior are relatively random and therefore deficient in the qualities that convey rules which can be internalized. (p. 193-194)

In this unstable environment, the children become action oriented and impulsive. They respond to the mood and event of the moment. Remember that the ability to delay gratification depends upon expectancy of fulfillment or upon trust. In an unpredictable environment, future rewards are never certain. By seizing available gratifications, the child is doing nothing more than adapting to the reality of his existence.

The Effects of Organic Brain Pathology

Cognitive factors play an important part in the normal development of self-control. The child must be able to understand the baffling array of prohibitions and directives which come his way and integrate them into meaningful guides to behavior. Thinking itself serves to delay motor activity, thereby aiding self-control. We might expect that pathological conditions or injuries to the brain which adversely affect cognition and thinking might also prevent the achievement of adequate self-control. Instead of summarizing the present status of research, we shall limit outselves to some general statements about organic brain pathology and self-control.[7]

We shall begin with a fact: there is no one-to-one relation between pathological alterations of the brain and deviations from normal self-control. Brain pathology can have a variety of consequences ranging from no discernible effect on behavior to

[7] For an exemplary evaluation and summary of the field, see Birch (1964). For a more general background, see White (1964). For a descriptive account of pathological conditions, see Kanner (1957).

amentia, paralysis, and death. Intelligence and behavior can be unaffected by the removal of one cerebral hemisphere, provided the remaining structure is normal; a hydrocephalic child can function normally; children with Down's syndrome (Mongoloid children), who are grossly defective mentally and physically, are reputedly quiet, good-natured, and well behaved, and they rarely indulge in temper tantrums, willful destructiveness, or masturbation. Thus, there are no inevitable, uniform behavioral consequences of brain pathology which justify the belief that the brain-damaged child is fated to have poor control and behavior problems. Rather, the effects depend upon factors such as the locus of the injury, the extent, nature, and distribution of the lesion, the developmental stage of the child at the time of injury, and even his personality characteristics prior to the damage.

The term "the brain-injured child" is frequently met in the clinical literature, and it has come to mean a child who is hyperactive, emotionally labile, distractable, and perseverative. However, this image is as much an oversimplification as the stereotypes of "the mentally retarded child" and "the physically handicapped child" (see Chapter 10). Brain-injured children, like normal children, vary widely in their personalities. As we have just seen, brain pathology can result in a variety of behavioral disturbances or no disturbance at all; ironically, even when all the behavioral signs of "the brain-damaged child" are present, neurological examination may fail to uncover any actual brain pathology.

With these cautioning remarks as a background, let us examine one constellation of behavioral deviations which is suggestive of some—as yet to be defined—pathological condition of the brain. One striking symptom is often labeled hyperactivity because the child darts from one activity to another. On closer inspection, however, the behavior turns out to be more a matter of short attention span than heightened motor activity. Unlike the normal child, certain organically damaged children cannot settle down to a task; they go from one thing to another in a hit-or-miss fashion. To illustrate: a typical eight-year-old, after entering a room for a psychological evaluation, would sit in the chair provided for him and wait for the examiner to present the test items. A hyperactive child might sit down, then immediately stand, go to the blackboard and scribble with chalk for a few seconds, walk to the window and briefly pull on the window shade, begin opening the drawers in the examiner's desk, rush to the door and look out into the hall, come back and begin turning the water faucet at the washstand off and on, only to abandon this for a brief episode of feeding a baby doll which is on the desk. This continual shift in activities can give way to the opposite extreme of perseveration, in which the child repeats an activity over and over again. Efforts to divert his attention or modify his behavior are met with irritation or angry outbursts. Emotional lability is another characteristic of children with brain damage. Their moods change swiftly and tend to extremes—excited delight can quickly give way to outbursts of rage or sullen withdrawal. Not unexpectedly, the children's social relations suffer. Their pace is out of step with the goal directedness and persistence of their peers, their moodiness and resentment of change alienate them, and their clumsiness precludes skillful participation in games.

It is their distractibility which interests us most. The children cannot keep their attention on a given activity. Paying attention is not as simple as it may sound. It requires both focusing on a limited segment of the environment and inhibition of response to a vast array of irrelevant stimuli. In order to read a book, one must attend to the words on the page and "unattend" to everything else—the people sitting nearby, the noises from the street, etc. It is just this ability to inhibit responses which certain organically damaged children lack. It is as if no stimulus is any more compelling than any other—each must be reacted to as it appears, and none is sufficiently dominant to sustain attention.

The difficulty in distinguishing relevant from irrelevant stimuli is evidenced in visual perception. Show a child a figure of a square intersected by two wavy lines and ask him to copy the square. The normal child can do this with no trouble. The child with certain kinds of brain pathology might begin correctly, but when he comes to the intersection of square and wavy lines, he is apt to be perplexed or to follow the wavy line. He cannot say, in effect, "square is relevant, wavy line is irrelevant." Expressed technically, he has difficulty distinguishing figure and ground. At any moment, a normal person automatically organizes the visual field into stimuli which are clear, prominent, and central, and others which are vague, dispersed, and peripheral. When one talks to another person, his face is the figure, his body and surroundings are the ground. When one is window shopping, store windows and their contents are the figure which stands out against the general background of buildings and streets and other shoppers. The process of ordering the visual field into figure and ground is so automatic that it goes unnoticed—until something happens to disrupt it. It is just this process which goes awry in certain instances of organic brain pathology. In a world without emphasis, the child is buffeted about by each new event that comes his way.

Just as the study of disorganized families has sensitized us to middle class environmental consistencies that we might have taken for granted, the scattered behavior of certain children with organic brain pathology sensitizes us to the importance of focused attention. Like the proverbial fish who were the last creatures to discover the water, we might have overlooked this essential factor in self-control because it is omnipresent in normal behavior.

An Example of Deviant Self-Control

The last study we shall present adds nothing new to our understanding but serves as a counterpart to our introduction to normal control. We noted that Anna Freud and Aronfreed agreed on the importance of certain affective and cognitive factors in the development of normal control. We shall now present a description of children in whom both factors were deficient.

Goldfarb (1945) compared two groups of twelve-year-olds, one of children who spent the first three years of life in a bleak and unstimulating institution before being transferred to foster homes, and one of children who had been raised in foster homes from early infancy. The institutionalized children showed many signs of pathologically weak self-control. They were hyperactive, disorganized, flighty, and unable to concentrate. They were easily angered, but their outbursts were unpredictable and brief. They had an insatiable need for love, attention, and food, but were incapable of forming a deep relationship. Their foster parents found that they were unteachable, since they were unresponsive to rewards and punishments and indifferent to normal motivations. They reacted to their own acts of hostility, unprovoked aggression, and cruelty with bland indifference rather than anxiety. Failure in school did not shame them, and even threats to remove them from the foster home did not faze them. They were heedless of warnings concerning their own safety and frequently wandered aimlessly into the street.

The behavior of the institutionalized children may well have been due, in part, to their lack of a strong attachment to a caretaker. In the formal language of social learning theory, Aronfreed says, "...a substantial positive attachment to an agent of nurturance, of the kind that emerges in varying intensities from the extended early dependency of the child, may be required as a foundation for the child's internalization of any form of socially acquired conduct" (p. 313). The psychoanalyst Fenichel says love makes a child educable (1945, p. 41). Since they never had an affectionate attachment, the institutionalized children had nothing to lose. The love of one person had no special reward value, and the threat of withdrawal of love held no particular terror. The primary motivations for achieving self-control were absent.

The cognitive components of self-control were also pathologically weak. The children had little memory of the past and limited foresight concerning the consequences of their actions. They had difficulty with concept formation; consequently, the ability to integrate discrete prohibitions and directives into general guides to behavior was low. Instead of healthy initiative and sustained curiosity, they evidenced the flighty, restless jumping from one task to another which characterizes the behavior of certain brain-damaged children.

Thus, point by point, the institutionalized children offer a deviant counterpart to Anna Freud's and Aronfreed's list of variables involved in normal self-control: they failed to form a deep emotional attachment to adults, their memory and ability to anticipate were poor, their grasp of cause and effect relations and their ability to integrate specific experiences into guiding principles of behavior were both weak. In light of these multiple deviations from normality, the resultant picture of weak self-control is understandable.

Normal Development of Judgment

Initiative, Willfulness, and Judgment

We shall begin with two examples. A mother of a four-year-old reported: "He was fiddling with the arm of the record player. I slapped his hand for the thousandth time. He stood there for a minute. Then he looked me square in the eye and reached over for the record player and started fiddling with it again." The boy was not testing his mother's firmness; he knew how she reacted. Nor was he trying to provoke her into losing her temper. He was saying, in effect, "I know what I want to do, and I know how you feel—and I am going to do it anyway." Next, we return to the account of the healthy adolescent deciding which college he will attend (see Chapter 3). An important element in that situation was the compatibility of his own educational goals with those of his parents. When there was a conflict, he either tried to find a compromise solution acceptable to both sides or stuck to his guns and dealt with parental reactions as best he could.

The adolescent's situation is vastly more complex than that of the four-year-old. The adolescent has arrived at the point where he can make a major life decision for himself. Further, his grasp of his parents' point of view and feelings goes far beyond the four-year-old's understanding and empathy. Yet, there is an essential similarity. In both instances the child knows what he wants to do, what his parents wish him to do, and the consequences of following either alternative. Then he chooses. The direction chosen is not important. The essential elements are that a choice is possible and that neither the child's inner needs nor the directives of others can preempt the outcome.

Judgment can be viewed as developing out of initiative and willfulness. In infancy, self-reliant venturing out is not socialized. The toddler, however, begins to encounter special kinds of parental reactions aimed at checking his unfettered expansiveness. He meets these prohibitions and directives with characteristic initiative and willfulness. Because he is enterprising, he strives to assimilate them in terms of their content and affective loading. This assimilation may involve testing the parent to determine what a directive means, how firmly it is held, and the cost of disregarding it. Such testing is the prototype for behavior in future confrontations with a power structure, whether it be a peer group, a school, or a work situation. The questions as to what is prohibited and what is allowed, what is fixed and what is negotiable, and what is the price of not conforming, are universally asked. As prohibitions begin to register, the child begins to entertain alternatives. His initiative and willfulness serve as counterforces to mindless acceptance, and he matches parental directives with his need to pursue his own interests and maximize his own pleasure. In the preschool years, there is a simple clash between his socializer and himself, since he has limited ability to deliberate in the sense of mentally weighing alternatives. However, even the preschooler can represent his own action with its consequences and a parental prohibition with its consequences, and choose one over the other. In normal development the choice will go now one way, now the other. The direction does not matter. What matters is the experience of entertaining alternatives and deciding. In this experience lie the seeds of the more sophisticated deliberation of middle childhood and, eventually, of mature judgment and responsible action.

The Development of Judgment

Gesell and Ilg (1946) offer a descriptive account of the development of judgment which is in keeping with the one we have proposed. With the naturalist's sensitivity to the complexity of behavior, they note that judgment is not a simple response which becomes progressively stronger; rather, it is contingent upon the child's awareness of himself, his awareness of others, and his grasp of principles to guide decision making. Each has an involved history. In like manner, the child's decision to follow his inclinations or to obey his parents has its own developmental rhythm, which is worth examining in some detail.

In the first twelve weeks of life a primitive precursor of judgment can be detected. At six weeks the infant's smile is egocentric. At eight weeks he smiles back when his mother smiles, thus evidencing socially responsive behavior. At twelve weeks he takes the initiative and smiles at the mother so that she will smile at him in return. (Gesell and Ilg's implication that initiative underlies judgment is consistent with our own derivation of judgment.) Here we have an "intrinsic self" phase, a "social reference" phase, and a "reciprocal self-and-social" phase.

At about thirty-six weeks, the infant becomes sensitive to disapproval as well as to the smile of approval, and by one year he is "caught in a complex web of smiles and frowns" (p. 405). He enjoys repeating behavior which others will laugh at and can begin inhibiting himself in direct response to disapproval. At fifteen months, willfulness appears, bringing with it an overriding assertion of the "intrinsic self" and a defiance of even benevolent social intervention. By two-and-a-half years the defiant *No* has given way to a desire to try both alternatives. With experience and with help (although Gesell and Ilg do not specify what kind), the two-and-a-half-year-old can begin to make simple situational choices. As early as three years, the intrinsic gratification of choosing and the social gratification of compliance blend with a general desire to please. This blending, aided by an increased intellectual capacity, makes the three-year-old a reasonable creature. The parent can bargain with him, and he will do something he dislikes if given a good reason. The happy wedding of choice, social responsiveness, and intellect is short-lived, however. By four years of age, the child is less anxious to please, less responsive to praise and blame, and more willful. However, his understanding of rules and restrictions has increased. At five he is once again tractable, enjoys approval, and is, from the parent's point of view, "an angel."

And so it goes. At six there is a return of the vacillation which marked his behavior at two-and-a-half, but once the child makes up his mind, he does not change it. Instead of being responsive to reasoning, he explodes. However, he will carry out parental prescriptions if he can be made to feel they are really his. His self-centeredness is seen in his need to win, to have his way, to deny guilt. But through it all, he is progressing. Prior to age six, he has primarily thought in situational terms and in terms of parental approval or disapproval. Now he begins a task which will occupy him for the next ten years or so—finding principles to guide his decision making which are valid over and beyond parental reactions. He is beginning to evaluate playmates' behavior in terms of whether it is good or bad—although he exempts his own behavior from such scrutiny! By eight years of age, he has reached a higher level of integration. He can make up his own mind, and he can change it; he knows what he wants, but he will listen to reason. Thus, he is flexible with himself and with the directives of others. He bases his decisions on a much clearer understanding of standards of good and bad behavior than he had at six, and he genuinely wants to be good. He has greater insight into the behavior of his peers and he is able to size up their assets and shortcomings. Within this framework, earlier behaviors still persist–he still likes to have his own way, is sensitive to criticism and teasing, loves praise and applause, and can be obstructionistic before complying.

We will abstract the factors which determine the development of judgment from Gesell and Ilg's description. There is a variable of self-centeredness and other-centeredness which has both an affective and a cognitive component. The affective component may be intense, as evidenced in the willful self-assertiveness of the toddler, or, to a lesser degree, the excessive concern over pleasing others which can be seen during middle childhood. In our account of healthy adolescents deciding where to go to college, the emotional intensity was moderate, however, and distributed between self-interest and empathy for the parents. The cognitive component of self- and other-centeredness involves the child's increasing understanding of himself, his parents, and his peers. In our discussion of initiative (see Chapter 3) we noted how the young child begins to define himself to himself though venturing out, and in subsequent discussions (see Chapters 5 and 6) we shall learn how he can perceive a situation from the viewpoint of others. Such understanding can be utilized to make decisions on the basis of a realistic grasp of himself and of others. Next, Gesell and Ilg hint at a factor of intellectual flexibility. The preschooler becomes capable of choice and derives pleasure from his newly acquired power. Later the child develops the ability to change his mind. This ability to entertain and successively select alternatives stands in contrast to the rigidity which results either from an intellectual inability to see alternatives and keep them in mind, or from a strong emotional investment which channels thinking along a single path. The final factor is the evolution of principles which the child can use to guide choice. In the willful toddler, such "principles" are primitive indeed and seem to involve no more than "me versus you,"–a tenacious clinging to what the toddler wants to do and a rejection of everything else. Next, situational considerations come into play, bringing a crude weighing of pleasure and pain. Parental rewards and punishments figure prominently here. The four-year-old in our original example was probably making this kind of situational assessment before he chose to defy his mother. Note, too, that the child's projections into the future do not go beyond the immediate consequences of his behavior. Yet, even as early as three years of age, the child can be reasoned with; that is, he will listen to reason if it is geared to his level of understanding. He can also learn that his behavior has consequences other than parental rewards and punishments: his physical well-being may be involved ("If you touch the heater you will get hurt") or property rights ("This is your sister's doll, not yours") or the feelings of others ("It hurts when you hit so hard"). His curiosity motivates him to understand the explanations of consequences, and the reasons why his behavior is evaluated as being good or bad. This effort at understanding will continue into adulthood (see Chapter 5). When confronted with decisions, he will be able to draw upon an increasingly varied repertoire of guides which he has assimilated. Some guides concern enlightened self-interest, others concern the rights or feelings or values or idiosyncracies of others–initially of the family, but then of peers and finally of abstract groups such as the community, the political party, or the nation.

Another feature of Gesell and Ilg's description should be underscored. There is no simple, quantitative increase in judgment, so that we can expect a child at a later stage to have "more judgment" than a child at an earlier stage of development. There is no reason to assume that the many variables involved in judgment follow the same developmental course, so that, for example, self-understanding will keep pace at every point with the understanding of rules for fair play among peers. In addition, Gesell and Ilg's descriptions indicate that a child may be capable of reasoning but may be unreasonable in his decisions because of the turmoil inherent in his developmental stage.

Perspective and Reflection. Piaget (1967) discusses the development of other-centeredness in terms of the ability to adopt the viewpoint of another individual and the development of the ability to reflect upon one's own behavior. He focuses on cognitive rather than on affective factors. The preschooler is egocentric, by which Piaget means that the child thinks his is the only point of view. (For a more extensive presentation of egocentricity, see Chapter 6.) He believes, for example, that he can say anything which comes to mind and others will understand him. A father of a four-year-old reports the following, baffling experience:

> We were having a good time walking along together, when Lee suddenly said, "You know Veronica, the one with the yellow hair; she is always sneaking up on this guy Harvey in the cartoon, and he jumps in this old car and really zooms away." I hadn't the slightest idea who Veronica or Harvey were, and, only after questioning Lee, found out that he was telling me a funny incident he saw that morning on TV.

The boy had assumed that, because he knew what he was talking about, his father would also know. It would never occur to him to put himself in his father's shoes, psychologically, and ask whether he was communicating sufficient information to be understood.

Piaget also maintains that the preschool child believes that the content of what he says is patently true and needs no defense. Left to himself, he would never challenge the veracity of his own ideas. Both forms of egocentricity are poorly suited to social living. The child must learn to adapt his speech to his listener in order to communicate. This necessitates taking the point of view of the listener. He also must learn to defend the veracity of his statements, since they do not go unchallenged. At around seven years of age, reflection replaces egocentricity. The child engages in "inner social discussions" in which he represents both his own and another's point of view. He challenges and contradicts his ideas in the arena of his mind, just as he previously had been challenged and contradicted by others in a social setting. Once capable of reflection, the child can juxtapose two viewpoints and decide on the basis of this juxtaposition. This ability to understand another person's point of view, to keep this point of view in mind, and to weigh it against his own in order to arrive at a decision, is exactly like the ability Gesell and Ilg describe in the eight-year-old.

Summary

Judgment is contingent upon the ebb and flow of self-centeredness and other-centeredness. Affectively, it involves feelings of self-worth and of empathy, defiant self-assertion in opposition to others and longings for their affection and approval. Cognitively, it involves the increased understanding of himself and of others which the child gains by broadening and deepening his experiences. Judgment is also contingent upon the child's increased ability to reflect in the sense of holding alternatives in mind—particularly alternative points of view—and testing one against the other. And, finally, judgment is contingent upon the evolution of criteria for choosing, starting with simple situational concerns over parental approval and culminating in general guides which are valid over a variety of situations and long spans of time.

We must now postpone further examination of judgment until we discuss the development of conscience (see Chapter 5) and of vocational choice (see Chapter 9). The reason for doing so is a practical one: conscience and vocational choice have captured the interest of psychologists while judgment has not. Even Gesell and Ilg placed their description of the development of judgment in a chapter entitled, "The Ethical Sense." Most of the relevant literature has a substantive focus which makes it inappropriate here. In discussing conscience we shall trace in detail how the three factors involved in judgment (self-centeredness and other-centeredness, reflection, and the evolution of guides) effect the production of a mature conscience; the same variables are involved in vocational choice, except here we deal with the self-as-worker confronting the world of work.

One observation will serve as a bridge to our future discussions. In this chapter we have distinguished between self-control as conformity to parental directives and self-control as the exercise of judgment. In the literature on conscience we shall meet a distinction between a rigid and a flexible conscience, the former producing an unquestioning adherence to parental standards, the latter fostering independent evaluation. Similarly, the vocational literature contrasts choices made primarily on identification with a parent (e.g., because his father is a doctor, the son decides he will be a doctor too) with decisions made on the basis on self-evaluation and an understanding of a variety of occupations. In both instances, the distinction is similar to the one we made between the self as executor of parental directives and the self-as-judge. We shall be particularly interested in discovering the conditions under which the self develops along one line rather than the other.

5 Values, Ideals, and Conscience

Self-control evolves from parental prohibitions typified by *No, Don't,* and *Wait*. Conscience develops from parental evaluations typified by *Good Boy, Bad Boy*. The two merge imperceptibly, but separate processes can be distinguished. *No, Don't*, and *Wait* inhibit or delay behavior. *Good Boy* and *Bad Boy* not only control but also evaluate behavior. They are judgments of the child himself.

Two Descriptive Features of the Adult Conscience

We shall begin by describing two characteristics of the adult conscience. To do this, we shall look at the fleeting thoughts of a hypothetical twenty-year-old as he shaves in the morning. The evaluations made by his conscience are in parentheses.

> I really wasn't hungry at breakfast but I managed to eat everything anyway. (Good boy). That newspaper story about the sniper killing all those people was terrible (Good boy). I didn't tell Tom I could see one of his hands in the poker game last night (Bad boy). He outscored me in basketball (Bad boy), so I am going to have to practice longer every day (Good boy). I really could go for his girlfriend (Bad boy), but I'd better lay off (Good boy). My car is dirty; I haven't washed it in two weeks (Bad boy).

The first thing to note is that conscience is composed of diverse values and a variety of affects. Some values are religious (Thou shall not kill), some are cultural (Be successful), some come from the peer group (Do not interfere with lovers), some originate in family themes (Be clean), and some grow out of parental idiosyncrasies (Eat everything on your plate). The affects can vary in quality and intensity. They include barely perceptible feelings of satisfaction or dissatisfaction, healthy pride or realistic concern, priggish superiority or shame, ecstasy or the deepest guilt. We should also note that affect does not necessarily become more intense as one goes from idiosyncratic values to moral values; an individual may feel as guilty over wasting food as over being dishonest.

The second descriptive feature of conscience is that it involves *thinking, feeling,* and *behaving*. All three elements are integrated in the mature conscience. The individual understands what is morally right and wrong, feels appropriate satisfaction or guilt, and behaves in a moral manner. Yet, there is no guarantee that integration will occur, and any one element may overshadow the other two: the Devil can quote Scripture; the mere feeling of self-approval or guilt is no guarantee that an act was moral; and moral behavior can be used to serve immoral or amoral ends.

The conscience we are describing is the human conscience, not the conscience of the moralist or the theologian. It is a puzzling mixture which deserves all the names it has been called: the surest guide to human conduct, the crowning glory of human evolution, the repository of bigotry and prejudice, the cruel, irrational tormentor, the sublime angel, the destructive devil. Our study of the development of conscience is concerned with understanding the complexities and paradoxes of this strange creation. But first we must examine the literature on values and on the related phenomenon of ideals, since both figure prominently in the development of conscience.

Values

Definition and Controversial Status

Values are more than mere interests. They involve the assigning of a high degree of worth to a pursuit. The status of values in psychology has been a matter of debate. Some psychologists have advocated banishing them because they are superfluous or because they are ethical rather than scientific concepts; others have given them a central place in personality. There is little agreement as to what values are, how they should be classified and conceptualized, or how they are related to other theoretical constructs. Despite all the controversy, there have been only sporadic research efforts. Developmental data are particularly meager.[1]

Those who champion the cause of values claim that values selectively sensitize a person to environmental events. After seeing the same movie, for example, a politically active student talks about the subtle use of Communistic propaganda, an aesthete discusses the deft camera work, and a divinity student objects to the amoral attitude toward responsibility. Each person may have been unaware of those aspects of the movie which were so obvious to the others. But values can also desensitize a person by

[1]For more comprehensive discussions of values, see Dukes (1955) and Thompson (1962). As an example of the discrepancy between opinion and fact, see Mitton and Harris (1954). And for a discussion of methodological problems in the objective study of values, see McClure and Tyler (1967).

actively interfering with his perception of environmental stimuli. For example, a woman in her thirties recalls her youth: "I was brought up in the European tradition which values modesty in adolescent girls. I was healthy and athletic, so I went to a lot of swimming parties. But it was only after I got married and began enjoying sexual relations that I realized men had bodies. For the first time I could actually look at their bodies and enjoy them instead of shutting them out." Finally, there is evidence that a person remembers information which is congruent with his value system; e.g., a political liberal will recall an article written by another liberal more readily and accurately than one written by a conservative. However, the objective research supporting the role of values in perception and memory has been challenged, defended, and challenged again, with no general consensus as to what is fact and what is artifact (Dukes, 1955).

The Development of Values

The developmental picture of values is a sketchy one up to adolescence. Some psychologists doubt whether children can be regarded as having values in the first three or four years of life. Other psychologists claim that early values do exist but that they are either very flexible or very fragmentary and unstable. Perhaps the issue is one of definition rather than conflicting data. However, five-year-olds have social prejudices which directly reflect family and neighborhood values, and the child of six respects the property of others. According to Thompson (1962), nonsocial values, such as those regarding specific kinds of music, clothes, and recreation, are acquired slowly from preschool through early adulthood, probably on the basis of approval from parents and peers. By contrast, values which guide behavior are acquired rapidly. We also know that the meaning of a value may change as the child's experience and understanding expand. For example, the six-year-old respects property because he is afraid of punishment; whereas, the high school student is concerned with the consequences for others if he violates property rights. When we discuss conscience development, we will see that this shift from concern with punishment to concern with the rights of others is part of an overall development in moral understanding.

Adolescent values have been studied in greater detail. The findings are interesting in their own right and are important to our subsequent discussion of the integration of conscience. In part, adolescent values reflect the ever-present theme of sex typing (see Chapter 7). Twelfth-grade boys subscribe to a standard of conduct which emphasizes work, achievement, and intellectual stimulation (Moss, 1955). A boy should attend to the business at hand, know what to do, get the job done on time, and not fool around. His discussions should be clear and to the point. He should support the team and, above all, he should not be shy. In these values we glimpse the adolescent's concern about his future which will revolve, in part, around his ability to work effectively. The valuing of clear communication represents an interesting shift. In middle childhood, one was suspect if he valued intellect. Boys emphasized athletic ability, assertiveness, and aggressiveness rather than excelling in school. The adolescent boy who is preparing for the serious business of adulthood turns toward ideas and away from athletics. At the same time, boys begin to outstrip girls in intellectual achievement—a fact which delights those who regard values as exercising a determining influence on personality.

Moss (1955) describes the adolescent girl as more socially oriented than the boy. She values kindness and consideration and friendliness; she does not like cliquishness. She is concerned about her appearance and her reputation. She believes that girls should not become angry in public or be sexually provocative. In general the adolescent girl values social sensitivity, friendliness, and a conventionally defined goodness.

Religious values undergo a significant change in adolescence (Kuhlen and Arnold, 1944). The adolescent becomes more tolerant and more questioning. For example, two-thirds of the twelve-year-olds believed that every word of the Bible is true and that God watches to see if children behave so as to punish those who are not good; only one-third of the eighteen-year-olds subscribed to such statements. Eighteen-year-olds wonder about death and hell more than twelve-year-olds. They do not reject the notion of life after death, but, rather, try actively to come to terms with the issue. Finally, more eighteen-year-olds than twelve-year-olds believed that Catholics, Protestants, and Jews are equally good. In general, the late adolescent has shifted from a concrete, absolutistic, authoritarian view of religion to a more abstract, humane, evaluative view.

Not all values undergo the pilgrim's progress which Kuhlen and Arnold suggest for religious ones. A comprehensive review of studies of ethnic relations by Harding and others (1954) indicates that other values, especially those regarding racial groups, may follow different developmental trends. Cognitive growth between the first grade and eleventh grade can result in a more accurate mirroring of the Negro stereotypes in the community rather than an increased understanding of the brotherhood of man. Greater religious hostility has been observed among ten- and twelve-year-olds than in younger children. And, finally, between grade seven and grade twelve there is only a slight change in what children regard as the right thing to do in situations involving different ethnic groups, while a steady increase in prejudice is observed.

The literature on religious values and prejudice is not necessarily contradictory. The adolescent is intellectually capable of moral understanding, and because he is making the transition from childhood to adulthood, he is in a unique position to re-evaluate the past and search for new values. As we shall see, this search may be essential to the formation of a mature conscience. Yet, such a favorable outcome is far from inevitable. The adolescent may be exposed to rigid, authoritarian, or prejudicial values which are part of the community's way of life or his family's theme. He himself may have rages, rigidities, or deep insecurities which lead him to hold tenaciously to narrow, autocratic values. Adolescence, like any time of ferment, can be a prelude either to growth or regression, and consistency across value systems should not be expected.

Nor should all adolescents be expected to share the same value system. Havighurst and Taba (1949) present a picture of sixteen-year-olds which suggests a continuation of the uncritical, authoritarian view of goodness typical of middle childhood. The adolescents they studied still subscribed to stereotypes and slogans, such as, "It is always a good policy to be nice to people," or "One has to sacrifice fun for honesty in life." Friendship was characterized by politeness, responsibility by punctuality and completing a job. They valued obedience and acceptance of standards and tried to avoid grappling with inconsistencies and deviations. Instead of the re-evaluation often associated with adolescence, they maintained a fixed hierarchy in which their values were subordinated, first to adults, and then to peers. In contrast, Stone and Barker (1939) found a significant difference in values in eleven- to fifteen-and-a-half-year-old girls, depending on whether they were pre- or post-menarcheal. The latter were much more favorably disposed toward petting and necking and going to dances. They were also less critical of smoking, becoming angry, using slang, and being divorced. This constellation suggests that they were becoming less prudish about sex, less priggish about morality. In short, they were becoming more "worldly" (see p. 170). Tryon (1939) also found a difference between twelve-year-old and fifteen-year-old girls. The younger girls placed a high value on being docile, conforming, quiet, and ladylike, while the older ones placed a high value on being active, lively, interesting, and a good sport.

These findings underscore our point: adolescents are not a homogeneous group in regard to values. An adolescent in a small community may retain the values of middle childhood longer than his counterpart in the large, heterogeneous city. The middle-class adolescent in the city will also be different from his counterpart in the slums. The timing of physiological maturation and environmental pressures for him to adopt new values or re-evaluate old ones will significantly affect both the rate and the extent of change.

The Acquisition of Values

Turning now to the question of how values are acquired, we find that relatively little is known (see Thompson, 1962). Theorists tend to fit values into their own frame of reference; consequently, psychoanalysts emphasize identification, learning theorists emphasize reinforcement, social psychologists emphasize conformity to a norm. It seems reasonable to assume that, up to a point, values are acquired and sustained in the same way (or ways) as any other personality variable, but it is not clear whether there are unique features to value acquisition and change.

One thing is certain: values are not acquired through teaching alone. Formal instruction in moral behavior, whether it be in the home, church, or scout group, has consistently shown little relation to the development of moral values. The personality of the teacher, however, does seem to be important. "Do as I say, not as I do" seems almost to guarantee that a child will adopt the undesirable trait that the teacher embodies rather than the desirable one he espouses. Values are acquired only if they engage the child's interest and are made relevant to his experience. After the child is so engaged, then discussions will help him to generalize from his experience. Values presented by parents or teachers become memorable, therefore, only if they capture the child's interest and are intellectually relevant (Mitton and Harris, 1954).

There is some evidence concerning the family's role in the acquisition of moral values. (We shall discuss the role of the group in Chapter 8.) Brown, Morrison, and Couch (1947) found a substantial relationship between affectional family relations and various aspects of "character." The highest correlation was between affection and responsibility and honesty, the lowest between affection and moral courage. Sharing of decisions in a democratic atmosphere, positive attitudes of the parents toward each other, and parental acceptance and encouragement of peer relations are aspects of the family interaction which were found to contribute most to character development.

Although most of the literature is concerned with family and group influences on the acquisition of values, Rogers (1964) regards the growing child himself as the ultimate and crucial evaluator. It is the child who has the highest stake in his own growth and who is in the best position to decide whether a new experience contributes to or impedes his development. He values positively those experiences which promote what Rogers calls self-actualization and values negatively those which hinder it. The child does not merely absorb values from the environment; the most enduring and fulfilling ones are those which he evolves for himself. In Rogers's views we find Emerson's "self-reliance" and Shakespeare's "to thine own self be true" along with a challenge to the learning theorist's emphasis on environmental influences. For Rogers, the child always knows more than he is taught.

Critique of the Literature

The status of values in psychology is uncertain. The term itself is difficult to conceptualize and to distinguish from other terms. Yet we noted in Chapter 1 that many terms used in personality theories are difficult to define and yet have proved fruitful. Our dissatisfaction with the literature on values arises from other sources. First, the answer to the question, *What*? (see Chapter 1) is fragmentary, since most studies are of adolescents. Next, the answer to the question, *Why*? is even more unsatisfactory. Discussions of values tend to be biased in the direction of what is ideologically desirable rather than objectively probing the issue of value acquisition and change. Our most serious concern is that values fail to emerge from the literature as an independent personality variable which has motivational and/or cognitive properties in its own right—they seem primarily to be resultants rather than causes. To illustrate: the congruence between parents' and children's values seems just another instance of the potency of socialization in the family; the sex-linked values in middle childhood and adolescence seem no more than another manifestation of sex role identity (see Chapter 7); the adolescent boy's shift to work oriented values may only reflect his increased orientation to the future, while his religious tolerance and questioning may only be one facet of his ability to achieve perspective on and to question absolute statements of any sort (see Chapter 6). Values have a "nothing but" quality which makes them seem hardly worthy of study.

Douvan and Adelson (1966) make a similar point in line with our concern that values are "nothing but" byproducts of other psychological variables. They speculate that the apparent ideological ferment in early adolescence is little more than a reflection of the adolescent's personal preoccupation with the major problems of sexuality and emancipation. If he suddenly shifts from conventional to unconventional values, he is doing little more than expressing his rebellion toward his family; if he becomes bohemian, he will conform slavishly to the dictates of his fellow bohemians because of his basic need for the security of peer approval. In fact, Douvan and Adelson conclude from their research that, aside from the modifications or realignments of values stemming from personal needs, there is very little change in values during adolescence. Once his problems in regard to sexuality, independence, and vocation have been resolved, he usually reverts to the values of his parents.

Yet, Douvan and Adelson are not willing to settle for a "nothing but" analysis. There is such a thing as genuine re-evaluation of values in adolescence, but it is rare. It probably does not come early in the period because personal needs are too strong and they bend values to their own purposes. But toward the end of the period the adolescent has a chance to make a genuine re-evaluation. His transitional status carries with it an openness to new values, and he is capable of flexible, abstract, creative thinking. Douvan and Adelson stress the ability to get along without the security offered by conformity to family, peer groups, and society. During this comparatively brief span of time, the adolescent becomes both sufficiently mature and sufficiently uncommitted to achieve what the authors call a "value autonomy" or "...the capacity to manage a clarity of vision which permits one to transcend customary structurings of reality" (p. 131).

Now we are no longer dealing with values as such but with the process of evaluating. If Douvan and Adelson are correct in their analysis, evaluating becomes a rare and special quality which flourishes during a brief time span. It is like creativity in that it is both precious and atypical. It also rescues the study of values from being nothing more than what parents, peers, and society teach the child. However, evaluating is also a vaguely defined and elusive phenomenon at present. To conceptualize it more precisely and to trace its developmental roots are not easily done and probably will not be done for many years. In the meantime, we must rely on our observations of the human scene and on our interpretation of human history for evidence that "value autonomy" is a viable and significant personality construct.

Ideals

Definition and Characteristics

Ideals are standards one sets for himself. Since they are only standards, ideals are independent of the content of behavior. One can strive for perfection in whatever he values most, whether the value is moral, nonmoral, or immoral. Suppose an adolescent girl resolves to lead a perfectly virtuous life; her value is virtue and her ideal is perfection. However, she could just as well have aspired to be the most dazzling Washington hostess, and the standard itself would be the same. The general characteristics of ideals will be only briefly mentioned since they are not our main concern. Ideals vary in their likelihood of attainment; some can be realized with increased effort, others are unattainable in their perfection. Ideals provide the child with a frame of reference for evaluating himself, and he judges his behavior as good or bad as he realizes or fails to live up to his standards. Ideals also have a motivating function, since they spur the child on to find ways of achieving the standard he has set for himself, although they give no clues as to which ways would be most effective.

We shall be concerned with only two kinds of standards: the first is called either the ideal self or the ego ideal, and the second is achievement motivation.

The Ego Ideal or the Ideal Self

The ego ideal is the individual's ideal picture of himself. It is the image of "me as I would like to be." Jacobson (1964) traces the developmental course of the ego ideal utilizing psychoanalytic theory. This theory addresses itself to the questions, How are ideals acquired? and How are the grandiose idealizations of earliest childhood reduced to realistic proportions? The infant-toddler regards his parents as omnipotent and omniscient. Such exaggerations are only natural in light of the vast discrepancy between the infant's and adult's power and knowledge and also in light of the infant's limited ability to comprehend the world around him (see Chapter 6). When his father lifts a suitcase which he cannot budge, why should he not believe the father is all powerful? When his mother knows the name and function of everything in his environment, why should the child not assume that she knows everything in the world?

While such grandiose idealizations must eventually be corrected, they serve an important function during the preschool period, according to Jacobson. The child's intense love of the opposite-sex parent makes him resent any denial of love and any rivalry for affection from the same-sex parent. This turmoil of love and hate, of irrational longings to overpower the rival parent and irrational helplessness in the face of that parent's power, reaches its climax in the Oedipal complex (see Chapter 7). It is the child's idealized image of the rival parent, compounded of admiration and exaggeration, which helps him maintain his attachment to that parent throughout this tumultuous period. While the little boy or girl is devaluing his same-sex parent, he or she still preserves the image of an ideally good father or mother. The Oedipal complex is resolved by identification with the same-sex parent. In effect, the child says, I am not your rival—I am like you. Through identification, the child now has an ideal self, an ideally "good me" which is modeled on the ideally good parent and which helps heal the narcissistic wound of being powerless in comparison with adults. Thus, the ego ideal protects both the child's attachment to his parent and his self-valuing. Jacobson implies that the preschooler needs an image of an ideally good parent to cushion the difficulties of the actual relationship between parent and child, and he needs an ideally good self to maintain self-respect in the face of realistic inferiorities.[2]

[2]While not accepting the psychoanalytic framework, Ausubel (1958) has a comparable progression from an omnipotent phase to an ego devaluation crisis and a "satellizing" solution to ego devaluation in which the child models himself on the parent.

Even after the Oedipal phase, the ego ideal possesses exaggerated, omnipotent qualities. Its values are primitive and nonmoral: "eternal happiness, glamour and wealth, or physical and mental power and strength" (Jacobson, 1964, p. 112). These values are slowly transformed into moral values, an ethical code, and a serviceable ideal for the self to emulate. The transformation occurs during middle childhood when the child is relatively free from the turmoil of the first six years and is in a more favorable position to understand human relations as they actually exist. In psychoanalytic terms, the ego—that aspect of personality which is concerned with understanding reality and achieving self-control—is strengthened. As the child broadens his contacts and increases his emotional ties outside the family, he gains a clearer understanding of who he is and transforms his image of exaggerated and idealized goodness into one composed of socially relevant values.

Objective studies on the development of the ideal self tend to support Jacobson's picture of middle childhood. Havighurst, Robinson, and Dorr (1946) found that children aged six to eight, when asked to write an essay on "The person I would like to be" named their parents most frequently. Children between the ages of eight and sixteen tended to describe either superficially glamorous adults, such as movie stars and comic strip heroes, or attractive, successful adults whom they had observed. The ideals of the oldest children tended to be composites of desirable characteristics from all the individuals identified with in childhood. Unfortunately, this study does little to further our understanding of the qualities represented by various adult figures or why different qualities were valued by children of different ages.

Pringle and Gooch (1965) studied a group of children when they were eleven and again when they were fifteen years of age. They were able to identify a Fashionable/Famous Ideal, who is admired for superficial qualities such as good looks, money, and fame, and a Moral Ideal, who is admired for qualities such as kindness, understanding, and fairness. The Moral Ideal was more prominent in the older children. The authors believe this change is based on a shift from fantasy idealization to reality. The Fashionable/Famous Ideals have no problems, no responsibilities, no negative traits or feelings; the Moral Ideals are real people trying to cope with the problems of real life. Interestingly, this trend also was related to scholastic achievement; children who chose realistic figures were overachievers, those who chose fantasy ideals were underachievers. Thus, development of an ego ideal is related to the academic standards the child sets for himself.

Bossard and Boll (1957) delved more deeply into the process by which an ideal self develops. They suggest that the child selects a limited number of adults who play the most important role in defining and evaluating experience for him. The particular adults chosen reflect the child's unique needs. A special emotional rapport develops in which the child not only values the admired adult, but the adult also values the child's feelings for him. The objective characteristics of the adult do not matter; what matters is the way the child views him. These adults come to personify traits, values, habits, occupations and behavior patterns which the child-adolescent more or less consciously tries to emulate. Incidentally, the child may also develop an antipathy for a given adult who then becomes the embodiment of everything the child does not want to be. Negative ideals have a motivational potency all their own.

Jersild (1952), studying children from fourth grade through college, found that they increasingly valued personality characteristics (moral qualities, inner resources, a sense of humor), social attitudes (friendliness, heterosexual interests), and intellectual abilities, while they decreasingly valued characteristics centering around home and family. His study, like Havighurst, Robinson, and Dorr's (1946), reveals a shift in orientation away from the family and toward social relations and the self. There is also some evidence (Perkins, 1958) that between the fourth and sixth grades the discrepancy between the child's view of his actual self and his ideal self—between "me as I am" and "me as I would like to be"—decreases. This change is not related to school achievement or peer acceptance, but it apparently can be facilitated by an understanding child-centered teacher.

In sum, the objective evidence generally supports Jacobson's description of a shift in the ideal self away from exaggeration, perfection, and a home orientation, toward realistic standards which are oriented toward society in general.

Achievement Motivation

Achievement motivation is concerned with evaluating performance in the light of a standard of excellence. The research literature concentrates primarily upon the nonmoral value of doing a good job (Crandall, 1963). Theoretically, a small discrepancy between the standard and performance will produce a positive evaluation (Good boy), while a large discrepancy will produce a negative evaluation (Bad boy).[3]

Achievement motivation obviously is not inborn. According to Heckhausen (1967), the first signs of achievement motivation can be detected between the ages of three and three-and-a-half years. The child by then is no longer totally absorbed in doing for the sake of doing. His self-concept is now sufficiently firm so that he feels pleasure when he is competent and shame when he is incompetent. By four years of age, the child has developed the defense of protecting himself from the shame of failure by denial, evasions, and excuses. The four-and-a-half-year-old begins to show consistent preferences for certain levels of difficulty. Between the ages of four and fourteen, persistence in pursuing goals and tolerance for failure both increase. As in other aspects of achievement, there is a strong sex difference—boys return to failed tasks more frequently than girls do. This may well be part of the work orientation in older boys which we have already noted.

Intellectual achievement is remarkably stable over time (Moss and Kagan, 1961). Preschool children who show a strong desire to master intellectual skills retain this drive through adolescence and into early adulthood. Such consistency in a personality trait is rare. In addition, increases in the Intelligence Quotient (I.Q.) between the ages

[3] For a review of the studies of achievement motivation in children, see Crandall (1963). Achievement motivation is closely related to Lewin's concept of level of aspiration. His penetrating analysis and classical studies (Lewin et al., 1944) are still worth reading, although they are not primarily concerned with development.

of six and ten also are related to strong achievement motivation. We do not know whether achievement in nonintellectual areas would show a similar long-range consistence (Crandall, 1963).

Personality correlates of achievement motivation have been studied. Although we will present the findings in generalized terms, the data are not sufficiently detailed to determine whether the relationships are general or age-specific. High achievers are more independent in problem solving, more pleased with success, more self-reliant in competition, and more willing to delay immediate gratification than are low achievers. They are also moderate risk-takers, in contrast to low achievers, who tend to either play it very safe or undertake spectacular ventures. This picture of the self-reliant, assertive, competitive achiever fits nicely into the nonmoral value system of the American way of life. However, there are some hints that the child pays a price for his success. Third-grade achievers are more guilty than their nonachieving counterparts, while seventh graders are more willing to be destructive in order to win out over others. The evidence is only suggestive, but it raises an important issue: as one ideal is realized, what are the effects on other values and on other aspects of personality? Because achievement is so much a part of the American dream, we should pay particular attention to all of its significant interpersonal and interindividual consequences.

We have already referred to the striking sex difference in achievement motivation. Crandall suggests that achievement is congruent with the sex-typed values of boys but not necessarily congruent with those of girls. A girl's motivation to achieve will be reflected in her academic standards only if she perceives high achievement as being properly feminine. If she regards it as unfeminine for some reason, her academic aspirations remain hidden or suppressed. In addition, achievement in girls is part of a larger trait of sociability. They accomplish educational goals in order to gain extrinsic social goals of approval and attention. Boys are more work oriented, and achievement is an end in itself.

The role of parents in regard to achievement is not altogether clear. Initial studies suggested that early independence training rewarded by demonstrative affection produced high achievement motivation in middle childhood. Boys between the ages of four and seven, whose mothers had encouraged them to stand up for their own rights, to try new things without help, and to feed and dress themselves, but who also hugged, kissed, and verbally praised them, were high achievers subsequently. However, later studies showed the situation to be more complicated. As one might expect, the sex of the parent and the child is an important variable, as are the area of functioning (intellectual, athletic, etc.), the timing of the parental influence, and the general parent-child relationship. A child may be intimidated by an extremely competent, perfectionistic parent of the same sex because he despairs of becoming as powerful or capable as his model; or again, premature pressure may undermine the child's motivation, even if accompanied by affection. Parental behavior itself may change over time; a parent who rewards early independence may disapprove of the adolescent's desire to be autonomous. Finally, peer group values may contradict those of the parents; e.g., the group's valuing of athletic achievement may override parental valuing of academic success.

Heckhausen (1967) gives what he considers to be the best solution to the problem of parental influence. The mother should be warm, but she should also stimulate achievement directly by positive and negative sanctions. She should not determine what the child does but should be genuinely concerned that he do it well. The father should play more of a background role, sympathetically encouraging self-reliance while serving as a model of independence and assurance. Thus, in this analysis, the mother is the stimulating, rewarding teacher of achievement, the father is the benevolent model of independence. If the parents are cold and/or pushing or if they are authoritarian, they run the risk of intimidating the child. If they pamper the child, unrealistically inflating his image of himself, and require little in the way of accomplishment, achievement motive will be weak.

Family structure also affects achievement. Firstborn children are more highly motivated to achieve, perhaps because they are given more responsibility at a younger age. In the middle class, the smaller the family the more highly motivated the child, as a rule. Broken homes, parental disharmony, and weak ties between the parents tend to impede the development of achievement motivation.

There is no agreed upon explanation of how achievement motivation is acquired. Learning theorists stress the role of parental rewards and punishments. The child, since he is vitally concerned with gaining parental approval, regards achievement as another means of obtaining this goal. If the parent is unconcerned with setting standards, the child will be, too. Heckhausen, while recognizing the importance of social sanctions, challenges the notion that they constitute the child's only reinforcement. Rather, the development of a conceptual self is essential to achievement motivation. As we have seen, Heckhausen claims that the toddler is merely "acting"; but the four-year-old has developed the concept of an "I" or a self which acts. Through his explorations, he also develops an idea of levels of activity which are neither too easy nor too difficult. Having established this frame of reference, he can have the feelings of success in attainment or failure in lack of attainment which are an essential element in achievement motivation. In sum, the child himself evolves his standards and is intrinsically rewarded with pride and self-esteem when he realizes them. Parental approval is not the inevitable goal but, rather, the goal is the measure of success which meets his own criteria of excellence.[4]

Values, Conscience Proper, and the Polyglot Conscience

From the descriptive account at the beginning of our discussion we learned that the adult conscience contains diverse values and a variety of affects. We now need to elaborate upon this general characterization.

In content, values can be either moral or nonmoral. Moral values are concerned with special ethical rules of conduct among men. The Ten Commandments and the Golden Rule are examples. Nonmoral values come from society (e.g., self-reliance,

[4]Heckhausen extends our discussion of initiative with its affect of healthy pride (see Chapter 3) to include standards of performance with their affect of self-esteem.

industriousness, achievement), from peer groups (e.g., fair play, leadership, cooperativeness), from family themes, and from parental idiosyncrasies. Family values, for example, are not moral but represent "the way we are." The Lansons and the Newbolds (see Chapter 4) have strikingly different value systems and both are convinced that theirs is the best; but neither would accuse the other of being immoral.

The affective loadings of values come from two sources. The first is the granting or withholding of parental love. *Good boy* comes to mean "I am loveworthy," *bad boy* to mean "I am not loveworthy." There is no generally agreed upon label for the former; we shall use the term self-satisfaction. The latter develops into the feeling of guilt. The second source of affect is the granting or withholding of esteem. *Good boy* comes to mean "I am esteemworthy," and *bad boy* to mean "I am not esteemworthy." The former affect will be called self-esteem, the latter inferiority. The purest examples of esteem are seen in peer relations, where loveworthiness is irrelevant. The child's standing in terms of athletic ability, sociability, intelligence, and the like determine the degree to which he is esteemed by the group. Of course, parents can also have esteem for the child, being pleased with or disappointed in his behavior, independent of his loveworthiness. Finally, the child can hold himself in high or low esteem as he lives up to or falls short of self-imposed standards. Self-satisfaction differs from self-esteem, just as guilt differs from inferiority, both in origin and in quality of experience.

For purposes of discussion we can make a rough division. What we shall call conscience proper contains moral values and the affects of self-satisfaction and guilt. Self-esteem and inferiority, on the other hand, are usually associated with nonmoral values; they occur when one succeeds or fails in being as bright or witty or powerful or healthy or skillful as one wants to be. Yet, this convenient pairing of content and affect does not do justice to the complexities of conscience and to its development. For one thing, nonmoral values can become foci for judgments of loveworthiness. For puritanical parents, idleness and playfulness are sinful, while work is a moral requirement. Their child feels guilty, not inferior, after going to his first dance or playing his first game of cards. Or in middle childhood a child may derive moral self-satisfaction from being fastidiously neat or polite. Second, moral values can be used in the service of nonmoral ones. For example, "Honesty is the best policy—I've tried all the others" shows a pragmatic turn of mind in which morality comes off second best. More seriously, a self-centered, manipulative child can use the show of morality in the service of nonmoral values such as personal success. Our neat dichotomy with moral values and loveworthiness on one side and nonmoral values and esteemworthiness on the other, with the two realms in harmony, may characterize the ideally mature adult, but it is not an accurate picture of many adults or of children.

We need a conceptualization broader than conscience proper which will include the two kinds of values and four affects described above, plus the relations among them. We have coined the term polyglot conscience to serve this need. The hypothetical twenty-year-old in the initial section of this chapter has a polyglot conscience composed of moral and nonmoral values, especially if he feels self-satisfaction and guilt over the nonmoral activities of cleaning his plate and not washing his car.

One might object that our term polyglot conscience confuses the issue, since conscience, by definition, concerns only moral values or ethical principles together with their affects of self-satisfaction and guilt. We do not deny the validity of this definition or the validity of studying the development of conscience proper. However, clinicians have confronted us with the fact that nonmoral values can be held with the same conviction of their ethical rightness and can arouse the same affects of self-satisfaction and guilt as moral values. An adult can believe that God ordained racial segregation, while another is tortured with guilt for months after forgetting to buy a birthday present for his mother. Regardless of formal definitions of conscience, we must account for the similarity between moral and nonmoral values in adults rather than treating them as independent entities. There is a more compelling reason for the concept of a polyglot conscience. The development of conscience, like the development of many important aspects of personality, is not a matter of quantitative change. The toddler does not have a weak moral sense which only needs strengthening to become a mature conscience. As we shall see, conscience development entails dramatic revisions in thinking and feeling. Inherent in this change is an interplay between moral and nonmoral values and between the affects of self-satisfaction, guilt, self-esteem, and inferiority. Even if we had chosen to study only conscience proper, we would have been forced to broaden our scope to include the variables comprising the polyglot conscience in order to understand how it came into being.

Before pursuing our developmental theme, we must clarify one point concerning the difference between the polyglot conscience in the adult and in the child. Adults can distinguish between moral and nonmoral values or have the potential for doing so; young children are incapable of making this distinction. Our hypothetical twenty-year-old might readily admit that eating breakfast had nothing to do with morality and that it was "silly" for him to feel self-satisfied. A four-year-old would not be so sure. For one thing, parents probably do not behave in discriminably different ways as they move from a moral to a nonmoral value system. Specifically, there is no evidence that they change their socialization techniques or affects as they go from an idiosyncratic statement ("I can't stand it when you whine!") to one involving a moral principle ("I can't stand it when you lie to me"). Quite literally, there is no evidence. Until there is, we are assuming an unsystematic relation between the source of a directive (whether it is a personal, familial, or moral value) and the technique and affect used to impress it upon the child. If a parent offers no consistent clues, the child is denied an important means of distinguishing sources of values. He has no way of telling whether the parent is saying, "You must do this in order to get along with me," or "You must do this in order to be an acceptable member of the family," or "You must do this in order to meet the moral demands of the society in which you live."

How, then, does the preschooler respond to parental evaluations? First, he reacts to the content of such evaluations, even though he is intellectually unable to grasp the nature of the underlying value system. His effort to understand content continues throughout childhood and his thinking undergoes a number of significant transformations before he disentangles moral and nonmoral values and reaches a mature level of moral insight. Next, the preschooler reacts to the affective loading of parental evaluations. He is concerned with the amount of distress he will suffer if he is "bad" (e.g., if he violates a prohibition) and with the amount of pleasure he will experience from being "good" (e.g., if he does an act he would not have chosen to do). Finally, he is concerned with the consistency of the evaluation in terms of its being immutable, transient, or negotiable. Some parental evaluations are made with finality and invariably applied to certain behaviors (e.g., calling the parents derogatory names may always be regarded as "bad"); other evaluations may or may not be applied to certain behaviors depending on factors such as the parent's mood or conviction about their importance. Finally, there are evaluations which the child himself can succeed in altering. He might learn, for instance, that a judgment of "bad" will be altered if he is contrite, or that, if Mother says "That's bad," Father might say, "There's nothing wrong in that." Remember that the preschooler can be counted on to want his own way, so he is constantly alert to means of bypassing or counteracting or revising adverse parental judgments.

We are now in a position to take stock of the basic stuff out of which the preschooler will fashion his conscience. There is the bond of love for the parents which is essential to self-satisfaction and guilt; there are sources of self-esteem and inferiority, some deriving from achieving or failing in what he sets out to do, others deriving from parental and peer reactions to his accomplishments. There is initiative and empathy which leads him to explore his social environment and enables him to be emotionally reponsive to it. There is an awareness of and a desire to assimilate evaluations intellectually, plus a budding ability to weigh these evaluations against his own inclinations and to exercise a crude kind of judgment as to which he will follow.

The Development of Conscience

Conscience proper is a developmental achievement which undergoes many transformations before becoming a dependable moral guide. Nonmoral values must be disentangled from moral ones; some values must be reversed (e.g., sexual behavior is "wrong" in middle childhood but "right" after marriage); parroting of ethical rules must be replaced by understanding; perfectionistic standards of goodness must be modified into realistic guides to moral behavior; and anxiety and guilt must give way to moral responsibility. We shall be interested in all these transformations and why they occur.

In this section we will examine the earliest manifestations of conscience and then trace the development of the emotional, cognitive, and behavioral components of conscience. These components can be presented separately because they develop somewhat independently. The harmonious integration of feeling, understanding, and action marks the culmination of moral developments, and our next section will be devoted to an examination of how this integration is achieved. The final section will be concerned with the development of a healthy or an unhealthy conscience. Since conscience can be either a constructive or a destructive force in adult personality, we shall try to understand the conditions under which it takes one path rather than the other. Unlike our discussions to date, deviant development will not be considered in a special section, since it more readily fits in with topics included in the second and third sections.

Freud's Concept of the Superego

We shall begin with the development of conscience in the first six years of life. The most influential ideas concerning this early stage come from one of conscience's most unsympathetic critics. Freud, in dealing with emotionally disturbed adults, was impressed with the ravages of irrational guilt. At best, conscience seemed an unnecessary burden; at worst, it immobilized the individual by its pitiless laceration for even trivial transgressions, of thought as well as deed. Freud labeled this tyrant the superego. The superego is immutable and irrational in its demands for moral perfection. It is the implacable enemy both of sexual and aggressive impulses and of the necessary compromises involved in adjusting to the exigencies of reality. Freud further maintained that this irrational adult superego is a perpetuation or fixation of normal developments in the first six years of life. He hypothesized that conscience is originally rigid and uncompromising in its demands for perfection and its harsh punishment for transgressions. Whatever moderation it ultimately shows is learned during middle childhood or later.

How could such a monster be spawned? Freud hypothesized that the superego of the young child is far more restrictive than the standards of conduct which the parents are trying to convey. In adopting parental standards, the child exaggerates them into a punitive caricature. Isaacs (1937) provides observations which support Freud's hypothesis. She first noted that preschool children were guilty over screaming, biting, wetting and defecating, breaking things, being greedy and messy, thumbsucking, and masturbating. The list seems realistic enough. But the punishments children devise for themselves often far exceed the seriousness of the crime. One little girl believed that if she were naughty in school, God would drown the world; a little boy talked of being killed by God. Other punishments involved being deprived of food, being starved, being shut out of the home, and losing the mother's love (all of which may be equivalent psychologically). When one boy broke a cup, he denied himself cocoa for some days, being much more severe with himself than his parents or a teacher would have been. The realistic fear of being caught and whipped can mushroom into a frightening fantasy of being pursued and eaten by a large animal. Isaacs, like many others, observed that such fantasies of severe punishment can be found even in children who have benign, nonpunitive parents.

Freud's idea that early conscience represents an extreme and distorted version of parental teachings has been generally accepted. However, his explanation is questioned. Before presenting it, we should note that Freud was concerned with understanding a person's inner life. In his psychology it is never surprising to find a discrepancy between reality (such as parental discipline) and a psychological phenomenon (such as conscience), because the child rarely reacts directly in terms of reality. Instead, reality is constantly distorted by the basic passions of love, hate, and fear. While learning theorists typically ask, What environmental event produced this change in the child's behavior?, Freud asked, What emotional need motivated this change in the child's psyche? Thus, his speculations concerning conscience centered on the emotional crises of early childhood.

Freud thought the precursors of the superego could be found in what he called the anal period (see Chapter 7). The first primitive notion of *Good boy, Bad boy* is connected with performing a bowel movement on request rather than at will. However, the main part of the superego is born of the rage and anxiety associated with the Oedipal complex. As we have seen, the child of four or five undergoes an emotional crisis. His possessiveness of the opposite-sex parent makes him the rival of the same-sex parent, and he is terrified that this powerful rival will retaliate and destroy him. He resolves the conflict by identifying with the same-sex parent. In his effort to be like the parent, he adopts parental ideas of right and wrong. *Their* standards literally become *his* standards. From that time on, he rewards and punishes himself as his parents have done in the past.

Identification is widely believed to be the basic mechanism in the formation of conscience. We shall soon examine it in more detail. The view that conscience represents the resolution of an emotional crisis is held only by the Freudians. One reason the superego is so strong is that it relieves the anxiety of the Oedipal situation. Children, like adults, cling tenaciously to maneuvers which protect them from anxiety (see Chapter 7). The superego also protects the child from the guilt he feels over his Oedipal rivalry with, and desire to do away with, the same-sex parent. He may even be much more obedient to his own superego than he was to parental directives in order to atone for this destructive desire. Not only must he be good, he must be above reproach.

Balint (1953) describes different ways in which four- and five-year-olds try to appear blameless. A nursery school child, asked who did a certain forbidden act, blamed another person, either real or imaginary; when confronted with his own naughtiness, he protested violently. A child who was told it is not nice to vomit said that she was blameless, since her tummy did it; when she stamped her foot in anger, she accused her leg of naughtiness. (In certain adults, the Devil serves the function of keeping the self blameless–forbidden feelings come from him.) Balint also describes the priggishness which develops a few years later, as the child scrutinizes parental behavior. Minor transgressions or lapses by the parents are met with reproof: "You promised to take me for a ride on Sunday, and you didn't. You lied to me"; "You told me not to litter, but you just threw a gum wrapper out of the car window"; "Don't say 'It's me'–say 'It's I.'" The child is getting his revenge, unconsciously, claim the psychoanalysts. He is now the Good Child condemning the Bad Parent.

Another important feature of the Freudian concept of the superego is that it divides the child against himself. Freud was particularly concerned with the child's alienation from his sexual and aggressive impulses. This split can hamper future personality growth and become the breeding ground of psychopathology. Although Freud was not clear on this point, he seemed to maintain that the alienation, once it occurs, is never outgrown, that it remains as part of the personality and may lead to irrational self-accusation, religious dogmatism, prejudices, and war. In mature individuals, however, the forces of reason eventually counterbalance the primitive superego.

Granting that the superego is strong because it protects the child from anxiety and guilt, why is it also a caricature of parental values? The answer lies partly in the nature of identification in childhood, which is nearer to mimicry than to understanding (Balint, 1953; see also our discussion of Piaget's concept of imitation in Chapter 4, p. 100). It is the child's literal interpretation of parental values which distorts them. Subtleties of meaning and distinctions between what is and is not important are difficult for him to grasp. To the four- or five-year-old, adult morality is an alien culture whose deeper meanings are beyond his ken. He can do no more than ape the manifest content. Here Freud is making an important general point about socialization. The child does not internalize actual parental values; he internalizes parental values as he understands them. It is a mistake to assume that the understanding of a four- or five-year-old resembles that of an adult. Rather, it is a mixture of fantasy and magic and naivete and the seeds of realistic thinking. Only when reasoning becomes the primary mode—and this may not happen until adulthood—can one assume that environmental messages are received without distortion. (Chapter 6 will elaborate this theme.)

Psychoanalytic theory points to other distorting factors in the production of the superego. The role of hostility is particularly important. The child in the Oedipal phase is extremely angry with his rival parent. He also tends to believe that everyone feels the way he feels, since he is intellectually incapable of grasping the intentions and motives of others (see Chapter 6). In his primitive way, he reasons, "If I want to destroy my parent, my parent must want to destroy me." He thus reads into parental prohibitions a destructive intent which may never have been there. Subsequently, when he identifies with this parent, he incorporates a punitiveness which, ironically, may have been of his own making. "The severity toward our self which we call conscience is indeed nothing but a turning back of sadism upon our own ego." (Balint, 1953, p. 121). The superego is also exaggerated because of the importance of parental affection to the preschooler. He is still at the stage where the parents provide most of his love and security. Unlike an adult who has a network of attachments and friendships, the child is heavily invested in two emotional attachments. Thus, *Good boy* and *Bad boy* have a life-or-death quality which magnifies the importance of compliance or transgression. Finally, the preschooler lacks the perspective of time. When his mother says, "You're bad for hitting your little sister," she knows that such "bad" behavior is temporary and will change. This is exactly what the child does not know. He is incapable of thinking, "This is just a phase and I will outgrow it." The judgment *Bad* is an absolute condemnation. The parental recognition of relativity is not conveyed and, even if it were attempted, could not be grasped by the child's immature understanding.

Another important contribution from Freud is his analysis of guilt, the emotional substrate of conscience. Guilt has its origins in love. It is love which gives the infant-toddler his earliest, most basic feeling of pleasure in himself. The well-loved child feels good all over. Concomitantly, withdrawal of love arouses the anguish of separation. The child feels that his world is going to pieces. As the preschooler internalizes parental values, his superego takes over the parental function of rewarding him with a feeling of pleasure or self-satisfaction when he is good and punishing him with anxiety when he is bad. It is this self-administered withdrawal of love, this "moral anxiety," which Freud calls guilt (Fenichel, 1945).

Freud's theory of the superego set the stage for much of the subsequent inquiry into conscience development. His ideas that conscience consists of parental values which become internalized, that the mechanism of internalization is identification, and that the early conscience is authoritarian and rigid, are generally accepted. His ideas about the relation of the superego to the resolution of the Oedipal complex and the special roles of hostility and anxiety in the formation and perpetuation of the superego have remained specific to psychoanalytic theory.

Identification

The concept of identification was developed by Freud and has continued to fascinate and frustrate psychologists ever since (Bronfenbrenner, 1960). The term is used for two related phenomena. The first is the increased similarity between two people, typically the child and the parent. When the child identifies with the parent, he becomes like the parent. Increased similarity can be evidenced at different levels. At times, Freud implied that identification was equivalent to what some psychologists would call modeling, since the child primarily apes the overt behavior of the adult. For instance, a four-year-old girl misses her mother who has just gone shopping; she goes to the door and exactly duplicates the mother's exit of waving goodbye and throwing a kiss. More frequently the term identification is used in connection with pervasive motives, attitudes, and values; e.g., a girl identifies with her mother's disgust over sexuality and subsequently becomes frigid, or a boy identifies with his father's strict, humorless attitude toward life and becomes a martinet. In psychoanalytic theory identification is not a transient mirroring of parental characteristics, but a set of widespread and permanent changes in the child's personality, or what psychoanalysts would prefer to call his psychic structure. After identification, the child regards the behavior or motive or attitude or value as intrinsic to himself. What was formerly "theirs" is now "his" as deeply as any aspect of personality is "his."

Identification also is used to describe a *process* as well as a product. Here we see Freud's preoccupation with motivation, his continual questioning as to why a psychological phenomenon exists (see Chapters 6 and 7). He proposed two motives for identification, thereby delineating two types of identification. The first is anaclitic identification. Anaclitic means leaning on, or being dependent upon, or being attached to someone. Recall that the infant comes to depend upon the caretaker to reduce distress and provide pleasures (see Chapter 2). Anaclitic identification is motivated by the anxiety which inevitably occurs when the caretaker is not present when needed. Through assuming the caretaker's characteristics, the child can, metaphorically, become his own mother and allay his anxiety. A hypochondriacal child, for example, might show the same exaggerated concern over minor aches and pains that his overly solicitous mother did, thereby protecting himself from the greater anxiety of separation from the mother.

The second type is identification with the aggressor, which is motivated by fear of the power and destructiveness of the parent, or, more precisely, fear of the power and destructiveness the child views the parent as having. In order to master his anxiety, the child himself becomes aggressive, changing from helpless victim to attacker—the child wages a "preventive war" to allay his fears of being attacked and destroyed. Identification with the aggressor, in which the boy takes on the attributes of the father, is the classical resolution of the Oedipal complex.

Freud speculated that identification in girls is primarily anaclitic, with some subsequent overlay of aggressive identification, whereas the situation is reversed in boys. That is why, according to Freud, the female superego revolves around loving and being loved, the male superego revolves around power. Women tend to be emotional and need social approval and acceptance, while men tend to be strict, objective champions of principles in their stance concerning right and wrong. Mother is warm, forgiving, extenuating, sensitive to hurt; father is stern, demanding, uncompromising.

While there is a certain validity to Freud's characterization of mothers and fathers (Stolz, 1966), subsequent investigation has not uniformly supported his hypothesis of identification with the aggressor (Bronfenbrenner, 1960). A number of revisions have been suggested. Mowrer (1953), for example, speculates that identification takes place against a background of both love and fear of the parent. The first emotional relation is affectionate, but, as the child is socialized, he becomes resentful and strikes out at the parent. The parent punishes him by withdrawal of love which, in turn, leads to intense anxiety. At the height of his distress, the child realizes that he can both satisfy his parent and relieve his anxiety if he accepts parental standards and values as his own. Mowrer's conceptualization accords better with empirical findings that boys who have strong positive feelings toward their fathers, rather than being fearful of them, are also strongly identified with them.

Other theorists emphasize the power and competence of the father as incentives for identification rather than his punitiveness. The father who gives commands is in command. He works and makes money. He can make decisions and act on them. Bronfenbrenner suggests that such qualities appeal to the child's drive for mastery or what we have called initiative. Instead of being the personification of a fearful authority, the father serves as a model of assured, effective action. Identification with him facilitates the child's growth toward mastery.

As has happened in past discussion, we are again left with diverse, plausible theories, but with insufficient empirical evidence to support one over the other or to decide whether they all have validity under given conditions.

The Role of Thinking

The intellectual component of conscience has been dealt with most extensively under the rubric of moral judgment. The broad developmental trend is from ignorance, to awareness, to understanding. Moral judgment, then, follows a route similar to that found in other areas of intellectual mastery. An infant learning to speak typically parrots words first and later discovers that they are means of communicating ideas; the college student studying statistics may learn formula after formula before the real meaning of experimental design strikes home. And with morality, as with speech and statistics, understanding opens the door to flexibility and improvisation, which were impossible when rules were learned by rote and followed to the letter.

Piaget on Moral Judgment. Piaget has given us many insights into the development of moral judgment (Piaget, 1948; Flavell, 1963). His research technique consists of telling stories which pose moral issues and asking children of different ages to make a moral judgment. In one story a child breaks a cup as he is stealing some jam, while another breaks five cups when he is helping his mother set the table; who is naughtier? A child steals a piece of ribbon for herself, while another steals a roll of ribbon to give to a poor and hungry friend; who is naughtier? Typically, Piaget does not accept the child's first response as the answer but continues to question until he is certain the child has revealed his true judgment and reasoning. The methodological point is important, because children, especially bright ones, are adept at parroting socially acceptable ideas which they do not understand.

In studying the child's understanding of transgressions, Piaget found the following progression. Young children, around four to five years old, judge the seriousness of a transgression in terms of its physical consequences or in terms of its deviation from the truth. For example, breaking five cups is naughtier than breaking one. The child reasons literally: if it is bad to break one cup, it is "badder" to break more than one. Likewise, if telling lies is bad, telling a big lie is categorically worse than telling a small lie. Only much later, around age nine or ten, can children grasp the subtle notion of intention and judge behavior in terms of underlying motives. Thus, an intentional bad act is naughtier than an accidental one; a transgression motivated by selfishness is naughtier than one done altruistically.

Piaget found other interesting developmental trends in regard to lying. For young children, the lie which fails to deceive is naughtier than the successful one, the unintentional lie with serious objective consequences is worse than the deliberate one with no serious consequences; lying is bad because it is punished, and lying to adults is worse than lying to a peer. In older children, all these judgments are reversed—lying is wrong regardless of whether it is successful or not, regardless of subsequent punishment, and regardless of whether it is told to an adult or a peer.

Piaget conducted an intensive study of children's concepts of justice. There is a developmental trend in the punishment deemed appropriate for transgressions from expiatory punishment to punishment by reciprocity. Young children have a categorical, unconditional, expiatory view; wrongdoing must be punished, and, in certain cases, the more severe, the better. The older child believes in a reciprocal type of punishment, in which the punishment "fits the crime"—its severity is contingent upon the misdeed and it should be accompanied by explanations which would help the child avoid similar misdeeds in the future.

Piaget also traces the development of distributive justice, which concerns the distribution of rewards and punishments among members of a group of children. Children six years old or younger believe that whatever an authority figure does is fair simply because he does it. Just as the adult who is a True Believer does not question God's judgment in the face of manifest injustices, the little child accepts favoritism or unequal punishments for similar crimes. To illustrate this belief that the adult can do no wrong: if a four-year-old is told a story about a boy who obediently watches his baby brother while his mother is away but then is given a spanking when the mother returns, he will claim that the boy actually was bad. The idea that the mother may be wrong is untenable. In the next stage the child becomes extremely equalitarian and insists that everyone must be treated the same regardless of the circumstances. His thinking has the same inflexibility as in the previous stage, but now the theme is, What is right for one is right for all. He believes in "an eye for an eye" in regard to retaliation; in regard to altruism, he believes in doing something nice for others so they will do something nice in return. In the final stage, the child is still equalitarian but takes extenuating circumstances and individuality into account. For example, the child in the second stage justifies retaliating when he is hit, regardless of who did the hitting; in the third stage, he still upholds the right to retaliate, but he would refrain in certain instances, such as being hit by a smaller child. He is more capable of of putting himself in the place of the recipient of injury and is also able to think in terms of giving without expecting anything in return.

Piaget conducted one other investigation which is relevant to our discussion. It concerns children's understanding of the rules governing the game of marbles. The toddler playing marbles is blissfully oblivious of rules. He does what he pleases, and the rituals he develops are his own invention. In the second stage, around four to five years, the child has grasped the basic rules of the game but regards them as sacred and inviolable. He claims that the rules have always existed and that they came from some parental or divine authority. Regardless of the wishes of the players and regardless of the possibility of making the game more interesting or more fun, rules cannot be altered. Only in the third stage, when the child is ten or eleven, do rules cease to have an inviolable authority of their own. Then they are seen for what they are—human inventions to serve the practical purpose of regulating behavior. However, it is not until the child is eleven or older that he indulges in improvisations. Now he truly knows that rules exist to guarantee fairness to all, to maximize skill and minimize chance, and to insure pleasure. If everyone agrees to a new rule, the game can be varied at will.

Piaget's choice of a game such as marbles is fortunate for our discussion of conscience. If he had only dealt with justice, honesty, and other moral values, we could still question whether his developmental stages apply to nonmoral activities, such as games. Piaget's data suggest that they do. However, we still have no evidence as to whether the child's understanding of family and idiosyncratic rules undergoes similar transformations.

Let us summarize the development of moral understanding from the preschool period to early adolescence. In the early years, an act is either right or wrong; naughtiness is judged in terms of an act's physical consequences or its deviation from an absolute standard of correct behavior; the authoritative adult is always right; punishment by the parent defines badness and is approved as an end in itself; rules are immutable, and the talion is justified. In later stages, the seriousness of a transgression is judged in terms of the transgressor's intentions; right and wrong depend upon circumstances, and authoritative adults can be mistaken; punishment should be in keeping with the seriousness of the transgression and should serve a corrective purpose; rules exist for social regulation and can be altered by common consent; fairness requires that an issue be considered from the points of view of all parties involved.

Note the parallel between Piaget's picture of the early conscience and Freud's concept of the early superego. Both consciences are absolutistic, unconditional, punitive, and immutable. Freud and Piaget both imply that the child does not accurately receive the messages which his parents send, no matter how clear and rational their directives might be. What message he reads will be determined by either his current emotional problems (according to Freud) or the stage of his cognitive development (according to Piaget). This similarity is all the more striking because Piaget conducted his studies with normal children rather than with disturbed adults and used an entirely different method of data collection. But there are also differences. Freud, the clinician, was concerned with the motivational forces which produce the superego and with the emotional toll it takes; Piaget, the epistemologist, was interested in cognitive factors underlying conscience. The two men, looking at the phenomenon from different viewpoints, may well have explored complementary variables. (For a comparison of their general theories, see Wolff, 1960.)

How does Piaget account for his findings? The child's early understanding of moral values and rules is determined by two factors: the stage of cognitive development he has reached and the nature of adult constraints placed upon him. The immature mind of the young child obviously cannot grasp the complexity of his physical and social environment and Piaget delineates the successive stages the child's thinking passes through in his efforts to comprehend the world. Around five or six years of age, the child's judgment is characterized by what Piaget calls moral realism. Realism seems a peculiar label until one realizes that it carries the connotation of literalness rather than rationality. The child uncritically accepts his immediate experience as reality. To take a simple example: during a moonlight stroll, the moon appears to move along with the individual. This naive experience is exactly what the child believes is happening in reality. Ask him what makes the moon move in the sky, and he will answer, "I do, by walking" (see Chapter 6). He accepts parental prescriptions as literally as he accepts his sensory experiences. He can grasp only the letter, not the spirit, of the law. He can see a broken cup; he cannot see an intention. When he learns that breaking a cup is bad, all he can understand is the manifest act of breaking and the resulting physical damage. He does not think of intentions because the concepts of "accidental" and "on purpose" are too inferential to grasp. Similarly, when he is told that lying is bad, he accepts the statement as literally and absolutely true.

The preschooler is also egocentric in that he cannot adopt a viewpoint different from his own. He cannot view the physical world from a different perspective. Show him a model of a city street and ask him how it would look if he were viewing it from the opposite end of the street, and he is stymied. He also cannot see the human world from any other perspective. He knows how many brothers and sisters he has, but he cannot answer the question, How many brothers and sisters does your brother have? Lerner (1937) told boys six to thirteen years old a story in which the hero violated the teacher's prohibition and helped a lazy student. Then each boy was asked, "How do you feel about this? How did the hero feel? How did the lazy boy feel? How did the teacher feel?" The six- to eight-year-olds typically gave the same response to all five questions. The idea that the same act could be viewed differently from the perspectives of the various people involved did not occur to the younger boys. Because the preschooler has only limited ability to shift his perspective, it is difficult for him to think of his behavior in terms of its effects on others.

Piaget maintains that the constraints which parents impose upon the young child reinforce his egocentricity. In order to impress the child with their importance, parents present moral values as absolute and judge behavior as categorically good or bad. They might say, "It's naughty to tell lies," or "You were naughty to lie to me." The parent rarely suggests that there might be extenuating circumstances, or explains the reason why a value is important in human relations, or even conveys his own sense that a childhood transgression is not, after all, a cardinal sin. Even if the parent did try to explain such abstract matters, the child probably would not grasp them. In his own way, Piaget is as fatalistic about early moral judgment as Freud is about the early superego. Because the five-year-old's stage of cognitive development is what it is, and because socializing parents must place constraints, a primitive understanding of morality is inevitable. In fact, Piaget labels this early phase the morality of constraint.

What, then, is responsible for the development of true moral understanding? Piaget's answer is simple: cooperation. He labels the mature stage of moral judgment the morality of cooperation. Cooperation alone can deliver the child from his egocentricism and counterbalance parental authoritarianism. Why? Because of the nature of cognitive development. The child's understanding progresses as his inquiring mind is forced to assimilate information which is incongruous with his present stage of understanding. He thinks the rules for marbles are immutable, but he sees older boys improvising–how can that be? He thinks it is bad to say "God" except in prayer, but when his friend is excited he says, "Oh God!" and sees nothing wrong–how can that be? The child receives maximum exposure to divergent points of view in the free exchange of ideas which peers enjoy while engaging in activities of mutual interest. In assimilating ideas which are incompatible with moral realism and egocentricity, he is forced to conclude that he is not the center of the universe and that his experience does not define social and physical reality. There are many points of view, many loci of thoughts and feelings. His self is only one of many selves, his values only one of many possible sets of values.

There is some independent validation of Piaget's theory. Fite (1940) observed the educative influence of friends and peers in a group of preschoolers who were caught in a double standard regarding aggression. Their parents had taught them one standard of behavior which was more strict than that of the peer group; specifically, parents discouraged the very fighting, toughness, and retaliation which were valued by the group. Although the child was confused at first, he began to achieve perspective on values. He began to make a distinction between "right and wrong" and "what my mother and father say"; parental rules represent not the only way of behaving but simply one way of behaving. Naturally, the preschool child was still too young to evaluate the relative merits of conflicting directives, so his behavior was a series of contradictions. He would stoutly defend his parents' rules one minute and then abide by a group mandate the next. Since he had no concept of overall consistency and no need for a guiding life plan, he was not unduly distressed over the fragmented nature of his value system. What he gained was an ability to stand apart from parental rules. Fite emphasizes the importance of friendships in enabling the young child to detach himself from parental values and in giving him emotional support as he experiments with new standards of conduct.

Unfortunately, other objective studies have not borne out Piaget's thinking. There is no conclusive evidence of a consistent development from an authoritarian to a democratic ethic in children, nor is there evidence that active participation in peer groups inevitably results in an understanding of intentions and a gaining of perspective (see Kohlberg, 1963). Kugelmass and Breznitz (1967) found that Israeli adolescents reared in a kibbutz, which minimized parental authority and stressed group participation, were no more advanced in their consideration of intentions in judging behavior than were family-reared controls. Lerner (1937) also makes the telling point that group pressures for conformity can be as coercive as parental demands for obedience. Commands "from above" can be replaced by majority rule which can also block the achievement of perspective.

Piaget's theory has met with other difficulties. In his own investigations Piaget was impressed with the variability of moral judgment, and subsequent research has shown that, indeed, it is not all of a piece. We should not assume that, because a child has reached an advanced level in one area of moral understanding, he is equally advanced in other areas (Johnson, 1962). Lerner also found that the child is not uniformly insightful in his perspective. While he achieves increasing perspective on his community and school between the ages of six and thirteen, perspective on the family levels off around eight years of age. Thus, the early adolescent realizes that his city or his school is no better and no worse than other cities or other schools, but he is still likely to judge his family as all good or all bad. Clinicians have made a similar point. Parents are perhaps the last persons to be seen as individuals in their own right, with their own dreams and pleasures and troubles and failings. Not infrequently, the child must become a parent himself before he can achieve perspective on his own parents. And the values of the family often are the last to be viewed as just one of many ways of getting along in the world. A person may be forty-five years old before he realizes that he does not have to eat everything on his plate!

A Modification of Piaget's Categories. Kohlberg (1964) has introduced a number of modifications within Piaget's framework. Like Piaget, he maintains that the development from amoral to moral thinking is as legitimate a cognitive dimension as the development from alogical to logical reasoning. A genuine moral judgment represents the ultimate standard for behavior, just as logic represents the ultimate standard of rationality. As Piaget puts it, "...morality is the logic of action" (1932, p. 404).

Kohlberg's reconstruction of the stages the child goes through in achieving a mature moral judgment are somewhat different from Piaget's. At the Premoral level, the child is concerned only with the punishment or reward that an act will lead to. At the next level, labeled the Morality of Conventional Role Conformity, the child adopts the conventional standards of "good boy"behavior in order to maintain the approval of others or to conform to some moral authority, such as religious principles. He is significantly more social and more conceptual in his judgments than he was at the previous level. At the final level, called the Morality of Self-accepted Moral Principles, the child judges behavior in terms of the morality of contract, of individual rights, of democratically accepted law, and of individual conscience. At this stage, he holds himself accountable to himself for his moral decisions. Between six and sixteen there is a gradual decrease in the first type of morality and a gradual increase in the second and third types, although only about 25 percent of sixteen-year-olds have achieved the highest level.

Like Piaget, Kohlberg maintains that the child's moral judgment reflects his level of cognitive development. If he first judges in terms of punishment, it is because this is the concept he is capable of grasping. Kohlberg, unlike social learning theorists, maintains that different parental practices in regard to punishment will have little effect on the child. "Bad equals punishment" and "Good equals obedience" are the only kinds of statements he can understand. And it does not matter whether parents or neighbors or teachers or policemen provide him with the equation. Changes in moral judgment, Kohlberg says, do not represent merely the internalizing of parental values, as both the psychoanalysts and the learning theorists maintain, and it has nothing to do with identification. Rather, it reflects the progressive steps in purely cognitive mastery.

What is responsible for the progression from lower to higher levels? Kohlberg maintains that it is not parental influence per se, nor, as Piaget claims, peer contact per se. Rather, it is the child's experience with a variety of conflicting values. Middle class children and popular children are more advanced in their moral judgment because they have more extensive social contacts and assume more responsibility in social situations. Kohlberg has no quarrel with Piaget's model of the inquiring mind assimilating incongruous information, but he maintains that the child needs a larger matrix of experience than peer cooperation.

A Supplement to Piaget's Concept of Cooperation. Cameron and Magaret (1951), like Piaget, maintain that the child achieves perspective through peer relations, but they also stress the importance of play. Offhand, play might seem too trivial to be a catalyst for the development of conscience, but this is because the label is misleading (see Chapter 9). The preschool child plays because he wants to play, because he needs to play, because he enjoys playing. At times his play is repetitive and ritualistic, at times it is inventive, but always it has the quality of intrinsic meaningfulness. What a young child learns through play he truly learns because the activity originates with him. It is the Sunday School lesson and the didactic moralizing which have little effect on conscience because they are alien products from an adult world.

As Cameron and Magaret point out, a prominent feature of play is that the child assumes many roles. The preschooler is an avid dramatizer. He plays house and store and doctor; he is a pilot, a pussycat, a policeman, a monster, a nurse; he comforts and feeds and arrests and kills. In assuming these diverse roles and functions, he is preparing to break out of the egocentricity which binds him to a single point of view.

During the preschool years, play becomes progressively more social. Children begin to act out dramas together, and a critical change takes place. The child is required to accommodate to his playmate. As long as he was playing by himself, he could go wherever his fantasies led him. However, once he begins to play house, where he is the father and a little girl is the mother, he must adapt his play to that of his companion. In order to make such an adaptation—and here is the crucial point—he must be able to understand the other child's behavior. If she chooses to be a nice mother preparing supper, he must recognize this role and act accordingly. If he is still so egocentric that he assigns her the role of a punitive mother, there can be no communication and there can be no play. The beauty of play is that children do understand their play roles so well. By contrast, the adult world is often alien and incomprehensible. Parents use rewards and punishments to induce children to do many things which are meaningful only in terms of keeping or losing love. But children playing together are sharing the same psychological world. The little boy may more readily understand the little girl playing mother than he understands his own mother. It is the desire to understand and the ability to understand the behavior of another which is the essence of perspective. And it is this willingness to accommodate to another person to achieve a mutually gratifying social interaction which contains the germ of truly moral behavior. Both of these factors are strong counterforces to the child's egocentricity.

Cameron and Magaret also discuss the parent's role in facilitating the growth of perspective. The parent, perhaps in reaction to the child's egocentricity, is constantly calling attention to the way other people or other living things feel. When the toddler pulls a dog's tail or hits his sister with a stick, the mother tells him that it hurts. The toddler may not realize that other living things have feelings, until the parent sensitizes him to the fact. Another example of sensitizing is the familiar phrase, "How would you feel if someone else did that to you?" Here the parent is the ally of the child's empathy and is engaging the child's intellect in a way which empathy alone cannot. In essence, the parent is confronting the child with the challenge of the Golden Rule. It is important to note the contrast between Cameron and Magaret's image of the sensitizing parent and Piaget's image of the constraining parent. Later on we shall see that these different images may make a great difference in the kind of conscience the child develops.

Alternatives to Piaget's Theory. All psychologists do not agree either with Piaget's emphasis on cognition in moral development or with his theory of cognitive development itself. Let us take a familiar bit of behavior. A five-year-old girl with permissive, loving parents is observed spanking her doll and saying, "Bad, bad." Kohlberg would regard this as evidence that the child has grasped the only moral idea she is capable of understanding, namely, "bad equals punishment." Balint, within the psychoanalytic framework, offers a different interpretation. The girl, after resolving the Oedipal complex by identifying with the same-sex parent, is split into a Good-me and a Bad-me. She is particularly desirous of being perfectly good because only recently she had been so bad as to view her mother as a hated rival. In spanking the doll, the girl is displacing her feelings about herself onto the doll, as children often do in play. Thus, the doll symbolizes her bad self which is punished by her new, good self. More generally, psychoanalysts might object to the view that moral judgment is determined by purely cognitive factors. The child's compelling emotional experiences determine what he will think to a greater degree than does his level of cognitive development. The most potent meaning of "good" and "bad" throughout childhood is "I am or I am not loveworthy." Loveworthiness undergoes a complex sequence of changes (see Chapter 7), and the child's moral judgments are significantly affected by his attempts to master the conflicting forces of love and hate within him.

Both Freud and Piaget would agree that there is an inevitable gap between parental messages and the child's comprehension of these messages. Aronfreed (1968), in the social learning tradition, objects. Moral development is orderly because the socializing parents automatically adjust their demands to the child's growing ability to comprehend such demands, not because of a fixed sequence of stages of cognitive unfolding. If, for example, the preschooler judges degree of badness in terms of physical criteria, it is because parents themselves are punitive in response to damage, mess, and disruptions; if the child regards rules as absolute, it is because parents often take an absolute, authoritarian stance. It is only as the child matures intellectually that his sensitive socializers begin to talk of intentions, consideration of others, and fairness, because they know that only now will the child understand. Because the child learns what his socializers teach him, the content of moral judgment can differ in children at the same level of cognitive sophistication. If society exploits the child's increased cognitive ability to reinforce authoritarian values or, as in the case of certain delinquents, antisocial values, the result will be an intellectually more adroit defense of authoritarian or antisocial values. Cognitive growth does not guarantee the appearance of higher levels of moral judgment, such as Piaget and Kohlberg describe; it only insures that the child will grasp these ideals if the environment provides them. In general, the child learns what he is taught within the limits of his ability to comprehend the messages he is sent.

Freud and Piaget stress inevitabilities—every child is fated to experience the Oedipal situation or to go through certain stages of cognitive development. Aronfreed stresses the importance of the environment in determining the content of moral judgment. Each argues persuasively with evidence to support his position. Choice of one over the other, at this point, is more a matter of accepting his general theoretical framework than of established superiority. For our purposes, we have learned much from the Piaget studies. They provide a richly detailed account of the change from a rigid, authoritarian conscience to a flexible, humane one, and they trace the child's growing awareness of the social consequences of his behavior and his feeling of personal responsibility in upholding abstract values and ethical principles. Piaget's picture of the inquiring child engaged in assimilating divergent values provides us with an important lead as to how the primitive conscience is transformed into a humanistic guide to behavior. He implies that it is essential to nourish the child's inquisitiveness and to keep the channels of cognitive growth open. Translated into our terminology, it is essential to preserve the child's initiative.

The Role of Feeling

Conscience is more than an intellectual understanding of values. It has its own affects which give it special qualities and potency. On the negative side are shame and guilt; on the positive side is the pleasure in doing something one considers to be good, which we call self-satisfaction. Empathy, while not a feeling characteristic of conscience, is important to its development. The ability to "feel with" another person is the counterpart of the cognitive ability to see another person's point of view, which we have just discussed.

Shame. According to Erikson (1950), shame occurs when the child is "visible and not ready to be visible" (p. 223). Shame is being exposed to public view while doing something one would prefer to do in private. When one is "caught with his pants down," the impulse is to hide, to vanish from sight, to sink into the ground. Shame is fostered by the child's feeling of being small and powerless compared to adults. It is more public and situational than guilt, which is an inner voice condemning the child, regardless of where he is or whether he has been detected. Shaming may be a technique of discipline used infrequently by middle class American parents; at least they do not engage in the kind of public embarrassment, mocking, and critical teasing found in other cultures.

Guilt. Guilt is the affect which accompanies the child's judgment of himself as bad. According to psychoanalytic theory, the child assumes the parental function of rewarding good behavior with love and punishing transgressions with withdrawal of love. Since his deepest need is for love and his deepest anxiety concerns loss of love, the child's self-condemnation, "I am bad, I am not worthy of love," reproduces the distress he formerly felt in response to parental condemnation. If the parent has been negligent or punitive and the bond of love weak, the child will not develop guilt feelings and their special kind of braking power. If the parent has been possessive and has placed too high a premium of love on conformity, guilt will be intense and pervasive.

There is an unfortunate split in the research on guilt. Clinicians have been concerned with its development primarily during the first six years of life, while objective investigators have concentrated on school-age children. Objective studies rely on a technique which involves verbal comprehension and communication of explicitly formulated ideas. Typically, the child is told a story in which the hero violates a rule, with no chance that he will be detected; e.g., a boy steals a catcher's mitt and no one sees him. The child is then asked to complete the story. Whether this technique taps the early, often unconscious, primitive guilt which Freud described is open to question.

However, the children's responses to the stories do serve to illuminate the subsequent development of guilt feelings.

Young school-age children tend to be concerned about punishment rather than feeling guilty over transgressions. Their anxiety over being caught is unrealistic, since the stories are all constructed so as to eliminate the possibility of detection. These youngsters, like those at Kohlberg's first stage of moral judgment and those with power-assertive parents (see Chapter 4), think an act is wrong because it involves the pain of physical punishment, and their main concern is with hiding or with being caught. They have an externally oriented conscience.

The theme of confession increases in frequency from the early school-age period. It is a half-way house between the externally oriented fear of punishment and an internally oriented guilt reaction. The child may be driven to admit to a wrongdoing, even when he could get away with it, because he is concerned over what others would think of him and because he is inwardly troubled by the transgression. Next, confession can relieve the anxiety of waiting to be punished. The child who has been caught repeatedly begins to live in fear that he might be detected at any time. This expectation is an intensely uncomfortable state. He confesses in order to be punished and relieve the anxiety of anticipation. Finally, confession can be used to placate a parent. An alert five-year-old quickly learns to say "I'm sorry" when caught red-handed. The parent may grumble, "Yes, but are you really sorry?" but his anger is checked by thoughts of his own guilt if he would punish a repentant sinner. From here it is an easy step for the child to admit transgressions spontaneously in order to dilute or avoid punishment or even be rewarded. (Parenthetically, religions which emphasize confession share the parent's concern that it can be used to bypass responsibility and smooth the way to the next transgression.) In sum, confession can be good for the soul. It can be motivated by a genuine concern over loss of love and a concomitant loss of self-love. In such cases, it is backed by the powerful motive to re-establish an inner sense of loveworthiness. However, confession can also be used primarily to relieve the anxiety of waiting to be punished or to divert punishment and gain approval.

True guilt, in the sense of an internal judgment and criticism of the self, is very rare in young children but is found frequently in twelve- to thirteen-year-olds. The early adolescent is primarily concerned with the damage done to his image of himself, with not being able to respect himself or to be proud of himself. Punishment from others, or even being worthy of the love and respect of others, is unimportant. The child must now live on good terms with himself. The potency of the need to regard oneself as good is seen in the fact that guilt is correlated with avoidance of delinquent acts and with resistance to temptation in experimental situations. Fear of punishment serves neither of these functions, perhaps because it often involves a concomitant desire to get away with as much as possible. Confession, true to its transitional nature, may or may not prevent succumbing to temptation.

Empathy and Sympathy. Empathy is feeling the same emotion that another person feels. In the 1920s and 1930s, when the heredity versus environment issue was in vogue, psychologists pondered whether empathy is innate or acquired (see Allport, 1924)

McDougall claimed that the perception of another, such as the facial expression, the bodily movements and voice of a panicky person, directly arouses the corresponding affect (in this case fear) in the observer; the connection is an innate one. Allport countered that empathy is learned. In essence, the situation that produces the affective response in the observed individual produces the same response in the observer. We empathize only when we have experienced, or have imaged ourselves experiencing, the event we now observe happening to someone else. The closer the observed situation to the experience of the observer, the greater the empathy. Thus, empathy is seen as a conditioned response. Allport's hypothesis is supported indirectly by a study of six- and seven-year-olds (Feshbach and Roe, 1968) in which empathy was facilitated if both persons were of the same sex. It seems reasonable to assume that similarity of experience is responsible for this result. Incidentally, the investigators also found that an intellectual understanding of another child's feelings does not necessarily involve empathy. Even at this early age it is possible to know what someone feels without feeling what he feels.

Empathy has not been investigated as intensively as sympathy, which is a concern for the plight of another. Murphy (1937) studied thirty-nine children between the ages of two-and-a-half and four years. Her purpose was to understand sympathetic behavior as such and its function in the child's total personality. She found that sympathetic responses are a visible and stable feature of social behavior at this early age, although their frequency is considerably less than one per hour in a nursery school setting. Typical sympathetic behavior includes helping a child in need of assistance, asking a child why he is crying, and making supportive comments. Friends, younger children, and favorites of the group are likely to receive relatively large proportions of these responses. Interestingly, boys and girls do not differ in sympathetic behavior at this age.

Murphy found that sympathetic behavior results from a variety of factors, ranging from temporary, situational ones to general personality traits and broad developmental trends. Situational variables include the age composition of the group and the size of the play area. A group with a narrow age range in a small play area will have fewer sympathetic responses than one in which older and younger children play in a large area. Sympathetic responses are less frequent in new situations than in established situations where friendships and group influences have had time to develop. Whether or not a child receives sympathy depends partly upon the frequency with which he is in distress. But sympathy is also related to general factors. It is moderately correlated with age and intelligence but probably is more closely related to the increased social responsiveness in the preschool period. Leadership and friendship, for instance, are on the rise and the child is becoming less egocentric. There is also a moderately high correlation between sympathy and hostility; the genuinely outgoing child is free to express both positive and negative feelings as they arise. (The finding that the child high on hostility may also be high on sympathy is an important one to remember when we examine the development of hostility in Chapter 7.) The relation of sympathy to other needs is highly individualized; e.g., one child may be most sympathetic when he is secure, another when he is insecure.

Murphy believes that it is not profitable to argue whether sympathy is innate or learned. Both heredity and environment are important. A child becomes more sympathetic as he is rewarded for such behavior and as he imitates the sympathetic actions of adults. But there is also a basic social sensitivity and awareness, a human responsiveness, which sensitizes him to the pleasures and distresses of others. Murphy's concept of social sensitivity resembles Adler's concept of "social feelings" (see Chapter 3), which are innate not in the sense that they appear inevitably but in the sense that a normal environment will elicit a vigorous, sustained, expansive response in the child.

Murphy's description of the forms sympathy takes early in life is illuminating. There is the well-adjusted child's cooperative, practical sympathy in response to the spontaneous need of a playmate. There is projected anxiety, in which a child's concern for others is primarily a projection of his anxiety about himself. There is conventional morality, or duty, in which manifestly sympathetic behavior is used for the ulterior purpose of being acceptable to adults and gaining their approval. There is even masked aggression, in which a child uses sympathy to express aggression in a manner which will not be socially disapproved. What is remarkable about Murphy's categories is the variety of meanings that the same behavior can express at this early age and how these meanings parallel those observable in adults. Murphy's study also underscores a general point about the complexity of personality: manifestly similar behaviors can have different meanings when underlying motivations are taken into account. Sympathetic behavior may be an expression of concern for others, but it also may have its roots in self-centeredness and fear.[5]

The Role of Action

There are people who, like Alice in Wonderland, give themselves very good advice but are seldom able to follow it. And there are those who genuinely feel guilty over the transgressions they continue to commit. The mature conscience must be consummated in moral behavior after thinking and feeling have done their work. It does not follow, however, that one can start with the act and assume the work of thinking and feeling has been done. As we have seen in our examination of confession and sympathy, the same behavior can arise from diverse motives and serve a variety of functions. Behavior alone cannot be regarded as the sole criterion of conscience development any more than can moral judgment or feeling taken by themselves.

The classical studies of moral behavior were done by Hartshorne and May (1928-1930) over forty years ago, and their findings are still sobering as well as enlightening. In one series of studies the abilities of school-aged children were first established under supervised conditions in which there was no opportunity to be dishonest. Subsequently, the children were put in a variety of situations where they could cheat; e.g., they were given a chance to correct their own examination papers or to take a

[5]For a review of the general topic of altruism, see Midlarsky (1968). For an analysis of empathy, sympathy, and altruism from a social learning theory point of view, see Aronfreed (1968).

test without supervision, or they were given opportunities to cheat in athletic contests or on homework. The discrepancy between supervised and unsupervised performance was taken as evidence for cheating. A roughly analogous situation would occur if the government announced one year that income tax returns would not be audited and then proceeded to determine how many people paid significantly less than in the previous year. Hartshorne and May showed similar ingenuity in studying lying and stealing. After evaluating all of their procedures, they found no evidence of consistency of behavior from one situation to another; e.g., a child who was honest in one situation might or might not be honest in another. Even lying, cheating, and stealing themselves were only loosely related, so that the child who did one might or might not do the others. The authors concluded that morality and specific components of morality are not all-or-none kinds of traits. Children cannot be divided into the "honest" and the "dishonest," nor can they be ordered on a continuum from high to low on a given trait. On the contrary , whether a child behaves honestly or not depends upon a number of specific factors of which the most important is the behavior of other members of the group. Cheating or not cheating followed class patterns, not individual patterns, since the individual tended to behave like his classmates. Honesty is also affected by what is to be gained by cheating. For example, cheating was greater on tests which counted toward a monthly grade than on practice tests. The likelihood of detection is a third factor; some children were unwilling to risk detection and so were more cautious than honest. Honesty was also found to be a function of general personality and social factors. It declined as intelligence, harmony in the home, and socioeconomic status declined.

Hartshorne and May concluded that the child has no autonomous need to be honest or considerate or self-controlled, nor do such values prescribe how he should behave. Instead of morality, we should think of moral behavior, which is a specific way of responding to a specific situation. If one wants to produce an honest child, he must teach the child to behave honestly in a variety of situations by making honest behavior more rewarding than dishonest behavior. Groups, as well as individuals, should be rewarded for behaving in conformity to moral values because of the group's potency in determining what the individual will do.

A subsequent re-analysis of Hartshorne and May's data (Burton, 1963) modified the original findings. Children's behavior is not entirely the result of situational factors; instead there is a degree of generality to moral behavior which allows us to regard one child as having more character than another. While this re-evaluation moderates Hartshorne and May's conclusions, it does not negate them. We still can expect moral behavior to be heavily influenced by nonmoral factors in school-aged children.

There is much in Hartshorne and May which concurs with points we have made in other contexts. We have seen how situational factors can support or undermine self-control (see Chapter 4) and we shall discuss the powerful influence of peer groups (see Chapter 8). Murphy's study of sympathy uncovered situational factors, as well as individual and general developmental ones. Our concept of the polyglot conscience involves the idea that moral values (Be honest) must constantly contend with nonmoral ones (Be successful, be popular). Yet, Hartshorne and May's extreme conclusion of situational specificity is not consistent with our picture of the school-aged child gradually achieving greater autonomy (see Chapter 4) and being increasingly able to use general principles as guides to behavior. Their emphasis on behavior bypasses the cognitive and affective variables we regard as essential to morality. We agree that moral behavior should be made rewarding to the child and to groups of children, since there is no reason to expect children to be good "for goodness' sake." (To be satisfied only with the rewards of one's conscience might prove too great a strain for many adults.) However, we would add that the integration of moral understanding and feeling—so that the child's grasp of moral principles is reinforced by the potent forces of self-satisfaction and guilt—is the essential ingredient in producing moral behavior. As we shall soon see, this integration may be fostered by the incongruity between moral and nonmoral values, between individual standards and social pressures, which Hartshorne and May delineate.

The Integration of Conscience

As we noted, thinking, feeling, and acting can be discussed separately because they develop somewhat independently. The preschooler has only fragments of a conscience. He can espouse parental values one minute and contradictory peer values the next with no sense of incongruity; he would not dare to touch a good vase, but he may be unconcerned over his merciless teasing of a peer; he may sympathetically hug a crying child one minute and push him down angrily the next. These fragments of a conscience must not only develop, they must coalesce as well. In the following discussions we shall raise two questions: At what point in development may we say that the child's conscience has been formed? and, Under what conditions do thinking, feeling, and acting become integrated?

The Controversy Concerning When Conscience Is Formed

Our first question concerns timing more than integration, but integration is implied in the literature we shall cite. The term "moral character" is often used, and the literature is addressed to the question of when the child becomes as moral—or amoral or immoral—as he is likely to be as an adult.

Theory and empirical evidence are not in agreement as to when the child's basic value system becomes relatively fixed. Freud postulated that the superego was established in the first six years of life, but he hinted at subsequent identifications which would modify the ego ideal and produce a reasonable and humane guide to conduct. Recently, certain psychoanalysts have become interested in exploring the fate of conscience during middle childhood and adolescence. While agreeing that the primitive superego is formed by age six, they claim that conscience continues to develop until maturity. As we have seen, Piaget and Kohlberg maintain that moral judgment changes progressively up to at least age sixteen. Thus, experts on the affective and cognitive facets of conscience come to the similar conclusion that there is a gradual development from the preschool years through adolescence.

By contrast, empirical evidence suggests that character has been set by age ten. Peck and Havighurst (1960), using general personality traits, such as rationality, self-control, and friendliness and global ratings of behavior by peers and teachers, found that the child's moral character was consistent between the ages of ten and sixteen. At one extreme, the amoral ten-year-old tended to become an amoral adolescent, and at the other extreme the rational, altruistic ten-year-old became the rational, altruistic adolescent. Sears, Maccoby, and Levin (1957) used different techniques and criteria and came to a similar conclusion. They regard the period between six and ten as critical in determining the individual's conscience.[6]

We can only speculate as to the reasons for these marked discrepancies. Perhaps they are due to experimental artifacts. When comparing children of different ages, one encounters the problem of relative versus absolute change. Children may maintain their relative standings with agemates as they grow older and yet change significantly in absolute terms. In the Peck and Havighurst study, we do not know whether absolute changes were taking place in moral understanding or emotional maturity while the children maintained their relative standings in the group. Another difficulty in evaluating the discrepancies arises from the use of different criteria of moral character. Peck and Havighurst were concerned primarily with behavioral manifestations of character, Piaget and Kohlberg with moral judgments, and Freud with inner experiences and feelings. Each approach is legitimate, but none contains any means of comparison with the others. We do not know what Peck and Havighurt's group would have done had they been given Kohlberg's tests of moral judgment; Kohlberg does not venture far into the realm of behavioral manifestations of character; and the psychoanalysts are only sporadically interested in overt behavior and pure cognition. The discrepancies in the evidence on the course of moral development can be resolved only by evaluating the same children in terms of multiple criteria of moral character.

Objective Studies of Integration

There are two approaches to understanding the integration of conscience. The objective approach involves determining the relations among the three facets of conscience—thinking, feeling, and acting—measured independently. The psychoanalytic approach involves a theoretical integration based on clinical data. We shall present both in order to gain the widest perspective on this crucial problem.

Objective studies generally have found modest rather than striking correlations among cognitive, affective, and behavioral variables (Kohlberg, 1964, 1965; Hoffman, 1964). Basic moral knowledge, in the sense of knowing the conventions of society,

[6] There is an historical irony here. Sears is an outstanding proponent of learning theory, which traditionally has championed the modifiability of human behavior. Classical Freudians have maintained that the basic features of personality are determined by the sixth year. Just as some psychoanalysts are beginning to explore the modification of conscience in middle childhood and adolescence, a learning theorist sounds more like a Freudian advocating early determinism.

is related to moral conduct as measured by resistance to temptation. If the child does not know right from wrong, or if he has been reared in a subculture which espouses immoral values, he cannot be expected to behave in a moral manner. However, expressed strength of belief in moral conventions is not related to behavior; the child who says he is strongly opposed to cheating is no more or less likely to cheat than one who says he is only moderately opposed to cheating. Finally, moral judgment as measured by Piaget's or Kohlberg's criteria is related to moral behavior. Thus, it is not how strongly a child says he believes, but how well he understands, which affects the level of morality of his behavior.

The role of guilt in producing moral behavior is somewhat ambiguous. Fear of being caught, which might be considered the absence of guilt, is ineffective in helping the child resist temptations and delinquent acts. Confession and restitution sometimes is, sometimes is not effective in protecting the child from temptation. Guilt, in the sense of a generalized distress reaction after a transgression, is not related to moral behavior. Only as guilt involves explicit self-criticism and remorse does it insure moral conformity. The child must consciously and painfully pass a judgment upon himself. As we have seen, such a judgment is rare before the ages of twelve or thirteen. In sum, it is not the "feeling bad" but the painful awareness of self, of standards of behavior, and of discrepancies between the two which holds the key to mature moral behavior.

The Process of Integration between Middle Childhood and Adulthood

While objective findings are descriptively interesting, they do not tell us how integration takes place. We must turn to the clinical literature for an account of the process of integration and, more specifically, to Jacobson's (1964) neo-Freudian psychoanalytic theory.

Jacobson assigns the ego a central role in conscience development. The ego serves many functions, but we shall discuss only those which are relevant to the topic at hand. First, the ego is responsible for learning the nature of reality and for providing realistic guides to behavior. The ego receives information about the environment, integrates the information, stores it in memory, and utilizes it in planning future actions. The psychoanalytic term for these functions is reality test. Next, the ego is a vehicle for identification, which is one of the chief mechanisms for the internationalization of values. Third, the ego has its own values, such as physical health, monetary success, prestige. We would classify these as nonmoral values. Jacobson claims that ego values are as important as moral values in a person's developmental history. Fourth, the ego performs an integrating function. It helps the child to satisfy demands for social conformity while preserving his autonomy and his right to enjoy personal pleasures. It protects him from the extremes of moralism, self-indulgent pleasure seeking, and infantile willfulness. In doing so, it operates on the principle of expediency, utilizing any maneuver which will reconcile or minimize friction between competing internal and external demands. Finally, the ego is the locus of self-esteem. Jacobson emphasizes the importance of love in producing self-esteem. The well-loved infant feels good all over, enjoying the maximum amount of physiological and psychological pleasure.[7] From this rudimentary notion of a "good me" will come the ego ideal, which, as we shall see, will assure the child of his own goodness, counterbalance the force of the superego, and, ultimately, serve as a mature moral guide.

Recall that the six-year-old child, according to psychoanalytic theory, has a superego which is absolute in its demands for perfection and an ego ideal which revolves around exaggerated notions of happiness, power, and glamor. His entire moral apparatus, therefore, needs to be humanized and attuned to the realities of human nature as well as to the realities of getting along in life.

During latency, or what we have called middle childhood, the child's family-centered values are supplemented by those of peers and new adult authority figures, such as teachers and group leaders. Since the child has little autonomy, new values conflict with old. Jacobson, like Kohlberg, regards the child's effort to integrate discrepant values as an invigorating experience which serves to change the orientation of his conscience from family to society. However, integration is more than a purely cognitive task. Values will lack potency unless they are conveyed through an emotionally significant relationship with a peer or adult. Jacobson also points out that the child can be baffled and distressed rather than stimulated by new values, especially if he is insecure in his interpersonal relations and has deep uncertainties about himself.

Jacobson lists three aspects of personality which must be integrated between middle childhood and adulthood. First, there are moral values, such as honesty and justice. Then there are ego values, or what we have called nonmoral values, such as physical health, sexual, vocational, and monetary success, personal and social prestige. Finally, there is the child's understanding of his capabilities and potentialities. For instance, a boy may identify with fair play (a moral value applied to sports) and winning (which is a nonmoral value) while he is discovering how good an athlete he is. The integration of moral and ego values—which is the key to moral growth—is not easily achieved. Fair play and success, equality and excellence can be harmonized, but they do not enjoy a natural fellowship.

Jacobson makes the interesting point that ego values are vulnerable to corruption, just as superego values can be rigidly perfectionistic. Prudent self-interest can become merely a front for aggressive and selfish motives; healthy competition can deteriorate into cutthroat rivalry; a desire to succeed can turn into a callous use of power. A good conscience requires an uncorrupted set of ego values as much as a humane set of moral values. She also notes that ego as well as superego values can produce painful affects. A successful but ruthless man may eventually be overcome by guilt when his moral values reassert themselves. This is the typical picture of the triumph of conscience. But a highly ethical man who achieves only modest success may reproach himself for being a weakling and avoiding competitive struggles. In this case ego values

[7] Recall that Heckhausen traces self-esteem to initiative and the achievement of self-imposed standards of performance. However, the two derivations are not incompatible, since self-esteem may have multiple roots.

are the cause of pain; they punish the man for being "too good" to enjoy the successes which are his due.

The integration of ego values and moral values is not possible without the changes which occur during the adolescent period. The adolescent is in the process of relinquishing emotional ties with his family and the values appropriate to being a child. He is searching for new modes of relating to other persons based on sexual intimacy. This search entails a major shift in values, since being good in childhood meant being asexual. Finally, the adolescent must forge an image of himself as an autonomous adult capable of making his own decisions and assuming adult responsibilities. These developments necessitate a concomitant change in conscience, from asexual goodness, dependence, and conformity, to autonomy, sexual liberation, and social responsiveness.

Jacobson describes the characteristics of the adolescent period which contribute to the growth of conscience. The early adolescent's personality is more disharmonious than it was during latency. At times his ego (or self-control) and superego are no match for the flood of sexual excitement demanding release. His ego may side with his impulse life, and he may glory in his sexual prowess, defy his parents and social conventions, and revel in the sudden achievement of adult status. He has arrived! Yet, the very libertine flavor of his sexual fantasies, the grandiosity of his self-esteem, and the irresponsibility of his defiance of convention betray his vulnerability. His superego reasserts itself and punishes him with pangs of guilt and shame, while his ego feels inadequate to assume the responsibilities of adulthood. He may then try to re-establish the parental relations he so recently defied while salvaging a modicum of self-esteem in the process. His ego ideal shows similar fluctuations between extremes. At times he glorifies the libertine and the rebel, at times the pure-hearted and the saintly.

The adolescent's temporary participation in delinquent and promiscuous activities and his experiments with new ego ideals can be a potent stimulant to growth. Just as the healthy latency child is broadened by contacts with values outside the family, the healthy adolescent now encounters those aspects of adult life which are callous and undisciplined. Such experiences introduce him to the "worldly" side of adulthood. They help him to realize that morality is more than prescriptions for good behavior.

The integration of worldly experiences is aided by another feature of adolescence which is a concern with philosophical and ethical problems. Initially this preoccupation has the same exaggerated quality as his attraction to glamorous ego ideals–he must solve the riddle of the universe, find the ultimate meaning of life. He champions causes, confronts parents and society with the irrefutable rightness of youthful idealism. His inexperience and the pressure to achieve instant maturity are responsible for the excessive and sometimes fickle nature of his ideals. Yet the underlying need is deep and abiding–the need to find a set of values which will constitute a serviceable guide through the problems of adult life.

The adolescent's rebelliousness is the final factor contributing to conscience development. Some adolescents seem to revert to the negativism of the two-year-old. Any authority, no matter how sweet and reasonable, is violently rejected. Parents are dismissed with sweeping condemnations and rules are regarded as personal affronts. However, this rebelliousness serves to counteract the conformity of middle childhood, the reliance on adult models, and, most important, the values which were incorporated uncritically through earlier identifications. It paves the way for the autonomous conscience of the adult.

A mature conscience depends upon moderating the exaggerated worldliness, idealism, and independence of early adolescence. Grandiose ideas of sexual freedom, the glorification of rebellion and callous power, the glamorizing of adult vocations must give way to a view in which sexual freedom and rebellion are part of a larger pattern of adult love and social concern, in which vocational fulfillment results from coordinating abilities with values concerning the worth of an activity, and in which the independence of early adolescence becomes a flexible autonomy, while stringent idealism is translated into workable principles of moral behavior. The development of such a mature conscience depends upon the strength of the ego. Jacobson emphasizes the ability to master aggressive and sexual impulses rather than being overwhelmed by or alienated from them; the ability to form deep, abiding attachments; the ability to participate in social, political, cultural, religious, and athletic groups; the capacity for assimilating new experiences into existing value systems; and the maintenance of a positive sense of worth. True to her Freudian heritage, she stresses maturational factors in ego development rather than specific environmental factors which might enhance or impede ego growth.

The ultimate harmonizing of ego and superego values produces a strengthened ego ideal. The individual does what he values doing, he does what he considers right and good, and these two sets of values tend to complement one another. The mature ego ideal thus supplies both self-esteem and moral self-satisfaction. Jacobson's developmental reconstruction also helps us understand the polyglot conscience, since there is no guarantee that the transformations necessary to produce a mature conscience will be made. Idiosyncratic and family values in early childhood may carry such a high charge of loveworthiness as permanently to produce self-satisfaction and guilt. They may continue to be obeyed blindly or may be stoutly defended as the child becomes more intelligent and articulate. The child may not be exposed to the diversity of values which would stimulate him to evaluate and re-evaluate his present system; or, if exposed, he may lack the emotional security necessary to relinquishing an accepted standard. Finally, the integration of ego and superego might be incomplete, so that moral values serve the self-interest and expediency of the ego. The adult conscience, like adult thinking (see Chapter 6) and adult sexuality (see Chapter 7), contains residuals from many stages of its developmental course.

Deviant Development. Jacobson both provides a model of normal conscience development and enlarges our understanding of deviant development. Deviation is usually thought of as the conscience being too severe or too indulgent, producing either a moralistic prude or a sexual or aggressive delinquent. (Delinquency is discussed in Chapter 7.) Yet adults can suffer from faulty ego ideals as well. They may have failed to outgrow the grandiose, egocentric models of childhood, the striving to be the most glamorous woman or the most irresistible lover or the most impeccable connoisseur of art or even the most immaculate of saints. Their self-centeredness warps their values and either corrupts their consciences so that they will do anything to maintain their image or subjects them to periods of agonizing shame and inferiority. In addition, failure to achieve a stable set of ego values prevents the normal development of conscience. Clinicians are familiar with adults who have "identity problems." They go from one role to another, with no sense of fulfillment. A woman may try being a homebody, a clubwoman, a mother, a mistress, an intellectual; she may be by turns promiscuous and prudish, heterosexual and homosexual, childishly dependent and manipulative—yet each new venture leaves her restless and dissatisfied. Such people have no psychological center. Their conscience is as fragmented and insubstantial as their sense of self. They may be subject to temporary waves of shame and remorse; they may become fearful of being caught or deeply suspicious of others; but neither the inhibiting nor the guiding function of conscience has any sustained and sustaining effect.

Evaluation. Jacobson's theory obviously is too abstract and complex to be readily tested by objective studies. However, objective studies point to a finding which is pertinent here: conscience development is related to a broad spectrum of variables in addition to affection, empathy, and the ability to understand the viewpoint of others (Kohlberg, 1964; Peck and Havighurst, 1960; Bobroff, 1960). It is related to intelligence and to foresight and anticipation. It is related to the various components of self-control—the ability to focus attention, persistence, the capacity to delay gratification, and the mature control of affect. It is related to the ability to evaluate oneself realistically, as well as to self-esteem. It is related to generally reasonable and consistent behavior. Like all major personality variables, conscience depends upon a network of interlocking variables. Whatever prevents the infant from forming an attachment, curbs his zestful exploration, hampers his self-reliance and judgment, weakens his self-control, or blocks his intellectual growth will adversely affect the shaping of a humanistic and serviceable conscience.

It is one thing to atomize conscience into its component variables and proceed to find correlates of these variables, such as parental attitudes or child rearing practices. The majority of the objective literature is of this nature. It is another thing to explore the interrelations among the variables and to order them within a sequential framework. Such an undertaking is rare, especially one which, like Jacobson's, attempts to do justice to the complexity of conscience. Her reconstruction is best regarded as an illuminating synthesis of a very fragmented area of investigation.

The Mature and Immature Conscience

Conscience has been called angel and devil; it can be either. When it first appears it is rigid, perfectionistic, punitive, authoritarian. It must become humane, flexible, and reasonable. What conditions are necessary to bring about the transition?

The Dangers of Rigidity and Valuelessness

Hemming (1957) sets the stage for our discussion by making a general point concerning conscience. A rigid, primitive conscience is not only a source of inner distress, as Freud claimed, it is also poorly adapted to the changing values of a changing society. The internalization model may possibly be adequate to explain how values are learned in static cultures or subcultures, but it includes no provisions for new adaptations later in life. The typical American must adopt a set of values which will change during his lifetime. Our country is becoming more urban and industrial; large, stable families are being replaced by small, mobile ones; the passing of political isolationism has made the nation more cosmopolitan, introducing a variety of new values to be assimilated—as if we must undergo as a nation the kind of confrontation with divergent viewpoints which Piaget regards as essential to the growth of moral judgment in the individual.

While overconformity is one impediment to adaptation, floundering about with no consistent value orientation is equally dangerous. The individual may be overwhelmed by diversity and fail to develop his own frame of reference. Moral growth involves the dual process of learning values and learning to evaluate values. Again, the internalization model neglects this second function.

The Humanistic-Flexible Versus the Conventional-Rigid Conscience

While we lack longitudinal data on the humanization of conscience from the preschool period to adulthood, we do have studies which concentrate on the early period and on adolescence. Hoffman (1964) is concerned with the development of what he calls a humanistic-flexible conscience. He maintains that there can be two quite different kinds of internally oriented consciences. The first is the conventional, rigid conscience which Freud described—the implacable censor of sexual and aggressive impulses, the enemy of reason and change, which rules the child by fear. But there is also the humanistic, flexible conscience which takes into account the interpersonal consequences of behavior, humanistic values such as trust, and extenuating circumstances. Hoffman argues that flexibility without standards can produce a shallow or easygoing moralism, while humanistic values can be rigidly applied by an authoritarian superego. Both standards and flexibility must be present.

Hoffman is critical of including both love and reasoning under the single rubric of love oriented discipline (see Chapter 4), since the two have distinct consequences. It is not love but reasoning, in the context of a relatively permissive atmosphere, which holds the key to developing a humanistic-flexible conscience. In fact, the fear induced either by power assertion or the threat of withdrawal of love prevents the child from grasping parental messages and sets the stage for uncomprehending obedience.

Socialization should proceed as follows. The attentive parent first introduces restrictions but always at a level the child can comprehend. In the process the parent also serves as a model for realistic and reasoning behavior. As the child develops, the parent adds a moral factor by explicitly pointing out the effects of the child's behavior on the feelings and well-being of others. He stresses the similarity of the child's own reactions to those of others and suggests constructive reparations to reduce guilt. In this way, the consequences of one's actions for others become the guide to moral behavior.

Hoffman explored the differences between the humanistic-flexible (h-f) conscience and the conventional-rigid (c-r) conscience in a study of seventh graders. The h-f children experienced guilt over the interpersonal consequences of transgression, whereas the c-r children felt guilty over having unacceptable impulses. The h-f children could entertain minor impulses to immoral acts, whereas the c-r group condemned all deviations from the "straight and narrow." While both groups upheld conventional morality, only the h-f children could modify their stands as the situation required. Finally, the h-f children were on good terms with their impulses, while the c-r children were more concerned with controlling all feelings, since feelings were viewed as alien and potentially disruptive.

The two kinds of conscience were associated with two different patterns of parental discipline. The parents of c-r children reacted to aggression toward them by withdrawing love and inducing guilt. The h-f parents focused on the issue leading up to the aggression and on reparations to be made. For parents of the c-r children, anger was bad in itself; for parents of h-f children, anger was bad because it was dysfunctional and disruptive. For example, the parent of a c-r child might send him to his room as punishment for an outburst of temper; the parent of an h-f child might send him to his room so that he can cool down sufficiently to discuss why he went into such a rage. The parent of the c-r child tended to make overall condemnations and his attitude typically was, "I can't love you; you are bad." The parent of the h-f child did not condemn as much as imply that the child could do better. Interestingly, the parents of the h-f child used power-assertive techniques more frequently than the parents of the c-r child, although they were not extreme in this respect. Apparently their benign and reasonable stance did not provide sufficient motivation without an occasional, "Do it now because I say so."

To summarize, Hoffman finds some evidence that withdrawal of love may produce self-control based on fear of being bad, fear of alienating the parents, and fear of expressing strong impulses, particularly hostile ones. Other-oriented discipline focuses on the causes and consequences of antisocial behavior and stresses reparations. Since there is no threat of loss of parental love, the child does not become alienated from his feelings. He also can identify with the constructive, interpersonal orientation of his parents.

The Rational-Altruistic Conscience and Four Deviant Patterns

Peck and Havighurst (1960), in an intensive, comprehensive study of seventeen-year-olds, distinguished five levels of character development. They were as follows, going from the lowest to the highest:

The Amoral type is hostile and immature emotionally. His self-control is weak. He reacts quickly and inappropriately with anger and negativism. He lacks any integrated system of internalized moral principles to guide him, and his guilt is ineffectual in controlling future behavior. He has little liking or regard for people and little concern with the consequences of his behavior. He readily gives vent to the feeling of the moment, retaliating irrationally against a harsh, confusing world.

The Expedient type is totally self-centered. His own pleasures, needs, and ambitions are uppermost in his mind. His relations to others and his moral behavior always have self-enhancement as an ulterior motive; e.g., he will be honest in order to maintain his reputation, but he can readily become dishonest if it suits his purpose. His "me first" orientation, although immature, at least gives him guidelines to behavior, which the amoral individual lacks. Thus, his life pattern may be more consistent and tolerable to society. He is kept in line by external rewards and punishments; when these are lacking, he quickly reverts to doing what he pleases.

The Conforming individual lives by the principle of doing what others say he should do. He accurately reflects the mores of his surroundings and at the same time caricatures them. He does all the right things without any understanding. He might be a regular churchgoer, but he might also be a Nazi herding Jews into the gas chamber. He feels shame rather than guilt because public exposure is more distressing than violation of inner principles. He is the end product of identification without reflection. His conscience is a series of parental *Don'ts*. He is depressed, dull, and unhappy, but unable to express resentment against the forces which made him so.

The Irrational-conscientious type is familiar. He epitomizes Freud's concept of the primitive superego and the conventional-rigid conscience described by Hoffman. He has a puritanical conscience which he follows to the letter, or else he pays the price of extreme guilt. He is loyal, kind, and honest, but out of a sense of righteousness rather than out of empathy and moral insight. In interpersonal relations, he is as demanding of others as he is of himself, and his rightous indignation only thinly veils the hostility he tries to deny. He illustrates the imbalance between an excessively strong superego and a weak ego.

The Rational-altruistic type is characterized by the combination of understanding and feeling which is the mark of a highly developed conscience. He has firmly established moral principles whose applicability to reality he continually re-evaluates. He has a high regard for others and is attuned to their feelings. He is self-reliant, constructive, and ethical in his behavior. He is spontaneous, natural, and free from irrational guilt. He is on good terms with his impulses and with society. Peck and Havighurst make the interesting point that, in observing adolescents of this type, one does not have the impression of a compartmentalized "ego" and "superego"—their consciences do not seem distinct from their general principles about getting along with others or from judicious self-interest. One is reminded also of Jacobson's (1964) point that in maturity the self and the conscience merge into an ego ideal which is at once practical and ethical.

Peck and Havighurst's empirical data revealed a number of personality factors which were related to these various character types. Not surprisingly, two factors become increasingly potent as one goes from the Amoral to the Rational-altruistic type: moral stability, or a tendency to comply with the conventional moral code willingly and with genuine pleasure, and superego strength, or the degree to which a person has an internalized set of moral principles which effectively guide his behavior. There was a slightly negative correlation between character type and hostility and guilt, suggesting that the Freudian type of superego is found in the lower character types, such as the irrational-conscientious, but not in the highest type. Complementing this finding was a positive correlation between character type and friendliness and spontaneity, indicating that a highly developed conscience produces not a dour, aloof individual but one who is on good terms with himself and with others. Finally, Peck and Havighurst found that character was related to a number of general personality variables which collectively they called ego strength: emotional maturity, internal consistency, rationality, accurate self-perception and assignment of moral responsibility, insight into others' motives, autonomy, and positive feelings toward peers.

Peck and Havighurst also found that character was shaped primarily by the parent-child relationship. They explored four aspects of family intereaction: mutual trust and approval, consistency of family life, democratic control, and severity of parental punishment. Each character type was related to a particular kind of family interaction.

Family relations of the Amoral type adolescents were inconsistent and rejecting. Some parents were severely punitive, others rather laissez-faire in their discipline. Some Expedient type adolescents had parents who were lenient and approving but in an indiscriminate, unthinking, inconsistent way. Others had parents who were highly consistent, severe, autocratic, and mistrustful. Thus, love without consistency and consistency without love produced similar types of character. The Conforming types had consistent, autocratic, severely punitive parents. Some experienced love and trust, others did not. However, the adamant authoritarianism of the family produced conventionalized conformity in the child. The family pattern of Irrational-conscientious subjects was somewhat ambiguous. The main theme was a high consistency coupled with severity of punishment. There was also a fair amount of variability in the dimensions of trust and affection, and autocratic versus democratic control.

The family pattern of the Rational-altruistic adolescents interest us most. It was characterized by consistency, trust, love, democracy, and leniency in punishment. Love, approval, and the absence of punitiveness suggest the affection which psychoanalysts claim is essential to socialization; democracy involves the opportunity to make decisions and to trade ideas freely with parents and peers, which Piaget and Kohlberg recommend as stimulants to cognitive development and Hemming regards as the basis of moral flexibility. These families were child-centered but not overly permissive. They had definite standards and reserved the right to make final decisions when necessary. At the same time they encouraged the child to make an increasing number of decisions for himself as his developing judgment allowed.

Peck and Havighurst summarize their findings as follows. Hostile, disorganized, uncaring parents produce a child who is undisciplined, hostile, and rejecting of others. Parental indulgence, in the sense of an uncritical permissiveness, produces a self-centered child who may temporarily conform in order to gain approval but whose self-discipline and restraint are too weak to generate dependable moral behavior. Autocratic parents produce children who are tightly bound to a literal interpretation of rules and regulations; tragically, they rebel in adolescence only because they want to become authoritarians themselves. Love, combined with consistency and reasonable standards, produce an adolescent capable of mature concern and affection for others.

One note of caution: One of the four Rational-altruistic type adolescents came from a family which was not particularly warm and consistent and which offered him little moral guidance. Thus, it lacked the positive features of other families of this group, although it did not have the negative characteristics typical of families producing other character types. We should also add that the study, intensive though it was, involved small numbers of subjects and should be regarded as furnishing leads rather than definitive evidence about the relation of family patterns to character.

Evaluation

Looking back over the studies of the mature and immature conscience, we see no glaring inconsistencies in theory or data. On the child's part, love, empathy, interest in others, an active curiosity about and involvement in social relations, experience in making decisions and taking responsibility for the consequences, a generally high level of self-esteem, and a sense of being on good terms with his impulses, are all important to the development of a mature conscience. As for the parent's part, affection, nonpunitiveness, and integrity, coupled with explicit help in understanding consequences, defining values, and promoting autonomy, provide the proper family setting for maturity.

This consistency is based on speculative reconstructions and meager data, and psychologists probably have only begun to give the issue of the mature versus the immature conscience the attention it deserves. However, it is encouraging to find evidence that socialization—that endless succession of *No, Don't, Wait, Good boy, Bad boy*—can be assimilated and utilized constructively and, even more important, congruently. Freud thought otherwise. So much of socialization goes against the grain: the compromise of autonomy, the moderation of willfulness, the control of hostility and sexuality, the incorporation of values which have no intrinsic meaning. Freud envisaged civilized man as divided against himself, forever harboring a demon who is unalterably determined to have his own amoral way and indulge his narcissistic pleasures (see Chapter 4). Certainly development can go that way, but there is evidence that it does not inevitably do so. Murphy (1937) and Hoffman (1964) have shown us preschoolers who are well socialized but at the same time spontaneous and well adjusted. Peck and Havighurst (1960) offer a similar picture of adolescents. All of which would not surprise Adler who maintained that man's basic nature is social not selfish. He would argue that we have every reason to expect personal fulfillment through learning to live with others. The only question is how to go about it.[8]

[8] Our final resume of conscience development involves material from subsequent chapters. However, it could be read at this point. See Chapter 11 on "The Mature Moral Act."

6 Cognition, Reality Contact, and Cognitive Styles

We have met thinking in our discussion of the development of self-control (see Chapter 4). Ideas serve the dual function of delaying and guiding action. As the infant-toddler becomes increasingly capable of thinking, he is increasingly able to delay his immediate reactions to stimuli; and, as his ideas reflect a greater understanding of his surroundings, he is better able to adjust his behavior to the demands of reality. We have also seen (Chapter 5) that both the toddler and the school-age child are incapable of grasping all the complexities and subtleties of reality. A child's thinking must undergo many transformations before it assumes mature form. These progressive changes and their effects upon personality are the subjects of our present inquiry.

Piaget will be our principal guide to the transformations which thinking undergoes as the child learns to understand his environment. Piaget is basically concerned (1) with the epistomological problem of how we come to know the world and (2) with tracing the developmental course from unrealistic to realistic thinking. He provides the richest account of the child's orderly progress toward rationality undisturbed by intense feeling (see Piaget 1967; Flavell, 1963). However, the young child's understanding is distorted not only by immaturity but also by emotions. No one knew better than Freud the ways in which strong affects of love, hate, fear, and guilt can force reason to serve their unreasonable ends. He was endlessly fascinated by psychopathological behavior, dreams, and the everyday irrationalities to be found in normal adults. Therefore, Freud (1953, 1963) will be our guide to understanding the nature of emotionally determined irrational thinking. By juxtaposing Piaget and Freud we shall be able to see the part played by thinking and by feeling in the development of understanding.[1]

[1] While Freud and Piaget had complementary interests in cognitive development, their theories are not readily reconciled (see Wolff, 1960).

Two General Features of Cognitive Development

Early cognitive development is difficult to present because we have lost touch with it completely. Emotions, by contrast, have a timeless quality. We can empathize with the infant's joys and distresses, his bewilderments and triumphs. But how can we appreciate the infant's struggle to understand an environment which is so obvious to us? Obviously we live in a world of permanent objects which occupy space; obviously a ball which rolls under one end of a couch will roll out the other; obviously a rabbit has two ears. All the infant has to do is look and see for himself! Actually, nothing—not even the most uncontested features of the physical world—is apparent initially. We wonder why the infant cannot immediately comprehend these uncontested features only because we have learned them so thoroughly.

In trying to appreciate early thinking, it is helpful to shift our perspective to the infant's point of view. Remember that the neonate knows nothing about the world he has justed entered. For the sake of convenience we can distinguish three realms which he must come to know: the inanimate world, the human world, and the self. The infant must learn a prodigious amount about these three realms. He must accumulate a staggering number of facts—water splashes when he hits it but stones do not; a soap bubble readily pops; an arm can be used to reach a toy; a bump hurts, a tickle pleases; people who disappear from sight will reappear—and on and on. He must learn categories and principles: certain kinds of objects are toys, certain kinds of animals are dogs; round objects can roll, blocks cannot; the speed of his tricycle is proportionate to his effort in pedaling—and on and on. Even this partial listing will help us appreciate the neonate-toddler's capacity for learning and the scope of his early intellectual achievements.

There are two general principles of cognitive development which should be emphasized. First, cognitive development does not proceed by the accumulation of increasingly large pieces of correct information about reality until, finally, a comprehensive understanding is achieved. The cognitive fog of infancy does not gradually lift to reveal now a rattle, now a face, now a room. On the contrary, understanding begins with primitive, erroneous ideas which are revised as the infant explores and assimilates his environment; but even the revisions may be in error. An accurate grasp of reality is achieved only after a series of revisions. Nor is it preordained that the child will come to know reality for what it is. Everyday life, as well as the clinic, reveals countless examples of immature thinking which has been carried over into adulthood.

The infant-toddler's ideas concerning reality are often fantastic and magical by adult standards. Again, our adult frame of reference misleads us and we fail to see that the child's thinking may be appropriate to his level of development and may represent an advance in understanding over the previous level. The growing child is constantly utilizing the information at his disposal to enlarge, enrich, or to modify his current ideas. At times his understanding may be faulty because his information is limited. To illustrate: until an infant has repeatedly encountered his crib, his toys, or his mother, he has no basis for inferring their permanent existence. Initially, all the information he has is: X was present and now X is not present. If this is all he experiences, how can he know that X continues to exist somewhere outside of his experience (see p. 186)? At times his understanding is faulty because he is not capable of engaging in certain kinds of thinking; e.g., abstract ideas, hypothetical propositions, or syllogistic logic are unknown in early childhood. However, if we accept the child at his own developmental level, taking into account what he knows and how he is capable of thinking, his fantastic ideas become reasonable. We have already seen examples of thinking which is magical by adult standards, but reasonable when put into its proper developmental context (see Chapter 3). The infant whose cries are consistently responded to by his attentive mother draws the obvious (to him) conclusion that his cry in and of itself makes his mother appear; the idea that she has an existence and a volition of her own and chooses to answer his cries lies wholly beyond his ken.

The second general principle of cognitive development is more controversial, but we shall accept it as valid in our discussion. Piaget maintains that the successive modifications in thinking occur in an orderly, fixed sequence, called stages of cognitive development. Each stage is qualitatively distinct and no higher type of thinking can evolve until the child has gone through all the preceding stages. Schematically, the child must have mastered stages a, b, and c if he is to think at stage d. The schedule can be speeded up or slowed down, but the order is always the same.

Early Cognitive Development

The Development of Basic Categories of Thinking

Let us now reconstruct the experience of the neonate as best we can. In all probability he experiences a succession of stimuli differing in quality, intensity, duration, and affective tone. The stimuli are unrelated and have no meaning. Only after repeated encounters with objective reality will he begin to structure his experience in terms of the basic categories of thinking.

The neonate must learn to structure his experiences in terms of the source of stimuli. Initially he cannot distinguish among stimuli which come from his empty stomach, from the blanket he grasps, from the mother's nipple in his mouth or from the mother's face. He must learn which stimuli emanate from his body, which belong to the physical world, and which to other human beings. In terms of the shorthand we shall use, he must distinguish "me" from "not-me," and, among the "not-me" stimuli, he must differentiate those belonging to things from those belonging to people. The infant must also learn to structure his experience in terms of time and space. He must grasp the fact that the sequence of stimuli he experiences is often orderly, and from these orderly successions he must evolve the idea of "before" and "after." Later he will understand that he himself has a past and a future and, still later, he will master the formal measurements of time. Concomitantly, the infant is learning to think in terms of space. From his fascination with moving objects he evolves the idea of space as a homogeneous medium through which all objects move; and, by experimenting with objects, he discovers the spatial relations among them, such as "on top of," "in front of," "inside of." The neonate must structure his thinking in terms of the distinguishing properties of objects or what we shall call object characteristics. Only then will he be able to separate the possible from the impossible, the realistic from the absurd. Is it realistic for the infant to expect that his rattle will give him milk and his mother will make an interesting sound if shaken? If not, then why not? What must he have had to learn in order to realize the absurdity? Finally, the neonate must grasp the concept of causality. We have already seen how the infant tends to equate temporal succession with causality (see Chapter 3) and the erroneous (magical) consequences of this equation. The growing child will have to correct a number of other erroneous ideas before achieving a realistic understanding of causal relations within and among the three realms of self, other people, and physical objects. While there are other categories of thinking the neonate must develop, these will be the ones which will concern us.

Our subsequent discussions will be organized in the following manner: First we shall give a more detailed developmental picture of the categories of time and of object characteristics. Then we shall examine Werner's theoretical account of the process of structuring experience. Next, we shall discuss the categories of source of stimuli and causality, which will serve to introduce us to a number of Piaget's theoretical concepts concerning cognitive development. Finally, we shall compare and contrast Piaget's image of the inquisitive child learning the nature of reality with Freud's concept of "reality testing" in which affect plays a more prominent role.[2]

The Concept of Time

Freud said that there is "no time dimension in the unconscious," by which he meant that primitive thinking is not ordered in terms of past, present, and future. While we described the neonate as "experiencing a succession of stimuli," the concept of succession does not exist for him. Following Freud's lead, we may look to dreams in adult life for an approximation of the timelessness of infantile experiences. In dreams, past, present, and future intermingle. A person long dead can appear in a contemporary setting or the adult dreamer can be a child talking to a Roman emperor. The dreamer does not protest the anachronism; the time sequence is not only distorted, but its fragmented nature also goes unnoticed.

We can only speculate about the development of the concept of time in the first year of life. The uniform relations among certain physiological processes, the stable characteristics of objects, the attentiveness of his caretaker, insure that the neonate-infant will experience regular successions of events. With repeated experience, the regularity begins to register: sucking milk—relief from distress; shaking rattle—interesting noise; cry—mother appears. This grasp of connections between events may constitute the first primitive notion of succession. Subsequently the appearance of one event leads to the expectation that the other will appear. As we have noted, the

[2] For more general and comprehensive reviews of early cognitive development, see Elkind (1967) and Stone and Church (1968).

sight of the bottle can quiet the crying infant. From this we infer that the bottle becomes a cue signaling the satisfaction of hunger which lies ahead. Such expectancies may constitute the first primitive notion of the future. With the appearance of expectancies, the infant's world is no longer limited to the experience of the moment. We can be less speculative concerning the development of the concept of time when the infant is one year old. If a toy is placed under a pillow, he can readily lift the pillow in order to obtain it. Here the ordering of a simple sequence of events has been learned (see toy disappear–lift pillow–get toy) and can be utilized as needed. Or again, the one-year-old can remember that he placed a toy behind him a few minutes ago and can turn around to obtain it. His ability to recall furnishes him with the rudimentary notion of past and present. In sum: the learning of sequences of events, the establishment of expectancies, and the ability to recall, begin to order the infant's experience into past, present, and future.

Now let us consider a bright, highly verbal, three-year-old. Up to now he has never consciously dipped back into his reservoir of experiences and revived them in the present. Suddenly he realizes he has such power–the concept of "remembering" has registered–and he begins to exercise his newfound ability. His standard introduction is "Remember when...," and his standard conclusion is "a long time ago." "Remember when I got a haircut a long time ago?" "Remember when we saw that fish in the water a long time ago?" "Remember when I ate cereal with raisins a long time ago?" The event may have happened an hour before, the day before, or a month before. While the past is now distinct from the present, the past is still homogeneous. This was Freud's point about primitive thinking: nothing is "more past" than anything else. Not until middle childhood is the child at home with the major ways of ordering events: personal time, by which he orders the events of his day and his life; impersonal time which is ordered in minutes, hours, days, and years; historical time, which orders the events of mankind's past.[3]

The Concept of Object Characteristics

The infant-toddler has yet to grasp the defining characteristics of animate and inanimate objects. Since he knows very little of reality, he does not know what to expect–anything can happen. There is no way for an infant to know that dogs do not walk on the ceiling. Only after he has observed dogs long enough to develop an idea of their characteristic behavior can he be surprised by a departure from their habitual pattern. There is nothing to tell an infant that a soap bubble will burst easily. Only after observing soap bubbles long enough to grasp their nature would he be puzzled to find a strong man unable to break one with a hammer. Let us look at the phenomenon from another angle. Why are we rarely surprised in dreams where the most fantastic things happen? Consider a typical dream: "I was in the living room, but it was really a swimming pool, and this big pink seal came over and began rubbing his wet

[3] For more complete discussion of the concept of time, see Flavell (1963), Werner (1957), and A. Freud (1965).

nose on my leg and saying some strange words in French." The events are accepted as being as natural as eating breakfast in the morning. Why? Because, in a dream objects lose their defining characteristics, and therefore expectancies are never violated. The same is true of early childhood thinking. Congruity must be established before incongruity can be noticed. As long as the child's grasp of the distinguishing properties and behaviors of animate and inanimate objects is incomplete, he lacks criteria for evaluating the probability of an event's occurring. If a living room and a swimming pool are still "large square things with people in them," then one can readily become the other, as in the dream.

We do not mean that all thinking in the first three or four years of life is alogical. The preschooler has acquired a remarkable knowledge of the world and can readily detect absurdities when he is on familiar grounds. However, there is so much that is unfamiliar, and there are so many stages of thinking still lying ahead of him, that the proportion of alogical to logical ideation is greater than it ever will be again. On the other hand, no matter how rational he is as an adult, his thinking will bear traces of its primitive origins–inevitably in his dreams and probably also in times of fatigue, illness, or emotional turmoil, and when he is under the influence of certain intoxicants and drugs.

Werner's Theory of Cognitive Development

We are now sufficiently familiar with the descriptive properties of early cognitive growth to present Werner's conceptualization of the changes which take place (see Werner, 1957; Baldwin, 1967). Werner regards all development as proceeding from a relatively undifferentiated state to a state of increasing differentiation and hierarchic integration. Utilizing this terminology, we can state that the neonate's experience is undifferentiated as to source of stimuli, time, space, object characteristics, and causation. Cognitive growth results from the dual process of differentiation and integration of this undifferentiated stream of experience.

Undifferentiated Thinking. Werner uses the term syncretic thinking when several ideas or concepts are merged which subsequently will become discrete in mature thinking. The inability to differentiate "me" from "not-me" and the inability to order events temporally are examples of syncretic ideation. Dreams are filled with such merging of discrete ideas; e.g., "I was in the living room, but it was really a swimming pool." In fact, Freud, through his study of dreams, arrived at a concept similar to Werner's, which he called condensation. In condensation, a single idea or image can represent several ideas or images. A young man in psychoanalysis dreamed of a middle-aged woman called "the Baroness." The title symbolized to him a haughty, commanding woman. It was also a pun on the word barren-est, which gave her the attribute of emotional and sexual sterility. Thus, the single dream image condensed three qualities that he imputed to the woman, who, not unexpectedly, represented his mother. In play, a child has no trouble being both horse and rider simultaneously, again condensing two discrete functions into one.

Another example of syncretic thinking is what Werner calls physiognomic perception. As we have seen, the infant-toddler does not consistently differentiate himself from physical objects. As he explores the world of things, his motor behavior and emotional reactions may be perceived as characteristics of the objects explored. A two-year-old picks up a box and shakes it with pleasure. He does not perceive the box as being completely "out there," while movement and affect are completely part of himself. Instead, the elements of the experience merge so that "effort–pleasure–box contour and color" are perceived as a unit. The world of things is not a dead, impersonal one at this stage; it is saturated with subjective factors. Physiognomic perception is the perception of animate qualities in inanimate objects. It differs from the fantasy play of the four-year-old or the conscious "let's pretend" attitude of middle childhood in that subjective factors are perceived as literally being characteristics of objects. A broken cup is literally perceived as distressed; a leaf in a stream is literally perceived as making an effort to break away from a rock. Only by age three or four does physiognomic perception begin to give way to make-believe which is less a literal perception of animation and more a fanciful use of inanimate objects. By six years of age, physiognomic perception is rare.

Physiognomic perception not only animates the inanimate world; it also concretizes the symbolic world of words. To most adults, words are sounds which are important only in terms of their functions of symbolization and communication. To the young child, however, words are more than mere sound-vehicles for meaning. First of all, sounds have a palpable quality, since the act of making them–the sensations from jaw and tongue and lips and cheeks–is an integral part of the sound itself. Incidentally, adult actors talk of rolling words around in their mouths or of the different way Shakespeare feels (not sounds) when translated into German. Werner also claims that sounds, as such, have a thinglike quality for young children. Preschoolers make up nonsense phrases, such as "gwol dump" and "doo lop" and react with pleasure to the sound. An adult counterpart would be the poet who orchestrates the sound of words as well as utilizing their meaning. The adult is probably nearest to the child's level when he is trying to learn a difficult-to-pronounce foreign language; at such a time, words can again take on their thing-like quality, since the act of making a sound and the sound itself have a compelling interest of their own.

Not only words but ideas also have a thing-like quality in early childhood. Werner, along with Freud and Piaget, maintains that early ideation is highly motoric and pictorial. The infant-toddler thinks in terms of sensations and images. An adult can think, "The boy is walking" at a purely symbolic level; if the toddler thinks, "The boy is walking," he is more apt to have a visual image and a motoric sensation, so that the idea is much closer to the actual event. As a result of animating the inanimate world and concretizing symbols and ideas, the toddler has difficulty distinguishing fantasy from reality. A case in point is the child's understanding of his dreams. The four- or five-year-old may believe his dreams actually happened. They are as real as life. Not until about age ten or twelve does a child regard his dreams as subjective and insubstantial (Piaget, 1951).

One final kind of undifferentiated thinking is what Werner refers to as *pars pro toto*—any part has the quality of the whole. Here Werner emphasizes the global, diffuse nature of early perception. The toddler perceives various "globs." He may be able to tell a "horse-glob" from a "dog-glob," but he lacks a detailed understanding of their distinguishing elements. Even a five-year-old may have trouble identifying what is missing in a drawing of a one-eared rabbit. He can label the drawing correctly, but the matter of the specific number of ears has not been firmly established. Details, instead of standing out clearly as elements in the whole, merge into the whole and more readily take on the properties of the whole. Werner cites the example of a child who, after being bitten by a spider became frightened of the spider web and everything associated with the setting in which she was bitten. Instead of singling out the spider, she perceived the entire experience as one in which she was bitten, so that any element of it aroused her fearfulness.

Hierarchization. By simple extrapolation from these examples of undifferentiated thinking, we can see how cognitive growth entails progressive differentiation. But mental development, according to Werner, is also a grouping together and an ordering or, to use Werner's phrase, a hierarchization of the elements. An adult's concept of himself is not a collection of discrete facts. The facts are organized in clusters (e.g., myself as a student, myself as an athlete, myself as a worker), and the clusters are given different weights according to their importance (e.g., my political opinions and religious beliefs are more central to myself than my ability to sing and play cards).

One form of hierarchization is in terms of essential and nonessential characteristics. Let us start with a hypothetical situation. A three-year-old sees a dog for the first time and notices that it has a head, a body, four legs, a tail, short hair, and a black spot over one eye. If this is the only dog he has seen, he has no way of knowing the relative importance of each detail. The black spot may be as much a part of his idea of "dog" as the tail or the four legs. Only after he has seen other dogs without the black spot can he recognize the black spot as a nonessential detail. In like manner, hair remains an essential characteristic, although its length can vary. Comparing many dogs enables the toddler to order the details in a hierarchy of importance and, incidentally, prepares him to develop the abstract idea of "dog" (see Sigel, 1964). A three-year-old asks, "When is Daddy coming home in (i.e., through) the door?" Every time he has seen his father return home, the child has been playing in the living room and has seen him coming through the glass door. Thus the door was an essential part of the homecoming. A two-year-old who had just learned to walk wheelbarrow fashion returned to the exact spot where the learning had taken place before she would do the trick again (Woodcock, 1941). The site was an essential part of the activity itself. Only with repeated experiences in which certain elements remained fixed while others varied or disappeared could these children learn the essential and nonessential features of these situations.

Werner shows how unserviceable thinking is before hierarchization occurs. Because no detail is more important than any other, a change in any one element may threaten the integrity of the whole. An infant who sees his mother wearing a hat for the first time may be genuinely distressed. His concept of "mother" gives hair the same weight as other features. When this detail is changed, the mother herself is altered. Werner claims that the toddler's need for sameness and his vulnerability to upset over minor variations is due to his failure to order details in terms of relative importance. But if a minor change can cause a major upset, on the other hand, a change in the essential nature of something may go unnoticed before hierarchization has taken place. In the preschooler's play and drawings, an object leads the life of a chameleon: a dog becomes a horse and then a pig in the very act of being drawn, or an airplane turns into a racing car or a rocket. Before their essential features are grasped, objects can readily merge into or change into one another.

As hierarchization stabilizes the child's world, it increases his security. There is no longer any danger that people and things will be significantly altered by minor variations in their appearance and behavior. There need be no rigid order imposed to insure continuity. While hierarchization decreases the fluidity of ideation which is so delightful in play, it also brings thinking into closer correspondence with the essential qualities of people and things and, hence, with reality.

The Concept of Independent Objects

Of all facets of cognitive developments, learning to differentiate the source of stimuli will concern us most because it is most relevant to personality. The basic task of separating "me" from "not-me" continues through middle childhood in one form or another. We shall discuss two aspects of this task. In this section we shall focus on the separation of the self from physical objects, so that the child can think of such objects as having an independent existence. In the following section we shall trace the development of the child's idea that causal relations among physical objects exist independently of himself. We shall be concerned with the child's encounters with the physical world because Piaget has studied these in greatest detail. The child's separation of himself from other human beings has not been studied as intensively and can only be sketched in.

The Object Concept. The neonate, as we have seen, experiences a succession of stimuli. He must learn to discriminate internal and external stimuli, and, furthermore, he must learn that physical objects have permanent, independent existences. There is no reason to assume that he has innate knowledge of object permanence, and there is no reason why he should immediately grasp it. All he knows is that certain patterns of stimuli are in his experience and then they are not. A rattle, for example, is experienced for a while and then is gone. If the infant has only his direct experience with the rattle to go on, how can he know that it has an existence all its own, even when he is no longer waving or patting or regarding it? That some stimuli continue to exist even when not experienced is an inference he must learn to make. To use Piaget's (1954) term, the infant must "construct" an objective universe.

Piaget offers a detailed account of the development of the concept of object permanence, or what he calls the object concept. Technically, the object concept includes the idea that an object exists in its own right, independent of the infant's activity, and that it is permanent. For the first three or four months, the infant gives no evidence of missing an object which is removed. At best he will orient himself in the direction of the vanished object and appear to expect that it will reappear. However, this expectancy cannot be equated with a realization that the object exists independent of his experience. Out of sight is not only out of mind; it is also out of existence. In the next stage, approximately between the fifth and eighth months, the infant's repertoire of responses has vastly increased. He is accustomed to following objects with his eyes as they move through his visual field and manipulating them in various ways, such as shaking, mouthing, feeling, and banging. Having begun a motor activity, he can sustain this involvement briefly, even if the object is not directly experienced. For example, if the rattle drops from his highchair tray to the floor, he will lean over and look for it; when he loses hold of an object, he will search for it with his hands. He can even begin an activity, abandon it briefly, and return to it. This is an advance over the previous stage, but it still falls short of true object permanence. It is as if the infant wants to continue what he started and therefore searches for the vanished object. The search is brief and is limited to the immediate area of the disappearance. If it is not successful, the infant reacts as if the object no longer exists. The burden of object permanence therefore rests primarily on motor involvement.

In the next stage, roughly between nine months and one year, the infant no longer has to be motorically engaged in order to perpetuate the object. Objects which are merely seen continue to exist even when removed from sight. For example, if the infant observes a toy being hidden behind a screen, he will remove the screen in order to obtain it. However, object permanence is still not totally independent of motor activity. If the infant recovers the toy from under a blanket and subsequently sees it being hidden under a pillow, he will continue to look for it under the blanket! The object exists where he last acted upon it.

In the next stage, between twelve and sixteen months, object permanence is no longer dependent upon the toddler's manipulation, since he can react in terms of visual cues alone. Now if the toy is hidden in one place and then in another or even a third, the toddler will immediately look for it where it last disappeared. But the rigorous Piaget points out that the toddler, at this stage, is as bound to visual perception as he had been previously to motor activity. He will continue to search where the object went out of sight, just as previously he had searched where he had last acted upon it. For example, while a toddler is watching, the experimenter picks up an object and closes his hand around it, moves his hand behind a screen, and deposits the object there. The toddler will subsequently search for the object only in the experimenter's hand, because this is where he last saw it, and he will be baffled when he does not find it there. Only in the final stage, beginning around sixteen months, is the object freed of its dependence upon visual perception and regarded as having an independent existence. The object is a thing apart—apart from the child's motor activities and perception. It is then a "not-me," with an existence all its own.

Incidentally, we have here our first illustration of the toddler's overcoming his tendency to believe what he sees. In the next-to-final stage, reacting to what he saw did not enable him to adapt to reality. Only as the toddler transcended the strictly visual evidence, only as he grasped the concept of object permanence, could he meet the challenges which reality presented. We can now appreciate more fully the title of Piaget's book, *The Construction of Reality in the Child*. The independent existence of physical objects, as well as the concepts of time, space, and causality, all must be constructed by the infant-toddler on the basis of his transactions with the environment. Once he has constructed a conceptual world, these concepts are used so naturally and automatically that it is difficult to imagine a time when they were not present.

The Object Concept and Attachment. The object concept is not a purely cognitive achievement; it affects the toddler's emotional life as well. In general, Piaget claims that cognition and affect are two complementary aspects of all behavior. Each cognitive advance is accompanied by an affective change. It therefore makes no sense to argue about whether behavior is determined by reason or by emotions. In earliest infancy the body is the source of the primitive emotions of distress and pleasure. As the infant becomes capable of exploration, his emotional life is concomitantly enriched: he experiences effort and fatigue as his interests wax and wane, pleasure and displeasure as he succeeds or fails in what he tries to do. However, emotion, or affect, merely accompanies activity at this stage; it is not directed toward any particular object. Directed affect becomes possible only after the object concept has been achieved. The toddler can then be angry *at* a tower for falling, not merely angry that the tower frustrated his activity of building; he can be delighted *with* a pushtoy, not merely pleased by the activity of pushing.

Most important, the object concept enables the infant-toddler to develop stable attachments to people. He can love a person and be anxious when that person is not present, because he can grasp the concept of independent existence. In Chapter 2 we mentioned the cognitive component in attachment; now we can see its role more clearly. The infant may love being with the mother and may react to her presence with intense pleasure, but he can love the mother—in the sense of having an object of his affection which is permanent and no longer dependent upon her presence—only after the mother's permanent existence has been grasped. Expressed mentalistically, the infant-toddler has the idea "mother" which is part of his storehouse of ideas, regardless of the presence or absence of the mother herself. The idea can arouse affect and motivate behavior; e.g., he can suddenly long for his mother while playing with his toys and search the house to find her.

Piaget, like Freud, stresses the importance of the mother as the first human object. Both claim that she is the affective object par excellence; Piaget also regards her as the cognitive object par excellence in the infant's environment: she engages his attention, piques his curiosity, and stimulates his exploration in many ways. She is a compelling visual and auditory stimulus. She can be sucked and smelled and tasted and grasped. Her rocking and caressing arouse kinesthetic and tactual sensations. And, as an animate object, she has that quality of variability within the context of sameness which never fails to excite human curiosity.

It is difficult to marshall definitive evidence concerning Piaget's ideas, since the purely cognitive effects of mother-as-object are impossible to isolate from other important variables in the development of attachment. Decarie (1965) hypothesized that, if cognition and affect are complementary aspects of behavior as Piaget claims, there should be a positive correlation between the development of emotional attachment and the development of the concept of the permanence of physical objects. Her scale for the development of object permanence follows Piaget's stages which we presented earlier. Her scale for the development of attachment begins with evidence of the infant responding to the mother during feeding; then comes an intermediate stage marked by the indiscriminate social smile, negative reactions to play interruption, and the ability to wait; and, finally, there is an objectal stage, which includes differential smiling to the mother, negative response to the loss of a toy, compliance with requests and prohibitions, and subtle discrimination of signs of communication from the mother. The scales were used to evaluate the behavior of ninety infants between three and twenty months of age. In general, Décarie confirmed the hypothesized relation between attachment and the object concept, though not conclusively. The results are confounded by the unsatisfactory nature of the criteria for attachment which includes many items which, at best, only indirectly reflect the infant's relation to the mother. While Décarie's book clarified a number of theoretical issues and points up the practical problems involved in testing Piaget's ideas, her study itself is best regarded as a preliminary test of her hypothesis.

Piaget's Theory of Development of the Object Concept. In order to understand Piaget's theory of the development of the object concept, we must examine his general theory of cognitive development. Piaget designates the first year and a half of life as the period of sensorimotor intelligence. Cognition develops in the context of early perception and motor activity. The infant cannot represent objects or actions symbolically. He cannot say to himself, "I will now reach for that rattle." He sees and reaches. He literally learns through doing. At first his sensorimotor repertoire consists of simple, repetitious acts—he sucks, he grasps, he looks. As he continues to act, however, each of these activities becomes more varied and versatile; e.g., his crude, reflex-like grasp develops into complex, coordinated patterns of holding and manual exploration. Piaget calls an organized, integrated pattern of behavior a schema. In the sensorimotor stage a schema is "an ensemble of sensorimotor elements mutually dependent or unable to function without each other" (1952, p. 244).

Schemata are not only organized; they are also continually expanding. Expansion takes place through the dual processes of accommodation and assimilation. Accommodation is the organism's adapting itself to a new stimulus in order to incorporate it. Assimilation is the mental "digesting" of the new information and integrating it into an existing schema. Through assimilation, the schema is enriched and expanded. To illustrate: When a student picks up a psychology book, schemata relevant to reading and to psychology are activated while other schemata remain dormant. Because of this activity, he can accommodate himself to the contents of the book. As he reads, he gains new information which must be understood or integrated with his existing knowledge of psychology. After the information has been assimilated, his knowledge is enriched and enlarged. Or, to take an example from infancy: The infant who has heretofore always ingested milk by sucking encounters a cup for the first time. The stimulus initially activates the sucking schema but, since the cup differs from a nipple, milk spills out of his mouth instead of flowing down his throat. The information concerning the new stimulus must be assimilated so that future accommodations will be more efficient. After new adaptations are devised, the infant's schema is enriched to include drinking as well as sucking, so he can accommodate to cups as well as to nipples.

Schemata also expand by assimilating other schemata. By about the fourth month, for instance, sucking and looking are integrated, and thereafter the infant makes sucking movements at the sight of the bottle as well as when the nipple is in his mouth. It is just this integration of independent schemata which facilitates the development of the object concept. The more schemata an object activates, the more likely the toddler is to grasp its independent existence. Let us first simplify the situation. The infant hits different objects in the course of his exploration. He discovers that some objects break, some jump, some roll away, some hit back, and some remain unaffected. The very fact that the same action produces a variety of results enables the infant to distinguish an act from its consequences. The former becomes part of the self, the latter part of the environment; actor is differentiated from object acted upon. Of course, the learning situation is never so simple. Instead, each object explored activates, as the result of learning, a specific constellation of schemata. For example a ball has special properties of being round, of rolling, and of bouncing, which will lead to the activation of a different constellation of schemata than, say, a block or a teething ring. The fact that the infant must form distinctive constellations of accommodations in order to adapt to different objects greatly facilitates the establishment of object permanence. As we noted in discussing attachment, the mother activates a rich array of schemata; now we see that this very quality of mother-as-object helps the infant grasp the idea of her independent existence.

Evaluation. Piaget has no equal when it comes to observing and explaining the infant's transactions with the physical world. He has shown only a casual interest in early social and emotional development, so that his contribution to our understanding of attachment is limited. If one juxtaposes Ainsworth's account of the stages of attachment (see Chapter 2) with Piaget's account of the stages of achieving the concept of object permanence, one finds very little overlap between the two. There is no obvious counterpart in the human world of the six-month-old's idea that an object's existence depends on his motor activity; and there is no obvious counterpart in the physical world of a human being as a cue for relief from distress or as an object which, of its own accord, behaves in such a manner as to pleasure the infant. How is the object concept affected as the infant learns the peculiarly human characteristics of the human object? We do not know.

Piaget's two theoretical ideas concerning attachment (the complexity of mother-as-object aiding the development of the object concept and the complementary nature of affect and cognition) are programmatic and poorly documented. In regard to mother-as-object, there is none of the ingenious detailing of integration of schemata which can be found elsewhere (e.g., the integration of looking and grasping which will be presented in Chapter 10). In a like manner, his theory of emotional development is too general to account for the appearance and interplay of delight, comfort, anguish, and anxiety which characterize the development of attachment.

This evaluation is not meant to be a criticism. Piaget did not intend to explore emotional and social development and he does not claim to have a comprehensive account of attachment. He brilliantly accomplished what he set out to do in terms of his interest in the question, How do we come to know the world? We are only pointing to the gap between his insights into the process by which the infant constructs the physical world and his insights into the process of forming an attachment and cautioning against regarding his account of one as a satisfactory account of the other.

The Concept of Causality

A woman believes that her unexpressed thought, "I wish my mother were dead," caused her mother to die of cancer; a college student will not speak during meals for fear her words will poison the food; a man always sleeps on his back out of fear that, if he did not, he would change into a woman. These severely disturbed individuals engage in a particular kind of thinking called omnipotence, since they believe they can produce events which are only partially in their control or which are out of their control altogether. Because they are adults, we say that their idea of causation is magical. Yet, all ideas of causation begin with magic; it is only the perpetuation of magical ideas into adulthood which is pathological. In this section we shall be concerned with the development and eventual dispelling of magical ideas concerning causality (see Piaget, 1930).

The Development of Causal Thinking. As we have just seen, the infant, between the fifth and eighth months, begins to grasp the concept of an object as being "out there" or spatialized, to use Piaget's term. At the same time, the infant begins to have a vague concept of one event being related to another. However, causality, like object permanence, is at first contingent upon his own motor activity. He dimly grasps the idea that motor activity A produces external event B, but he cannot conceive of an external event C producing external event B. For example, he can understand that his hitting will knock down a block tower; he cannot understand that a ball can also knock it down. The belief that external events are determined exclusively by the individual, when seen in adults, would be called magical thinking. Recall the illustration we gave in Chapter 3: as the infant cries and the mother appears and comforts him, he comes to think that his behavior causes the mother's appearance. His inference is understandable because he knows only the beginning and the end of the sequence. He does not know that his cry is heard by his mother, who has an existence, personality, and volition of her own, or that she could choose either to come or to ignore him, and chooses to come. Since he has not even grasped the concept of object permanence, this understanding of his mother is totally beyond his ken.

By the end of the sensorimotor period, the toddler has identified self and objects as separate entities and realizes that both can be causal agents. Causality is no longer contingent upon his activity or his perception. Why, then, is this not the end of magical thinking? Because the toddler now enters a new phase of cognitive development called the pre-operational stage, which extends roughly from the second to the seventh year. Piaget contends that the child entering the next stage must relearn many lessons from the previous one. The concept of causality does not transfer automatically; on the contrary, the insights must be rewon in terms of a new vehicle of thought which is at the toddler's disposal. This new vehicle is language. As the toddler learns language, he no longer is tied to doing and perceiving. He can symbolize. He can represent an entire sequence of actions rather than having to perform them. Instead of reaching for a ball that he sees, for example, he can say or think, "I will reach for the ball." Both the ability to remember past events and the ability to anticipate consequences in the future are greatly facilitated with the addition of language to the child's repertoire.

But the fact that the preschooler now has verbal symbols available to him does not mean that he uses them for realistic or logical thinking. Realism and logic must be achieved over a period of years. The preschooler does not lose the insights of the sensorimotor period; as long as he can function at an action level, he behaves with understanding. But when asked to explain what he did (and thus forced to go to a symbolic level), he will relate thoughts which are as fantastic as those in the initial phases of the sensorimotor period. One sees a proliferation of exotic distortions in the preschooler's thinking, all due to his tendency to view the world of objects in human terms.

In the earliest phase of thinking, roughly between two and three years, confounding of the subjective and the objective—the "me" with the "not-me"—is at its height. Piaget borrows a time-honored label from psychiatry—autistic—to describe thinking which is more like daydreaming than mature logic. He points to early symbolic play (see Chapter 9) to illustrate the toddler's disregard for reality and logic, his ability to conceive of something as being so just because he wishes it to be so, his arbitrary ordering of events predicated on his need at the moment rather than on reality, and his obliviousness to incongruities. A bit of wood can readily become a gun and as readily turn into a fighter plane and then a rocket, as the child fantasies he is now shooting a monster, now dropping a bomb, now going to the moon. (This example is more typical of a preschool child's play, and is only meant to illustrate wishful, need-determined thinking.) Piaget concludes that autistic thinking and magical thinking are two sides of the same coin.

The period of autistic thinking ends when the preschooler begins to ask *Why*. The question itself indicates that he has run up against an external world which has refused to act according to his wishes. This resistance must be assimilated and understood. Yet the leap from autistic thinking to reality is too great to be made all at once. The preschooler, like any individual, must struggle to comprehend the new in terms of the schemata he has available. At this particular point, he reverts to his earlier egocentrism and revives his error of modeling the external world on himself. He conceptualizes the physical world in terms of animism—"objects have life and consciousness"; in terms of finalism—"objects exist in order to serve human needs and purposes"; and in terms of artificialism—"humans, or a highly humanized God, made all things." Thus, the world becomes conscious, purposeful, and industrious, just as he himself is. Piaget calls all three of these forms precausal thinking.

The preschooler's tendency to conceptualize objective events in terms of himself is revealed by his responses to questions concerning the reasons for natural phenomena: e.g., Why do clouds move, Why does water flow? Around five years of age, children give reasons which are animistic and magical. Natural objects move because an agent (such as a person) commands or wills them to move, and because of an object's own will to acquiesce. The external agent is most frequently a humanized God who pushes the clouds with his hand, for example, or pulls them by a thread, or presses a button to make them move. The explanation is in terms of divine artificialism. Children also believe that people cause natural objects to move when they themselves move. This explanation is particularly interesting because it is another example of the child's believing literally what he sees. While walking, one does see trees, roads, fences, clouds moving as he moves. The child, like the astronomer who believed the sun went around the earth, must transcend his naive perception in order to grasp reality.

In subsequent stages, the child no longer believes that a human or human-like agent causes movement in nature; rather, a natural object is considered responsible. This is a step toward disentangling the self from the environment, but the explanation is still animistic, since the object is humanized: the sun makes the clouds move by pushing them with its rays, or the air makes them move by chasing them. Only by about eight or ten years does the child completely divest the physical world of human qualities and conceptualize causality in purely physical and mechanical terms. Thus, it requires five years or so for him to cease using himself as a model for constructing the physical world. Again, we see how long it takes to grasp an "obvious" feature of reality.

Piaget is more explicit in depicting the stages of cognitive development than he is in accounting for progress through the stages. He suggests that it may be the child's encounters with toys and mechanical objects which force him to reject his subjectivism. The physical world stubbornly refuses to verify his belief that he can control it by his will alone. In assimilating this resistance, the child is forced to reconceptualize objective reality.

Objective validation of Piaget's principal ideas concerning precausal thinking in general and his explanations of physical movement in particular is found in an extensive study by Laurendeau and Pinard (1962). Certain details of Piaget's theory were not corroborated, but the trend from anthropocentric to a mechanical conceptualization of the universe was clearly supported by the data. The book also contains a worthwhile discussion of the methodological problems involved in testing Piaget's concepts.

Comment: Omnipotence in Adults. Omnipotence is developmentally appropriate in infancy but psychopathological in adults. Why? Because the infant is incapable of thinking in any other way. He has not grasped the independent existence of objects and persons, much less their defining characteristics. Omnipotence, in its proper place in the developmental sequence, represents the infant's best utilization of the information and of the kinds of thinking available to him. Disturbed adults are capable of thinking in terms other than omnipotence and do so in areas unaffected by their pathology. They have separated "me" from "not-me" and can grasp the nature of causality within and between the realms of self, people, and objects. In their psychopathological state, this separation is lost; thinking which a normal adult would recognize as wish or idea or daydream becomes part of the disturbed adult's perception of reality. It is the discrepancy between their levels of functioning–with magical thinking side by side with rationality–which marks adults as disturbed.

Egocentricism. We have already noted that the failure to distinguish "me" from "not-me" is a theme common to the object concept and causal thinking. In the sensorimotor period the infant believes that the existence of physical objects is contingent upon his motor activity or his perception. Only at the end of this stage does he undergo a "miniature Copernican revolution" and conceive of himself as merely one entity in a universe of permanent objects. In the precausal stage, he continues to use himself as a model for the physical world, regarding physical objects as having life, goal-directedness, and purpose. Piaget uses the term "egocentricism" to describe the condition in which the child conceives of the world exclusively from his own point of view. Characteristics of the self are used to define or interpret characteristics of the objective environment.

Egocentricism marks the child's early transactions with the social world as well as with the physical world. Just as he cannot conceive of physical objects as different from himself, he cannot conceive of other persons holding viewpoints different from his own (see Flavell, Botkin, and Fry, 1968). He regards people as nothing more than "other me's." This egocentricism affects both his ability to communicate with others and to reflect on his own ideas (Piaget, 1926, 1928).

The child around four years of age believes that his ideas are immediately comprehended by others. He says whatever comes into his mind and takes it for granted that his audience knows what he means. If other people are "other me's" there is no necessity to explain what he means, any more than (in a roughly analogous way) an adult feels the need to explain his private thoughts to himself. The preschooler engages in long monologues when he is by himself, and several children can carry on monologues in each other's presence. In these collective monologues, as they are called, the children may seem to be conversing, but there is no real exchange of ideas. Each child is merely expressing his own view. Piaget claims that a child is about seven years of age before he is capable of having a true discussion with other children, thereby indicating that he has grasped the idea that they have other points of view.

Incidentally, Piaget (1966) points to another consequence of lack of differentiation between self and others, namely, the child's tendency to attribute his own private ideas to others. Before he understands that other persons are originators of their own thoughts, opinions, and viewpoints, he believes that his ideas are also theirs. This is similar to what the psychoanalysts call projection, which we shall discuss later in this chapter.

Another facet of the child's oblivousness to others is his belief that something is true just because he has said it. The child under seven feels that his assertions require no justification. He is startled when asked to prove one of his fantastic statements because he cannot differentiate his own point of view from that of others. It is only as he runs up against uncomprehending or challenging or contradicting reactions from others that he is forced to relinquish his exclusive monopoly on truth and wisdom. Only when he begins to challenge himself as he has been challenged by others, and to question his own statements as he has been questioned, does he begin to reflect. Reflection is primarily an internalization of discussions first carried on with others (see Chapter 4).

Both Werner (1957) and Anna Freud (1965) also characterize early childhood as egocentric, but they stress the affective rather than the cognitive components. For them, egocentrism refers to the child's tendency to evaluate events in terms of their hedonistic value to himself. The behavior of others is good if it brings him pleasure; it is bad if it brings him pain. A father is good to the extent that he admires a block tower, fixes a toy car, or puts money in a candy machine; he is bad to the extent that he pays attention to the baby brother, does not let the child hold his pocket watch, or orders him to go to bed. In essence, the child says, "Anything that you do which makes me feel good is good; anything that you do which makes me feel bad is bad." It does not occur to the child that persons have feelings of their own or that they are governed by motives which have nothing to do with him.

Early Cognition and Deviant Development

First let us look at developmental timetables. The child's grasp of physical and social reality remains faulty up to the time he begins formal schooling. Ideation, which should serve as a guide to reality, is still heavily saturated with magic. Yet, during this same period, the child has had a number of crucial experiences. He has formed an attachment to his mother, he has been toilet trained, he has known violent jealousy, he has been required to control his destructiveness, sexual self-stimulation (see Chapter 7), and curiosity, he has been introduced to concepts of right and wrong, and he has discovered the pleasures and frustrations in social participation (see Chapter 8). One may not agree with Freud that the first six years determine the fate of personality development, but it would be difficult to match them in terms of fateful beginnings.

All these things happen during a time when the child is incapable of understanding them in realistic terms. Inevitably his world is distorted. For example, after he has formed an attachment, the infant reacts to the mother's going away as rejection, regardless of how she may feel toward him. The mother may be ill or involved with a family crisis, but all the egocentric infant knows and feels is that she is not there when he needs her. We also have a better appreciation of why toilet training enhances the sense of power in the toddler, whose thinking is naturally inclined to be omnipotent (see Chapter 3). Finally, the child's tendency to be literal in his understanding of rules for behavior is a variation on the tendency literally to believe what he sees, which distorted his understanding of object permanence and causality (see Chapter 5).

While Piaget supplies us with the most detailed picture of early cognitive distortions, particularly of the physical world, the psychoanalytic literature is rich in descriptions of how the child's omnipotent thinking distorts social reality. In the Oedipal situation, the boy of four or five is caught in a rivalry with his father and wishes his father were dead (i.e., that he would disappear from the home). Suppose the father had to be absent for a long period of time, due to military service or a new job in another city. The son is likely to believe that his wish caused the father to disappear and to react with guilt. If the father actually dies, the child's irrational guilt is even greater. The mother may explain the situation and the boy may accept the explanation at a conscious level, but the primitive part of his personality which contains both the wish to destroy and the omnipotent notion of causality may make him feel responsible for and guilty over the father's death.

Examples of omnipotent thinking need not be tied to specific Freudian concepts. Let us suppose the parents of a five-year-old are going through a stormy period and are constantly quarreling. Let us also assume that as sensitive adults, they take care not to involve the child directly in their problems. How can the child understand the reality of their situation—that the father is attracted to another woman, or that the mother thinks the father is not making enough money, or that they disagree over what to do with a senile grandparent? Such issues are beyond his comprehension, so the child understands the situation as best he can in terms of his own perceptions and needs. Since he does not understand the real causes, his omnipotence may lead him to think that he is the cause of his parent's unhappiness. The particular interpretation he constructs will depend upon his particular personality. The following quote from a young man in psychotherapy, indicates one possibility:

> I remember my parents fought a lot. Now that I think of it, they tried to hide it from us kids because they always went up to their room and closed the door. I remember one night—I guess I was six or seven—and Effie (the eldest sibling who would take charge of the others) sent me upstairs to get her scissors. Mom and Dad were fighting like they always did. I remember feeling like I couldn't stand it. I crawled under my bed and said, "Please God, don't make them fight. I promise I'll be good."

This young man was deeply disturbed by hostility when he was a boy. In his bargain with God he assumed the burden of ending behavior which he had no part in producing. In his primitive way he reasoned, "My badness caused bad things to happen; my goodness will make the bad things good." We can reconstruct the situation in less mentalistic terms: the socialization process builds up a strong association between the child's being "bad" and the parents becoming angry; if parents become angry for other reasons while the child is locked in his egocentric mode of thinking, he may conclude that his badness was the cause.

There are two general conclusions we may draw from Piaget's and Freud's theories. A simple socialization model in which the child's personality is regarded as the product of parental behavior has serious limitations. It is the child's understanding of his environment which shapes his growth, and this understanding can differ markedly from reality. Both Freud and Piaget claim that there is not one reality but a series of evolving realities. Environmental events must always be translated into terms which are consonant with the child's level of understanding, if their effects on his personality are to be understood. Psychologists in clinical settings are accustomed to discrepancies between aspects of reality as viewed by adults and by children. A boy may be terrified of a father who is no better or worse than any other father; he may respond to reasonable discipline with unreasonable defiance; he may cling tenaciously to parents who are models of self-reliance.

The second point concerns deviant development. The first six years of life is a period when affect is particularly intense and understanding is particularly faulty. Therefore, it is regarded as the period most fertile for breeding severe adult disturbance. Because the young child is so passionate, and because his feelings are concentrated on a limited number of people (usually those in the nuclear family), his early misperceptions are highly charged with emotion. If he is hindered in passing through subsequent developmental stages, he is in danger of becoming psychopathologically disturbed; he may live in terror of being rejected, or have blind, destructive rage toward authority, or a deep disgust over sexuality, or a debilitating guilt over minor transgressions. We can also understand the tragic irony of certain adult disturbances. An adult may be psychologically crippled all his life by a distorted view of reality which originated in childhood. A man may be terrified of authority throughout his mature years because of the destructiveness he attributed to his father; a woman may be sexually frigid because of a moral rigidity she attributed to her mother; in reality, the man's father may have been somewhat violent and the woman's mother somewhat prudish, but neither may have been significantly different from most parents. It was the child who manufactured an exaggerated image of a destructive or prudish parent out of his needs and immature thinking. Similarly, to suspect that behind every problem child there is a problem parent oversimplifies the nature of human development. One of Freud's significant insights was that a child does not have to undergo a real trauma in order to develop a neurosis. The trauma may have been a common event (such as a father going to war or a mother having to care for a terminally ill grandparent) which triggered an irrational overreaction because of the state of the child's emotional life and his magical thinking. It is just such fantastic creations of early childhood which can become the nucleus of a neurosis that prevents the personality from achieving full maturity.

Cognitive Development in Middle Childhood and Adolescence

We have discussed the preschooler's tendency to believe literally what he sees. Piaget uses the term intuition to denote the kind of thinking which is based on the mental representation of sensory experience rather than on logic. Between five and six years of age, a child will say that a line of eight blue discs is the same as a line of eight red discs, if there is a red disc under each blue disc. If, before his very eyes, the experimenter moves the row of red discs so that it is no longer exactly under the blue but a little to the side, the child says the rows are no longer equal. To the child, whose thinking is intuitive, things which look equal are equal; things which do not look equal are not equal!

Because the child believes what he sees, his world is fluid. A perceptual change means a conceptual change. The triumph of middle childhood is the substitution of reason for perception as the basis of conceptualizing the physical world. The child learns that objects remain constant, in spite of perceptual variations. When a ball of clay becomes a snake, you do not have more clay; when water is poured from a wide, squat glass into a tall, narrow glass, you do not have more water. Piaget uses the term conservation for the cognition that properties of objects remain constant regardless of perceived variations. Between eight and twelve years of age, children master first the conservation of matter, then of weight, and finally of volume.

The child's literal approach to objects also shows up in the moral realm. The young child interprets parental rules literally and rigidly. Just as he does not reflect on the validity of his own assertions, he does not question the validity of parental dictates. Right is right and wrong is wrong. There are no extenuating circumstances and no consideration of intentions. In middle childhood, the morality of obedience gives way to the morality of mutual respect. As the child becomes capable of understanding other points of view, he comes to perceive rules as ways of regulating behavior. Morality then becomes functional rather than absolute; rules can be changed by mutual agreement; goodness and badness are judged in the context of intentions and extenuating circumstances. (These developments were discussed in Chapter 5.)

Piaget's ideas concerning the role of cognition in social development and in the development of the self are particularly interesting. As we have seen, egocentric language and thinking disappear around seven years of age. Social reality forces the child to realize that everything he says is not inevitably understood by others. He now knows that he has his ideas, other people have theirs. If he wants to communicate, he must adapt his language so that it will be understood, since communication requires the speaker to take cognizance of the listener. Much of the time adults make this adaptation unconsciously. However, when they speak to young children, or to foreigners with a limited English vocabulary, or to severely retarded individuals, they become acutely aware of the need to modify their vocabulary in order to communicate.

As the child becomes capable of communication and reflection, he also becomes capable of cooperation. Piaget would not deny that sympathy and fellow-feeling might exist in the preschool child, but he would say that a true working together demands that each child be able to cognize the viewpoints of the other children and to coordinate his thinking and action with theirs. Piaget makes another point we have already discussed: As the child becomes more reflective, he also becomes less impulsive (see Chapter 4). Instead of saying and doing whatever comes to mind, he can delay and engage in an internalized social discussion and so adapt more readily to the demands of a social situation. Thus, the child becomes both more considering and more considerate.

Finally, Piaget implies that the child begins to define himself at the same time that he is defining others. The two processes are complementary. If the child could always say and do whatever he liked, he would never advance beyond the stage of egocentricity. Only as he is forced to take other points of view into account is he made aware of his own views. And only as he is forced to reflect on his ideas is he made aware that they are his and no one else's. Whereas, in the sensorimotor period the child's body and bodily experiences were the essence of his concept of "me," in middle childhood his ideas and his point of view become part of his concept of "me." His psychological self has then taken a significant step forward.

By the end of middle childhood the child can think realistically. He no longer confounds aspects of the physical world with characteristics of the self, and he no longer confuses the social world with his own view of it. The realms of self, people, and things are conceptually separate. However, his thinking is still tied to concrete reality, to the situation immediately at hand. The cognitive development in adolescence is the ability to derive common principles from specific instances and to fit specific instances into general systems and theories. General ideas and abstract construction flourish. Piaget labels this the period of hypothetico-deductive thinking, since the adolescent can now draw conclusions from pure hypotheses as well as from actual observations. What is more, ideas are intriguing in their own right; they do not have to mirror concrete reality. From around age twelve on, the child can go wherever his thoughts lead him. And he is heady with his newfound powers. He discusses, he writes, he ruminates. He creates a philosophy of life or a political system or a theory of aesthetics. He wants to explain the universe. Finally, the adolescent is able to evaluate his own thinking. In middle childhood he was able to differentiate his ideas from those of others, but he could not reflect on their quality. Now he is capable of being truly self-critical.

Yet, like previous stages in cognitive development, this one is marked by an initial egocentrism. The adolescent believes that ideas alone will win the day and that his ideas hold a special key to solving the world's ills. He enters adulthood with the dual postures of self-aggrandizement and selfless devotion to others. This egocentrism is overcome through contact with the realities of adult life, just as the egocentrism of the preschooler is tempered by the reality of social relations. The adolescent discovers that he cannot impose his answers on the world. His idealism must be moderated if it is to have any effect. Instead of intruding himself into the lives of adults, he must work cooperatively with them. He realizes that the purpose of reflection is not to reorder experience but to predict and interpret it.

Affect and Cognition

Piaget's theory of cognitive development is the most comprehensive extant. However, his treatment of interpersonal relations is too limited and his theory of affect too sketchy to satisfy our special need for viewing cognition in the context of personality development. We shall therefore turn to Freud in order to add scope to our discussion.

Freud and Piaget

We shall start our comparison with a caricature. Freudian theory has been called a psychology without intelligence, Piaget's theory a psychology without feelings. Obviously, both statements are oversimplifications. Freud's theory postulates an ego, one of whose functions is to learn to understand the nature of reality. Piaget states that behavior always has both a cognitive and an affective component, the affective being the energizing, motivating factor, the cognitive the guiding, adaptive factor. However, a good caricature distorts by highlighting certain aspects of the truth, and there is a difference between the official stance of a theorist and the manifest content of his work. As we have seen, Piaget is primarily interested in the development of cognition, and his treatment of affect is superficial. Freud's primary interest was in understanding abnormal behavior, especially the potentially disruptive drives and emotions. His theory of the ego, the part of the self which gives rise to rational, adaptive behavior, was sketchy and unsatisfactory.

There is another important element in this picture of Piaget the epistemologist and Freud the clinician. Piaget's subjects were typically normal children in normal environments. Freud's subjects were emotionally disturbed adults; he studied neither normal individuals nor children at first hand. Now let us look at a partial list of emotions which Piaget includes in *Six Psychological Studies* (1967). In the sensorimotor phase we find fear, pleasure, and pain; later come joy and sadness, interest and disinterest, sympathy and antipathy, moral sentiments and respect, feelings of inferiority and superiority. Certainly there is nothing here which could drive a man mad or torture him to death or make him beat his child unmercifully. Yet, it was the mad, the tortured, the uncontrollably violent whom Freud was trying to understand. The only emotions Piaget acknowledges are the genteel ones seen in normal individuals; Freud, on the other hand, was dealing with the furies. Piaget gives a developmental picture of emotion and cognition traveling hand in hand toward the ultimate goal of rational behavior. Freud gives us a psychology of conflict, in which man is at odds with himself and his environment, struggling to be rational but continually thwarted by primitive impulses or an equally primitive conscience.

When we turn to the cognitive theories Piaget and Freud evolved, we find a good deal of overlap. Both men conceive of cognition as progressing from primitive, irrational, magical thinking to rational thinking, and both regard development as extending throughout childhood. Freud's description of early thought as having no logic or time orientation is congruent with Piaget's thesis that logic and time are developmental achievements. Both agree on the relation between cognitive development and control: thinking serves to check impulsive actions and makes behavior more adaptive to reality (see Chapter 3). However, Freud's delineation of developmental stages is crude in comparison with Piaget's. He refers to primitive ideation as primary process thinking and logical ideation as secondary process thinking. The infant has only the former, while secondary process thought appears and gradually increases in proportion and strength during childhood. Since Freud was continually exploring a wide range of personality variables, his cognitive theory does not have the richness of concepts, the substantive scope, or the detailed documentation of the developmental process, which characterize Piaget's theory.

In spite of the general similarities, there is a difference in Piaget's and Freud's theories which reflects the difference in their interest in cognition. Piaget is most ingenious when dealing with the structure of thought; Freud was most insightful when dealing with its functions. Piaget asks, How do children think? Freud asked, Why do they think so? Piaget delineates the forms that thinking assumes at various developmental levels—the earliest cognizing of the world via sensory and motor behavior, the subsequent appearance of intuitive thought, precausal thinking, and hypothetico-deductive thinking, just to mention a few high points. However, the underlying processes he postulates—assimilation and accommodation in particular, and adaptation to reality, more generally—provide only a very general account of the forces responsible for cognitive change.

The situation is reversed with Freud, who was constantly asking the question, *Why?*; that is, What motivates this behavior? and What function does the behavior serve in maximizing pleasure and minimizing pain? Why does a normal adult forget his friend's name? Why does a psychotic adult believe he is Jesus? Why do we dream? Why, indeed, do we think at all? Typically, Freud accounted for thinking in terms of the function it served. In fact, he went so far as to claim that the ego originally comes into being for the purpose of impulse gratification. We become cognizant of reality in order to find socially acceptable ways of gratifying our needs and in order to minimize the distress inherent in socialization itself. Suppose, for example, a jealous two-year-old hits his baby sister and thus gratifies a primitive impulse by expressing his rage. His mother spanks him, and the pain and humiliation of the punishment outweigh the pleasure he gained from hitting. The ego, looking for a way to express his rage without incurring punishment, says, "Next time you want to hit, wait until mother is out of the room." By heeding the ego and refraining from impulsive action, the toddler can both have the pleasure of hitting and avoid the pain of punishment. It is only because the ego can lead the way to more pleasure in the long run that the infant follows its reality oriented advice.[4]

[4]Neo-Freudian ego psychologists have reformulated Freud's theory, making the ego autonomous in infancy (see Hartmann, 1951). However, they are expanding the ego by giving it more independence from the id, rather than by denying its impulse gratification function.

Fixation and Regression. As we have seen, Freud confronts us with the questions: How is thinking affected by intense emotions? What happens to cognition in states of high sexual arousal, rage, guilt, ecstasy, and despair? Freud's own answer is clear: passion can become the master of reason. It can make the most fantastic distortions of reality seem like manifest truth. Recall our examples of magical thinking in disturbed adults (p. 191). Here bizarre ideas represented either a perpetuation into adulthood of early omnipotent thinking, which Freud terms fixation, or the return to such omnipotent thinking under stress, which Freud terms regression. If Freud is only telling us that bizarre ideation in adults was appropriate in early childhood, we do not need to change our model of cognitive development; we need only to add the possibility of fixation and regression. In fact, both Piaget and Werner recognize that the child does not inevitably advance to higher cognitive levels and that, having advanced, he may return to lower levels. The concepts of fixation and regression are common to many developmental theories.

However, Freud means more than fixation and regression when he refers to emotion distorting cognition. In order to appreciate his ideas, we must first understand his concept of personality structure and his concept of mechanisms of defense.

Freud's Concept of Personality Structure

Freud conceptualized personality structure as consisting of the id, the ego, and the superego. The id contains what Freud called instincts or what we would call drives. Although Freud revised his ideas concerning instincts, he always thought in terms of a limited number, such as hunger, thirst, and sex. As we shall see (Chapter 7), he regarded the sexual instinct as the crucial one for personality development. The id is primitive not only in its content but also in its operation. It functions according to the pleasure principle; that is, it strives to maximize pleasure and minimize pain, and it never changes. The most mature adult personality has this infantile core, although it may be observable only under stress, such as a severe illness or personal tragedy.

The superego contains moral precepts and ideals. As we have seen (Chapter 5), it can be as primitive in its way as the id. It is absolute and uncompromising in demanding perfectly righteous behavior, and it punishes the individual with guilt when he transgresses. A mature adult evolves a more reasonable conscience, but the primitive superego continues to exist, if only at an unconscious level.

It is the ego which interests us most. One of the ego's basic tasks is that of adapting to the environment. Adaptation requires knowledge, and the ego employs a number of psychological functions in order to obtain this knowledge. Through perception it gathers information concerning the environment. It uses memory to store this information. It exercises judgment by comparing new information with that stored in memory. Finally, through voluntary motor activity it explores the environment in order to gain new information. The ego also has a special function called reality testing which involves distinguishing environmental stimuli from wishes and impulses. To illustrate reality testing we shall return to the omnipotent infant who has concluded, "All I have to do is to want Mother badly (i.e. cry), and she appears." It is the ego which questions, "Is this really true? Is reality so constituted that it is inevitably obedient to needs and wishes?" The ego then begins to test the idea of omnipotence by gathering more information about reality; e.g., it might observe and remember occasions when the mother did not appear when needed. As this information fails to confirm omnipotent thinking, such thinking is replaced by more realistic ideas. Reality testing thus corresponds to Piaget's concept of dispelling egocentrism by differentiating the objective from the subjective. Only through repeated checking of primitive ideas against information in the objective world can autistic or primary process thinking be replaced by realistic thinking. To use Freud's terminology, the ego operates on the reality principle, which means that it is concerned with arriving at an accurate understanding of the self, of other human beings, and of the physical environment. This concern with accurate understanding is in marked contrast to the id's inalterable need to maximize pleasure.[5]

Mechanisms of Defense

The ego, in addition to testing reality, is in charge of the mechanisms of defense.[6] The term refers to a group of psychological maneuvers designed to defend the child against anxiety, since the primary purpose of all defense mechanisms is the reduction of anxiety. (Note again Freud's concern with function.) Anxiety in childhood can arise from a number of sources (see Chapter 7). The superego may condemn a child for sexual or aggressive impulses and punish him with guilt, which Freud regarded as "moral anxiety" (see Chapter 5). Next, there is objective anxiety in which a realistic threat causes the child to become frightened of an impulse. Some parents still use actual or disguised castration threats; e.g., a little boy might be terrified by a parent who says he will cut off the boy's hands if he masturbates, or who predicts that the boy will "go crazy." Finally, the ego can fear the id itself. The intensity of the id's primitive impulses, its relentless demand for immediate gratification, and the contradictory, alogical nature of its functioning pose a threat to the moderate, reasoning ego. The feeling of getting out of control, of having more lust or rage than one can manage, is always frightening to that part of the personality responsible for self-control and management of impulses.

We shall not attempt to draw up a comprehensive list of specific defense mechanisms or a critique of them. Rather, we shall describe a few of the most common ones as they bear on cognitive development and reality testing.

Repression is usually regarded as the most basic defense. In repression, the individual banishes an impulse and the ideas and fantasies associated with it from consciousness. In essence he says, "What I am not aware of does not exist." Let us imagine a five-year-old boy who becomes anxious over his wish that his father were

[5] For a more complete account of Freud's theory of personality structure and of the function of the ego, see Brenner (1955). For a discussion of the pleasure principle and the reality principle, see S. Freud (1963).

[6] For a summary of defense mechanisms, see Brenner (1955) and A. Freud (1966).

dead. He bars the hostile impulse (the wish to kill or destroy) and the thoughts accompanying it ("I wish you were dead") from conscious awareness. Then he can feel with conviction that he does not hate his father.

Freud maintained that the anger continues to exist in the boy's unconscious and continues to exert an influence on him. The boy's behavior does not necessarily become pathological; rather, it may become unrealistic in very subtle ways. He may be too nonchalant in situations where he has a right to resent the father's punitiveness, or he may be a shade too polite with all older men, or he may be unusually awkward with a coach who pressures him to "fight harder." It often is difficult to distinguish behavior which is due to an unconscious conflict from that which is a person's special way of managing life's problems.

Repression should not be confused with more superficial means of "saving face." It is not a transparent rationalization or a "sour grapes" reaction (a game we play to convince ourselves we really did not want something after we have failed to get it). In all such self-deceptions, the individual readily admits, if pressed, that his underlying feelings are at odds with his manifest behavior. In contrast, the boy who says, as a result of repression, that he never really hated his father is as convinced of this as one who never had a deep hatred. As we shall see, this subjective sense of certainty is an important characteristic of an established mechanism of defense.

In addition to banning an impulse from consciousness, the individual can sometimes defend himself further by feeling, thinking, and acting in a manner diametrically opposed to the anxiety-provoking impulse. This defense is called reaction formation. The most common examples involve excessive expression of love as a reaction to anxiety over hostile impulses. An over-solicitous mother who is continually concerned for her child's health and safety may be defending against a desire to be rid of him. Excessive hate may also be a reaction formation against a threatening attraction. The rabid moralist may be overreacting to an unconscious fascination with the lasciviousness and fornication he condemns. In this vein, an Eastern religion claims that the God-lover must go through seven steps in order to attain heaven, while the God-hater must go through only one. The most common example of reaction formation in childhood is the toddler's marked disgust with feces after toilet training, which replaces the positive interest in them which preceded toilet training.

In projection, another type of defense, the individual attributes his unacceptable impulse to others. The adolescent girl who claims, "All boys think about is sex," may be projecting, especially if she regards herself as being "above that kind of thing." Belief in the devil can also be a means of preserving one's innocence by attributing unacceptable ideas to an external force. Preschool children are notorious for projecting. They people the world with wild animals and monsters which represent their own aggressive impulses. Quite often they believe such fancied creatures actually exist and become terrified of them. A three-year-old may be afraid to go to bed because he is convinced there is a wolf hiding under it.

Not all defenses banish the unacceptable impulse from consciousness. In displacement, for example, the impulse is expressed but directed to a different object. A familiar example is the man who is afraid of expressing anger toward his boss but comes home and shouts at his wife (or his children or his dog, depending upon which is the safest target). If the displacement is a true defense mechanism, he is unaware of what he is doing. When we feel that a person's reaction to us is excessive, that he is being ingratiating or fond or provocative or attacking far beyond the emotional intensity we usually evoke in others, then there is a good chance that the person is reacting to stimuli other than those we present; for example he may be displacing onto us his anger toward another person.

Anna Freud describes an early use of defenses in childhood in which the subjective sense of certainty is not absolute. If a child finds a given aspect of reality painful, he may deny that it exists and take refuge in compensatory fantasy. She cites the case of a boy who felt painfully weak and developed an elaborate fantasy concerning a tame lion who was his friend and who would do his bidding. The fantasy gave him the feeling of power which he lacked in reality. Anna Freud calls this protective maneuver denial in fantasy. Although the child was totally absorbed in the fantasy while it lasted, he did not really believe that it was true. In adult life the shock of losing a loved one may bring an initial denial—No, it is not true!—but the reaction is fleeting. A genuine denial of reality always indicates a severe emotional disturbance, because it means the ego is no longer performing its basic function of maintaining contact with the real world. Expressed in Freudian terminology, it means that reality testing has collapsed and wishes are regarded as objective realities.

The Loss of Reality Contact

We are now prepared for Freud's answer to the question, What happens to cognition when affect becomes intense? We shall begin with the subjective sense of certainty which accompanies a well-established mechanism of defense. Freud tells us that our most firmly held convictions concerning the kind of person we are and the nature of our relations to others may be erroneous. Defense mechanisms distort our understanding of ourselves or others and, in certain instances, of the physical world, and they make these distortions indistinguishable from accurate information concerning reality. Here we are at the heart of the matter. Freud assigned the ego two important functions: reality testing in order to achieve an accurate understanding of reality and the utilization of defense mechanisms in order to protect the individual from anxiety. These two functions are incompatible as far as the cognitive element in reality testing is concerned. Every defense entails an alienation from reality (A. Freud, 1946). As the child becomes too frightened to perceive an impulse or idea for what it is, he inevitably distorts his perception of himself and others. Repression, for example, represents a loss of contact with basic facts about the self and about the relation of the self to significant others. The fact "I am angry" is replaced by the fiction "I am not angry"; the fact "I hate you" by the fiction "I do not hate you." Each defense involves its special form of distortion.

Let us now review our comparison of Piaget and Freud. Both view cognition as developing from primitive to realistic ideation. Before reality is correctly understood, the child goes through a series of erroneous conceptualizations which, nevertheless, are appropriate to his developmental level. Thus, he thinks as realistically as his stage of development permits. A child must have direct access to information about reality, since he progresses in understanding through the confrontation of erroneous concepts with realistic data. He may still distort this information, but the distortions themselves will be challenged by subsequent information. Eventually, the child's inquiring intellect will achieve an accurate conceptualization of self, others, and the physical world.

Piaget essentially goes no further than this. The affects which complement cognitive development are never of a nature or an intensity to deflect the child's steady progress toward realistic thinking. However, for Freud, the child's inquiring intellect is never free to follow its own bent and his cognitive development is never only a matter of successively correcting errors in light of new information about reality. At any point cognitive development may be blocked by the appearance of anxiety and the consequent sacrifice of reality contact to diminish it. Any future information will be distorted in accordance with the particular defense the child employs; e.g., a child who projects hostility onto his parents will incorrectly perceive their behavior and be incapable of altering his erroneous concept to make it congruent with reality. Note also that we are not talking about a developmentally appropriate error when we talk of the cognitive distortion inherent in defense mechanisms. The child can have a realistic grasp of the terrifying impulse and correctly understand the meaning of the anxiety-provoking thought. "I hate you" is well within his ken. In sum: the necessity of reducing anxiety results in the loss of reality contact essential to the ultimate development of realistic thinking. Piaget says that during the developmental process, reality is distorted because there are thoughts a child cannot think; Freud agrees but adds that reality is also distorted because there are terrifying thoughts a child can but dare not think.

Defenses and Patterns of Thinking

Do defense mechanisms involve a new kind of thinking or does anxiety activate thought patterns already in the child's repertoire? Both Piaget and Freud imply that the thought processes are the same. While Piaget rarely addressed himself to psychopathology, he does regard the stages of cognitive development as uniform and universal. He might defend this stance unless presented with clear instances in which psychopathological thinking could not be embraced by his theory. Freud also regarded the difference between normality and pathology as a quantitative one; the thinking of a disturbed individual is not unique or without a counterpart in normal development.

From our discussion of early ideation we can see how certain thought patterns can become cognitive bases for various defense mechanisms. For example, both Freud and Piaget state that the toddler and preschooler attribute their private ideas to others as a consequence of failing to differentiate the subjective from the objective; projection can utilize this primitive thinking. What about the thinking involved in reaction formation, where feelings and ideas are turned into their opposite? One possibility is that the transformation is the result of repression. Every strong attachment is ambivalent, or a mixture of love and hate. If the child becomes anxious over the expression of one aspect of the ambivalence, he may repress it and allow the other to remain in consciousness. An alternate hypothesis is that the child might utilize an existing pattern of thinking when he turns a thought into its opposite. If so, he would have to be at a rather sophisticated level of cognitive development. "Opposite" is a logical concept, one which is not found in primitive ideations. In primitive ideation one thing can readily become another; e.g., recall our example of the dream in which a living room "really was a swimming pool." Changing a thing into its opposite, however, requires knowledge of a unique type of relation between two things; e.g., changing "I hate you" into "I love you" requires the understanding of the unique relation of love to hate. To return to the general issue of defense mechanisms and thinking, we can conclude that defense mechanisms utilize various patterns of thinking drawn from different developmental levels. Some of these patterns may be primitive, others advanced, but the need for defense does not seem to generate new ways of thinking.

However, it is important to remember that neither Freud nor Piaget studied psychopathological thinking in children. While followers of Freud have found his concepts of primary and secondary process thinking adequate to account for psychopathology, they have seen only a small, select sample of disturbed children. Followers of Piaget, who have a richer conceptual scheme at their disposal, have been relatively uninterested in studying extremely disturbed children. Whether intense affective states such as rage, sexual arousal, and guilt, produce qualitatively different ways of thinking in children, and whether all psychopathological thinking in childhood can be accounted for by Freud's or Piaget's theories, remain open questions.

Cognitive Style

We have already encountered the concept of style in our discussion of coping styles and family styles (see Chapter 4). Individuals and families have different patterns in handling stress and fulfilling the basic requirements of living. Some of these patterns can be ordered on a continuum of maturity or adaptiveness; but some represent nothing more than individual differences. A child can be reserved or effusive in a new situation; a family can be constrained or demonstrative in expressing affect. Each style has its special assets and limitations; all are equally defensible.

Cognitive style has attracted investigators with different theoretical backgrounds and appears under various labels. Generally speaking, cognitive style refers to the way one thinks rather than to the content of thinking. The concern is with the question *How?* (see Chapter 1). A person can think clearly, vaguely, impulsively, superficially, and so on. Obviously, style and content are related; we do not expect an impulsive person to come up with the same ideas as a thoughtful one. More formally, cognitive style is defined as a stable, individual preference in mode of perceiving and conceptualizing the environment. Individuals do not take in and process information in the same manner. Confronted with the same data, they approach and organize it in different ways: one person might skim over all of the information and be satisfied with a superficial impression, another may go thoughtfully from one bit of information to the next, determined to arrive at a comprehensive understanding. Cognitive style is of interest to us because it is an expression of personality. While the aphorism, "The style is the man," may be exaggerated, the *How* of thinking furnishes important insights into the kind of personality an individual has.

Diffuse and Articulated Thinking

Werner's development theory has been one of the most fruitful sources of concepts concerning cognitive style. Recall that Werner sees development as a process of differentiation and hierarchical integration. The child expands his behavior repertoire and becomes increasingly able to organize individual elements into complex patterns. Within this general framework, Werner's theory of self-differentiation is particularly relevant to the development of cognitive style. Self-differentiation involves the body, interpersonal relations, and self-control, all of which we shall describe before proceeding to the cognitive style literature itself. Instead of "environment" we shall use Werner's term "field" which refers to the child's perception or interpretation of events rather than to the objective events themselves.

As we know, the infant does not at first distinguish inner experiences from environmental events, the "me" from the "not-me." He must learn to differentiate himself from the psychological field. One step in this process involves differentiating the body from the rest of experience. The most primitive concept of me is "my body"—the gradually developing idea of what one's body feels like from the inside and looks like from the outside. We said "gradually developing" because the infant must learn what his body is like, just as he must learn what people and physical objects are. The concept of the body is not present from birth or achieved by a sudden insight but evolves over a period of years (see Chapter 10). Because the concept develops gradually, there are degrees of body differentiation. There is a good deal of objective evidence that one person may have a strong sense of body separateness, while another does not differentiate his body clearly from the field. The first individual may have a body concept which is richly detailed and well integrated. The latter individual may regard his body as a diffuse, poorly articulated "glob."

Interpersonal relationships play an important role in self-differentiation. After attachment, the infant is merged with the mother in the sense that her prescence is essential to his feelings of security and her actions and precepts are his primary guides to behavior. Subsequently, he becomes less dependent upon her and relies more upon himself. As his repertoire of realistic and proper behavior grows, he is able to evolve his own standards for appropriate action. The self becomes differentiated as an autonomous entity, just as his physical body becomes differentiated. Werner describes the self which is differentiated, integrated, and clearly separated from the environment as field-independent. A diffuse self which is still merged with the social environment is referred to as field-dependent. In the former instance, the "I" is firmly established and clearly the locus of decision; in the latter, the "I" is uncertain and the individual still looks to others for emotional support or for clues as to how to behave.

Finally, self-differentiation is aided by the development of self-control. Affective responses in infancy are diffuse and unmodulated. As self-control and defenses develop, emotional reactions are tamed so that they are no longer distressing and disruptive of interpersonal relations. Instead of being totally engulfed with the feeling of the moment, the child can observe himself. He has sufficient "psychological distance" to realize what feelings he is expressing and what effects he is having on others. Werner also adds that flexible use of a variety of defenses represents a higher level of self-differentiation than use of a few defenses which massively inhibit all emotional expression.

Witkin and his colleagues (1962) summarize a considerable body of data supporting Werner's thesis that the personality goes from being global and diffuse to being differentiated and articulated; and from being field-dependent to being field-independent. They also describe their own research on ten-year-old boys and their mothers. The children were interviewed about everyday events, and their accounts were evaluated in terms of the degree of articulation or the extent to which information was detailed, structured, and assimilated, at one extreme, or hazy, confused, and not well assimilated, at the other extreme. Highly articulated children described themselves, others, and events with a high degree of cognitive clarity, as compared with those low on articulation. Articulated children tended to be accurate, analytic, and detailed in their thinking; the thoughts of poorly articulated children tended to be global, impressionistic, and general. Other research showed that boys tended to be more analytic, girls more global; this sex difference was significant but not as large as the sex differences found in other intellectual and personality variables.

Children whose thinking was highly articulated had a strong sense of their own identity. They had a stable, individualized self, a well-defined psychological core. Consequently, their need to look to others for support or guidance, their dependence upon others for cues to values, attitudes, and behavior, was much less than in children whose thinking was poorly articulated. In Werner's terminology, the self was separated from the interpersonal field rather than fused with it. Even the body concept of articulated children was detailed and integrated, in contrast with that of children low on articution, who characteristically had a "global mass" body concept. In regard to self-control, the articulated boys had more highly developed defensive structures and could better modulate aggressive impulses than the boys low on articulation.

However, Witkin and his co-workers point out repeatedly that the degree of differentiation should not be equated with adequacy of adjustment, which involves a number of additional factors. The highly differentiated child can be overly controlled and isolated, just as the diffuse child can be overly dependent, helpless, and have an inadequate sense of identity. In summary, then, cognitive style—perceiving and thinking in an articulated, analytic way or in a global, diffuse way—is significantly related to the self, the body concept, and defenses and control. (For a companion study and a critique, see Kagan, Moss, and Sigel, 1963.)

Witkin and his associates also explored the relation of these two different cognitive styles to personality characteristics of the mothers and to aspects of the mother-child interaction. Mothers who facilitated differentiation in the children were high on self-assurance and self-realization. They allowed the children to be separate from them and to develop on their own. However, they also set firm limits for the children, thus promoting self-control. By contrast, mothers who impeded differentiation lacked self-assurance and self-realization. They complained of being worn out, unable to cope with problems, and dissatisfied with themselves, their husbands, their children, and the life they were leading. The child was seen by such a mother as contributing to her lack of self-realization through his failure to live up to her standards of academic achievement, appearance, sociability, and self-control. These mothers prevented differentiation in their children either by being overprotective or by coercing the children to conform to their models of appropriate behavior. Controls were too lenient or inconsistent or were enforced with anxiety and punitiveness. Finally, there was some evidence that extent of differentiation in the mother was correlated with extent of differentiation in the child. The authors stress that the highly differentiated mother, like the highly differentiated child, should not necessarily be regarded as optimally adjusted. She might be emotionally cold or impose exacting standards of behavior. Differentiation, then, is an important aspect of maturation, but it does not account for all aspects of the mature personality.

Creative Thinking

Creative thinking both intrigues and baffles psychologists. Since our understanding of personality would not be advanced by an extensive exploration of the perplexing and controversial literature generated by the topic, we shall limit ourselves to the presentation of a single study. Wallach and Kogan (1965), in reading introspective reports of creative adults, were struck by a phase in the creative process of free-flowing ideas and playful contemplation unhindered by concern for logic or correctness. In order to tap this kind of thinking, they presented certain tasks to fifth-graders as games. For example, they asked the children to list all the ways in which a newspaper or a cork or a shoe could be used. The context was always game-like rather than oriented toward giving the correct responses. Their criterion for creativity was the number of unique but relevant ideas the child produced. An obvious response would not qualify as creative; neither would one which was irrelevant.

Wallach and Kogan found that children differ in the extent to which they produce creative ideas and that creativity is independent of intelligence. When they studied the children with various combinations of extreme creativity and intelligence, they discovered a number of personality correlates. The highly creative, highly intelligent children were outstanding in every way. They had the highest number of reciprocated friendships. They were both aesthetically sensitive and sensitive in their reading of human behavior, that is, in grasping and reacting to subtleties of feelings expressed by others. They had some anxiety but not so much as to be crippling nor so little as to make them indifferent. Of all the children, they had the least doubts about themselves, and their work. They engaged in various disruptive activities in the classroom, probably because they were too inventive to stay within the bounds of routine. Children low in creativity and high in intelligence were excellent workers but uniformly cautious and reserved. Far from being disruptive in class, they hesitated to express opinions at all. They responded to overtures of friendliness with reserve or aloofness. They seemed fearful of making a mistake or taking a chance. Although capable of being creative upon request, they preferred to stay within the confines of conventional ways of viewing the physical and social worlds. They were also the least anxious of all the children, perhaps because they were model examples of the bright, conscientious, well-behaved child. The highly creative children with low intelligence were least able to concentrate in class, were lowest in self-confidence, and had the strongest feelings of hopelessness about themselves. These children were the most isolated; they both avoided others and were shunned by others. In general, they seemed out of joint with themselves and their environments. They could not succeed in the conventional academic pursuits, and their talents for originality were a disruptive influence in the group. Their fear of being evaluated was particularly high. The low creativity, low intelligence children presented a more harmonious, consistent picture than the preceding group. Although not rewarded for academic performance, they were more outgoing and more self-confident. They also were reasonably at ease in meeting the social requirements on the playground and in the classroom, even though they were particularly poor in reading human behavior.

Incidentally, Wallach and Kogan's finding that the consistently high and consistently low groups were better adjusted than the high-low disparate groups is interesting in itself. It raises the general issue of the relation between the individual's internal consistency and his adjustment. Both clinical and objective studies point to the importance of a consonant development of the many facets of personality. When there are gross discrepancies, the child is at odds with himself and/or with the environment and is subject to an unusual amount of distress.[7]

One final caution. A majority of studies of cognitive style are correlational. This means that certain personality variables are correlated with certain measures of cognitive style. They reveal nothing about the direction of causation. In the Wallach and Kogan study, for example, we do not know whether a child is self-confident because he is creative, or whether he is creative because he is self-confident, or whether some third factor causes both.

Conclusion

We have discussed three different ways of viewing thinking. Piaget's theory represents a pure distillation of cognitive development, an idealized account of the child's progress from infantile distortions of reality to adult accuracy in understanding reality. Theoretically, emotional development complements the cognitive changes, but, for all practical purposes, it plays a minor role and is never disruptive. Freud was also concerned with the development of realistic thinking, but he described the many ways thinking can be used to serve powerful emotional needs for security and relief from anxiety. A realistic understanding of self, others, and the physical world is achieved in the face of tremendous odds, and even the most mature individual has intellectual

[7] For a summary of the literature on creativity, see Golann (1963). For a discussion of creativity and its relation to Witkin's studies of differentiation, see Bloomberg (1967).

blindspots and distortions. Cognitive style describes how a child thinks independently of his particular stage of cognitive development; e.g., a child can be either articulated or diffuse in his animistic response to Piaget's question, What makes the clouds move? Cognitive style is neither pure cognition (like Piaget's stages of cognitive development) nor thinking distorted by the need for security (like Freud's defense mechanisms). It is a distinctive way of taking in and assimilating information about reality; yet, it also is related to a number of personality variables such as self-control and autonomy. How the child thinks is an important index to the kind of personality he has.

7 Anxiety, Sex, and Aggression

In our discussion of the development of self-control (see Chapter 4) we were not concerned with the kinds of behavior being controlled. It would be a mistake to conclude that the behaviors themselves do not matter to personality development, since those related to sex and aggression are pre-potent in their influence. Their socialization involves an inescapable, frequently affect-laden confrontation between parent and child. The outcome of this confrontation will determine, to a significant degree, whether the young adult is on good terms with his impulses or whether he will be terrified of or overruled by them. At one extreme lies the danger of excessive inhibition and neurosis; at the other lies the danger of impulsivity and violence. Socialization also forces primitive sexual and aggressive behavior to assume multiform disguises which, in turn, affect many of the young adult's most important interpersonal relations.

Before proceeding to sex and aggression, however, we must round out our discussion of anxiety. As we have seen, anxiety is a major force in socialization, one which is sufficiently strong to counter and contain the child's sexual and aggressive impulses (see Chapter 4). But anxiety is also a double-edged sword—it can get out of control and twist behavior into the grotesque irrationalities which we call psychopathology (see Chapter 6). We shall inquire into the various origins of anxiety and into the factors which make it both a constructive and a destructive force in personality development. Then we shall trace its course through childhood and adolescence and, finally, discuss its relation to psychopathology.

Anxiety

Origins of Anxiety

Signal Anxiety. We have already presented the concept of anxiety as the anticipation of pain (see Chapter 4). The concept originated with Freud (1936) who used the term signal anxiety, and it was subsequently translated into the terminology of learning theory (Mowrer, 1939). Having once experienced pain in a situation, the child need not re-experience it repeatedly. He comes to anticipate the noxious event–components of the pain response become anticipatory, so that distress is experienced but at a less intense level. As the child develops cognitively, he also can become intellectually aware of imminent danger. The moderately painful anticipatory response, either by itself or in conjunction with the intellectual warning of "Danger ahead!" constitutes anxiety. Forewarned, the organism can take proper measures to avoid re-exposure to the noxious situation. As we have seen, signal anxiety plays a central role in socialization. Here pain is either physical punishment by the parent or withdrawal of parental love. Anxiety warns the child that his behavior will have physically or psychologically noxious consequences. If the distress of anticipated punishment outweighs the pleasure of achieving a desired goal, the child will refrain from doing the forbidden act.

Fear of Strangeness. Anxiety has origins which are unrelated to pain. There is a purely cognitive phenomenon called fear of strangeness or fear of novelty. Hebb (1946, 1949) claims that anxiety is aroused when an ordinarily familiar stimulus is presented in conjunction with unfamiliar elements. The clearest examples come from animal research. Chimpanzees react with fear when familiar attendants exchange lab coats or when they are presented with a plaster cast of a chimpanzee head. A totally unfamiliar stimulus would not have had such an effect; the disruption of an established expectancy is essential. Bronson (1968) claims that stranger anxiety in the second half of the first year (see Chapter 2) may arise because the infant has developed a reasonably clear image of the mother and is distressed by adults who are similar yet also dissimilar from her. Valentine (1930) also observed fear in one- to two-and-a-half-year-old children when a familiar stimulus was distorted; e.g., his children were terrified of him when he crouched in an animal position, his daughter was frightened of her brother dressed in a costume, and she developed an aversion for her favorite doll after it was dismembered.

The fear of novelty is an interesting addendum to our discussion of curiosity (see Chapter 3). Novelty not only excites and lures, it can also frighten. Psychologists know little about the differential conditions which produce one or the other kind of reaction. Unexpectedness of stimuli seems to enhance anxiety, while repetition of a novel stimulus seems to make for adaptation, exploration, and even enjoyment (Ellesor, 1933). Ainsworth (1967), in her study of attachment, notes that the infant uses the mother as "a secure base" from which to explore the world. This suggests that physical or visual contact with the mother reduces the infant's fear of strangeness.

Intense Stimulation, Helplessness, and Constitutional Predisposition. Anxiety has been linked to intense stimulation. One of Freud's theories of anxiety states that birth is traumatic because the neonate is "flooded" with intense physiological stimulation. Having had a placid intrauterine existence, he is overwhelmed by the strength of his physiological needs. Although Freud revised his theory, the notion has persisted that intense physiological stimulation in the vulnerable neonate readily exceeds his level of tolerance and produces a reaction of distress (Benjamin, 1961; Bronson, 1968).

Some theorists have emphasized the helplessness of the infant, his total inability to relieve distress, as the critical factor in engendering anxiety. Clinical studies strongly suggest that the neurotic is most terrified, not when he is anxious, but when he does not know what produces the anxiety and what he can do to diminish it. Anecdotal evidence also supports the relations between helplessness (in the sense of enforced inactivity at critical times) and anxiety; e.g., soldiers' anxiety is often greatest in the inactive period before the fighting begins, and the typical picture of the expectant father has him pacing nervously and helplessly in the waiting room.

Finally, infants have been observed to differ widely in the strength of their initial reactions, and clinicians have speculated that neonates who are more totally distressed will be predisposed to more intense anxiety reactions in later development. Therefore there may be a constitutional predisposition to intense anxiety reactions.[1] Parenthetically, constitutional vulnerability creates a practical problem in studying the etiology of anxiety. An infant may be so strongly predisposed that even the caretaking of a good mother brings insufficient relief; on the other hand, his initially normal reactions may be intensified to a pathological degree by lack of adequate mothering. Dissimilar environmental conditions produce the same end result. Knowing only that a toddler is highly anxious, then, we can say nothing definitive about the condition being due to constitutional factors or faulty caretaking. In Chapter 10 we will examine other complications which constitutional factors introduce into the study of personality variables.

Anxiety in Infancy. By assuming that all of the above concepts of anxiety have validity, and by relying on the few clinical and objective studies which have been done, we can reconstruct the development of anxiety in infancy. (For a more comprehensive review of studies of anxiety in children, see Sarason, 1960.) The initial encounter with anxiety may come at birth when the immature organism is "flooded" with intense physiological stimulation which the neonate is helpless to reduce. Since neonates differ widely in the strength of initial reactions, there may be a constitutional predisposition to intense anxiety. The environment also plays a crucial role in the fate of anxiety in early infancy. Bronson presents evidence showing that physical contact, especially being rocked, reduces anxiety. Similarly, Kessen and Mandler (1961) argue that the newborn has periodic distress reactions (anxiety) which can be inhibited by certain internal stimuli, such as sucking and ingesting food, or by the kind of external stimuli provided by caretaking. In sum, caretaking must do for the infant what he is helpless to do for himself—namely, reduce anxiety.

[1] There is also some clinical evidence that initial hypersensitivity to stimuli and overexcitability may produce psychopathological results (see Bergman and Escalona, 1949).

Signal anxiety and the fear of strangeness obviously depend on the development of anticipation and on cognitive development. Until the infant experiences environmental regularities frequently enough to associate two events, one cannot become a warning signal for the other. To illustrate: one classical signal anxiety, according to psychoanalytic theory, is separation anxiety, which cannot appear before the infant has associated the mother's presence with comfort and her absence with helplessness and distress. Similarly, the infant cannot be distressed by strangeness until he has experienced objects frequently enough for them to become familiar. Stranger anxiety may be the kind of cognitive fear Hebb (1946) delineates in which a stimulus is both familiar and unfamiliar. Both separation anxiety and the fear of strangeness are lessened by contact with the mother, either as a haven of safety or as a secure base from which to explore the world.

This reconstruction also helps us in our search for the conditions under which anxiety aids or blocks development. In the neonatal period there may be both innate predispositions to anxiety and innate relaxors, such as sucking, ingesting food, being rocked, and the external stimuli provided by caretaking. The balance between the intensity of the initial distress reaction and the relaxation provided by caretaking may be one determinant of subsequent intensity of the anxiety reactions. Caretaking continues to be important in the latter half of infancy when using the mother as a secure base from which to explore the world helps allay the fear of strangeness. Thus caretaking throughout infancy may serve to prevent anxiety from becoming unduly distressing.

Signal anxiety, with its crucial protective function, depends on the development of expectancies. We can infer that anything which adversely affects the child's ability to anticipate will impede the development of signal anxiety. Anticipation, in turn, depends upon a number of variables essential to self-control (see Chapter 4)—an orderly environment, attention, memory, the ability to see cause and effect relations, plus (the psychoanalysts insist) the emotional security which comes from being loved. If the ability to anticipate is weak, one might expect both signal anxiety and general self-control to be weakened also.

The Developmental Course of Anxiety

The Preschool Period. According to Freud, the nodal anxieties between two and six years of age are castration anxiety and guilt (which he calls "moral anxiety"). In the nonpsychoanalytic descriptive literature, Jersild (1954) states that the fear of noise, strangeness, pain, falling, and unexpected movement decline between two and six years of age, while fear of the dark, imaginary creatures, being alone, physical harm, and wild animals increase. There are striking individual differences in the stimuli which elicit fear, just as there are a variety of ways normal children cope with fear.[2] Anxiety can have either a facilitating or a disruptive effect on behavior, the usual interpretation being that a moderate degree mobilizes the child to make the most

[2] See, for example, Moriarty's (1961) rich account of the ways preschoolers manage the stress of being tested.

effective use of his resources, while an intense degree results in disorganization. Girls tend to have more fears than boys, bright children more than average or dull ones. Finally, there is evidence that children's fears are similar to those of their parents. In general, the descriptive picture is one of diversity in regard to the stimuli eliciting anxiety and its behavioral effects.

There are diverse theoretical explanations for anxiety in the preschool period. Castration anxiety and guilt both signal impending trauma, the former involving physical mutilation, the latter, loss of love. Freud also maintained that a strong drive which has no socially acceptable outlets can bring about a recapitulation of neonatal trauma with its combination of intense stimulation and helplessness. The arousal of strong sexual and aggressive impulses which the child cannot master terrifies him and shatters his self-control. Learning theorists deny the importance of Freud's nodal anxieties and account for developmental changes in terms of objectively verifiable principles of learning. Anxiety spreads on the basis of classical conditioning. Neutral stimuli present at the time of anxiety can come to elicit it. A child who is attacked by a dog may come to fear not only the dog but also a number of incidental stimuli, such as the physical features of the place where the attack occurred. Anxiety can spread on the basis of stimulus generalization; e.g., the child comes to fear all dogs or even all animals resembling dogs. The empirical finding that children's fears are similar to those of their parents suggests that imitation is also involved in the learning of anxiety. However, there is one descriptive fact which is difficult to explain. This is the phenomenon of incubation, in which the strength of anxiety increases with the passing of time, even though the child is not re-exposed to the frightening situation. In our example, the child may be even more afraid when he returns to the scene of the attack after a period of a few weeks than he was initially. Instead of diminishing with time, therefore, anxiety actually can increase.

Mechanisms of Defense. The preschool period is the time in which mechanisms of defense are elaborated. Their purpose is to protect the child from the pain of intense anxiety (see Chapter 6). Since we have already discussed the function of defenses and described a number of them, we shall turn to the more general topic of their role in facilitating or impeding personality development.

For the psychoanalysts, the fundamental problem of adaptation is that of obtaining maximum impulse gratification in a social setting which inevitably curbs the free expression of primitive impulses. Since socialization entails punishment, anxiety is inevitable, and, since no human being chooses to live in a state of distress, some use of defenses is essential. In the sense that everyone must use them, defenses are normal. Anna Freud (1946) goes a step further in saying that defenses are essential to ego development. The child's increasing ability to control himself through the use of defenses facilitates his adaptation to social reality. Restricting the expression of impulses and adapting to reality are merely two sides of the same coin. While repressed hostility toward a sibling may remain forever in the unconscious, the resulting ability to get along amicably with siblings and peers is an essential step in the child's personality development. Responsibility and conscientiousness may represent the preschooler's reaction formation against the primitive impulse to mess whenever and wherever he pleases, but, nonetheless, they are important personality assets.

But the consequences of defense mechanisms are never simple. While they may protect and strengthen the ego, they also can prevent further development. That is why Anna Freud emphasizes the ability to abandon defenses once the threat has passed. If they continue to operate, they may trigger a chain of events leading to neurosis or they may significantly impoverish the individual's life. Recall our example of the boy who was frightened of his destructive wish toward his father and, as a defense, became overly compliant. The defense which initially protects him from excessive anxiety may prevent him from learning that a certain amount of aggressiveness with men is tolerable and even desirable. He may be as subservient at twenty or forty as he was at six, even though his talent is exploited and he fears making any move that might antagonize his superiors. Defense mechanisms may prevent the child from using reality testing to determine whether his fear is still justified. By depriving him of openness to new experiences, defenses can block the growth of personality.

In sum: Anna Freud distinguishes between defenses which generate socially adaptive, growth-promoting personality traits from those which offer protection at a given developmental stage but which should be relinquished when they are no longer appropriate to social reality.

Under what conditions do defenses help the child master anxiety and under what conditions do they make him more vulnerable to anxiety? Anna Freud does not agree with some psychologists who claim that defenses are either healthy or unhealthy to the extent that they permit impulse gratification. By this criterion, displacement would be healthy because it allows some expression of the impulse, while repression would be unhealthy because it does not. She suggests three criteria for judging the healthy utilization of defenses. The first is a developmental one. Defenses have their own chronology—denial and projection come early in childhood; repression and reaction formation come later—and a significant deviation in timing is suspect, such as a three-year-old who is correct and fastidious in his behavior or an eight-year-old who believes malevolent people are conspiring against him. Second, the healthy individual is capable of using a variety of defenses and is able to relinquish them once the threat has passed. Ultimately, however, pathology is not a matter of the kind or flexibility of defenses, but a quantitative matter of the amount of impulse inhibition. This is Anna Freud's third criterion. Excessive inhibition of impulses, no matter what the mechanisms, inevitably generates more anxiety than the child can master and produces neurosis. The child who is totally alienated from his sexual and hostile feelings finds that his defenses have been self-defeating. (The rationale for Anna Freud's position will become clearer when we discuss the psychoanalytic theory of sex and aggression in the next sections.)

We have been discussing defense mechanisms from the psychoanalytic point of view. There are many personality theorists who regard the mechanisms of defense as significant variables but who reject the Freudian conceptualization in varying degrees. Dollard and Miller (1950) define defense mechanisms in the language of learning theory, replacing Freud's mentalistic account with behaviorally defined concepts which can be objectively validated. Swanson (1961) not only rejects Freud's mentalism but also argues that defenses are created by interpersonal forces. He pictures the child facing the following dilemma: On the one hand, he has persistent but socially unacceptable impulses pressing for expression; on the other hand, if he expresses such impulses, he risks losing the rewards of love, respect, and approval, and suffering punishment and censure. The child is forced to block impulse expression at least partially, in order to maintain essential interpersonal relationships. Defense mechanisms represent techniques for managing impulses while staying on good terms with the social environment.

Swanson further speculates that particular social situations produce particular defense mechanisms, thus departing radically from the Freudian theory that they are created by the ego. Reaction formation, for example, occurs in close-knit, secure families whose members feel obligated to contribute to one another's comfort and to anticipate one another's needs. Disruptive actions and even disruptive thoughts must be replaced by supportive ones. In essence the child says, "I cannot be hostile because that would undermine the basis of family relationships. This would be too upsetting to others and too threatening to me. I must be loving instead." Another defense involves turning anger against the self. Instead of saying, "I hate you," the child says, "I hate myself" and feels worthless, guilty, and inferior. This defense occurs when the child wants to gain rewards but is faced with standards of behavior which are so demanding that he is bound to fail. For example, a mother may require her child to be completely nonaggressive. The child tries to comply but inevitably fails. His anger over being given an impossible task would jeopardize the mother's love, so he focuses on his repeated failure to meet her expectations and regards himself as worthless.

Swanson's account of the origin of defense mechanisms is specific but incomplete; Freud's theory is comprehensive but insufficiently specific. Swanson is most adept in accounting for defenses against hostility; it is not clear how erotic impulses would be defended against, so that, for example, a strong attraction would become a strong hatred. Thus, Swanson's program for delineating social antecedents for defenses is not fully realized. Freudian theory is versatile in that it accounts for defenses against both hostile and erotic impulses, yet it fails to explain the actual process by which various defenses originate. The theory lists ingredients—impulse, anxiety, the general level of personality development—but it is never clear how or why these ingredients combine to produce one defense rather than another. Too often the explanation is in terms of a personified ego which "defends itself" against anxiety. One day we might find that both Freud and Swanson teased out important elements in the formation of defenses, since both ego functions and social constellations are involved. At present, we are left with a Hobson's choice in regard to the relative merits of the two theories of etiology.

Middle Childhood. The descriptive literature in middle childhood indicates that fears become more general while their content involves imaginary, supernatural, and improbable dangers. The former finding is understandable in light of the intellectual development and increased integration of the school-age child. Instead of living in a world of specific reactions to specific situations, he begins to think in terms of generalities and overall plans. The second finding is unexpected and more interesting. In middle childhood the cognitive separation of external world and internal ideas, feelings, and fantasies is achieved (see Chapter 6). The separation has not reached its ultimate sophistication, but the child has a firm, workable grasp of reality. A three-year-old might be afraid to go upstairs because he literally believes a wolf is waiting there, but a ten-year-old would regard such an idea as ridiculous. Yet, children as old as nine to twelve fear unrealistic happenings (being attacked by a lion, or even by witches and ghosts) more strongly than realistic dangers (being hit by a car). One would expect that the child's fears would be exorcised by his realization that fantasy figures do not exist and that certain events are highly improbable.

Generally speaking, we should not be surprised at a lag. Development is not a grand forward march on all fronts. On any given day a child functions at different levels, for instance, becoming more immature when he is tired or hungry than when he is alert. The overall rhythm of development is one of advance and retreat and advance again, while certain areas of personality develop more rapidly than others. However, knowing that the general pace of development is erratic does not satisfactorily account for the specific phenomenon of unrealistic fears in middle childhood. We can only speculate as to the cause. It may be that anxieties in the toddler and preschool periods are so much more intense than other experiences that they are particularly resistant to change. Or it may be that the malevolent animals, witches, and ghosts symbolize the punitive parent of the preschool period and the preschooler's fear of retaliation for transgression. With the development of conscience, this fear would be internalized as the child becomes his own punitive parent (see Chapter 5). Since conscience in middle childhood is strict, and the child strongly desires to be good, the fear of transgression would be strong and would perpetuate the image of retaliating monsters.

Objective studies of anxiety in children are rare in light of its importance, and the research tends to be piecemeal. An exception is Sarason's (1960) investigation of test anxiety in children between eight and eleven years of age. Sarason is not interested in test anxiety per se; he chose it on the practical grounds that it is a clearly defined, frequently encountered experience of anxiety and on the theoretical grounds that it reflects basic conflicts within the child.

Why should a child be anxious about a test, especially in those cases where he has repeatedly done well? Sarason argues that the child is concerned about being evaluated by his teacher and more especially by his parents. What is being tested is not only the intellect but also the child's loveworthiness. He fears parental disapproval and the implicit threat of loss of love if he does not meet their standards. What, then, makes for an unusually high degree of text anxiety? Sarason has a number of hypotheses which were tentatively confirmed. The mother of the highly anxious child is uncertain about her own capabilities, and to compensate for her lack of firm inner values, she strongly supports conventional ones concerning right and wrong. Instead of being attuned to her child's individuality and age-appropriate needs, she is invested in public, external standards of "good" behavior. More important, she makes affection contingent upon the child's conformity to her standards of "good" behavior.

The child reacts with hostility to the demand that he surrender his individuality and autonomy and become subservient to maternal standards. Yet he fears the mother will abandon him if he expresses his resentment. Therefore he tries to deny his anger and becomes overly conforming in order to assure himself that he will be loved. The denial does not make his anger vanish; rather, it is turned back upon the self. Instead of shouting "You're bad!" at his parents, the child regards himself as bad for wanting to defy them. He also somaticizes this self-aggression and becomes overly concerned about damage to his body through injury or illness. Being tested in school, then, becomes just one more instance of a generalized anxiety over the consequences of not meeting parental standards.[3]

The study is interesting in its own right, and also because of its similarity to a number of clinical investigations of childhood disturbances centering around anxiety. The general theme is as follows: The parent, usually the mother, has a personality problem and becomes threatened when the child's developmentally normal demands touch on her area of vulnerability. She reacts by making love contingent upon the child's serving her own security maneuvers. The child is enraged, but feels anxious because his anger jeopardizes maternal affection. Subsequently, he sets up his own defensive maneuvers to keep his anger from becoming conscious and from being expressed. The particular maneuver will determine the particular kind of deviant behavior he will display. Anything which increases the strength of his unconscious anger will cause a concomitant increase in anxiety and may necessitate further defenses.

Let us put Sarason's research in another context. Recall our discussion of love oriented parental discipline (see Chapter 4) and its effect of producing a responsible, conforming, intropunitive child. Here we have a normal variation on Sarason's theme. The parent uses the granting and withdrawal of love to inculcate socially desirable behaviors. The affects aroused in the child are anxiety and guilt—anxiety over losing love, guilt over the pain his oppositional behavior might cause his loving parent. He turns his resentment against himself, regarding himself as bad rather than accusing others. As long as the balance between behavior inhibited and love received is maintained, the anger over being inhibited is kept in bounds. The child can develop into an acceptably socialized adult. However, as the parent loses touch with the needs of the child and begins to require massive inhibitions as the price for love received, the anger and the ensuing anxiety and guilt become unduly strong. The anxiety may be expressed in a test situation and still be considered within normal limits, but it also may result in a school phobia or other neurotic disturbances. The difference between normal and disturbed behavior, therefore, is only a matter of degree.

Adolescence. During adolescence fears become developmentally appropriate. Some reflect the heightened self-consciousness resulting from the adolescent's indeterminate, transitional status; e.g., he fears making a mistake, looking ridiculous or naive, and being teased or ridiculed. Adolescents fear loss of prestige with the group through

[3] In a subsequent publication, Sarason recommends that more attention be paid to factors within the school itself which contribute to a high level of anxiety (see Hill and Sarason, 1966).

failure to live up to group-prescribed behavior. Boys, in particular, are concerned about the future, especially in regard to their ability to assume responsibilities and to find a suitable job. Sexual fears take the form of anxiety over physical or behavioral deviations from the norm, both of which raise doubts as to sexual attractiveness or general sexual adequacy. Probably fears of being on display, such as reciting in front of the class, are derivatives of general sexual anxiety.

Anxiety and Psychopathology

The clinical literature on anxiety in middle childhood and adolescence is heavily weighted with studies of various psychopathological conditions. Implicitly or explicitly, it generally supports Freud's contention that anxiety holds the key to understanding psychopathological behavior. As we have seen, defense mechanisms can distort reality and block the development of appropriate ways of dealing with the anxiety-producing situation. The individual is apt to be labeled deviant or emotionally disturbed as his distortion becomes extreme and his behavior grossly inappropriate to his developmental stage. The hysterical young woman, for example, can utilize the defense of repression and deny she has sexual feelings while dressing in an obviously provocative manner and complaining that men are always making passes at her. The compulsive adolescent may live in terror of dirt and disorder and may expend a disproportionate amount of his energies in nonutilitarian actions, such as washing his hands, adding a column of figures again and again, repeatedly returning home for fear he has not turned off the stove or the lights or shut his closet door. Under the extreme pressure of anxiety, the adolescent can regress to the primitive level of autistic thinking; e.g., in his omnipotence he can believe he is the Son of God, or, in his confusion between inner feelings and external reality, he can be convinced that the world is coming to an end.[4]

Conclusion

Anxiety may be unique among personality variables in its constructive and destructive potential. Without it, the organism might not survive, yet it can perpetuate the most primitive and bizarre behavior. On the positive side, anxiety raises the level of the child's motivation and enables him to make maximal use of his resources. As a warning signal, anxiety protects the child from re-exposure to painful and destructive situations. Certain defenses against anxiety can engender socially adaptive, growth-promoting personality traits. However, intense anxiety can block personality development in a given area, can perpetuate defenses which are no longer appropriate to the reality of the individual's life, and can force thinking and behavior to assume bizarre, primitive forms in which omnipotence and magic abound and inner fantasies are confused with external realities. In this last instance, the individual clings to his psychopathological distortions out of the sheer terror of what would happen if he relinquished them.

[4] For more detailed accounts of the development of deviant behavior, see White (1964) and A. Freud (1965).

We have also seen that a variety of factors determine whether anxiety will be moderate and manageable, or intense and overwhelming. There may be an innate predisposition to react with intense anxiety which makes some children more vulnerable than others. Too much or too little control both favor the production of overwhelming anxiety; in the former case the child attempts to repress all expression of strong drives and becomes the helpless victim of intense stimulation when these drives are aroused; in the latter instance, anticipations are not firmly established and the child has no reliable protection against re-experiencing traumatic situations. Mothering may reduce neonatal anxiety and, later, counteract the fear of novelty. However, if the mother subsequently demands that the child sacrifice his autonomy and individuality in order to retain her love, she is apt to create a high level of anxiety over unexpressed resentment. Thus, the question, How can the child avoid overwhelming anxiety? has no simple answer, since it involves constitutional factors, self-control, attachment, and parental discipline.

Sex

Freud shattered forever the Victorian image of childhood as the age of innocence. Far from being nonsexual—and, for that matter, nonhostile—children have both impulses. Only the demands of society force children to check, moderate, and disguise their primitive nature. If all goes well, they will develop into adults who are both civilized and on good terms with their erotic and aggressive feelings. Failure to master their impulses may be a prelude to adult psychopathology. While Freud opened psychologists' eyes to the fact of early sexuality, his interpretation of the facts in terms of a sexual instinct created a storm which has not subsided. We shall turn to his theory after presenting an alternate, less controversial one. This approach states that the infant must (1) learn the concept of classification by sex and (2) learn the special set of behaviors and feelings society prescribes on the basis of his being a boy or a girl. Our terms for these two facets of sexuality are gender typing and sex role identification.

Gender Typing and Sex Role Identification

Description and Influence. The facts of gender typing and sex role identification are not too obvious to mention—they are worth mentioning because they are so obvious. The neonate must learn that he is a "boy" or a "girl," since he has no innate knowledge of the matter. Gender is one of the most significant labels he will come to apply to himself and, barring exceptional circumstances, it will last a lifetime. He will also learn that, as a boy or a girl, society prescribes behaviors and feelings which are appropriate and inappropriate. These prescriptions define his sexual role and extend to many areas of functioning. The child who learns his sexual role well will be rewarded by social approval; one who does not conform will be punished by ridicule, disparagement, or isolation.

Objective research (see Kagan, 1964) testifies to the clarity of society's prescriptions concerning sex role. Boys should be dominant and seize the initiative; girls should be nurturant and sociable. Boys are expected and even encouraged to be aggressive when threatened or attacked by another boy; girls, by contrast, are required to inhibit both physical and undisguised verbal expressions of aggression. Girls can express feelings of tenderness and fear; boys are expected to be unsentimental and stoic in the face of pain and anxiety. Boys should be pragmatic and work oriented; girls should be domestic. In physical appearance, the ideal boy is large and strong, the ideal girl small and pretty. As adolescence approaches, the boy is expected to be sexually assertive and the girl to refrain from open display of sexual urges. The male gravitates to the acquisition of money and power, the female to domestic skills and child care.

If one reacts to this list of differences by saying, "I don't need to read a book on psychology to tell me that"—that is just the point. Society makes sure that everyone knows the prescribed sex roles; parents know, advertising men know, movie and television directors know, gym teachers and shoe salesmen know. The message is redundant and inescapable. Society does not take chances by being ambiguous or subtle.

And children receive the message. Between three and seven years of age they gradually learn sex identification for themselves and realize that people can be categorized according to sex. The association between maleness and aggression is firm and pervasive; preschool and school-age boys are both more aggressive than girls in their overt behavior and in their fantasies, while fathers are seen as more punitive and dangerous than mothers. For their part, adults regard men as more aggressive than women and expect more overt aggression from boys than girls. The same consistency between generations is seen in the areas of dependency, passivity, and conformity. Girls have more of these qualities in their overt behavior and in their fantasies, just as parents expect them to.

The influence of sex role does not stop here, however. It spreads to games and toys. Boys prefer sports, machines, and activities capitalizing on speed and power; girls prefer games and activities involving babies, home, personal attractiveness, and fantasy themselves as nurses or secretaries. Boys' preference for active games starts as early as three years of age and becomes increasingly strong, while girls are more variable in their preferences until nine or ten years of age.

There is another current running here: Girls more frequently express a desire to be a boy or to be a daddy when they grow up than boys express preferences for the female sex role. Kagan (1964) notes that the association of the male role with aggression, dominance, and independence, and the female role with passivity, nurturance, and emotionality has remained constant in spite of increasingly liberal attitudes toward sexual behavior in women and an increasing number of working mothers during the last twenty-five years. Behavior can change while sex role standards remain constant.

Sex role also affects intellectual functioning. Preschool boys ask more "how" and "why" questions; girls are more concerned with social rules and correct labels for objects. The stage is set for males to be more analytical in their thinking, females more conventional. Starting at three years, boys also tend to be more autonomous and persistent in their problem solving behavior. These sex differences increase until adolescence, when females typically feel inadequate in the face of problems involving analysis and reasoning. Mathematics, science, and mechanical reasoning are viewed as being in the masculine domain, and girls consistently perform less well in such areas. Once again, the male is rational and objective, the female is social and feelingful.

Sex role identity may also be related to the reversal in academic achievement in which girls typically outperform boys during the early school years but lag behind them in adolescence. The primary grades are "a woman's world." Typically the teacher is a woman, and she values the feminine qualities of docility and obedience and emphasizes feminine activities, such as coloring and singing. Rote memory, giving the correct answer, and quiet concentration are important to learning basic academic skills. The aggressive, independent, active boy is out of his element. As adolescence approaches, however, the boy begins to become concerned with his future and his ability to acquire power and prestige. Academic success is now seen as instrumental in achieving vocational success. Moreover, practical knowledge which is useful in achieving success (such as the sciences) is differentiated from impractical knowledge (such as the humanities, language, and history) and the former is valued more highly than the latter. The adolescent girl's future is rarely as work oriented as the boy's, since it revolves around domesticity and maternity. The pressure to achieve is less. In addition, the adolescent girl may hestitate to compete actively with a boy, since this runs counter to her prescribed role of submissiveness.

Sex role identity is not only extensive in its influence, it is also stable over time (see Kagan and Moss, 1962). Childhood behavior between six and ten years of age is predictive of behavior in adolescence and adulthood, provided that it is congruent with the sex roles defined by society. Passivity and dependence in childhood, for example, are predictive of adult behavior for girls but not for boys; aggression and sexual assertiveness, on the other hand, show continuity for males but not for females.

Theoretical Explanations. It is easier to describe sex role identification and gender typing than to account for them. A learning theorist might explain them in terms of reinforcement and imitation. Parents, peers, and society in general differentially reward desirable responses and punish undesirable ones in order to inculcate appropriate sex role behavior and attitudes. The child's need for love, esteem, and status make reinforcement a potent technique for effecting behavior change. In addition, parents and certain other adults serve as models of male and female behavior. Through imitation the child learns more than the parent consciously teaches. (Since we have discussed both reinforcement and imitation in Chapter 4, we will not elaborate them further at this point.)

Kagan suggests that once sex role standards have been grasped, the child becomes self-rewarding or self-punishing as he perceives that he is realizing or deviating from the sexual ideal. As the boy becomes a competent athlete or the girl becomes socially poised, for example, feelings of masculinity or femininity are strengthened; conversely, as these attributes fail to develop or are lost, the child becomes increasingly concerned over his failure to realize the ideal. Kagan's hypothesis, while it falls within a learning theory framework, goes beyond simple reinforcement. It involves an ongoing process of self-monitoring. The child is constantly scanning his social environment for information concerning what is expected of him. As we have seen, the social environment is particularly explicit in its prescriptions for acceptable behavior. The child then compares these expectances with his actual behavior. If the match is close, he is pleased. If there is a discrepancy, he experiences various negative affects, such as inferiority, anger, dissatisfaction. The positive affects motivate him to continue his appropriate behavior; the negative ones motivate him to change his behavior in order to bring it into line with social expectations. If the negative affects are too distressing, however, they can produce a defiance of convention or socially unacceptable defensive behaviors designed to protect the child from distress. In sum, reward or punishment is contingent upon the congruence between two sets of information, one from the social environment and one from the self. Cognition thus plays an important role in the maintenance of sex role identity.

Kohlberg (1966) makes cognition even more central in his discussion of gender typing and sex role identity. He disagrees with learning theorists who maintain that sex role identity represents an internalization of sexually prescribed behavior by means of reinforcement and imitation. He also takes issue with the Freudian emphasis on a biologically based sexual drive (see p. 234). The factor responsible for sex role identity is cognition and the stages of cognitive development (see Chapter 6). Kohlberg begins by depicting the child as curious, enterprising, and self-reliant, very much as we did in discussing initiative (see Chapter 3). Two of the most significant concepts the child must master are those of gender and sex role. His initial grasp will be faulty and crude because his capacity to understand is limited; as his understanding deepens and broadens, his concept of sex role is concomitantly enriched and expanded.

We shall concentrate on that segment of Kohlberg's theory concerned with the development of sex identity, or the self-categorization of "boy" or "girl." A typical three-year-old has grasped the idea that "boy" or "girl" is one of the labels which can be applied to him and can answer the question, "Are you a boy or a girl?" However, he does not comprehend the real meaning of the label or the principle of categorizing people by sex. Only half of the children tested at three years of age could correctly label father, mother, boy, and girl dolls as "boy" or "girl," and they blithely go around saying, "Johnny is a boy, Mommy is a boy, Daddy is a boy." By four years of age, the child can label the sexes correctly, usually guided by external cues of clothing and hair style. While he has mastered classification on this spurious but useful basis, the continuity of sex role behavior is still beyond his grasp. He does not know whether he will grow up to be a mommy or a daddy, and he believes boys can change into girls if they wish or if they dress like girls and play girls' games. Only by six or seven years of age does he realize that one's sex remains constant even when external appearance varies, and that it is permanent and immutable.

Kohlberg maintains that awareness of genital differences plays a role in sex typing but disagrees with Freud that the role is central. The enlightened four-year-old who knows that boys have penises and girls do not is just as uncertain about sex identity as his more ignorant counterpart. Sexual information cannot clarify the issue for him because he is cognitively incapable of assimilating it. The four- to five-year-old has a clearer grasp of cultural stereotypes of sex roles, such as size, strength, aggressiveness, and power, than he has of genital differences. It is only around six to seven years of age that the child assimilates the fact that genitals are the crucial factor determining sex type.

Kohlberg contends that this progression in understanding sex type exactly parallels cognitive growth between three and seven years. A wide array of concepts concerning the animate and inanimate world also undergo a similar clarification. The child comes to realize that a cat cannot become a dog at will, nor does a cat become a dog when it wears a dog mask; its essential "cat-ness" remains constant in spite of apparent change. In short, the seven-year-old, unlike the three-year-old, no longer literally believes what he sees but can infer permanence underlying changes in appearances (see Chapter 6).

Once sex type permanence has been cognitively grasped, a change of sex is as inconceivable as the idea that cats or any other object in the world can change their natures. As evidence, Kohlberg cites the admittedly controversial studies of hermaphrodites assigned to one sex at birth because of external genital characteristics but later assigned to the opposite sex on the basis of internal reproductive organs. If the change occurred before the child was three to four years old, his subsequent sexual adjustment was satisfactory; reassignment at a later age resulted in intense conflict over sex identity.

Kohlberg does more than explore the cognitive factor involved in gender typing. He makes the daring assumption that cognition determines the values a child will assign to experiences. Once having achieved sex identity, or, better, because the child has achieved sex identity, male or female activities and objects become differentially valued. In essence, the child says, "Because I am a boy, boys' things are rewarding." This reaction is part of a general tendency to value positively experiences which are consistent with or like the self. The self-image thus becomes a potent source of positive and negative affective reactions, while the need for affection and approval, emphasized by learning theorists, is secondary. As evidence, Kohlberg cites studies showing that social reinforcement is ineffectual in changing preferences for masculine over feminine toys in kindergarten children. Thus, Kohlberg turns reinforcement inside out: boys and girls do not learn appropriate sex roles because their behavior is differentially rewarded and punished; they first grasp the idea that sex type is an essential characteristic of the self; having done this, they are attracted to and find gratification in sexually appropriate activities.

Kohlberg's radical reformulation is too recent to have been tested satisfactorily by objective studies. His speculation concerning the development of the concept of sex types is his most valuable contribution to our understanding of sex. However, the cognitive component does not exhaust the phenomenon. Just as there is more to moral behavior than moral insight, there is more to childhood sexuality than the cognitive grasp of gender and role. After examining the nature of masculine identification, we shall explore other facets of this complex phenomenon of childhood sexuality.

Masculine Identification. A review of theories and objective studies of masculine identification by Biller and Borstelmann (1967) is worth examining in detail, since it makes two important points. First, it highlights the complexity of a concept which seems simple and commonly understood. "Everyone knows" what a strong masculine identity is, until he is forced to conceptualize and operationalize the term (see Chapter 1). Next, the article concludes that the development of a masculine identity is contingent upon the interaction of a host of factors rather than upon a simple relation between parent and child. As we shall soon see, clinicians have reached a similar conclusion on the basis of clinical data.

Masculine identity is conceptually complex. First, there is sex role orientation (a variable similar to Kagan's concept of sex role identity) which corresponds to the individual's basic view of himself. It is not necessarily conscious, but it represents the fundamental perceptions of the maleness or femaleness of the self. Sex role preference is the perception of one sex role as more desirable than the other or the desire to adopt one sex role over another. Finally, sex role adoption refers to the overt behavior of the individual in regard to sex role. The three concepts are related but not identical. A boy may have strong feminine components in his sexual orientation and may even prefer the feminine role, but social pressures may force him to adopt masculine behavior.

Objective studies of masculine identification have begun to tease out the many variables which influence its development. Clearly, constitutional factors and cultural expectations are important, but the authors concentrate on parent-child relationships. Here the availability of the father, his masculinity, nurturance, punitiveness, and power are all important. The power and overprotectiveness of the mother, as well as her encouragement of masculine behavior, are also significant determinants as is the role played by siblings and peers.

One interesting finding is that the relationship between father and son is as important to identification as the characteristics of the father. Specifically, nurturance seems to act as a catalyst. In itself it does not bring about a masculine identification, but a warm father-son relationship greatly facilitates the boy's identification with his father's masculinity, limit setting, and punitiveness. Father-absence is more important in retarding the development of masculine orientation and sex role adoption than of masculine preference. Maternal overprotection, a matriarchal home, or the discouragement of masculine behavior by the mother, hamper the development of the son's masculinity.

Biller and Borstelmann's review is valuable in its underscoring of the complex interweaving of influences, each of which changes as the others change. The effects of father-absence, for example, are difficult to determine in their own right because an absent father often significantly alters both the mother's behavior and the mother-son relationship. Or again, the effects of father-absence are determined, in part, by the availability of older siblings or peers who could serve as masculine models and encourage masculine behavior. Thus, the question, What influences masculine identification? cannot be answered in terms of simple, one-to-one relations; "masculine fathers have masculine sons," for example, is an oversimplification. What we need is a series of contingency statements which will delineate different patterns of variables; e.g., if the father is masculine and warm, and if the mother is not dominating or subversive in regard to the father, and if sibling and peer relations support masculine behavior, and if the boy's physical development and constitutional endowment are within normal limits, then the chances are that the boy will develop a sound, masculine identification. Development itself is an important, complicating variable in such contingency statements. The potency of all influence may change from infancy to adolescence. Both Kohlberg and Freud, for example, regard the preschool period as crucial to the development of identification; if this is so, an adverse event (such as the father being sent overseas in the army) or an adverse relationship (such as the boy being caught up in the marital disharmony of the parents) would be more of an impediment during the preschool period than it would be subsequently. Or again, the relative potency of different influences may change over time; e.g., peers may be relatively unimportant during the preschool period but very important during adolescence. While developmentally oriented contingency statements are difficult to arrive at, they more accurately reflect the complexity of the phenomenon of masculine identification.

Erotic Pleasure and Sexual Curiosity

Up to now we have dealt with sex as a prescribed social role but have had little to say about sex as erotic pleasure. Kagan (1964) states that sexual behavior is like other social skills, and the child's primary concern is whether he is performing up to standard. Even in adolescence, the biological urge is of no special consequence, the boy's sexual ventures being similar to his practicing foul shots in the driveway in order to close the gap between ideal and actual performance. Kohlberg, like many cognitive theorists, has difficulty integrating intense needs and feelings into his system. At best, a child can be extremely interested in and attracted to another person, can value an activity highly, and can enjoy the pleasures of competent, effective functioning. But there is nothing which suggests that sexuality involves ecstasy, jealousy, rage, or terror. For these affective components we must turn to an examination of sexual feelings themselves and ultimately to Freud, who was continually concerned with understanding the primary passions of love, hate, and fear.

Sex in Early Childhood. Since Freud's theory of sexual drive is special and controversial, we shall begin with a more conservative account of normal development. Certain precursors of sexual behavior can be seen in early infancy. The unsuspecting mother may be startled when she begins to change her infant son's diaper and finds herself face to face with a tiny, erect penis. The toddler may derive brief, sporadic pleasure from stimulating his genitals, and during the preschool period masturbation is frequently practiced as a source of pleasurable sensations.[5] During this period the child is also curious about sex. He asks questions concerning anatomical sex differences and the origin of babies. He may want to touch his mother's breasts, and he is interested

[5] In one study, half the mothers reported their preschoolers engaged in sex play or masturbation; this is probably a conservative estimate (Sears, Maccoby, and Levin, 1957).

in looking at the genitals of adults and peers as well as exposing his own (Gesell and Ilg, 1949). While curiosity might not seem to be evidence of erotic feelings, studies of more permissive cultures show that it can be a prelude to sexually oriented genital play in preschoolers. The implication is that the American middle class culture forces the child to inhibit his exploratory behavior before it is fully satisfied; otherwise it would culminate in sexual behavior.

Sexual feelings in the four- to five-year-old can become part of his affection for his parents. As we have seen (Chapter 2), the mother in her caretaking stimulates intensely pleasurable reactions. Parents continue to take delight in delighting their children; they tickle and tease, they jounce and roughhouse, so as to produce a high level of pleasurable excitement. It is understandable, then, that the pleasurable excitement of genital stimulation would generalize to the child's relation to his loved parent. A mother may become concerned when her preschooler wants to lie close to her in bed and masturbate, or a father may realize that his little girl is becoming too excited by "riding horsy" on his foot. The erotic element in attachment may also be a factor in the jealous rage at siblings or even at the same-sex parent who interferes with exclusive possession of the parent of the opposite sex.

Socialization of Early Sexual Behavior. Middle class attitudes toward early sexuality have been significantly liberalized since World War I. The enlightened parent no longer terrifies the child with crude threats to cut his hands or penis off if he masturbates or with the myth that masturbation will cause physical weakness, moral irresponsibility, sexual impotence, or insanity. He also avoids excessive shaming and conveying an attitude that masturbation is dirty. In answering sexual questions, the parent strives to be neither evasive nor overly detailed but factual and sensitive to the nature of the preschooler's interests and level of comprehension. However, open masturbation and sexual curiosity are not condoned; the supermarket, the playground, or even the living room, are not approved settings. *No* and *Don't* inevitably appear with their demand for self-control. While the enlightened parent may accept the naturalness of childhood sexuality, he still feels there is a proper time and place for it. Ideally, the parent wants to curb socially disapproved expressions of sexual behavior without alienating the child from his feelings and curiosities.

Even under normal circumstances this ideal is difficult to achieve. Strong emotional reactions die slowly, and having to deal with the child's sexual behavior can revive the parent's own childhood feelings of fear or disgust. The message which manifestly states that such behavior is natural may be betrayed by an unnatural air of concern or casualness. Or the message itself can be contaminated. One enlightened mother said, "I told my little girl it was perfectly all right to masturbate but that she should wash her hands afterwards," unaware that she combined adult reasonableness with a residual of childhood disgust.

Another limitation to enlightenment involves the child's ability to understand sexual information no matter how clearly presented. As we have seen (Chapter 6), a child can grasp a new idea only in terms which are meaningful to him at his particular stage of cognitive development. The unfamiliar can be assimilated only in terms of the familiar. Because of his limited experience, the preschooler may significantly distort parental explanations. To illustrate: a five-year-old is told that babies grow "inside Mommy's tummy" or even "in a special place inside Mommy." His understanding of the workings of the inside of the body is limited (see Chapter 10). What he knows for a certainty is that things get inside by eating. In grappling with the problem of how the baby got inside the mother, he might understandably conclude that the mother ate something. Psychotherapists claim that an adolescent girl's extreme anxiety over eating may be due to the persistence, at an unconscious level, of this childish concept of oral impregnation. The mother does not help clarify the mystery of how the baby got inside by gliding over intercourse with a pleasantry such as "Daddy plants the seed"; but, more to the point, even if she gave an objective description of intercourse, she may only be adding another incomprehensible fact which compounds the mystery. A bright child can parrot her explanation, but in his private efforts to assimilate it, he is more likely to ignore the information or assimilate it into his understanding of familiar bodily processes. As he develops he will return again and again to the same sexual questions, each time being better able emotionally and intellectually to understand them. It is essential, therefore, that he has the freedom to return in order to correct and enrich his understanding, rather than being blocked from inquiry by fear or shame.

Sex in Middle Childhood. Information concerning sexual behavior in middle childhood is sparse. As far as we can tell, sexual interests and activities continue, but the overall pattern of the child's life makes them less pronounced. The preschooler's intense involvement with the family is now supplemented by more moderate involvements with peers and a more varied pattern of interests centering around school and social activities. In this expanded psychological environment, sexual feelings are out of place. In addition, society continues to condemn and inhibit overt sexual exploration and experimentation. The study of other cultures tells us that, if our society did not, we might expect a continuous increase in overt heterosexual behavior through this period. Middle childhood is also a time of strong sex cleavage, with boys and girls setting their own limits on sustained contact with the opposite sex.

Yet, sexual interests and activities persist. In their gangs, boys talk and joke about sex and sometimes engage in experiments involving simultaneous or mutual masturbation. Girls talk more about love and, while they may not engage in as much experimentation as boys, they have powerful sexual fantasies (Stone and Church, 1968). In studying eight- and nine-year-olds, Gesell and Ilg (1949) note a continuation of sexual curiosity as evidenced by an interest in peeping, telling dirty jokes, writing sex words, talking about sex information with a friend of the same sex, seeking books with pictures relevant to sexual organs and function, using sexual words in swearing, and writing sex poems. It may be that the child goes as far in sexual behavior as his uncongenial environment will allow.

Sex in Adolescence. Puberty ushers in physiological maturity and we may now regard the child as having a sexual drive. The period extends from around eleven to fifteen years of age for girls, and from around twelve to sixteen years of age for boys. A brief account of the facts of sexual behavior during adolescence will be helpful to our subsequent discussion (see Mussen, Conger, and Kagan, 1963; Horrocks, 1969). The picture is strikingly different for boys and for girls. In early adolescence the main source of orgasm for boys is masturbation, and most boys masturbate. Between sixteen and twenty, masturbation is gradually replaced by intercourse, but it still is a common source of orgasm. The lifetime peak of orgasm is reached between sixteen and seventeen years of age and begins to taper off around thirty. All manifestations of the sexual drive—petting, masturbation, and intercourse—are less frequent in the teenage girl. By twenty years of age, only half as many girls have masturbated; only 40 percent have engaged in intercourse, as compared with 71 percent of men. Equally important, only half have achieved orgasm by twenty and the lifetime peak comes at age thirty-five. The phase difference between males and females is indicated by the fact that only 30 percent of females have reached orgasm by the time of marriage while almost all males have; on the other hand, females reach a peak at age thirty-five—the time when the male begins to taper off.

Behaviors derived from the sexual drive are legion in adolescence. Sexual curiosity is intense, especially among boys, who are concerned about sexual techniques and reproductive anatomy. They trade information and misinformation with their peers, read literary and scientific accounts, occasionally peek directly at the female anatomy, but more frequently look at "girlie" magazines or pornography. Conversation among peers reveals their preoccupation with sexual feelings. Early in this period the topics concern how to get along with that erstwhile enemy, the opposite sex: what to talk about, what to wear, whether to go out alone or on a double date, whether to kiss. In later adolescence the topics are more serious: the implications of premarital intercourse, the nature of love, the problem of marrying someone with a different racial, religious, or cultural background. Diversity of opinion reflects the diversity of sexual behavior itself, which among college students runs from no physical contact through petting to intercourse. Petting—a kind of sexual behavior which stops short of intercourse—is prevalent in adolescence. The participants are not only exploring one another's bodies and trying out techniques, but they also are exploring their feelings about themselves as actors and recipients. Finally, there is evidence that differences in sexual behavior are accompanied by attitude differences between boys and girls. Boys are expected to take the initiative and to become sexually vigorous and accomplished. The majority of girls, on the other hand, disapprove of girls who are "fast" or promiscuous or who display an open interest in sexual behavior.

The Complexities of Adolescent Sexuality. This background is helpful in understanding some of the complexities of adolescent sexuality. At the simplest level there is the matter of obtaining accurate information. Before adolescence, all information, no matter how accurate, was hypothetical; now, it is vital. The adolescent does not merely want to know, he must know. Since society rarely provides him with ready access to factual material, he is apt to accumulate both correct information and misinformation and have areas of uncertainty and ignorance. Next, the adolescent is highly motivated to discover personally gratifying sexual techniques. Instructions in lovemaking must be adapted to the personality of the individuals. Equally important, sexual techniques, unlike foul shots, cannot be practiced in the driveway—they are intrinsically interpersonal. Thus, the adolescent must accommodate to the individual who is his partner. We are not dealing here merely with the awkwardness of inexperience, which can be easily remedied; we are dealing with individual differences which make the same technique exquisitely pleasurable to one partner and deeply repugnant to another. Thus, the uncertainty concerning technique occurs in a highly charged emotional context. Adolescents cannot be expected to have a relaxed tolerance of one another's insensitivities; because the urge is great, frustration readily becomes rage, insensitivity touches off anxiety or disgust. Each sexual venture also involves the question, What kind of sexual man (or woman) am I? This is a significant part of the more general question, What kind of man (or woman) am I? The adolescent knows that society will judge him, and he in turn will judge himself, in terms of the success of his ventures. But there is even more at stake. In middle class American society, sexual behavior is closely connected with psychological intimacy and, ultimately, with the choice of a lifetime partner. Sexuality is part of the questions, Whom can I love? and With whom can I share my life? These are the issues which concern the future oriented adolescent. They lie beyond questions of information, technique, or even social criteria of adequacy, and they involve areas of the personality which are both highly idiosyncratic and highly charged emotionally. When we discuss Freud, we shall give a more detailed accounting of such areas. Regardless of whether one accepts his theory, Freud's observations of the complex and personalized meanings of sexual intimacy are unique contributions to our knowledge. For the present we must be satisfied with the generality that the adolescent is searching for a physically and psychologically satisfying relation with a person of the opposite sex under the pressure of irresistible demands for periodic sexual gratification.

As we have seen (Chapter 3), one of the greatest aids in mastering a new challenge is a background of successfully managing similar challenges. Ironically, at one of the most crucial developmental junctures, the adolescent finds that much of his past is taboo. His chief attachments have been to his parents and, at a less intense level, to his same-sex peers. Closeness to parents is a particularly sensitive area because the erotic component in early attachment now threatens to stimulate manifestly sexual feelings. There is a double meaning to the adolescent's protest, "I'm not a baby anymore"; he is asserting his independence, but he is also protecting himself from the anxiety of eroticized intimacy. Sexualization of peer relations also carries the threat of a homosexual attachment, since the preadolescent has been psychologically close to peers and may even have shared some sexual experimentation with them. While not absolutely taboo, masturbation is also not serviceable to the adolescent, since it affords him no practice in the techniques and sensitivities necessary for heterosexual behavior.

Fortunately, certain aspects of the past come to the adolescent's aid. Sexual curiosity and experimentation flourish, since being "good" no longer means being asexual, as it did in middle childhood. While the adolescent is not given free rein, sexual activity is regarded as natural. Next, parental identifications and familial values continue to serve as guides to mature behavior. There is general agreement that a strong identification with a parent who is secure in his own sexual role is a great asset to a boy or girl. In spite of stormy confrontations with parents, the adolescent still is guided by the middle class values he learned from them; e.g., a boy wants his future wife to know how to cook and keep house; a girl wants her future husband to be a good provider; and both sexes value dependability, trustworthiness, a sense of humor, and compatible interests. Sex role identity also continues to guide and reward appropriate behavior. As we have seen, specifically sexual prescriptions are added in adolescence, so that the boy is expected to be sexually assertive and dominant, while the girl is expected to be nonassertive but sexually attractive and receptive. Finally, if we follow Sullivan (see Chapter 7), the mutuality inherent in healthy peer relations is an essential preparation for the accommodation and sharing which is part of mature heterosexual relations.

Achieving sexual maturity, therefore, depends upon a host of contingencies: if the toddler-preschooler has not been alienated from sexual feelings and curiosity by threats and shaming, if he has not become involved in strong attachments or resentments toward his parents, if his parents have served as proper models of masculinity or femininity, if he has successfully met the requirements of his prescribed sex role, and if his peer relations have developed his capacity for mutuality, then his chances of achieving maturity are good. (Our list of contingencies does not pretend to be exhaustive.) No one element is critical. However, the more deviant any one element in the pattern, or the more the total pattern goes awry, the greater the chances that sexual maturity will not be achieved. Anna Freud (1965) points out that each step in the sequence is necessary, but each one permits excesses which may lead development astray in some way: the toddler's attachment to the mother is an indispensable first step in relating closely to another person, but it can also be an exclusive bond which overrides adult sexual strivings; a boy's contempt for girls and delight in comradeship may advance his masculinity during middle childhood, but it also can lay the foundation for adult homosexuality. An intricate web of forces are at work, one of them being chance events—the death of a parent or a severe illness coming just at the time the child is least able to cope with it, or the accidental meeting of an adult or peer who becomes a friend and counteracts anxieties created in the home. While we may be able to discover general landmarks and common factors in sexual development, the fate of an individual is too multi-determined to predict.

Freud's Concept of Libido

For Freud, sex was an irresistible drive to experience intensely pleasurable body sensations. Although its compelling power is evident from puberty on, Freud made the daring hypothesis that the sex drive appears soon after birth. There is no doubting the fact that infants respond with intense delight to certain kinds of body stimulation such as stroking the lips, tickling the stomach, being playfully jounced. Freud contended that these are not situation-specific responses; rather they are expressions of a biologically determined characteristic of the human organism to react with pleasure to certain kinds of stimulation. Moreover, once having experienced these erotic sensations, the infant develops an intense craving to re-experience them. The need is continuous and irresistible. Puberty is only the climax of a lifelong history of deriving pleasure from various parts of the body. Freud called this biologically determined drive to obtain erotic bodily sensations libido. Libido exists because of the biological character of the human organism; just as our body is constructed so that we can see, hear, and smell, it is constructed so as to afford intense pleasurable sensations when stimulated. Thus, biological factors are basic in Freud's theory.

Although the entire body is a source of erotic sensations, stimulation of the mouth, the anus, and the genitals produce the most pleasure. In Freudian terminology, these organs are highly endowed with libidinal energy, while the organs themselves are called erogenous zones. Finally, Freud hypothesized that libido was highly mobile, so that, when one source of gratification was denied, a substitute would be sought; e.g., the infant who is suddenly weaned from the bottle will suck his thumb vigorously. (For a further discussion of displacement, see Chapter 4, on substitute gratification.)

From his study of disturbed adults, Freud theorized that the focus of libidinal investment moved in a fixed, perhaps biologically determined sequence from the oral to the anal to the genital erogeneous zones in the first four or five years of life. This means that stimulation first of the mouth, then the anus, and finally of the genitals provides the child with the maximum amount of erotic feelings. As the focus of gratification moves from one zone to the next, distinct psychological developments occur. Freud labeled his theory of early personality development psychosexual to emphasize the complementary relation of psychological and erotic progression. He termed the three psychosexual stages, the oral, and anal, and the phallic. Finally, Freud introduced the concepts of fixation and regression. No stage is completely outgrown. Even in the mature personality one can find residuals of oral, anal, and phallic activities. However, if the child receives too much or too little gratification at any particular stage, an undue amount of libido becomes fixated and healthy progress may be blocked. The child may continue to function at an immature level, or, expressed in Freudian terminology, he may be fixated at a certain psychosexual stage. Or the child may progress but return to the fixated stage during times of severe stress. This reverting to an earlier developmental stage is called regression.

The Psychosexual Theory of Development. We will present an outline of the psychosexual stages of development. (For a more detailed exposition, see Munroe, 1955.)

i. The Oral Stage. The infant derives erotic pleasure first from sucking and later (when he has teeth) from biting. In the context of being fed by the mother, he forms his first emotional attachment or, to use the Freudian term, his first *object relation*. Instead of being exclusively bound to bodily sensations, he derives pleasure from being with the mother and from the mental image of the mother he evolves from repeated contact (see Chapter 6). This image of a loving caretaker helps him master the anxiety of neonatal helplessness while serving as an essential entre to subsequent positive human relations (see Chapter 2). At a primitive level, "mother" and "being fed" are forever associated. If the infant is not well cared for, his image of mother will be saturated with feelings of rage and anxiety.

Normal fixations retained from the oral period include the adult's pleasure in kissing, sucking candy, chewing gum, and smoking. In religion, the Madonna preserves the image of the all-loving, all-powerful mother figure, while, in literature, the witch epitomizes the angry, destructive one. The need to be psychologically and physically nurtured also persists. "The way to a man's heart is through his stomach" expresses the Freudian idea that being well fed is one basis for adult intimacy, regardless of the sexual attractiveness of the cook. The association of love with food is also seen in the adult who claims that there is no apple pie (or spaghetti or noodle soup) like mother's apple pie (or spaghetti or noodle soup). Similarly, poisoned food symbolizes maternal caretaking which has been contaminated by anger and destructive feelings.

ii. The Anal Stage. The toddler achieves erotic gratification from retaining and evacuating feces, or from manual manipulation of the anus. With toilet training, control of the anal sphincter epitomizes the achievement of self-control in general. If the mother rewards the toddler with love, she compensates him for the loss of pleasure and power that he experiences from defecating whenever and wherever he wishes. If the mother is punitive, unloving, or coercive, the toddler becomes rebellious or stubbornly resistive or anxious and overly compliant (see Chapter 4).

The normal residue of the anal stage retained into adulthood is the cult of regularity; the laxative business is as enduring as the life-saver business, and there are many adults whose daily mood is contingent upon a daily evacuation. Just as children disparage one another by saying, "You old B.M." or "You old pooh-pooh," adults use four-letter equivalents of feces when cursing their contemporaries. In addition, a number of behaviors can be located along a continuum which goes from "clean" to "dirty"; at one extreme, being personally clean, neat, and orderly, being "clean-minded," being dutiful (which harks back to the mother asking, "Did you do your duty?"); at the other extreme, being messy and slovenly, being a dirty dealer, having a dirty mind, and telling dirty jokes, (interest in dirt is only a thin disguise for interest in feces).

iii. The Phallic Stage. Masturbation and curiosity about anatomical differences are at their height in this stage. The desire to peek and to show, to look and to exhibit, run high. The period is also marked by expansiveness, assertiveness, initiative, and an intoxicated sense of power. The child seems to feel he is (or soon will be) "big" and "grown up."

A persistent residual of the phallic stage is curiosity about the human body, sexual techniques, and intercourse, which is expressed as avid interest in scholarly descriptions, "girlie magazines," and pornography alike. A Freudian would also claim that an intense desire to look and learn is a displacement of sexual curiosity, even though the object might be the motor of a jalopy or a new virus under the microscope. The desire to show off also harks back to the phallic period, and is evident in the "muscle man" at the beach, the glittering glamor girl, the actor who tries to dazzle his audience, the teacher who tries to dazzle his class, and the conspicuous sufferer who proclaims how pitiable he is.[6]

[6] For a more comprehensive list of behavioral manifestations of the various psychosexual stages in children, see Ruttenberg, Dratman, Fraknoi, and Wenar (1966).

The Oedipus Complex and Castration Anxiety.[7] The Oedipus complex climaxes early psychosexual development for boys and was regarded by Freud as the prime determinant of adult sexuality. As we have seen, even nonFreudians concede that there is an erotic component in the boy's attachment to the mother; Freud claims that this component is critical. It represents the boy's deepest feelings, his most passionate longing, and for Freud, such longings and passions always determine the course of personality development. When a four-year-old says, "Mommy, when I grow up I want to marry you," or "Wouldn't it be nice if Daddy didn't come home so we could always be together?" he is not merely being cute; he is sending a heartfelt message which adults refuse to take seriously.

The boy's passionate attachment is accompanied by a desire for exclusive possession of his mother and intense jealousy of any rival. Of all rivals he resents the father most and quite naturally would like him to disappear. Typically he wishes the father were dead because, at this early age, death means disappearance. The desire to possess the mother and destroy the father is called the Oedipus complex. The rivalry with the father precipitates a crisis, partly because the boy loves his father and the father's absence would entail a loss of love, but primarily because of the boy's terrifying fantasy that his powerful rival will cut off his penis. This castration anxiety is an inevitable result of the Oedipal situation. It will be instructive to examine the reasons why this is so.

Castration anxiety has some basis in reality. Parents occasionally do threaten to cut off a boy's penis if he masturbates. Such threats do not have to be statistically frequent to be psychologically potent. Remember, the boy in the phallic stage derives his most intense feelings of pleasure and pride from his penis, and consequently is hypersensitive to anything which threatens to damage it. Even in a benevolent environment, however, the fantasy is inevitable. The little boy had assumed previously that everyone was anatomically alike and therefore had a penis. This assumption results from his general tendency to model the world on himself (see Chapter 6). When his sexual explorations lead to the discovery that girls and women do not, he assumes that they once had a penis but it was taken away. He concludes that what happened to them could happen to him. Castration anxiety is inevitable for another reason. As we have seen (Chapter 6), the four- to five-year-old is unable to understand the feelings and motives of others; instead he interprets the world in terms of himself. He assumes that if he wants to destroy his father, his father also wants to destroy him. Consequently he becomes terrified of his own wish which has been projected onto the father.

We have dwelt upon castration anxiety not because it is more important than other aspects of psychosexual development, but because it illustrates the basic "inward" orientation of Freudian theory. Learning theorists generally regard personality as developing from the "outside in"; parental standards and societal prescriptions are learned by reinforcement and imitation, and are adopted as the child's own. To understand personality, one must examine the characteristics of the child's social environment. For Freud, on the other hand, personality develops from the "inside out." Freud focuses on intrapsychic forces, or the interplay of conscious and unconscious

[7] See Munroe (1955) for a discussion of the Electra complex which the girl experiences.

thoughts, wishes, and impulses. The essence of personality lies not in behavior but in what the child thinks, wants, and feels. Moreover, the child's mental life is far from an accurate reflection of his environment. Especially in the early years, his thinking is determined more by his longings ("When I grow up I will marry Mommy") and rages and fears ("If I want to hurt Daddy, he will want to hurt me") than by logic and reason. To understand a child, one must understand his feelings and fantasies and the strange transformations that these perform on reality. Castration anxiety, like other aspects of personality, results from the combination of strong feelings and a faulty grasp of reality which is characteristic of early childhood, regardless of how benign and reasonable the environment might be.

To return to the Oedipal situation: rivalry with the father results in anxiety over loss of erotic pleasure through castration and loss of phallic pride in being a "big boy"; it also produces painful feelings of inferiority when the boy compares himself to his powerful father. The typical resolution is for the child to renounce his erotic feelings and his claim on the mother and to identify with the father. Through identification he says, in essence, "I am not your rival, I am like you," and he can say to himself, "I am not weak and helpless, I am strong like Father." Identification is such a potent factor in the boy's life because it resolves an emotional crisis and salvages self-esteem or what Freud calls narcissism. Meanwhile, the renounced erotic feelings do not cease to exist; the boy merely represses them and refuses to admit them into conscious awareness.

While the Oedipus complex is inevitable, each child experiences it differently because of his unique history of psychosexual development and his unique relationships with his parents. While parents cannot prevent the Oedipus complex from occurring, they can make it easier or more difficult to resolve. In general, the mother should be warm and loving, so that the boy comes to regard women as desirable love objects. If she is cold, he may fail to develop a strong attachment to her or he may react with rage to her failure to gratify his need for affection. If she is seductive (as by bathing, dressing, and taking naps with the boy, calling him endearing terms, or unconsciously trying to gain from him the pleasures she fails to find with her husband), the attachment may become too intense. A submissive boy may passively accept his subservient role, while an active one may react with rage to being tantalized by the unattainable lure of being his mother's lover. In general, the father should be strong so that the boy has a strong model to identify with. Moreover, the father's strength reassures the son that his childish fantasy, "I wish you were dead," is just that and has no chance of being realized. The weak, uncertain father offers no such assurance, and the boy is terrified that his death wish might become a reality. Too powerful a father runs the risk of intimidating the son unduly and magnifying his castration anxiety; the boy may become passive, placating, and fearful of authority. "I'm afraid to stick my neck out because someone will come along and chop it off," is a thinly disguised castration anxiety carried into adulthood. It is important to understand that even these generalities should not be taken as absolute prescriptions for a healthy resolution of the Oedipus complex. Constitutional factors and special developments in the prior psychosexual stages require different parental characteristics for a positive outcome.

Latency and Adolescence. Freud described middle childhood as a period of sexual inactivity, perhaps brought about by a biological decrease in libido but certainly caused by the child's renunciation of erotic feelings at the climax of the Oedipal struggle. Because the child is relatively undisturbed by intense feeling, he is free to expand in other areas, such as academic learning, friendship, group activities, and social interests. Freud did not investigate latency intensively and contributed little to our understanding of middle childhood.

Physiological maturation at puberty again focuses the child's attention on irresistible, biologically determined demands for erotic gratification. It also revives all the difficulties of psychosexual development experienced in the first six years of life. The "storm and stress" of adolescence, the erratic swings from babyishness to negativism to slovenliness to idealism, represent revivals of previous developmental difficulties and attempts to master them. Aside from this emphasis on recapitulation, however, Freud's theorizing was meager and his insights were few when compared with his study of early childhood.

Maturity. In the mature sexual act, the individual utilizes the earlier stages of psychosexual development to heighten the pleasure preceding mutual genital union. Kissing, biting, and sucking, stimulation of breast and buttocks, looking at and exhibiting the naked body, become part of the forepleasure, while the climactic erotic experience comes with orgasm. According to Freud, then, the mature sexual act draws upon a lifetime of erotic experiences.

Freudian theory also states that there can be no psychological maturity without sexual maturity. We are now in a position to understand the meaning of this deceptively simple criterion of maturity. Sex, or libido, is the drive which spearheads the development of basic personality variables—affection for others, self-control and responsibility, pride in achievement, assertiveness, attitudes toward authority, self-esteem, as well as sexual identity and specifically sexual attitudes. In learning to master the sexual drive, the child learns to master the passions of love, hate, fear, and guilt. We also can understand that the criterion of maturity has little to do with mere performance of the sexual act. As we have noted, it is never behavior but the underlying feelings and fantasies which are essential. A man fixated at the phallic stage, for example, has intercourse primarily to exhibit his masculinity, with little regard for the woman who happens to be his partner; an anally fixated man might find out how frequently the average person has intercourse, decide what techniques he will use and in what order, and then religiously follow his sexual agenda with the same feeling of "doing his duty" that he had years ago during toilet training. True maturity must transcend the egocentric dependency, demandingness, ritualization, hostility, and exhibitionism of the earlier psychosexual stages, just as the genital union transcends the oral, anal, and phallic activities of the forepleasure. Maturity requires a true mutuality, a genuine willingness to accommodate to the partner in the interest of maximizing mutual pleasures. The mature sexual act is still a psychosexual phenomenon. Freud claims that it is the ultimate fruition of human love.

Freud's theory of libido and psychosexual development was, in its day, and still is the controversial part of his complex psychology. Efforts to validate his theory by objective approaches have been inconclusive; some studies confirm, others fail to confirm predictions.[8] In reviewing the evidence, there is always the nagging question concerning the aptness of techniques. How effectively do objective procedures reveal that inner world of feeling and fantasy which is the special domain of Freudian psychology? All too frequently the essential intrapsychic orientation is lost when the investigator utilizes the available techniques of objective research. The task of validation, while not impossible, is formidable. Regardless of its ultimate validity, however, Freud's psychosexual theory, because of its comprehensivness, its insights, and its impact on psychology, stands as the single most important contribution to the study of personality development.

Deviant Sexual Development

The Problem of Defining Deviation. For adults, heterosexuality defines the norm; nonheterosexuality is deviant. In childhood socially prescribed roles for boys and girls are clear and fixed, so that one can judge how "typical" a given child's behavior is. However, other aspects of sexual behavior have a remarkably diverse history. Even a nonFreudian might include the following as part of normal development of a boy: masturbation as the primary source of erotic stimulation, a strong, partially erotic attachment to the mother, disdain for the opposite sex, preference for the company of the same sex with the likelihood of some sexual experimentation. Thus, normal childhood development includes behaviors which would be regarded as perverse in the adult because they are no longer developmentally appropriate. Anna Freud's (1965) caution should also be repeated (see p. 234): sexual development is so fluid and dependent upon such a diversity of factors—including chance—that one cannot be certain of its ultimate outcome until patterns of sexual behavior begin to solidify after puberty. In short, there is no single, fixed criterion for normal sexual behavior in childhood.

There is a more general problem involved in understanding deviant sexual behavior in childhood: the same behavior can result from different antecedents. Let us take homosexuality in early adolescence as an example. A homosexual attachment may represent a transitory phenomenon—a temporary clinging to the safety of same-sex relations due to a fear of heterosexuality. However, it may also be due to a problem in identification resulting from an early childhood fear of an overpowering father or resentment of a possessive, seductive mother. In the most extreme case, a homosexual attachment may represent the sole human relationship a chronically isolated, withdrawn adolescent can tolerate and thus becomes the only positive human contact he has been able to establish. If all we know is that an adolescent is engaging in homosexual activity, we know very little, since different developmental histories cast the same behavior in a different light.

There is a final complication arising from the interrelatedness of personality variables. Deviant sexual behavior may have little to do with difficulties in managing sexual

[8] For a resume of research on the oral and anal stages of psychosexual development, see Caldwell (1964).

feelings as such; rather, they may result from disturbances in other areas. Masturbation is a case in point. Once the four- to five-year-old has discovered the pleasures of solitary, autoerotic stimulation, he may use it to reduce anxiety arising from any number of developmental problems. Or again, homosexuality may represent an unconscious means of getting revenge on parents for setting unrealistically high goals and offering little love in return; its motive is anger, not sex. With these cautions in mind, we shall briefly describe deviant development.

Extremes of Control. Sexually deviant behavior can result from too little control which produces impulsivity or from overcontrol which produces inhibitions. Impulsive sexual behavior may be due to any of the factors which affect the development of self-control in general (see Chapter 4): the bond of affection between parent and child may be weak, the child may be unable to assimilate parental directives because of a cognitive deficiency, the parents may be lax in teaching control or may serve as models for uncontrolled behavior, and so on.

The greater danger for the middle class child is that he will become frightened by, ashamed of, or disgusted with his normal sexual urges. These negative feelings may represent reactions to overly punitive and threatening parental socialization, although a Freudian would claim that the child is capable of creating his own monsters, even if his parents are benign. Sexual anxiety and disgust is particularly burdensome because it alienates the child from an inescapable physiological drive and from an omnipresent feature of his social environment. Anxiety also prevents him from being open to the continual experimentation and cognitive reappraisals of sexuality which are essential to correcting his initial misconceptions.

Deviations Resulting from Fixations. Freud's psychosexual theory provides the most comprehensive framework for conceptualizing deviant development. As we have seen, a certain amount of fixation at the various psychosexual stages is normal, but an extreme degree significantly distorts development. Either excessive or insufficient libidinal gratification produces this pathogenic fixation; e.g., an infant who has an unduly prolonged period of sucking is reluctant to relinquish its pleasures, while the infant who has been incompletely gratified continues to long for oral pleasures and strives to obtain them. True to the psychosexual nature of the theory, fixations affect general psychological characteristics as well as specifically sexual behavior.

Fixations lay the foundation for the following character traits and sexual perversions:

i. The orally fixated individual retains an infantile dependency, expecting to be taken care of while doing nothing in return. He may be irresponsible and foolishly optimistic, always believing that "something will turn up" to solve his problems. He latches on to others and is always looking for something for nothing. In school his teachers may become irritated by his desire to be "spoonfed"; i.e., to learn with no effort on his part. The orally deprived individual, on the other hand, is sour, bitter, pessimistic, and depressive. He chronically expects the worst to happen and is suspicious and mistrustful of others. Oral perversions are found primarily in severely disturbed children and adults who use sucking, mouthing, and biting as a primary source of erotic gratification.

ii. The anal character is stubborn, pedantic, stingy, perfectionistic. A stickler for order, neatness, and rules, he upholds the letter of the law and is oblivious to the spirit. He believes in a place for everything and everything in its place. Ordinary spontaneity upsets him, and he substitutes ritual for passion. His self-righteousness may give him the appearance of strength, but he is basically uncertain of himself and mistrustful of others (see Chapter 3).

In anal perversions, the individual derives a significant amount of his sexual gratification from stimulation of the anus; e.g., through anal masturbation or defecation. As one adult male expressed it: "Anyone can have intercourse, but a good bowel movement is the greatest pleasure known to man." (Incidentally, this particular individual was not in psychotherapy, but was leading a socially acceptable life. Some sexual perversions are known only to the individual.)

iii. The phallic individual is the self-centered exhibitionist who must be the first, the most, the biggest. He sees others primarily as competitors or admirers. He may have a captivating vitality and capacity for materialistic accomplishments, but he is basically egocentric and vain. Voyeurism and exhibitionism are the perversions of the phallic period. The voyeuristic male may have an irresistible compulsion to look at nude females and may engage in "peeping Tom" activities, while the exhibitionistic male has an equally irresistible urge to display his genitals.

iv. The deviations deriving from an unresolved Oedipus complex are as varied as the distortions of sexual identification and the irrationalities of love. Our description of the Oedipus complex cited some conditions which may lead a boy to be uncertain about or fearful of his masculinity. He may protect himself from his intense fear of the powerful father by adopting a feminine identification in the hope of avoiding the dangers of rivalry. If his mother is possessive and seductive, he may identify with her as the stronger parent or, in his rage he may spitefully become effeminate in order to frustrate her attempts to use him as a lover. Again, we can do no more than sample the voluminous psychoanalytic literature on the psychopathological effects of an unresolved Oedipus complex on identification.

The distortions of heterosexual love are equally as important and as varied as those of identification. Sexuality comes to be viewed as a sterile experience or as a source of rage, anxiety, or intimidation. Any one of these distortions can be elaborated in a variety of ways. The boy who is enraged by his seductive mother may grow up to be a woman hater; but he is equally apt to make himself attractive to women, luring them into loving him and then deserting them, tantalizing and enraging them as he himself was tantalized. The nymphomaniac and the frigid spinster may try to master their common fear of sexual intimacy by very different behaviors. Also remember that psychoanalysts regard perversion of love as a matter of fantasy and feeling, not necessarily a matter of behavior. While engaging in the accepted forms of intercourse, a woman may fantasy that she now has power over the man and will exhaust and emasculate him; her feelings and fantasies define her deviation.

Understanding Irrational Behavior. The psychosexual theory of development is an important aid in understanding the irrational behavior of adults. Freud's insight that irrational behavior originates in childhood and, in certain instances, is developmentally appropriate revolutionized the study of adult psychopathology. (His thesis was illustrated in the cognitive realm in Chapter 6.) The adolescent girl's unconscious fear that she will become pregnant by eating becomes understandable when placed in a developmental context (see p. 231). To take another illustration, let us juxtapose two incidents, one from a mother of a two-year-old, another from a psychotherapist of a schizophrenic woman.

> We had just moved into our new summer home, and a neighbor dropped in to get acquainted. We were talking in the living room when Teddy (her two-year-old) entered naked as a jaybird, holding a pottie with a B.M. in it for me to see. He was bursting with pride as he always is, but I could have dropped through the floor. You know, I usually make such a to-do when he has a B.M. in the pottie, and take him up and hug him and everything; he looked so proud and happy that I didn't know what to do.

> I was treating this severely withdrawn schizophrenic woman. She was so terrified that she did not say a word for the first couple of months. She just sat looking down at the floor. However, I would make casual conversation and felt friendly and comfortable with her. Gradually I sensed that she was changing. She seemed to want to look at me or to make some kind of overture. The crucial event came one day when she opened her purse and shyly took out a perfectly clean white handerkerchief with something wrapped inside. She handed it to me. I opened it up and found a bowel movement.

This behavior, which might otherwise be dismissed as a bizarre, disgusting act of a "crazy woman" now becomes meaningful. She is offering her therapist a present, and in her primitive way she is making a poignant bid to be loved and to be valued for being a good girl.

Aggression

Let us start with a working definition: aggression is behavior which has injury or destruction as its goal and anger or hatred as its accompanying affect. We shall see later that aggression is as difficult to define as anxiety or sexuality. But first we shall give a descriptive account of how aggression is expressed at different ages and what situations elicit it. Then we shall turn to the matter of conceptualizing aggression and, finally, to the issue of its control (see Hurlock, 1964; Jersild, 1954; and Goodenough, 1931).

The Natural History of Aggression

Origins of Aggression. Psychologists often disagree about when various emotions appear on the developmental scene. In part, this is due to differences in definition, as we shall see. In part, it is due to the fact that emotions have preludes. Some experts are willing to label an affect in a preliminary stage, while others prefer to wait until it has fully blossomed. Disagreements also arise as a result of differing degrees of willingness to interpret early behavior. Some psychologists, who regard neonatal and early infantile behavior as ambiguous, constantly warn against reading adult meanings into it; others feel free to identify specific affects, intents, and fantasies. Where one expert notes that an infant "sucks vigorously," another will claim that it "clearly is attacking the mother's breast with hostility." For the present, we shall follow a conservative course.

In her classical descriptive study, Bridges (1932) was able to identify in the neonate only a generalized emotional excitement which is differentiated into distress and delight around three months of age. Anger, fear, and disgust develop out of the distress reaction around six months of age. In expressing anger, the infant cries and uses random, overall body movements, such as kicking, flailing his arms and legs, arching his back, and twisting his body.

The period between one and four years of age is the high-water mark for unvarnished outbursts of rage, although it also sees the beginning of more socially acceptable modes of expressing aggression. Temper tantrums, which include kicking, biting, striking, and screaming, reach a peak around three-and-a-half years of age and gradually decline thereafter (Macfarlane, Allen, and Honzik, 1954). Directed expressions of anger are negligible in the first year of life, but gradually increase until about a third of the emotional outbursts of children between four and five years of age are of this sort; for example, behaviors which Goodenough (1931) categorizes as retaliation—defined as a directed attack on an offending person—increase rapidly. In this same period, anger is increasingly expressed in verbal form, such as refusals, threats, name-calling, and arguing. Episodes of verbal aggression are still brief; the typical preschool quarrel lasts less than half a minute. Thus, we see a trend from explosive, undirected outbursts of temper to directed attacks and from physical violence to symbolic expressions of aggression. Goodenough's examples also reveal a progressive refinement of retaliation, which includes deliberately doing the forbidden (the earliest examples of which occurred in an eighteen-month-old child), doing what the child knows will annoy the parent, and subtle vindictiveness like that of a seven-year-old who, after being punished, remarked airily, "I wish I had a mother like Mary's." Finally, self-directed aggression is not uncommon in this early period. During a temper tantrum a child may hit himself, pull his hair, or, on rare occasions, bite himself.

These descriptive findings can be conceptualized in terms of the developmental trends we have met in previous discussions. The infant in the first half-year of life is not equipped to be aggressive *toward* anyone or anything. He lacks the necessary motor coordination for an attack; he has no clear concept of himself as a causal agent until around six months of life (see Chapters 3 and 6); he cannot intend to do harm because he cannot intend any but the simplest behaviors; finally, he is unaware that people and objects have independent existences (see Chapter 6), so he cannot direct his behavior toward a person or object. He may be angry as a consequence of being frustrated, but he cannot be angry *at* anyone or anything. These limitations in the first months of life make aggression primarily an uncoordinated, explosive release of affect. Some might prefer to say that we are dealing with the precursors of aggression rather than with aggression itself.

Aggression changes in form as the infant gains control of his body and as he is capable of purposive actions directed toward people and objects whose independent existence he has grasped. As his understanding of others continues to increase, his aggression becomes more complex. By the third year of life, he grasps the idea that an attack hurts, so now he can intend to hurt. Subsequently, hurting becomes less physical and more psychological. At the same time, his caretakers are facilitating his progress by defining how he can aggress (he may hit the body but not the face, he may hit with his fist but not with a stick, and he may not kick) and toward whom he may aggress (never to parents, rarely to siblings, sometimes to playmates, frequently to toys).

Before continuing with our descriptive account, it is important to put aggression into a broader developmental context. If we concentrate only on aggressive behavior, we might conclude that the toddler and preschooler are generally vicious and violent. But see what happens when we broaden our perspective to include other data. The same behavior takes on new meaning as new variables are added. Macfarlane (1954) notes that, while tantrums and destructiveness peak at three-and-a-half years, so do fears and timidity. Gesell and Ilg (1949) characterize the ages two-and-a-half, three-and-a-half, and four as times of general disequilibrium, with extremes of aggression and withdrawal, quarrelsomeness, and fearfulness. On the positive side, there is evidence that aggression is related to sociability and to friendships in the preschool years (Muste and Sharpe, 1947; Green, 1933). Vigorous social participation and the formation of mutual friendships increase the incidence of aggressive behavior. Thus, aggression may stem from either general emotional instability or social involvement. Taking its content into account may alter significantly one's evaluation of aggressive behavior. Aggression as an impulse to destroy may be regarded as "bad"; aggression as part of a general emotional instability may be viewed primarily as symptomatic rather than good or bad in its own right; aggression as part of a general sociability may be viewed as "good."

We would also like to make an interpretative point. The transition from physical to verbal aggression is marked by a decrease in primitive destructiveness as defined by inflicting physical harm. A mother is rightfully concerned over her toddler's physical attacks on a newborn sibling; his striking out may never again be equaled for its sheer, uninhibited expression of rage, except in the most severely disturbed adults. Yet it would be a mistake to conclude that the change from a physical to a symbolic mode per se diminishes aggression. Symbolic attacks are not inherently less destructive than physical attacks. The same child who says, "Sticks and stones may break my bones, but words will never hurt me," may come running to Mother in tears and protest bitterly, "Janie called me a dirty liar, and I'm not a dirty liar!" A blow to self-esteem can be as shattering as a blow on the face, a painful humiliation can be more damaging to a child than a painful beating. Thus, the shift from physical to symbolic expressions of aggression does not imply any moderation of destructive intent.

The situations evoking aggression show an interesting developmental trend between infancy and age four. Goodenough (1931) found that the most frequent category of situations was social or what we would call interpersonal, since it includes both parents and peers. In infancy, a frustrated desire for attention, usually from the mother, rated high as a cause but rapidly dropped to insignificance. Possessiveness (i.e., having to share an interesting object with another child or not being able to appropriate an object belonging to another child) was negligible in infancy and peaked at the three- to four-year period. Feeding and going to bed provoked anger in the first year of life and toileting in the second, especially if the toddler was forced to remain on the toilet after he wanted to leave. Being refused permission by an authoritative adult peaked as an anger-evoking situation in the second year of life and continued to be important subsequently. Failure to accomplish what the child set out to do was a rather infrequent but persistent source of anger. Thus, the trend is from anger engendered by a desire for attention and by caretaking and training, to anger over having to share and over authority, with anger at his own ineptness remaining low but constant. Aggression can be viewed as a response to any situation which counters the strong vested interests of a developmental stage.

Middle Childhood. In middle childhood, the physical behavior utilized in expressing anger is more directed as the child is increasingly capable of coordinated hitting and fighting. However, aggression continues to become less physical and more verbal and psychological, e.g., hurting another person's feelings begins to replace hitting him. The frequency of retaliation also continues to increase; the child is now concerned with getting even with another child and with paying him back in kind (Goodenough, 1931). The child's repertoire for expressing aggression proliferates: bickering, quarreling, teasing, and swearing abound, along with bullying, prejudice, cruelty, and antisocial acts (Jersild, 1954). Angry outbursts come to have aftereffects in the form of resentfulness or sulkiness; whereas the toddler's reactions were limited to the immediate situation, the school-age child is capable of being troubled subsequently and of holding a grudge.

In general, after the preschool period, aggression declines as a crude physical attack in reaction to the immediate situation. The child's behavior is now more intentional, more retaliatory, and more symbolic. From our previous discussions (see Chapters 5, and 6) we can assume that the child is increasingly capable of feeling guilt, since he is troubled by his outbursts. His increased capacity to represent situations symbolically and think about them has a dual effect: it helps him control his impulse to strike out immediately, but it also enables him to hold a grudge and retaliate later. As we have already noted, the ability to symbolize per se does not exercise a civilizing influence: the process of thinking creates a period of delayed action which can be used either to find means of avoiding aggression or to prolong anger and insure retaliation.

Anger-producing situations are more numerous in middle childhood than in the preschool period. The following rank high on the list: cheating and lying; scolding, whipping, and punishment, especially when it is unjust; teasing, bossing, faultfinding, lecturing; slight or neglect by others; ineptness, mistakes, poor grades in school. Siblings produce anger when they take another's property and when parents hold them up as models. As was true of the previous period, the list of "What makes me angry" reads like the counterpart of the list "What I feel deeply about." Concern with strict justice is pronounced in middle childhood, as is the growing desire for independence; the child does not want to be ignored but does want to be valued in his own right; his pride dictates that he should not be pushed around or mocked or treated less than he is; and he retains from earlier days his resentment of siblings and irritation at his own shortcomings. Note also that anger, in reflecting current developments, is different from anxiety which prolongs the fears of the preschool period into middle childhood.

Adolescence. The young adolescent experiences aggressive behavior as part of a general pattern of affective instability (see Hurlock, 1955). The more infantile modes of expression reappear (stamping feet, kicking, throwing objects, and crying) but disappear later. Verbal expressions of anger predominate, and include sarcasm, name-calling, swearing, ridiculing, and humiliating. Sulking continues after an outburst of anger. The situations which evoke anger resemble those of middle childhood–unfair or cruel treatment, encroachment on rights, refusal of privileges, being treated childishly, and being incapable of achieving a desired goal.

Additional Variables and the Interaction among Variables. We have been primarily interested in placing descriptive findings concerning aggression into a developmental framework. In order to complete our descriptive survey, we would like to mention a few additional variables which affect aggression. First come situational factors such as time of day or population density; e.g., a child is more likely to be angry before mealtime and when he is in a crowded as opposed to an uncrowded playroom. Physiological well-being is important: fatigue and physical illnesses increase the incidence of aggressive behavior. Boys are uniformly more aggressive than girls and more physically aggressive, while girls rely more on verbal forms of expression. The child's social class and his culture markedly affect the amount and kinds of aggressive behavior.

This inventory makes us aware of the many factors influencing aggression, but tells us nothing of the interaction among variables. One of the few objective investigators of interaction, Fite (1940), studied parental and peer influences on aggression in preschool children. By three or four years of age, both individual friendships and group pressures are sufficiently potent to influence aggressive behavior. The group, which values winning and toughness and accepts retaliation as a necessity, generally accepts aggression better than do parents and other adults. Highly aggressive children seek out each other's company and form close friendships in spite of constant bickering and quarreling. Only the child who is aggressive out of insecurity, and who becomes violent, hectic, and unreasoning, is shunned by individuals and the group alike. The preschooler is subject to a double standard (one from parents, the other from peers), and events at home may significantly affect his behavior at nursery school. In one instance parents made a little girl anxious over expressing aggression and she sought out more inhibited children to be her friends at nursery school. There is also an interaction of variables within the group itself; e.g., when her best friend began to tire of her, one child, who had been secure in expressing aggression and had been accepted by the group, became insistently and unreasonably aggressive, thereby jeopardizing her standing in the group. Thus, aggression is not a situation-specific or a "person-specific" reaction; instead, what happens in one setting or relationship may significantly affect the expression or inhibition of aggression in other settings or relationships.

Theories of Aggression

Although aggression is a primary concern to all students of human nature, psychologists' understanding of it is limited and their conceptualizations contradictory and unsatisfactory. The three conceptualizations we shall present represent three points along a continuum, with aggression as an innate response at one extreme and aggression as a learned response at the other. (For a more extensive presentation of conceptualizations of aggression, see Berkowitz, 1962.) In Freudian theory aggression is an instinct which man must learn to manage but which he never can eliminate. Man is biologically fated to be aggressive, as he is biologically fated to be sexual. However, Freud had a prior and different concept of aggression which, as adapted by a group of American learning theorists, became known as the frustration-aggression hypothesis (see Dollard et al., 1939). Here aggression is considered as a consquence of frustration, rather than as biologically determined, and theoretically, the amount of aggressive behavior can be diminished by decreasing frustration. Bandura and Walters (1963), in the social learning theory tradition, take an extreme environmental stance which denies any innate factors in determining the amount of aggression and sees it as no different from any other learned behavior. Its strength depends on the reinforcements (rewards) the child receives when he acts aggressively and on the aggressive models available for him to imitate. Extrapolating from their position, one can infer that aggression would all but disappear if reinforcement and modeling could be properly controlled.

The theoretical positions have important practical consequences. As we shall see, the conditions which would minimize aggression according to one position, would augment it according to another. This is particularly true of the Freudian and the social learning theories. The frustration-aggression hypothesis is derived from Freud and is nearer to the Freudian spirit. If aggression is innately tied to frustration, and frustration is an inevitability of human existence, then we are humanly fated to aggression, even if not biologically fated. Each theory can marshall evidence in support of its view; the psychoanalysts rely primarily on the clinical approach and the learning theorists on objective evidence. Instead of one answer to the question, What can be done about violence?, psychologists offer a number of conflicting answers, and constructive dialogues between opposing points of view are rare. Fortunately there are also areas of agreement which are impressive just because they grow out of different theories, independent investigation, and different kinds of data.

Freud's Concept of the Death Instinct

Freud formulated his concept of a death instinct comparatively late in his career and it failed to win acceptance even among psychoanalysts. Our presentation of a modified version, specifically addressed to child development,[9] will skirt an issue essential to an instinct theory; namely, the biological basis of aggression. At times aggression appears to be a purely psychological phenomenon, and research concerning its biological origins seems an effort to answer a nonexistent question. But the question keeps returning and the quest to discover the psychobiology of violence continues. The quest is part of a general hypothesis that, for every extreme psychological deviation there must be a biological deviation. The evidence neither proves the hypothesis is valid nor that it should be abandoned; it only tantalizes.

What will concern us primarily is the parallel Freud draws between the aggressive and the sexual instincts or drives. Both are somatically determined quantities of energy pressing for expression. Society can no more tolerate the raw, heedless destructiveness of primitive aggression than it can condone primitive sexuality. Realistic demands for control force the energy into socially acceptable channels, such as assertiveness, domination, and competition. Just as with the sexual drive, direct and disguised gratification is essential to healthy development; massive repression lays the groundwork for neurosis.

Psychoanalysts have provided a rich clinical literature to substantiate their claim that aggression, like sexuality, undergoes innumerable transformations and disguises under the pressure of socialization. One obvious maneuver in managing aggression is displacement: the object of anger is changed in order to reduce the fear of retaliation (see Chapter 6). The familiar example is the man who is too frightened to express anger toward the boss for not giving him a raise, but who shouts at his wife for spending too much money, or punishes his child for spilling milk, or whips his dog, depending on which member of the household is the least threatening. A mother who reports, "Every time I spank my (three-year-old) daughter, she kicks her highchair," implies that the innocent chair bears the brunt of the attack meant for the powerful mother. Although displacement can be obvious to an observer, the individual frequently is unaware that he is using this defense.

Both the affect and the destructive intent of aggression can be sublimated, i.e., sufficiently modified to make them socially acceptable. Athletics provides numerous examples: anger becomes a "fighting spirit," and the intent to destroy becomes a determination to defeat; the football cheer is "Yea, team, fight, fight, fight." In fact, the war correspondent's report and the sports writer's description of a game may use

[9]We rely mainly on A. Freud (1949) and Hartmann, Kris, and Loewenstein (1949).

the same verbs—"devastated," "slaughtered," "annihilated." The difference is that one is reality and the other is play—and this, quite literally, makes all the difference in the world. Psychoanalysts become concerned with brutality and corruption in athletics and with cutthroat competition (again, the label is significant), all of which are throwbacks to primitive expressions of aggression; the more primitive athletics become, the less they are able to perform their civilizing function.

Aggression can assume more subtle disguises. Instead of being active, it can become passive. An exquisite form of revenge is dawdling in which the child does not oppose the parent but complies in slow motion. The parent is both exasperated and robbed of his right to punish open defiance. What more could a child wish for? Various kinds of ineptness may also have an underlying hostile intent. The bright, underachieving child unconsciously may wish to distress his intellectually overambitious parents; the physically awkward boy may unconsciously enjoy the humiliation suffered by his pressuring, athletic father; even the immature wife who constantly burns the dinner may be expressing her anger at having to care for her husband when she herself still needs to be cared for. Note the careful use of the word "may" in these clinical examples; any given behavior can result from a variety of motivations, and we do not mean to imply that aggression is invariably the motive behind underachievement, awkwardness, or burned dinners.

Aggression can utilize guilt induction to serve its goals of harming others, since guilt is such a painful affect. A loving relation between parent and child is particularly vulnerable to this perversion since, as we have seen (Chapter 4), love oriented parental discipline tends to induce guilt in the child. The long-suffering mother who tearfully and constantly reproaches her adolescent child with, "How can you do this to your mother who loves you so, who fed you and cared for you and thought of nothing but your happiness?" is surely making her child suffer. She would vigorously and rightfully deny that she felt angry, since all she experiences is her own hurt. Yet, her anger may be unconscious, and in her own way she is inflicting pain. From the child's point of view, he might have difficulty choosing which kind of parent is more destructive—the sadistic father who beats him or the masochistic mother who makes him responsible for all her suffering.

Finally, suicide itself, which seems furthest removed from aggression, can be motivated by an unconscious desire to inflict pain on another. Not infrequently, the relationship to the other person has been a highly charged mixture of love and hate, rather than a purely hostile one. The child expresses the situation to perfection when he says, "I'll die, and then you will be sorry." Suicide can represent an unconscious implementation of this childish threat to punish through guilt.

This clinical evidence does not have to be explained in terms of an aggressive drive; but to the psychoanalysts, finely tuned to the myriad disguises of sexuality, a conceptualization in terms of drive is natural. Aggression, like sex, will out; deny it expression here, and it turns up there, parading as morality or love or indifference or silence. Even if one rejects the psychoanalysts' concept of an aggressive drive, their clinical observations richly illustrate the diverse and pervasive manifestations of the desire to injure another.

The Developmental Course of the Death Instinct. The psychoanalytic theory of aggression includes a developmental sequence which closely follows the model of the psychosexual stages.[10] Erotic and destructive elements constantly interact in personality development (see A. Freud, 1949). Aggression, like libidinal gratification, is concentrated on the body in earliest infancy. The behavioral evidence is the self-inflicted pain, such as head banging, or hitting the self, which, as we have seen, is not uncommon in infancy. In order to avoid self-destruction, the object of aggression must be diverted from the body to the outside world, a transition which is affected in the oral period. Sucking becomes charged with aggression, and psychoanalytic descriptions of nursing can include phrases such as "attacking the breast with devouring greed." With the eruption of teeth, the infant derives erotic pleasure from biting; pleasure and destruction thereby unite to form sadism. In the anal period the mixture of love and hate is so obvious that even antiFreudians recognize the picture of the possessive, clinging, demanding, tormenting toddler who exhausts the beloved mother and who treats his pets and toys with the same mixture of intense affection and destruction. In the phallic period, aggression assumes a more adultlike form, and becomes a desire to impress, dominate, and subdue the loved one. The development of conscience at the climax of the Oedipal phase is intimately bound up with aggression. In identifying with the same-sex parent, the child internalizes prohibitions and punishes himself for wrongdoing. Concomitantly, the child turns aggression against himself. Instead of saying, "I hate you for the bad things you do," he now says, "I hate myself for the bad things I do." This self-directed aggression is one factor which makes the early superego so vicious in its punitiveness (see Chapter 5).[11]

Characteristically, psychoanalysts have little to say regarding latency, except that successfully sublimated aggression adds vitality to the intellectual and social expansiveness of the period. Self-assertiveness, a determination to master academic subjects, leadership in groups and in athletics are some of the fruits of its successful management.[12] Adolescence typically is regarded as a period in which difficulties in the first three stages of psychosexual development are reactivated under the impact of physiological maturation. The affective instability of adolescence is a later edition of the

[10] There is an interesting paradox here. Psychoanalytic theory, which emphasizes the innateness of aggression, has the more elaborate account of its changes through childhood, while learning theories, which stress malleability of aggressive behavior, pay relatively little attention to development. Their main concern is to provide a universal model which will be applicable to all aggressive behavior at any time of life, and they tend to neglect the fact that the infant is flailing his arms and legs and the teenager is telling a malicious lie about his rival. The psychoanalytic school has addressed itself to describing and accounting for changes in form, while learning theorists have concentrated on universal principles of behavior and have been relatively unconcerned with the form of the behavior.

[11] Hartmann, Kris, and Loewenstein's (1949) genetic picture is somewhat different and more in keeping with our earlier descriptive account of aggression. They maintain that aggression in earliest infancy may be merely a massive release of energy rather than specifically directed toward the infant's body. In the so-called "oral sadistic" phase, they claim that the infant derives pleasure from aggressive behavior per se, but that he is not sadistic. To be truly sadistic, the infant must understand that he is inflicting pain on another; he is incapable of such understanding until after he has established an attachment to the mother and has perceived her as an independent individual.

[12] For an alternate conceptualization of latency, see White (1960).

affective instability of the first six years of life. However, some analysts (A. Freud, 1958) have ventured beyond a recapitulation approach. The crucial developmental task in adolescence is that of transferring libidinal attachments from the parents to opposite-sex peers, since parental attachments now threaten to be implemented by a mature sexual apparatus. The adolescent experiences intense loneliness when he realizes that he must give up the love of his parents while being unable, as yet, to replace it with meaningful heterosexual relationships. Ideally the transition between the old and the new proceeds gradually; but intense anxiety or loneliness can produce more extreme reactions. The adolescent may abruptly flee emotional involvements with parents and become passionately attached to the gang leaders, or he may develop crushes on older persons, or he may become involved in unrealistically demanding heterosexual or homosexual affairs. More to the point, he may defend himself against his earlier attachments by a "reversal of affect," in which love becomes hate, dependency becomes revolt, respect and admiration become contempt and derision. Bound to his parents as strongly by hate as he formerly was bound to them by love, the adolescent does not dare to relinquish his anger because then he would have to face the loneliness of being unloved as well as the fear of mature loving.

"Reversal of affect" adds a new element to our understanding of aggression, since we see that aggressive behavior can serve other motives. Hurting others can become a means of achieving nondestructive goals, such as protecting the child from anxiety and depression. ("It is better to be wanted for murder than not wanted at all" is an illustration of this use of aggression.) As we shall see, learning theorists have the same concept, which they call the instrumental use of aggression.

The psychoanalytic theory of development contains two points about aggression which will figure prominently in the comparison with learning theories. First, certain forms of aggressive behaviors are normal at given stages of development. Next, aggression, properly managed, contributes significantly to the child's strength of character by energizing self-assertion, competitiveness, mastery, and conscience. While one may or may not accept their theory, the psychoanalysts pose two challenging questions: Can aggression be considered normal? and, Can aggression promote the growth of personality?

Thus far we have stressed the similarity between the aggressive drive and libido; however, psychoanalysts recognize differences between the two. For our discussion, the most important difference is in the goal, or what the psychoanalysts call the aim. Libido is pleasure seeking, and, since the infant's caretaker is a source of manifold pleasures, libido is the force which links him to a human being. From the earliest mother love to mature heterosexuality, libido energizes the deepest human attachments. The aim of aggression is destruction. It is manifestly dangerous because it is potentially life-endangering. But, for the psychoanalysts, it is equally dangerous because it is love-endangering. Destruction of the loved one poses a unique threat to all of psychological development, because a positive attachment is essential to all subsequent steps in socialization—the capacity to tolerate frustrations, self-control, self-trust, self-esteem, concern for others, morality, and ideals—to mention only a few. The destruction of the loved one strikes at the heart of all civilizing forces.

Deviant Development. As was true of the sexual drive, anxiety or guilt can alienate the child from the aggressive impulse itself, and prevent him from utilizing developmentally appropriate outlets. Not only does he repress the sanctioned expressions of aggression, such as fighting, quarreling, name-calling, and swearing, but often the healthy derivatives also, such as self-assertion and competitiveness. The resulting picture is familiar to the clinician—the child who is too good in his daily behavior and who has a high charge of unconscious rage. A vicious cycle starts in which fear of expressing anger prevents the child from experiencing its realistic consequences and learning techniques for its control. Instead, he perpetuates the terrifying fantasy that, if once he were to give vent to his explosive feelings, he would destroy everything and everyone.

In another kind of deviation the child's self-control is weak and he is aggressive in a manner inappropriate to his developmental level. He is violent in his attacks on others or on property, and he commits the kinds of antisocial acts which society labels as delinquent or criminal. Among his behaviors one might find pleasure in or indifference to inflicting suffering, cruelty to animals, senseless or wanton destruction of property.[13]

The overly controlled child is constantly punishing himself for his destructive impulses and fantasies. He is at war with himself, immobilized by a strong need to act aggressively and an equally strong anxiety about acting. The child whose self-control is weak is at the other extreme—he cannot help giving vent to anger which the environment cannot tolerate. In psychoanalytic terms he is acting out his impulses.[14] Punishment is meted out by people who try to curb his behavior, and he is deterred primarily by the fear that he will be caught in the act. In sum, the social environment must supply the controls he cannot exert over his impulses.

Redl and Wineman (1951) in their study of severely disturbed, aggressive children, present a detailed clinical account of pathologically weak control. The children's attention is directed inward to their impulses rather than outward to the people and events in the environment. Preoccupied with their anger, restlessness, and pleasures, they have little ability to size up situations in an objective manner or to empathize with the feelings of others. Their ability to delay is pathologically low, and waiting for a meal which is not served when expected or even waiting for a traffic light to change can throw them into a spasm of impatience. Events often seem to evaporate after they happen, so that the children fail to profit from experience. In therapy, for example, such children must experience the good intentions of a therapist or the pleasure of an activity innumerable times before it begins to register. The boys are particularly slow to realize their responsibility for their own antisocial behavior, and they sincerely believe the most blatant distortions of reality; e.g., a thirteen-year-old blamed his stealing on his victim by claiming the victim should not have been carrying

[13] Technically, delinquency and criminality are defined by law and society, and are not psychological terms. Like any deviant behavior, delinquency and criminality can have a number of roots, and we do not imply that all delinquents and criminals have poorly controlled aggression. For a more comprehensive discussion, see White (1964).

[14] For a detailed exploration of acting out behavior in childhood, see Rexford (1966).

so much money. The impulse-ridden boy is not only trapped in the present, however; he also is magical in his thinking. In spite of experience to the contrary, he never really expects to be caught. Because he wishes to see himself as invincible, he believes that he will be. In sum, we see in the impulsive delinquent much of the emotional and intellectual egocentricity which characterizes the first few years of life (see Chapter 6).

Impulsive behavior results when the processes responsible for the development of self-control are disrupted (see Chapter 4). The psychoanalysts particularly stress the importance of identification with parents in determining the degree of self-control. The concept of identification helps us understand two different origins of delinquent behavior. If the parents themselves have antisocial values, the child may identify with these values and be regarded by society as delinquent. The process of identification is no different here than it is for a law-abiding child. An impoverished widow, for example, may teach her children that they are justified in stealing and lying and may herself be a model of such behavior. The children will incorporate her antisocial values and believe they are doing right when they steal. They will not have defective self-control. True impulsivity results when the process of identification itself is disrupted, thereby preventing the development of normal self-control. The parents may be indifferent or punitive or rejecting; or family unity may be destroyed by divorce or death or war. Whatever the reason, the bond of love is weak and cannot serve as a catalyst for identification with parental values. Lacking the stability which integrated guides bring to behavior, the child reacts with anger to any provocation in the immediate situation.

While failure to identify lays the groundwork for poor impulse control in general, hostile acting out is particularly apt to occur under two conditions, according to Anna Freud (1946). First is a mismanagement of aggression during the anal period. The toddler's high charge of aggression, evidenced by tantrums, willfulness, and provocativeness, must be counter-balanced by an equally large supply of libidinal gratification. If, for example, the parent is threatened by the child and reacts by withdrawing love or becoming harsh and retaliatory, the child's normal hostility will tend toward pure destructiveness; he will become quarrelsome, ruthless in his acquisitiveness, and basically angry in his attitude toward others.

The second condition is a more generalized version of the first—control of aggression is contingent upon its being counterbalanced by love. Psychoanalysts have a rather odd notion that when the two impulses are balanced, they can fuse and produce "neutralized" energy which is neither love nor hate, but which is available to energize a variety of constructive activities. We need not accept the concept of fusion and neutralization, but we should remember Anna Freud's point that the impulsively aggressive child is basically a child who is inadequately loved.

It follows that in order to help the impulsively aggressive child, the therapist must supply him with the love he lacks. This is an exceedingly difficult therapeutic task. The child is deeply mistrustful and fears he will be trapped by the show of friendliness and subsequently betrayed and deserted. Not infrequently he becomes even more aggressive in order to test the therapist's sincerity. Few therapists can tolerate both the child's slow progress and his provocative defiance of law and of life. Yet it is essential for the therapist to maintain an attitude which is sympathetic but firm. To be only sympathetic would run the risk of condoning the delinquent behavior and of being regarded as "soft." To be only firm would make him like all the other adults who, angrily or punitively, only want the child to behave himself. The therapist must help the child realize that he requires control because he values the child and is concerned about him. It is only as the child feels that he is valued and that the therapist will not desert him in the face of provocation that he will begin to incorporate the therapist's standards of conduct as his own.[15]

The Frustration-Aggression Hypothesis

The frustration-aggression hypothesis, as originally stated by Dollard (1939), postulates that aggressive behavior "always presupposes the existence of frustration" and frustration "always leads to some form of aggression" (p. 1). "Always" means that the connection is innate, and this issue of innateness has been a source of controversy to the present day. A more cautious rendering of the hypothesis is that the frustration produces an instigation to aggression, but that the occurrence of aggressive behavior depends on a variety of factors. Before reviewing the controversy, we will discuss the hypothesis itself and present some of the objective studies it has stimulated. Regardless of its ultimate fate, it represents one of the most fruitful collaborations of the clinical and the objective approaches. (Recall that the original idea was Freud's.)

Frustration is defined by Dollard and his colleagues as an interference with the occurrence of a goal response, while aggression is behavior whose goal is injury to a person. When scrutinized (see Berkowitz, 1962, and Lawson, 1965), the concept of frustration proves to be quite complex, but we will accept the definition at face value. Examples of interference with a goal response are limitless—playpen bars keep a toddler from reaching a toy; a mother refuses to let her child see a TV program at bedtime; duty forces a boy to study when he would rather play ball. Thus, frustration can originate from the physical world, the interpersonal world, or from within the child, but, whatever the source, it blocks the attainment of a desired goal. Moreover, the stronger the motivation to attain the goal, the more intense the frustration and the subsequent instigation to aggression. The aggression need not be overt; thoughts and fantasies, symbolic as well as direct attacks, qualify.

Noting that both anger and cognition are neglected in the formulation (probably due to the learning theorists' suspicion that feeling and thinking are not sufficiently objective for their behavioristic approach), Berkowitz proposes that the hypothesis be expanded to state that frustration produces anger, and that it is anger which serves as the instigation of aggression and the motivating force in the sequence. Cognition determines what will and will not be regarded as a frustration. Although there may be universal frustrations, most depend on an individual's perception and

[15] The classic application of psychoanalytic theory to the treatment of delinquent children is found in Aichhorn (1935). See also, Eissler (1950).

interpretation of situations. A gum-chewing waitress may be an affront to a woman's dignity, but be hardly noticed by a man.

Increasing and Decreasing Aggressive Behavior. According to learning theory, reinforcement (reward) should increase the likelihood of the occurrence of aggressive responses, while nonreinforcement and punishment should decrease the likelihood of occurrence. The presence of aggressive and nonaggressive models for the child to imitate is another variable which should significantly affect his aggression. A number of studies verify predictions made on the basis of learning theory. Goodenough (1931) noted that children whose parents give them their own way subsequently have more temper tantrums than children whose parents do not cater to them. In general, permissive parents who benignly accept and condone aggression or who might vicariously enjoy it, have highly aggressive children. Patterson and his co-workers (1967), perhaps suspecting that a preschooler's aggression usually outstrips parental reinforcement, focused on peer relations. As predicted, they found that a child, successful in his aggression toward a victim, would tend to repeat the same aggressive activity to the same victim; if his aggression produced painful consequences, the likelihood of repetition diminished. A more important finding disclosed that if an originally nonhostile child who was frequently victimized by others successfully counterattacked, his incidence of aggression would rise dramatically. Thus, peers can both sustain aggression by giving an already aggressive child what he wants and can actually produce more aggression in a nonhostile child who can make a successful counterattack.

The effects of punishment on aggressive behavior are more complex than the effects of reward. Punishment teaches the child that his aggressive behavior can have painful consequences. Subsequently, his tendency to be overtly aggressive will be a function of both the intensity of the instigation to hurt others and the intensity of anxiety or guilt which has become associated with hurting them. But the situation is not so simple. Punishing aggression is in itself a frustration, since it prevents the child from achieving his goal of hurting. Moreover, the angry, punishing parent serves as a model of aggressive behavior for his child to imitate. On two counts, then, punishment should increase the instigation to aggression while inhibiting specific aggressive acts. In the definitive study of this proposition, Sears and his colleagues (1953) found that children with highly punitive mothers were less aggressive with peers and adults in a nursery school setting than children with nonpunitive or mildly punitive mothers. However, the instigation to aggression had not disappeared. In a doll play situation, where the atmosphere was permissive and the targets were dolls rather than people, the children of highly punitive mothers exhibited the greatest amount of aggression. In learning theory terms, inhibition of aggressive responses generalizes to situations similar to the one in which punishment occurs; however, if the situation is dissimilar, the anxiety will be sufficiently weakened so that the aggressive response will appear.[16]

In sum, there is objective evidence that both extremes of permissiveness and of punitiveness increase aggressive behavior. The implication is that parents should never

[16] The same phenomenon is called displacement by the psychoanalysts, as we have seen. For a more detailed presentation, see Berkowitz (1962).

condone aggression but should rely on nonpunitive techniques in managing it, such as diverting the child, suggesting constructive means of dealing with the frustrating situation, trying to understand why anger flared up. Not only will the parent help the child substitute constructive reactions for destructive ones, but he will also be serving as a model of self-control for the child to imitate.

Instrumental Aggression and Nonaggressive Responses to Frustration. Learning theorists recognize that the need to harm others is only one of the needs which aggression satisfies. The child who realizes that his temper tantrums always bring mother running to investigate may well begin to scream and kick and throw things when he feels neglected; he uses aggression as an attention-getting device. As we have just seen, antisocial acting out can be used by a delinquent to test the sincerity of the therapist's professed concern, while an adolescent can cling to a pattern of hostile opposition to his parents to protect himself from the loneliness of relinquishing his attachment. These are instances of instrumental aggression, or the use of aggression to serve other than destructive purposes. However, as Feshbach (1964) aptly notes, instrumentality does not necessarily rob an act of its destructive sting. The four-year-old who hits his younger sibling may be motivated primarily by a desire to attract his mother's attention, but he is also gratifying his need to harm the sibling. A single bit of behavior may serve multiple purposes.

Just as aggression can come to serve other than injurious ends, frustration can produce behavior other than aggression. It can lead to regression, or behavior at a developmental level lower than the child is capable of. In their classical study Barker, Dembo, and Lewin (1941) observed thirty nursery school children with standard play material for half an hour and scored their play for constructiveness. Then the children were put in a situation in which the standard play material was incorporated into a set of unusually attractive toys. After they had become involved, the experimenter interrupted them and lowered a wire screen between the child and the attractive toys. The children once again had to play with the standard material, with the more attractive but unobtainable toys in full sight. They reacted in a number of ways, including aggressive attacks on the wire screen, and attempts to leave the room. More to the point, their play was significantly less creative and constructive than it had been before frustration; there was also evidence that the most frustrated children were the ones who regressed most.

Frustration does not inevitably elicit primitive behavior such as aggression or regression. Under certain circumstances, it can facilitate socially desirable acts. Wright (1943) found that when a pair of children who were strong friends were exposed to the Barker, Dembo, and Lewin frustration situation, they showed less immature behavior than did a pair whose friendship bonds were weak. In a second experiment, close friends became more cooperative in their play and more outgoing socially in response to frustration. They were less aggressive toward one another and more aggressive toward the frustrating experimenter than were weak friends. In the Patterson study (1967) peer relations significantly affected behavior by weakening the regressive pull of frustration and producing greater solidarity and cooperativeness.

The Issue of Innateness. Now let us return to the controversy over the innate connection between frustration and aggression. If frustration can produce behavior other than aggression, is the concept of innateness necessary? Bandura and Walters (1963) claim it is not and cite empirical studies to prove that aggressive behavior is learned on the basis of imitation and reinforcement. As we have seen, there is evidence that rewarding aggression will increase the frequency of its occurrence. There is also equally convincing evidence that, after observing an aggressive adult model or a film showing an aggressive adult model, children will behave more aggressively. The model serves both to increase their repertoire of aggressive responses and to facilitate the use of already acquired aggressive patterns. If a preschooler beats on a toy after observing an adult beating on a toy, there is no reason to assume that he has been frustrated.

To the evidence of diverse responses to frustration, Bandura and Walters add studies showing that the particular response chosen is a function of the child's previous training. Davitz (1952) rewarded half of a group of seven- to nine-year-olds for making competitive and aggressive responses and the other half for making constructive, cooperative responses. Then all the children were frustrated when a movie they were watching was interrupted at the climax and when candy which they had been given was taken away. In a subsequent play session the children in the first group displayed more aggression, while those in the second play group displayed more constructive behavior.

Bandura and Walters hypothesize that frustration produces a heightened motivational state which has no determining effect on the specific kind of behavior which will be energized. The frustrated child is mobilized to act, but what he will do depends on his existing repertoire of responses which in turn has been established through imitation and reinforcement. There is just as much chance for a child to behave cooperatively or constructively or submissively or assertively as there is for him to behave aggressively.

The camel's nose of innateness does get at least part way into Bandura and Walter's environmentalist tent. Frustration results in a temporary intensification of responses, and vigorous or "high magnitude" responses, such as hitting, kicking, pushing, are often regarded by others as aggressive. For example, a toddler is hitting a stick against the cement driveway and against a wooden porch as part of his exploratory behavior of discovering the different sounds it can make. For some reason he becomes frustrated–perhaps he wants the stick to make a different sound than it actually does make against the wood–and he starts hitting the porch vigorously. The innocent exploratory response will now be judged as aggressive, and an observer would say the child is angry at the stick and wants to break it. Bandura and Walters are careful to add that high intensity responses are not unique to frustrating situations but can be learned in nonfrustrating ones; e.g., a child is taught to punch or kick vigorously in the context of play. When frustrated subsequently, he has a repertoire of high intensity responses which can be activated. The child will be judged as behaving aggressively when such responses appear in the context of a frustrating situation and when they are directed toward the frustrating agent. In sum, an innate connection between frustration and intensity of response, and the tendency for high magnitude responses to be judged as aggressive, is all that is left of the once absolute connection between frustration and aggression. Note that Walters and Bandura are like Dollard and his co-workers in their avoidance of affect and cognition. Anger plays no part in their conceptualization of aggression, nor are they concerned with the child's interpretation of a situation.

The original frustration-aggression hypothesis is no longer tenable. The question is not whether aggression is always due to frustration or whether frustration always leads to aggression; the question is whether the child can learn any response when frustrated or whether aggression is the response which is most likely to occur initially. A crucial test would be difficult to devise. Bandura and Walters's studies of imitation, for example, do not disprove a modified innateness hypothesis. They show that all aggressive behavior cannot be regarded as a reaction to frustration, but they do not disprove the contention that, when frustrated, the aggressive response is more likely to appear than any other. The evidence concerning different reactions to frustration resulting from differential reinforcement of aggressive and cooperative behavior is also not crucial. Berkowitz readily acknowledges the importance of learning other responses to frustration, while still maintaining that anger with its instigation to aggression is the response which is most likely to occur initially. When a nonaggressive response to frustration appears, we cannot tell whether it has merely been reinforced (as Bandura and Walters maintain) or whether it has been reinforced to the point when it is stronger than the tendency to respond with anger and aggression (as Berkowitz contends).

The Developmental Perspective. The innateness issue and the learning of aggressive behavior have rarely been put into a developmental framework. As we have already noted, learning theorists have been more concerned with general principles of behavior than with the developmental course of a given behavior. Our impression is that the first six months of life is a promising period for studying innate responses to frustration and that the period between six and eighteen months is equally promising for studying the development of aggressive behavior relatively uncomplicated by imitation or instrumental use. (The earliest instance of imitated aggression Piaget (1962) noted occurred around eighteen months of age, and many of Goodenough's (1931) situations eliciting aggression in the six- to eighteen-month period involve a direct reaction rather than an instrumental use of aggression.)

We are not suggesting that observing young infants will provide an easy answer to the question of innate responses to frustration. For the moment, let us accept as valid the observation that some infants react with intense rage to frustration while others are "good babies" and seem unperturbed by delays in feeding, physical restraint, unexpected switches from one activity to another (see Chapter 10). What does this mean? It may mean that "good babies" are not being sufficiently frustrated. Remember that Dollard and his co-workers postulated that the intensity of aggression is a function of the intensity of the thwarted response. It is possible that the "good baby" is placid and moderate in the majority of his activities and typically is not in a heightened motivational state. Only if one could observe him being frustrated during times of intense motivations could one test his susceptibility to anger. Another possible explanation of these early individual differences is that infants are innately different in their characteristic responses to frustration. For some, anger is high on the list and is readily elicited; for others, shifting of attention or withdrawal might be high. Such a finding would tip the scales in favor of the innate camp, since it suggests unlearned temperamental differences (see Chapter 10); however, it would reduce the innate connection between frustration and anger from a universal principle to the status of a special case.

Now let us turn to the six- to eighteen-month period for leads concerning the development of aggressive behavior. A significant amount of aggression in this period involves a striking out at the perceived source of frustration. However, this "striking out" should not be equated with Dollard's concept of aggression as behavior whose goal is injury to a person. Such a goal requires that the toddler understand that his behavior has caused pain or distress to another. We are not sure when he is capable of understanding the feelings of others and the effect of his behavior on them. The earliest evidence of provocative (i.e., deliberately annoying) behavior Goodenough noted was at eighteen months of age, and our best guess is that, before then, the toddler seldom begins a sequence of behavior whose goal is injury. Developmentally, the distinction between striking out and injuring is a crucial one.

Feshbach (1964) elaborates on this distinction. Aggression originates in the emotion of anger, which is the innate response to restriction and discomfort, and is evidenced by loud vocalization and vigorous flailing of the limbs. Following both Darwin and Freud, Feshbach assumes that this motor behavior serves to reduce tension. In adult terms, the infant is "blowing off steam." As the infant becomes more coordinated and controlled, he is capable of directing his activity toward the frustrating agent. The most typical motor activity is hitting. If hitting succeeds in removing the frustrating agent, the infant will tend to use it more frequently, and the link between frustration and hitting will be strengthened. However this angry hitting is not an intent to injure the frustrating agent.

How, then, does the intent to injure develop? Feshbach's answer is primarily speculative. He notes that man is the only animal which habitually intends harm and is reinforced by inflicting harm; other animals inflict injury in the process of satisfying nonhostile needs. In order to intend harm, the toddler must first grasp the idea that his actions have noxious effects on others. In part he is aided by empathy, since he is capable of feeling distress merely by observing another child who is the victim of aggression; in part, he is helped by the socializing parent, who repeatedly tells him that he hurts when he hits. But primarily he is taught by being punished. Hurt to others is followed by punishment to himself; the more vigorously he hits out, the more vigorously he is hurt in return. At this point imitation of the parents helps him bridge the gap between hitting and hurting; if the parents hurt him when they hit, he must hurt when he hits. Anger continues to play an important role either in initiating hurting behavior or making it more vigorous, but anger and intentional hurting can also be independent: at times the child can primarily be "letting off steam"; at times he can intend to hurt without a high charge of anger.

We are not concerned with a critique of Feshbach's speculation, since our interest is in his general thesis: namely, goal-directed aggressive behavior is a product of development. As we have seen, an infant cannot be angry *at* anything until he has achieved a certain amount of control over his body, until he has experienced regularities of events sufficiently to anticipate that B will follow A, until he has some awareness of himself as an agent, and until he has grasped the concept of object independence. Only then can he intentionally hit out at the environment. The intent to injure requires an even more advanced understanding of how people respond to his actions. Cognition and empathy may play a prominent role in the development of his understanding of people, along with other as yet unspecified variables.

If we assume that behavior whose goal is injury cannot appear until around fifteen to eighteen months of age, we must recognize that a tremendous amount of learning has preceded its advent. Such learning does not depend primarily upon parental reinforcement or imitation of models, as learning theorists have assumed, since it involves cognitive factors which learning theorists have ignored. Only after aggressive behavior has developed are parental reinforcement and imitation effective in determining its fate.

Observing the development of aggressive behavior in the first eighteen months of life may be the best means of discovering the role frustration plays. We suspect that the innate versus learned issue will have to be broadened to give early temperamental differences (see Chapter 10), cognition, and rage a more prominent part than they have been assigned in the past.

The Management of Aggression

The issue of innateness is of more than academic concern. It affects one's philosophy concerning the management of aggression. The difference is most readily seen by contrasting Freud's drive theory with a thoroughgoing learning theory. In the former, developmentally appropriate outlets for expressing aggression are essential to mental health. The aggressive behaviors in the psychoanalytic genetic theory, supplemented by those from descriptive psychology, are normal not only in the sense of being statistically frequent but also in furnishing the child with age-appropriate outlets for his innate destructiveness. Parents should be loving and apt models so that the child can learn to manage his hostile impulses, not so that aggression will atrophy. While poorly controlled and regressive forms of aggression are suspect, so is the absence of age-appropriate expressions.

A thoroughgoing learning approach would claim that the amount of aggression a child displays merely reflects the degree to which his social environment has rewarded and supplied models for aggressive behavior. Normal aggression is only normal statistically. Societies have different degrees to which aggressive responses are strengthened. A society in which parents serve as models of love, firmness, and reason would produce children who are constructive in response to frustrations and provocations. There is no reason to believe that such children would be inhibited or that they would be seething volcanoes of pent-up anger waiting to destroy others or themselves. Any theory which requires a child to behave aggressively in order to develop normally is sanctioning a level of aggression which is unnecessarily high and may be blocking the achievement of a peaceful society.

One final point which is pragmatic rather than theoretical. The psychoanalytic concept of an aggressive drive has been regarded as "pessimistic," while the learning theory approach appears "optimistic." However, rejecting the drive concept does not simplify the problem of management of aggressive behavior. Since learning theorists have paid relatively little attention to the descriptive aspects of development, it often is not clear how they would deal with many substantive issues. From our descriptive survey we learned that aggression can be an element in friendship or in active social involvement and does not necessarily jeopardize either intense attachments or strong social feelings. Would the constructive context of such aggressive behavior make it acceptable, or would learning theorists maintain that any reinforced aggression brings the child that much closer to antisocial or asocial violence? Or again, would learning theorists control realistic aggression while permitting aggressive fantasies in stories, plays, movies, and on TV; or would every exposure be regarded as a potentially fatal beginning? Would multiform disguises of aggression be recognized, such as teasing, maliciousness, passive hostility, or guilt induction, and would these be controlled, or would there be a distinction between and differential treatment of direct and disguised aggression? Does learning theory require an absolute program of nonviolence, or would it allow for expressions which are developmentally appropriate and statistically within the range of normal frequency? We are not saying that learning theory cannot answer such practical questions, but we are saying that there is no obvious answer flowing from their environmentalist approach. We are also cautioning against confusing the simplicity of their explanatory concepts with a simple solution to the substantative issues involved in managing aggression.

While theoretical differences among psychologists lead to contradictory conclusions concerning the management of aggression, there are also many areas of agreement—the more surprising in light of their origins.[17] Learning theory and psychoanalytic theory agree on most major points concerning the reasons for poorly controlled aggression. Insufficient gratification of the need for love (which learning theorists conceptualize as a dependency need) by cold, rejecting, punitive parents, is frequently related to a high degree of aggressiveness in both delinquent and nondelinquent children. Like the psychoanalysts, learning theorists regard affection as important to identification with both parents and emphasize the role of the parents as models of self-control. Anxiety over receiving affection is also regarded as an obstacle in establishing a therapeutic relationship (Bandura and Walters, 1959; McCord, McCord, and Howard, 1961). More explicit in regard to parental discipline, learning theorists state

[17] In the discussion of sex, we saw how theories explored different facets of sexuality. The same holds true for aggression. Once again, learning theory concentrates on behavior and pays scant attention to affect and cognition. Psychoanalysts concentrate on the affect of rage and the motivating properties of hostility. Cognitive variables fitfully appear and disappear. Fortunately, there are attempts to integrate the behavioral, affective, motivational, and cognitive facets of aggression, notably Berkowitz (1962) and, in a more schematic form, Feshbach (1964).

that love oriented techniques and consistently prescribed standards for behavior favor the control of aggression, while punitiveness and inconsistency undermine control. Psychoanalysts, because of their concern with intrapsychic factors, have not systematically explored the effects of parental discipline, but the learning theorist's list is congruent with their emphasis on parental love and understanding.

In many of their recommendations concerning management, the theories agree or complement rather than contradict one another. The concept of instrumental aggression is common to both, and both recommend attention to the problems underlying such aggressive behavior—the fear of being unloved, the humiliating sense of insignificance, or even the self-loathing which lies behind the child's destructiveness. Both theories recommend helping the child to focus on the source of his anger rather than continually giving vent to his feelings.

Learning theorists recommend rewarding competing responses as a technique for managing aggression. Feshbach (1964) argues that the affect of anger is basically a heightened state of arousal. It energizes destructive behavior only as such behavior proves successful in reducing tension. Anger seeks release; it does not seek to destroy. Therefore it can be used to energize behavior incompatible with inflicting harm on others; it can motivate the sprinter, the satirist, the social reformer, as well as the delinquent. A study by Hoffman and others (1960) is consistent with Feshbach's point. They found that punitive, threatening parents did not produce an aggressive child if the child was high on autonomy. Rather, the child was outstanding for his assertiveness, leadership, and friendliness. The hostility engendered by the parents reinforced the child's initiative and produced a high degree of self-confidence. While psychoanalysts would not accept Feshbach's theorizing, the idea of rewarding incompatible responses is consistent with their idea of helping the child sublimate his aggressive drive—by becoming a sprinter, a satirist, or a social reformer, as Feshbach suggests.

In sum, there are important theoretical differences among psychologists concerning the nature of aggression, and from these arise differences concerning the need for aggressive behavior in normal personality development and the consequences of trying to eliminate such behavior. However, those looking for guides concerning the constructive management of aggression will find much agreement between the theories and complementary rather than contradictory suggestions.

8 Social Relations

The story is told that Freud, toward the end of his life, was asked what a normal person should be able to do well. The questioner expected a lengthy discourse, but Freud replied in three words: love and work. Adler considered the basic life problems to be love, occupation, and social relations. That additional factor epitomizes the difference between the two men. No one explored the consequences of passionate intimacy as exhaustively or with greater insight than Freud. While the topic of work did not figure prominently in his theory, his life itself serves as a model of devotion to intellectual inquiry. Yet, between the passions of intimacy and the stern demands of work lies a vast area of human relations which includes friendship, cooperation, status within a group, and that sense of being a member of the human race which Adler called social interests or social feelings. Significantly, Freud concentrated on the first half-dozen years of life when passionate attachments within the family are at their height. He had little interest in middle childhood, nor did he assign it much importance in personality development; yet it is in this middle period that peer relations come into their own. Moreover, Freud could find no reasonable answer to the question, "Why should I love my neighbor as myself?" Adler claimed that such love holds the key to optimal personality development.[1]

It would seem that Freud has won the day, since his genius has dominated the academic as well as the clinical scene. Learning theorists have translated his scheme of early personality development (i.e., his psychosexual theory) into terms of child-rearing practices and have produced an impressive number of studies of feeding, toilet training, and sex identification.[2] The Freudian approach has riveted clinicians' attention on the parent-child or even the mother-child relationship. Friends and group

[1] See Ansbacher and Ansbacher (1956), for a telling contrast between the two men and a detailed presentation of Adler's theory.

[2] We know now that this initial research reaped a meager harvest of knowledge and can ponder the reasons why, but the failure itself could not have been anticipated. For a review and critique of the studies on infant care, see Caldwell (1964).

adjustment were of secondary importance—if they were considered at all. If the early home life were good, subsequent peer relations would be good also. The well-loved, emotionally secure child with a healthy sex identification was considered prepared to meet life's problems.

This Freudian influence was buttressed subtlely by the American value system which also stresses the importance of home and mother above teachers and peer groups in the child's basic character development. To delegate the responsibility for character training or to think of personality as being molded by forces outside the family is culturally uncongenial.

Yet, victories in psychology tend to be temporary. Changes within American society itself, rather than in the academic and clinical community, are causing a shift away from the Freudian stance. The dramatic increase in day care centers, preschools, and nursery schools for children between two and five years of age, as well as the educational programs for disadvantaged preschoolers, mean that more children are spending more of their young lives away from the family. The child no longer waits until he is six to face a group of his peers under the guidance of a nonfamilial authority. As the preschooler has been forced to adjust to a group setting, those who care for him are being forced to face problems of what values to cultivate and how best to realize those values—problems which are Adlerian in spirit rather than Freudian. Should middle class values be impressed upon lower class children? Is intellectual stimulation as important as learning to cooperate with others? Should individuality or teamwork be encouraged? Social changes themselves seem to be forcing psychologists to be more mindful of man's relation to society.

But the case for Adler can be made in more concrete, more human terms. In an exchange between two mothers, one was complaining about the sibling rivalry among her sons. Her complaints were the usual ones of continual fighting, teasing, selfishness, and jealousy over parental attention, and she concluded in humorous despair, "I really think that children should be limited to one per customer." The other mother replied, "That's not the answer. After all, there's nothing more important for a child to learn than how to get along with another human being." That is the essence of Adler's message.

Early Peer Relations

Description

The infant has an innate need for environmental stimulation which is as basic as his need for nourishment (see Chapter 3). Some environmental stimuli have an intrinsically high interest value and are particularly apt to elicit a response of delight. The human being is first among such patterns of stimuli. The human face, with its combination of mobility and stability of structure, the melody of the human voice, the free and varied movements of the body, fascinate the infant and bring him pleasure.

Adult human beings may constitute more intriguing stimuli than fellow infants because adults consciously set about to make themselves so by distorting their faces and emitting special sounds to attract the infant's attention and elicit a positive response. They are also available for longer periods of time. However, there is evidence that, in the first two years of life, infants are both interested in and emotionally responsive to one another.[3] Bridges (1932) observed the following development between six months and two years of age: At six months of age, infants watch one another through the bars of their cribs, at times laughing in response to one another's movements. Recall that this is a time of general social responsiveness (see Chapter 2) when the infant indiscriminately enjoys other human beings. At eight to nine months of age, infants show a more exploratory interest in another infant, as well as in their own reflection (see Chapter 10). By ten months of age there is a more sustained interest in contemporaries, expressed in mimicking, patting, banging, and imitation of laughing. Social involvement reaches a climax of sorts toward the end of the first year, when one infant will make another laugh so that he himself can enjoy his companion's laughter. By fifteen months of age, affection appears, with toddlers taking one another by the hand, sitting close, patting, and smiling at each other. At eighteen months of age, amicable, nonsensical jabbering is added, and individual differences in overall friendliness are discernible. By the end of the second year the toddler can sustain a friendship for several weeks and can "parent" younger infants by directing their actions and pointing out errors.

Bühler (1933) enriches this early developmental picture by delineating three different types of social behavior observable in the second half-year of life. The socially blind infant behaves as if his companion did not exist; he looks at him without interest or emotion and proceeds to go his own way. The socially dependent infant, by contrast, is concerned constantly with the effect of his behavior on his companion—at times copying, obeying, showing off, or demonstrating objects in order to arouse interest. The socially independent child is also aware of and responsive to his companion but is not dependent upon him. He may join his peer in play or encourage or console him, but he is always ready to go his own way whenever he pleases. Bühler believes that these behaviors reflect innate dispositions rather than learning.

Finally, Maudry and Nekula (1939) traced the development of social behavior when two infants are alone, face to face, with various toys available to them. Between six to eight months of age, there are as many positive as negative responses, but the quality is bland and impersonal. From nine to thirteen months of age, hostility is at its height, not because of an increase in personal animosity, but because the infant is interested in playing and his companion interferes. Fourteen to eighteen months of age is a transitional period in which interest shifts from play to the partner. By nineteen to twenty-four months of age, positive social responses, which include cooperation as well as the more primitive responses of looking and smiling, predominate. Games are more frequent and the child is able to modify his play in response to the

[3] For more comprehensive reviews of early social development, see Adams (1967) and Bühler (1933).

behavior of the partner. To this sequence, Bühler adds the observation that two years of age marks the beginning of the child's ability to maintain contact with two other children instead of with a single companion.

Although the inventory of early peer relations is impressive, they should be viewed in the context of the general level of personality development. Woodcock (1941) admirably captures their flavor. Exploration, initiative, and willfulness are the themes which dominate much of the toddler's behavior. At the same time, his short attention span, his limited ability to use language to communicate, and his crude techniques for controlling the behavior of others, place realistic limits on his ability to initiate and sustain social interactions. The two-year-old has not completely differentiated animate from inanimate objects; e.g., he can say "Hello" to the bacon and earnestly shout "Dinner, wait for me!" when washing before a meal. The same youngster can react to peers as if they were objects to be explored, felt, pushed, and bent in the middle. His attempts to direct and prohibit other children are part of a general desire to be in command, which flourishes at home in the "battle of the spoon" and the "battle of the wills." While there are fleeting games and episodes of imitation, the child is most interested in simple activities, such as sitting on boxes and kicking heels in unison with other children or trailing others around in a circle with a wagon—activities which he can join or leave at will. In comforting a crying child, he often apes the behavior of an adult, but consolation can turn quickly into angry hitting and shoving. Thus, social behavior has neither the depth and richness of attachment, nor the intrinsic and compelling interest of exploration.

Peer Relations and Attachment

Certain aspects of the development of peer relations resemble the development of attachment. Bridges describes a progression from interest and delight in human stimuli, to imitation, to initiative in eliciting pleasurable behavior, to affection, which is reminiscent of Sander's (1962) description of stages of the early mother-child interaction. Progression is also from egocentricism to reciprocity—from eliciting a laugh because the elicitor himself enjoys it to becoming a playmate and adapting to the behavior of a companion. Finally, social relations become more individualized and sustained. However, these trends in peer relations are not as striking as they are in the case of attachment and affection. The timetable is slower, since attachment to the mother takes place around nine months of age, while peer affection does not solidify until fifteen months of age. The behaviors themselves are less intense and stable. We can observe full-blown mother love in the one-year-old, whereas the peer relations of the two-year-old are only a foreshadowing of what is to come.

There are several reasons why peer relations lack the intensity of parental relations. Peers are incapable of ministering to one another's physiological needs; adults or responsible older children must assume this function. Thus, peers do not become associated with relief from the distress of bodily tensions. This may be one reason why peer relations lack the life-or-death quality, the security of closeness, and anguish at separation which characterize attachment. Next, the meager amount of peer contact in infancy contributes to slow and uncertain progress in social relations. However, the quantity is not as important as the quality of the contact. Parents have a strong emotional investment in their infant's happiness because their own self-esteem and pleasure often are at stake. An infant who eats well, sleeps soundly, who is bright-eyed and rosy-cheeked, who smiles and laughs readily, is, quite literally, the parents' pride and joy. An alert mother will tickle the infant's stomach, echo his early vocalizations, present him with brightly colored objects, offer her fingers for him to pull himself from a prone to a sitting position. Either by trial-and-error or by experience, she will know how to capitalize on his interests and pleasures as they develop. In the process she insures her own pleasure in being a good mother. Finally, parents tend to love the infant because the infant is theirs. Over and above manifest behavior, or, more important, in spite of manifest behavior, the infant is intrinsically loveworthy. Peer contacts have none of these qualities. Although the toddler is capable of sympathy, concern, and parenting, such feelings and behaviors tend to be fleeting. Even when affect runs high, he has only a limited repertoire of techniques for expressing it, and even a more limited grasp of how to adapt his behavior in order to make it meaningful to his companion. His displays of affection and concern seem arbitrarily imposed on the recipient rather than finely attuned to him. Peers are never loved for what they are. The "spoiled" preschooler who thinks that his mere presence in the group will cause others to do his bidding will be shaken badly by his reception. Peers are valued because of proven utility.

Why Peer Relations Develop. In light of the many qualities peers lack, one might wonder what binds them together in meaningful relationships. Far less is known about the reasons for peer relations than about the reasons for attachment, so we must be more speculative. Adler regarded social feelings as innate, not in the sense that they will appear inevitably, but in the sense that the human young is so constituted as to respond vigorously and positively to social stimulation. Just as he has the capacity for intelligent behavior, the infant has the capacity for social feelings, although the environment must cultivate this natural predisposition.

There may be more specific factors involved in peer relation than Adler considers. Peers have one advantage which adults lack—they are similar to one another. Recall that a stimulus is most intriguing when it is partly familiar, partly novel; if it is too familiar, it is boring; if it is too unfamiliar it does not register (see Chapter 3). Especially as the infant becomes a toddler and a preschooler, adults must work at achieving this quality of similarity, since it requires a willful suspension of mature behavior. To sit on the floor and roll a ball back and forth, to play endless rounds of hide-and-seek, to delight in TV cartoons are beneath the dignity of some adults, threatening or tedious for others. Peers who are at the same physical and psychological level gravitate toward common interests. One child does what the other does or is potentially capable of doing, but he is always sufficiently different to pique his companion's curiosity. If peers lack techniques for communicating and relating, in a sense they do not need them: their behavior itself has that quality of novelty in the context of the familiar which is best calculated to intrigue. In sum, peers attract because they naturally do mutually interesting things; sharing interests and activities is one of the cornerstones of peer relations.

Since sharing mutual interests is different from love, we should not expect social feelings to be a dilute replication of parental affection. Rather, social feelings are important precisely because they add to personality new dimensions of mutuality, friendship, and prestige. Parents are loving and guiding superiors; peers are sharing equals. Both are important because each brings something different to personality development.

Friendship

The Earliest Friendships

Friendship, that special relationship which stands halfway between intensely personal affection and group-engendered popularity, has been observed as early as two years of age, although it is not typical of the toddler. Woodcock cites as reasons for compatibility an increasing ability to communicate, understand, and share, a need to be protected, and a need to protect. A less able child may gravitate to one who is larger and more competent, just as a small, inept child may arouse special parenting behavior in a large, competent one. Woodcock describes a child who offered to dress another and take him to the toilet (while lacking the skills to do either), and a little girl who would cheer up a boy whenever he began sulking. Such attachments can also have an impeding or even paralyzing effect. One child would persistently try to force play material on another who was only fitfully interested in it; a large, slow-moving boy developed a strong dependency on a tiny, alert girl to the point of becoming extremely anxious when she was not nearby and crying whenever she cried. Woodcock notes that such early friendships generally are not reciprocated; one child seems strongly drawn to the other who is significantly less involved.

These early friendships are different from a general sociability in that the children seem to be particular in their choice. One child is selected from the group to be the object of attention, concern, or affection because of some "intangible quality of personality" (Woodcock, 1941). In the above examples, the child who tried to force toys on his companion continued to show an intense concern over her for the next two years, while the slow-moving boy's dependency suggests that the tiny girl might have been uniquely suited for gratifying some deep need. Freud and Burlingham (1944), observing toddlers living in a residential home, noted similar signs of individualized attraction. Playmates were not indiscriminately chosen in certain instances, but playing with a particular partner seemed as important as the game itself.

Incidentally, this and other studies suggest that friendships are more intense and enduring when toddlers are in a group setting and have little opportunity to develop strong attachments to adults. The jealousies and rivalries so often described in the literature may well be a product of the close bond with the parent rather than inherent in peer and sibling interaction. Recall Adler's concern that parents not cater to emotional greediness for fear it would jeopardize constructive social relations (see Chapter 3)

The Preschool Period

Gesell and Ilg (1949) observed the following progression in children between three and six years of age. The three-year-old shows preference for and a beginning attachment to one companion, often of the opposite sex. He begins to use the word "friend" and by four years of age may have a friend of the same sex. While there is much quarreling and fighting with his friend, he will share or play cooperatively at times. By six years of age there is a marked interest in making friends, having friends, and being with friends. In spite of this, friendships readily come apart at the seams during play, since each child still wants his own way. We do not know how stable preschool friendships are over time, but there is evidence (McCandless and Marshall, 1957) that they are reasonably consistent over twenty-day intervals. Constructiveness in cooperative activities and high energy output in physical activities are important in boys' friendships; girls' friendships are based more on participation in social activities or general gregariousness (Challman, 1932). We may be seeing here the beginning of the energetic work orientation which is so important to the adolescent boy and the sociability which is so important to the adolescent girl. Preschool friendships have two other characteristics which will figure in our subsequent discussion. First, friendships are unrelated to general sociability; i.e., the number of friends shows no correlation with the number of children played with. Next, quarreling, instead of being inimicable to friendships, is a necessary part of it. Differing and then resolving differences, even at this early age, is an important feature of a close social relation (Green, 1933). As we noted (see Chapter 7), hostility takes on a different meaning when it was viewed as one manifestation of a high degree of social involvement. Individual differences in choice of friends incline some children to prefer those very much like themselves and others to prefer dissimilar companions (Hagman, 1933).

Fite (1940) documents the potency of friendships in children as young as three to four years of age. Friendships give the child a valuable feeling of sharing in the problems of assimilating the parents' baffling requirements for acceptable behavior. Preschoolers become aware of the fact that group standards of behavior differ from those of parents; e.g., the group condones a higher level of aggression. Friends support one another in a new standard of conduct which, in going counter to that of the parents, runs the risk of jeopardizing parental approval and love. Fite makes the familiar point that exposing the child to values different from those in the home facilitates moral development (see Chapter 5); but, in addition she observed how friends give one another the emotional support necessary to accepting a different standard of behavior.

Middle Childhood and Preadolescent Friendships

By eight to nine years of age, "best friends" begin to appear in a relationship which is close and demanding, according to Gesell and Ilg (1949). However, anger does not lead to disruption as it did at six years of age; the friendship itself is now important enough to be sustained in the face of difficulties. Instead of being egocentric, the child is beginning to want his friend to be happy and is concerned about the friend's attitudes toward him. Gesell places this increased mutuality in the context of a general increase in evaluativeness around eight and nine years of age. The child, much less prone to label behavior absolutely and rigidly, is more apt to wonder, ponder, and consider complexities. His friendships are beginning to show the deepening and humanizing trend so important to moral development (see Chapter 5).

We can also make an analogy to the effect that the object concept has on attachment. Before the infant grasps the independent existence of the mother, he can love being with her but he cannot love *her* (see Chapter 6). In a like manner, preschool friendships are little more than what their manifest behavior reveals them to be; they *are* the activities of sharing and laughing and quarreling and fighting. The label "friend" has appeared but not the concept of a friend. In middle childhood the concept of friend–of an independent individual–begins to register. The child now can want to be with his friend and he can have ideas and feelings about his friend apart from sharing activities. He no longer merely has fun when he is with his friend; his friend is the kind of person he can have fun with.

The ten-year-old luxuriates in friends who are often adapted to specific occasions, such as a best friend to play ball with, a best friend to work with, a best friend to talk to (Gesell, Ilg, and Ames, 1956). He knows a good deal about his friends and likes them because they can be trusted. Rotation of friendships is also found at twelve years of age, so that a girl can go with "nice girls" when she wants to be good and "horrid girls" when she wants to wear lipstick. Such a picture suggests the expansiveness and egocentrism of the six-year-old, except that now the child is more purposively selective and individualized in his choice, more cognizant of his friends' attitudes toward him, and more capable of moderating his feelings in the service of sustaining the relationship.

Parenthetically, we have here another example of Gesell's concept of personality development, in which periods of unstable expansiveness alternate with periods of stable consolidation. Each cycle serves to advance the child another step toward maturity. Friendship is not a matter of a child becoming more deeply attached to a smaller number of peers; rather, periods of relatively indiscriminate expansiveness are followed by others of more focused and deeper attachment. With each cycle the child is increasingly capable of the sustained mutuality which is the hallmark of friendship.

Objective studies[4] are generally congruent with Gesell's picture of increasing stability of friendships between five and eighteen years of age, while furnishing some revealing concrete details. The instability of sixth-graders' friendships is shown in the

[4] For a succinct presentation of objective studies, see Ausubel (1958), Horrocks and Buker (1951), and Austin and Thompson (1948).

finding (Austin and Thompson, 1948) that only forty percent listed the same three "best friends" after a two-week period! This rapid turnover might be the result of having a wide pool of friends (Thompson and Horrocks, 1947). While the same behavior in adults would rightfully be regarded as flighty and childish, a significant increase in friendship fluctuation over the expected rate is one of the concomitants of emotional instability or disturbance in middle childhood (Davids and Parenti, 1958). The dependence of early friendships on physical proximity is another example of their insubstantial nature. For sixth-graders, out of sight means out of mind, and lack of recent contact is one of the main causes of dissolving friendships (Austin and Thompson, 1948).

Sullivan: The Significance of Friendships

Sullivan (1953) agrees with Adler on the prime importance of peer relations and friendships for achieving psychological maturity. He regards the appearance of the chum, when the child is around eight to ten years of age, as a pivotal psychological event and directly relates the capacity for friendship to the capacity for mature love. To appreciate this point of view, we must first know something of his theoretical position.

Sullivan assumes that the most basic human needs are social or, in his terminology, interpersonal. In this respect he is like Adler but differs from Freud who, as we have seen, regards man's basic needs as biological (see Chapter 7). In Freud's theory, the infant comes to value adults as they reduce physiological distress or increase physiological pleasure. The mother has to "seduce" the infant into loving her. For Adler and Sullivan, human life is intrinsically interpersonal; the need for human contact does not have to be derived from a biological substrate. Nursing is significant, not because the infant is satisfying his hunger and deriving pleasure from sucking, but because he is having his first contact with another human being.

Sullivan's theory is based upon the need for interpersonal intimacy and the painful feeling of loneliness when this need is not gratified. In earliest infancy, there is a generalized need for physical and sensory contact with other human beings. Subsequently, the infant develops a need for tenderness and for sensitive care, which last through the preschool period. At the same time, a need for adult attention appears along with one for participation in verbal play. (Within the category of play, Sullivan includes many of the activities we have discussed in terms of early exploratory behavior and speech.) Around six to eight years of age, the child experiences a need for companions followed by a need for acceptance and a fear of ostricism or exclusion from the group. Preadolescence brings a need for intimate exchange, friendship, and love of another person. In adolescence, these same three needs come to involve a member of the opposite sex. Heterosexual love, therefore, is the final transformation of the need for intimacy and is the principal bulwark against loneliness.

Friendship in middle childhood stands midway between the need for protective care and the need for heterosexual intimacy. It is an essential bridge between the infancy-preschool years and adulthood. Put negatively, friendlessness during middle childhood seriously jeopardizes the child's chances of achieving a mature love relation. Friendship is the beginning of a new type of interest in another person, a subtle but momentous shift away from the prevailing egocentricity of the first eight or ten years of life. The child no longer asks, What should I do to get what I want? but, How can I contribute to the happiness, or well-being, or sense of worth of my friend? Sullivan's term for this new kind of relationship is collaboration.

Sullivan implies that the shift from "me" to "we" is highly likely among peers, highly unlikely in the parent-child relationship. The child's egocentricity is due, in part, to the fact that an intense relation with parents is established at an early age when emotions are strong and when the infant-toddler is cognitively incapable of grasping the point of view of another person. In part it is due to the vast gulf between the child's world and the adult's world. A loving, sensitive, and resourceful parent can dip down periodically into the child's world and share it, but inevitably he returns to his own realm of work, love, and friendship. How can the child experience "we" under such circumstances? And, finally, socialization itself mitigates against mutuality. So many adult values and expectations are foreign to the child; so many responses of parental pleasure and anger have nothing to do with the child's own experiences of interest, happiness, and distress. The "vertical" nature of the parent-child relationship almost precludes the development of mutuality in middle childhood, whereas the "horizontal" nature of peer relations fosters it. The child literally and figuratively must look up to the superior adult, whereas he looks directly at his peers.

Friendship gives the child a new feeling of personal worth. He is liked for what he can do. Children are utilitarian in their relations, rather than sentimental or devoted as the parents tend to be. There is no place for false praise or ulterior motives. If they play-act fantastic adventures together or make doll clothes or throw rocks into the water, it is because play-acting and clothes-making and rock-throwing are the most interesting activities in their lives at the time. Sullivan maintains that this sharing of intrinsically valued activities adds a new dimension to the child's sense of his own worth.

Friendship increases self-knowledge. Its intimacy exposes the child to the scrutiny of his friends, and both the clinical and objective literature agree that friendships are marked by uncompromising frankness. Just because closeness has no life-or-death quality, each partner is free to say what he thinks of the other. In the context of being criticized, the child begins to see himself as others see him. He is able to counteract the illusion of specialness which Sullivan and Adler regard as so detrimental to personality development. He finds that his supposed superiorities leave his friend unimpressed, and that his fears and concerns are not so different from those of his companion. Instead of regarding himself as a child apart, he begins to think of himself as a child among children. In Adlerian terms, his social feelings are enhanced; more simply, he begins to join the human race.

Douvan and Gold (1966), essentially agreeing with Sullivan, point out that the child can choose his friends on the basis of mutual attraction, whereas he has had no choice of a family. Family relations tend to involve the emotions of love-engendered guilt and power-engendered anxiety, while friendships tap the less torturous feelings of respect, loyalty, and tolerance. The mutuality of friendship allows each partner to reveal and expose himself without fear of ridicule or loss of respect—a sharing which becomes particularly important as adolescence approaches and the child is faced with so many disturbing impulses from within and with confusion concerning his role in the world.

We take issue with Sullivan and Douvan on only one point. They assume that, while children grow, parents do not. They picture parents who are fixated at the early stages of socialization and who secretly long to make their child a baby again. Although we know far less about adult growth than child development, there is no reason to assume that maturity is a time of stagnation. Some parents are relieved when the demands of early childhood are over and they look forward to middle childhood and adolescence with special pleasure. A similar point can be made in terms of research methodology. Clinicians and objective psychologists often obtain a child's-eye view of parents. Such a view is apt to be distorted. A child is poorly equipped to understand his parents because of his emotional needs and his cognitive limitations. It may even be that, in all the world, parents are the last people children come to see as the human beings they are. Thus, when Douvan claims that the early adolescent is "driven to friendship" because he dare not become entangled in old affectional ties and power struggles, it may be the adolescent who has failed to outgrow his image of the parents rather than the parents who have failed to develop beyond their early need to love and socialize the child.

Sullivan's data are primarily on boys. A vivid picture of friendships in girls is provided by Deutsch (1944) who, because of her psychoanalytic orientation, stresses the sexual aspect of friendship and its role in the parent-child relation. Around ten to twelve years of age, the young girl enjoys a vigorous and relatively carefree life. Although still dependent and uncertain, she wants to see herself as grown up and independent, particularly in relation to her mother. The affection ordinarily directed toward the mother finds outlets in attachments to an older woman at times but more frequently to a girl like herself. The typical picture is of two girls, in the privacy of a locked room, giggling and tittering and sharing secrets with each other. Each element of the picture is important. The locked door serves as a No Trespassing sign to adults and proclaims the children's independence. By being in league against the mother, each child helps counteract her underlying sense of inadequacy; together they feel stronger than each could feel alone. Secrets are essential. In earlier childhood, the girls were aware of a multitude of adult secrets which they could never learn. Part of the gulf between generations is the child's sense of being excluded from what is most intimate, exciting, and valued. Now the girl turns the tables and excludes the adult from her world. Moreover, the secrets are especially exciting because they usually concern sexual ideas—about intercourse, menstruation, and childbirth. Because the girls are prepubescent, they are more carefree about sexual matters than they will be when they are sexually mature. With their flare for dramatizing, they will stuff pillows under their dresses and pretend to be pregnant or paint their faces and walk around in their version of a prostitute. However, there is rarely any directly sexual activity, such as mutual masturbation or expressions of intense tenderness. Thus, friendships provide emotional warmth, strengthen the feeling of independence, and provide opportunities to share ideas on taboo subjects.

Adolescent Friendships

Horrocks (1969) outlines some of the distinctive features of adolescent friendships. Adolescents tend to be more selective and less interested in having "lots of friends" than they were in middle childhood. Friendships are more sustained and less dependent on propinquity. Friendship choices are markedly peer-bound, since adolescents are at a stage in which peer values are of prime importance. Adolescents are also more frank and critical of their friends than they were as children or than they will be as adults. In fact, adults are often dismayed at the crudeness of the labels they use and at the unvarnished, unsolicited advice adolescents give one another. Adolescents, along with certain children and psychotic adults, have the ability to size up character with deadly accuracy. They can give a teacher a nickname which epitomizes all the things he is but does not want to acknowledge in himself. And, like pitiless critics, they make the faults stand out in bold relief while the virtues or the extenuating circumstances, the overlooking and covering up, all vanish.

Objective studies corroborate the above picture. Sister Lucina (1940), for example, found that adolescents did not regard their best friends as perfect or even as superior to other friends; in fact, only about two-thirds regarded their best friends as a good influence. Getting angry and making-up was a common description of the relationship. Thus, from three years of age on, friendship and fighting go together. Sullivan might well point to this as an example of the down-to-earth quality of friendships. The friend's critical stance also provides a mirror which enables the adolescent to see himself from a different perspective. Being confronted with himself by someone he values is a priceless growth experience.

Another important trend is the deepening of friendships in adolescence. Bühler (1933) notes that in middle childhood friends are companions who have positive qualities of being "nice" and "good." Sharing interesting activities with a generally pleasant companion suffices, for friendship is a congenial doing-together, not a significant being-together. In adolescence, girls in particular are seeking a single individual to share confidences, understanding, and affection. Interestingly, Buhler makes the same point about friendship that we made previously about moral judgment and which we shall make concerning the adolescent's understanding of the adult sex role (see Chapter 10); it is the point which Piaget made about imitation and psychoanalysts make about identification (see Chapters 4 and 5). In the beginning, form predominates and only later does understanding emerge. Concrete and literal manifestations are, naturally enough, what the child is first capable of grasping. The development from appearing to being, from the letter to the spirit, from copying to comprehending, is a gradual one. The preadolescent girl is concerned with showing others that she has a friend. Friends like to wear the same style of clothes and hairdo, to use the same expressions, to be seen at the same events. Only in adolescence does the essential ingredient of intimacy enter. For girls in particular, the special, secret sharing becomes the essence of friendship, and overt display no longer matters. Since friends become individuals in their own right, there is no need to copy overt behavior.

Douvan and Adelson (1966) take up the theme of intimacy. Because the peer group is too impersonal and too stringent in its demands for conformity, the adolescent, particularly the girl, turns to friends to share forbidden and disturbing feelings—especially sexual ones. This sharing relieves the onus of being a lone sinner and helps the adolescent see herself as "more human than otherwise" in Sullivan's famous phrase. In listening to the experiences of others, the adolescent vicariously enlarges his understanding of how people behave when feelings run high; in telling about himself, he gains that special perspective which comes when experiences must be made explicit in order to be communicated. If friendship can relieve guilt and define experiences, it also can establish limits. In his state of high emotional arousal, the adolescent is often concerned with the question, Did I go too far? Was I too promiscuous, too cruel, too defiant, or too selfish? His friend may be openly critical or he may grow uneasy and withdraw, indicating that the behavior was too dangerous for comfort. One adolescent looking to another for control might appear to be like the blind leading the blind. However, the adolescent's fear of "going wild" may be even greater than that of his parents. Although inexperienced, he may have an intuitive insight that checks behavior which threatens to get out of bounds. Finally, friendship teaches the adolescent the responsibility and respect of intimacy—tact, tolerance, a balance of giving and receiving, knowing the limits of expression and confession which can be tolerated.

Sex differences in friendships are uniformly noted and frequently attributed to the earlier social and sexual maturation in girls. Douvan and Adelson, however, propose an explanation in terms of the different psychology of the two sexes. The girl's psychology revolves around love, feeling, and interpersonal relations. Being loved or unloved are the central concerns of childhood, and she looks forward to becoming a loving wife and mother. She cultivates her emotional responsiveness and sensitivity as they are relevant to gaining affection and uses interpersonal relations as a source of love, reassurance, and support. The boy's personality revolves around assertiveness, autonomy, and authority. Unlike the girl, he does not value sensitivity or empathy, but, on the other hand, he has no inclination to be gossipy and snobbish. Dispensing with parental control is an essential step in becoming his own man, and he is concerned about his ability to control himself and manage his own life effectively. Friendships furnish companionship, support, and the strength of unity. They have something of the air of "The Three Musketeers." (Note that there were three, not two; even in fiction men cannot be too close.) The boy is also more invested in the group than is the girl. In his uncertain efforts to become autonomous, he needs both the rules and the support which come from being part of an organization.

Douvan's picture of the girl as basically love oriented and the boy as authority oriented harks back to the two kinds of parental discipline discussed in Chapter 4. It also recalls Freud's hypothesis that the girl's superego was predicated on love, the boy's on self-assertiveness and fear (see Chapter 5). While there is no definitive evidence that Freud was correct in regard to the origins of conscience, he may have been descriptively accurate. The themes of love and power (or authority) come up so often in our discussions of personality that both must be fundamental to its development. It may not be by chance that Freud regarded love as basic while Adler stressed power. Both may have been correct.

Friendship and Heterosexuality

Sullivan was concerned more with development of heterosexuality than with tracing the fate of friendship in adolescence. The adolescent must change from seeking intimacy with someone like himself (as in the friendship of middle childhood) to seeking intimacy with someone significantly different from himself. This shift is necessitated by sexual maturation or what Sullivan calls the "lust dynamism." Maturity is achieved when the lust dynamism is integrated with the intimacy need. Here sexual gratification is united with mutuality, sharing, intimate exchange, concern for the partner, steadfastness, and all the other values of friendship. If the fusion does not take place, the individual's personality is immature or, to use Sullivan's term, "warped." Sullivan, like Freud, distinguished the sexual act from heterosexual maturity (see Chapter 7). Impeccable sexual performance can be used to satisfy immature, selfish needs; or to use Sullivan's picturesque language, intercourse can be nothing more than "instrumental masturbation."

Douvan and Gold elaborate on Sullivan's sketch of the change to heterosexual relationships. The preadolescent period is one of strongly sex-segregated friendships and groups. A boy is teased unmercifully, for example, if he is interested in a girl or in the things girls do. Then girls and boys begin to make heterosexual overtures. Because they need doubly to protect themselves against closeness, initial contacts are loose combinations of same-sex groups, so that the individual can think of "us" instead of "me" as being interested in the opposite sex and can retreat easily to the security of the group if the new venture threatens to become too intense. Around fourteen to fifteen years of age, boys and girls begin dating; by seventeen almost all girls are dating and thirty percent are going steady. Predictably, initial dates are hesitant and heavily protected by ritual. Neither boy nor girl is capable of committing himself deeply and is satisfied with such bland virtues as cheerfulness and being well behaved. As time goes on, anxiety lessens and, with girls at least, there is an increased awareness both of boys and of their girl-friends as individuals. Being a good dancer or a good conversationalist becomes less important than sensitivity and understanding. The rewards of dating change; whereas extensive or intensive dating was used as a means of gaining group prestige, now the relationship is valued or rejected in its own right. Thus, conformity to the outward show of heterosexuality is replaced by an effort to understand and relate to others.

The Basis for Attraction in Friendship

Objective studies of friendship in middle childhood and adolescence have been concerned with the question of whether similarity or difference is the basis for attraction. Is it true that birds of a feather flock together or do opposites attract? The evidence is in favor of the former hypothesis (see Ausubel, 1958). Similarity in sex, age, intelligence, and socioeconomic status, as well as in interests and personality traits, have been shown to correlate positively with choice of best friends. The relationship is not positive for all these variables at all age levels, and we lack a well-documented timetable indicating when each waxes and wanes in importance. Neither is it clear which factors are essential to friendship formation and which are concomitant; e.g., age, per se, is probably not a determinant of friendship, but same-aged children are likely to have common interests and values which are the basis of closeness.

The characteristics which form the basis of friendships are uniformly high on social desirability. Some could have been taken from the Boy Scout manual—cheerfulness, kindness, cooperativeness, generosity, and honesty (see Austin and Thompson, 1948); some reflect adult criteria of good behavior or child oriented virtues of fairness, sportsmanship, dependability, and inspiring others to do better. Even the reasons for changing friends are often the mirror image of social acceptability; e.g., the erstwhile friend was conceited, disloyal, quarrelsome, dishonest, noisy, or unkind. In one way these results are puzzling. In adulthood we think of a friend as being someone special, someone who is distinctive among all our acquaintances rather than someone who is socially acceptable. Observers of preschool children also sensed a special quality in friendships. Perhaps the child between eight and twelve years of age requires nothing more than social acceptability of his friend, since his own values tend to be conventional and his behavior conforming. As we have seen, a friend becomes an individual with a distinctive personality only in adolescence. However, it also could be that the objective studies have not tapped the more personalized reasons for friendship choices and have not probed beyond the layer of socially acceptable responses which children this age are apt to give.

Fortunately, psychologists interested in friendships in adults have begun to explore the problem of personal attraction from a developmental point of view.[5] "Similarity," it turns out, is not a simple dimension. People can be similar in a variety of ways, ranging from physical (height, weight) and demographic (age, socioeconomic status) characteristics to values and emotional responsiveness. In addition, actual and perceived similarity can be distinguished. The former are characteristics arrived at independently of the other friend, such as by administering personality inventories; the latter are the characteristics which an individual attributes to his friend. This distinction raises the question, Do friends have to be similar by some objective criterion, or is it enough that they regard one another as similar? The evidence is meager and inconclusive. Davitz (1955) found that, with children between the ages of six and one-half and thirteen years, there is a positive relation between perceived similarity and valuing of others. In fact, friends are perceived as more similar to the self than independent criteria show them to be. However, this study does not tell which came first—whether the exaggerated perception of similarity was present from the beginning and served to catalyze the friendship, or whether it developed as liking turned into friendship.

[5] Newcomb (1956) and Izard (1960) provide a general background. For a developmental study, see Byrne and Griffitt (1966).

While many of the complexities of similarity have not been explored with children, Byrne and Griffitt (1966) have found that individuals between the ages of nine and twenty years conform to what they call the law of attraction: attraction to a stranger is a positive linear function of the proportion of that stranger's attitudes which are similar to attitudes of the individual subject. Byrne thus makes attitudes an essential element in the formation of friendships. In their research, the investigators obtained children's attitudes toward a variety of topics such as sports, studying, comic books, and the relative intelligence of boys and girls, and adolescent's attitudes toward topics such as smoking, strict rules, and studying. He then asked them to rate liking for a hypothetical stranger who shared these attitudes in different degrees. Regardless of age, liking increased with the proportion of attitudes shared.

Another insight into attraction comes from Biber's (1942) informal observations of seven-year-olds. Like Austin and Thompson, she found that most of the children chose companions on the basis of "good" qualities, but the choice was not universal. What is more, having "bad" qualities did not mean that the child would be ostracized, since some of the worst children could be chosen as companions for eating or playing. This seeming inconsistency was resolved by studying the reasons individual children gave for their choices. When they preferred a "bad" child it was because they thought of themselves as being bad also; e.g., "Jimmy is my friend because he's bad like me." Biber implies that it is the "like-me-ness," rather than the social acceptability, which tips the scale in favor of friendship. Her tentative findings suggest a variable of self-evaluation which might be added to Byrne's list of attitudinal similarities.

But why should similar attitudes facilitate friendship? Byrne's explanation is in terms of cognitive factors. The child's curiosity and initiative lead him to explore the social world, and he strives to build logical and accurate belief systems about interpersonal relations. These belief systems may be simple at first—babies are pests, girls are silly, boys are noisy—but they become increasingly complex. The child feels a special pleasure in encountering similar beliefs in another child because he can say to himself, "I am right." The more similarities he discovers, the stronger the pleasure of consensual validation. Since belief systems come to include opinions, values, and ideals, the reaction of "I am right" can tap ideas and ideals which the child cherishes deeply and identifies as his own. Thus a friend bolsters his self-confidence and self-esteem.

Byrne's theory fails to account for the fact that friends can be critical of one another. Dissimilarity and dissent are part of the relation, just as are similarity and confirmation. Adolescents, in particular, disconfirm one another's beliefs; yet friendships are valued in spite of the implicit message "You are wrong"—or perhaps even because of it. Byrne's theory may be incomplete rather than in error; friendship may involve both the assurance of similarity and the challenge of disconfirmation.

Summary

From the beginning, friendships differ from general sociability, and a friend may be selected because of some elusive but special quality. Friendship is a mixture of sharing and antagonism: preschoolers play and quarrel, adolescents are supportive and critical. Friendships become more sustained with development; in the preschool period and in middle childhood, friendships are limited by propinquity and atrophy when frequent contact is not possible; only in adolescence can friendships persist over time and space. Friendships progress from egocentricity to mutuality and respect for individuality. The preschooler knows the label "friend' but not the concept, and his friendly relations cannot compete with egocentric needs to win, to be first, and so on. In middle childhood he has grasped the concept of "friend" as an independent individual, but the readiness to shift friends suggests a superficiality to the relationship, while a public show of congeniality seems more characteristic of friendship than true feelings. Only in adolescence does a friend emerge as an individual who is valued for his personal qualities. The adolescent can accommodate to and be loyal to his friend; in shifting from self-concern to concern for another's welfare, he is preparing himself for mature heterosexual relationships. Friendships help correct the child's faulty perceptions of himself and give him a realistic understanding of his actual worth. For girls, friendship brings a special intimacy in which they are free to reveal feelings and express secret or forbidden ideas. Of all the social relations we shall discuss, friendship contains the fewest potential dangers to the child's personality development.

Sociability and Social Acceptability

Sociability in the Toddler and Preschooler

There are positive peer relations which are more general than friendships in that they are dispersed among many peers and involve less intense affect.[6] Social interests and social interactions in the first two years of life have already been discussed. The years between two and five bring a striking proliferation of social behaviors. Woodcock's (1941) inventory of social responses in the two-year-old consists of the simplest kinds of behaviors: an interest in naming, watching, touching other children and exploring their physical properties, plus an impulse to play with the same kinds of material and go through the same actions as another. Such a general gregariousness is much weaker than the involvement in exploring physical objects, since the toddler is more thing oriented than peer oriented. Between two-and-a-half and three years of age, the child begins to show a keener understanding of others and a wider range of social behavior. He notes the absence of another child, is aware of and interested in another child's satisfaction, attends to his needs, and offers support. He explains his actions, includes another person in his plans, and wants a peer to witness what he can do. Most of all, he begins to share imaginative play, a development which we will discuss in detail (see Chapter 9).

[6] For a resume of the theoretical, conceptual, and methodological difficulties in objectively evaluating social behavior, as well as for a review of objective studies, see Thompson (1960, 1962).

In the preschool period, relationships become less impersonal. One child is no longer as good as another; instead, congenial children tend to pair off and play in groups, shunning the outsider because he is too aggressive or passive. Murphy (1937) points out the contradictory behaviors in the three-year-old—cooperation and competition, aggression and sympathy, selfishness and love, conflict and friendliness. She notes that all kinds of social behavior are on the increase between two and four years of age, regardless of whether they are socially approved or not. Sympathy, leadership, and friendships increase together with resistance, aggression, and negativism. Swift's (1964) review of objective studies confirms and extends Murphy's observation of a marked increase in all forms of social interaction. The older preschooler relies less on physical contacts and more on verbal techniques, while showing less dependent behavior toward adults. Another consistent finding is that children who rank high in one type of social activity rank high in others, suggesting a general personality trait running from social activity to social passivity.

Just as the preschool child is expanding and diversifying his social contacts, group phenomena are undergoing dramatic changes. The quantitative picture is not impressive; in the early preschool years, the group consists of two to three children and, in the later years, it has expanded to three to five. It is the quality of the change which is striking. Parten (1932), in her classic study, distinguished the following kinds of social participation: Onlooking, in which the child observes peer activity and even talks to or questions the participants but does not actively involve himself; Solitary play, in which the child pursues an activity without reference to others in his environment; Parallel play, in which the children play beside one another with the same objects or materials but do not play with one another; Associative play in which members are loosely organized around a common activity and share interests and material; and Cooperative play, in which there is a marked sense of belonging, control is in the hands of one or two leaders, and there is division of labor and assigning of different roles in order to achieve a common goal. Parallel play predominates in the two-year-old, e.g., children in a sandpile play with the same toys, but each goes his own way oblivious of the others. By the time the child is four years old, cooperative play has increased markedly. A number of children can band together to play house; roles of mother, father, baby, are assigned specifically; the leader makes sure that the roles are properly executed (the baby cannot cook supper, for instance) and excludes outsiders who might want to join in.

There is also information concerning social acceptability in preschool children. Social isolation is negatively correlated with acceptance. The child who plays alone and refuses the overtures of other children tends to be unpopular with his peers; acceptability is a matter of participation. Dependency on adults is another obstacle to acceptability. The dependent child may be unpopular because he interacts less with peers and is therefore as isolated as if he were playing alone. He may lack social skills for dealing with his contemporaries and feel more at home with adults. More ominous emotional entanglements have also been suggested: the child may be overly attached to adults or he may have a hostile-dependent relation which drives him constantly to seek attention and then reject it once it is given (Dunnington, 1957). However, using adults to further one's own activities, i.e., instrumental dependence, does not affect popularity adversely.

While involvement is necessary to acceptability, it is not sufficient to insure acceptance. The quarrelsome, combative, or dictatorial preschooler is likely to be rejected by his peers, as well as is the negativistic, obstructionistic, dawdling one. Preschoolers are attracted to qualities which facilitate social interaction: cooperativeness, respect for property, compliance. They gravitate to the child who is constructive and adapts quickly to the social demands of the situation. The highly aggressive child is not accepted, but preschoolers are less concerned about hostility than they will be later. Just as quarreling and friendship go together, fighting and tattling seem to be accepted as part of social involvement rather than as disruptive forces.

One of the most interesting findings, which we shall discuss in detail later (see Chapter 9), is that attraction is a function of facility in fantasy play. Play is the natural language of the preschooler, the activity which is least affected by adult pressures and values, the activity which frees him to ideate and act in accordance with his own interest and level of expression. The child who is adept at speaking this special language attracts others to himself, while one who has only meager fantasy play generates less interest in his peers.

Clearly, the preschool period is rich in social development. Although most of the topics we shall discuss have their origin in this period, it has been relatively neglected by psychologists. The meager data and the limited theorizing can be understood historically. Under the impact of Feud, attention has been focused on feeding, toileting, sex, and aggression, while early peer relations have receded into the background. Yet, the change from parallel to cooperative play, as outlined by Parten, may be as significant as any development in the preschool period. We know almost nothing of the factors which determine this transition and nothing of the conditions which facilitate or impede its progress. The same can be said of both the dramatic upsurge of social behavior and its highly complex forms. As Thompson (1962) notes, a brilliant beginning at conceptualizing social development was made by Lewin (1954), but interest shifted to adults. The subsequent conceptual and methodological advances in group dynamics have not been applied to personality development in children.

Because of the paucity of conceptualizations, we must be content with a few generalizations. Anderson (1938-39) suggests imitation as the first of three factors which condition social behavior. Parents are important as models for a wide range of social behaviors and attitudes, such as sportsmanship, prejudice, organizing ability, and cynicism. Parental behavior either sensitizes or hardens the child to social relations; e.g., quarreling parents not only serve as models but, if they upset the child, they also increase his sensitivity to overtly angry interactions. (Incidentally, Anderson could have included imitation of peers as an important factor in social development. Imitation can be the initial step in becoming a member of a group, and the child's social repertoire is constantly being enlarged by observing and copying the behavior of his contemporaries.) The second factor is the observed consequences of the child's own behavior. Anderson states that, in general, children are keenly aware of the social consequences of their behavior; just as they are endlessly intrigued with the effects of their exploratory behavior on the physical environment, they are constantly monitoring the effects of their social behavior on the human environment (see Chapter 3). The child is endowed with a rich repertoire of responses, some of which will produce favorable results, others not; for example, whimpering may be successful for one child, smiling for another. By successful, Anderson means more than gaining parental approval. He includes any social consequence which is favorable from the child's point of view. By this expanded criterion, making a parent angry may be highly rewarding for the child. A final set of factors involves the meagerness or richness of social experiences available to the child. Social behavior expands as the child encounters a variety of situations which must be mastered. In an impoverished setting, his social techniques will be limited.

Isaacs (1937), in her psychoanalytically oriented account of sociability among preschoolers, sees the desire to protect younger children and newcomers as an instance of identification with the good parent. Sociability also involves sympathy and empathy arising from identification with the feelings of another child. But there is a new, deeper kind of identification with others which gives rise to a sense of oneness or togetherness. Taking turns, for example, reflects the child's growing awareness of reciprocity, as does his occasional pride in reporting what "we" did rather than what "I" did. Finally, and most important, sociability depends upon the capacity for giving to another child, whether it be tangible things or helpful services. Children value gifts not because they are materialistic and greedy for things; on the contrary, gifts prove to children that they are loveworthy. The prototype of being given to is being fed by a loving mother, for food is the original gift and symbol of love (see Chapter 7). The child's capacity to give is a sign that he is secure in the knowledge that he is loved; his generosity is the overflow of the feeling of his own goodness. If he feels unloved, uncertain, or hateful, he is no longer able to give to others and his social relations suffer as a consequence.

Sociability in Middle Childhood

Sullivan (1953) discusses the child's need for playmates in middle childhood, which, in turn, exposes him to a new world of social experiences. Brought face to face with the many different ways in which peers conduct themselves, he observes and absorbs new information, and evaluates it as well. Some of his social encounters are positive and enjoyable; others are negative and uncomfortable, but he must live with them anyway. Sullivan's term for this special kind of social learning is accommodation. Instead of believing that he is unique and the center of the universe (as he tends to feel within the family), the child begins to learn how to get along with others, to think of himself as a social being and to evolve his own set of interpersonal values. He learns what needs motivate his relations to others (such as prestige, cooperativeness, leadership) and under what circumstances they are important. In short, the child begins to know his way around in the social environment—he begins to know what he wants and how to go about getting it, what setbacks he can take without experiencing damaging feelings of failure or inferiority, his areas of comfortable functioning, of uneasiness, of exciting venturing out. Sullivan, like Adler, maintains that socialization counteracts the tendency to be spoiled, to expect unlimited attention and service, to fret and demand if not catered to.

White (1964) echoes Sullivan's point. For the child to be loveworthy in the home involves a significant amount of obedient, affectionate behavior. On the playground the child must be respectworthy, which is quite different, since it is contingent upon the child's competence. He must expose himself to comparisons with other children in regard to athletic ability, courage, manual skills, sociability, friendliness. If he fares well, he will earn the acceptance and respect of his peers and his self-esteem will be raised. If he fails, he will suffer humiliation, ridicule, and rejection by the group, and his self-esteem will be lowered.

In sum: a high degree of positive involvement with peers or a high degree of sociability in middle childhood prepare the child for social participation in adulthood, just as friendships prepare him for the mutuality of heterosexual love.

Objective studies give a detailed account of acceptance by peers in middle childhood.[7] Before summarizing this research, we should note that social acceptability is not equivalent to Adler's concept of social feelings or Sullivan's concept of the need for playmates. Both authors are using a broadly defined variable, which ranges from a general gregariousness (liking to be around people) to a deep emotional investment in interpersonal transactions. Their concern is with tracing the process by which the child first becomes interested in, then involved with, and finally committed to the welfare of his contemporaries. Objective investigators, with their emphasis on precision, have preferred to deal with more specific variables, such as friendliness, acceptability, and popularity. Their studies have explored the empirical interrelations among these variables (showing, for example, that sociability and popularity can be unrelated at certain ages), rather than assuming an interrelatedness which may not be warranted.

In their concern for taxonomy, objective psychologists have distinguished various degrees of social acceptability. At the low end is the reject who is actively disliked or rejected by others. There is some evidence that he has socially negative characteristics, such as being hotheaded, noisy, and boastful, or that he is gauche and ineffectual in his social techniques. Next comes the social isolate who is more or less ignored by other children. Social isolation in some children represents one aspect of a general detachment from all deep human relations. Other isolates are loners, either because circumstances have placed them with children who do not interest them, or because of a general lack of interest in peers. The subjective isolate might have attempted, at one time, to establish positive contacts but have been continually rebuffed because of his race or religion, for example, or the lack of a valued skill, such as athletic ability. While neither asocial nor a loner, the child becomes prisoner of his idea that he is unwanted and no longer makes social overtures. At the positive end is the popular child who is above average in social acceptance and, at the very top is the star who is the preferred companion of almost all members of a group. However, even stars experience their special kind of isolation, since they do not reciprocate many friendship choices,

[7] For a more extensive review, see Campbell (1964) and Hurlock (1964).

either because they are not interested or because they are afraid of becoming friendly with less popular children. Finally, there are fringers who are constantly fearful of losing what acceptability they have by doing something which others would not like; and there are climbers who want to gain the acceptance of the socially more favored peer groups.

A given individual rarely occupies a single position on the continuum. In his daily life, he is confronted with a variety of groups serving diverse functions, and his acceptability varies as he moves from one to another. A football star may not be a star in the classroom or at parties. However, there is a sour note in this tune that anyone can be acceptable to someone—namely, everyone cannot have the degree of acceptability he desires in the group he chooses. As older children begin to acquire reputations, their acceptance or rejection becomes more and more dependent on the image their peers have of them, rather than on their actual behavior. The personality characteristics underlying acceptance are difficult to change; for example, it is difficult for a dour person to become cheerful and for a phlegmatic one to become vital. The image the child develops of his social self can also box him in. A girl who regards herself as inept behaves in keeping with this inner image when she meets a new group; unless the group is unusual, it will treat her as she presents herself. Finally, social settings become increasingly stable. In early childhood the home is supplemented by the classroom, the playground, and a variety of special interest groups spring up; by middle childhood it becomes increasingly difficult to find a setting novel enough to tap a new set of skills and resources through which a child may gain acceptance. Thus, a host of factors prevent the child from seeking and finding acceptance in settings where he would desire it and from changing his status if he would like to. Social acceptance is remarkably stable over time and rivals intelligence in this respect.

The reasons for acceptability remain essentially the same in middle childhood as they were in the preschool years. Some social nonparticipants are quiet, shy individuals who are uninterested in others, some are self-centered and selfish. There are the children who are actively involved but rejected because they annoy others by their tattling, bullying, showing-off or making trouble. The accepted child is often the success oriented one who is both vitally involved in social relations and has a variety of positive techniques with which to implement his enthusiasm. However, a friendly, sympathetic, loyal child who is "goodness" oriented can also be popular, even though he does not have the assertive vitality of his success oriented counterpart.

Social Acceptability and Middle Class Values

What gradually unfolds as one reads the objective data on social acceptability is an ideal picture of the personality traits of the middle class American child: he is vigorous, tolerant, dependable, cooperative, neat, clean, thorough, and sensitive to feelings; he has good will toward others, provides new experiences for his peers, is above average in intelligence and scholastic achievement, and is emotionally stable. The child low on social acceptability is retiring, is low in tolerance of others and adaptability to them, is undependable, negativistic, individualistic, quarrelsome, overbearing, and tends to be socially insensitive or socially maladjusted. The popular boy is one who is most boyish—strong, agile, active, aggressive, daring. The popular girl is the most girlish—sociable, neat, docile, pretty. Between the first and fifth grades the "little lady" quality tends to be replaced by admiration for more assertive, vigorous, and sportsmanlike behavior, although the girl still could not be quarrelsome, aggressive, or sexually forward and be popular with peers. Being a real boy is consistently important to acceptability among boys; effeminacy and timidity are significant handicaps. The popular boy, unlike the girl, can quarrel and fight, but bossiness and blatant aggressiveness are not acceptable.

Children in middle childhood accept one another on the basis of possessing the same personality traits for which they were accepted by their parents in the preschool period; when they themselves become parents, they will accept their own children for the same traits which were the basis of parental and peer acceptance. The point is that we are now dealing with relatively permanent values of middle class American society. If we were sociologists, we would focus our discussion on such social class variables. While our concern is with the individual, we should not lose sight of the potency and pervasiveness of middle class values, since they give continuity and direction to the lives of middle class children and serve as the broad frame of reference which affects their daily transactions and long range goals. Such values are not intrinsic to human development or to human nature. Whether they are the best ones for the preservation and growth of middle class society or for the individual are questions which lie beyond the scope of our discussion. However, we must constantly guard against the tendency to assume that the values are best because they are ours.

The fact that children in middle childhood are so finely attuned to middle class standards raises several questions. Given the freedom inherent in peer relations, why do children not evolve diverse standards of acceptability? If they suffer from parental disapproval at home, why do they uphold parental standards when they are on their own? Concretely, why do not slovenly, dawdling, and morose children band together and regard themselves equally as acceptable as their neat, responsible, cheerful counterparts? Since peer rejection is so painful, why do they continue to adhere to a system which penalizes them? We can only speculate about answers to these questions. Our discussion of moral development (see Chapter 5) indicated that in middle childhood rules and regulations are both rigidly defined and strongly championed. The "good boy, good girl" aura of middle childhood arises from a deep conviction of its basic rightness. If children bring the same absolutism to their standards of esteemworthy behavior, one might expect them to have a more narrowly defined criterion of acceptability and to be more cruel in their rejections than their parents have been. Drawing a parallel with friendship, the ability to accept—and even enjoy and value—diversity in personality may not appear until adolescence.

Supporting this speculation concerning conformity is the evidence that, in making judgments, young children are more susceptible than older ones to the opinions of others. Berenda (1950) asked a child to compare the length of three straight lines with a single, standard line. When he made his judgments alone, the child was usually accurate. Later he was asked to repeat the procedure with eight other children who were "stooges" instructed to give the wrong answer. Confronted with the fact that his peers were giving judgments which ran counter to what he saw with his own eyes, the child tended to agree with the group. In another procedure the child tended to agree with a teacher who was the stooge. However, the tendency to agree decreased with age; children from ten to thirteen upheld their correct judgment more than children between seven and ten. In middle childhood the attitude seems to be: if peers or authorities say so, it must be right. Only toward preadolescence does the child begin to think that each person is entitled to his own opinion—and that his is as good as anyone's.

If children champion social conventions so early, where does this leave Sullivan's concept of the enriching effect of peer contracts? How can playmates teach the child to get along with others unless there is diversity among them? The studies of acceptability inform us that the child's new experiences with peers still occur within a traditional context. He does not enter a world of vastly discrepant values in which anything goes. It would be difficult to know how cultural continuity could be preserved were this the case. He is exposed to variations on a common theme and different embodiments of common values. He can observe and learn from these variations both about others and about his own role in the middle class scheme of things.

Competition, Compromise, and Cooperation

Compromise and Integrative Behavior

Sullivan singled out competition and compromise when discussing the significance of peer relations in middle childhood. The former stimulates the individual to be the best in the eyes of his peers; the latter enables him to integrate his contribution with those of others. Balanced in healthy growth, compromise checks the danger of a selfish drive to be first regardless of consequences to others, and competition prevents the child from lapsing into a mindless, valueless yielding to others or into the pursuit of peace at any price.

Sullivan's concept of compromise has its counterpart in Anderson's (1937a, 1937b) concept of integrative behavior. Domination, compromise, and integrative behavior are possible outcomes of a confrontation between two people. The one who dominates by use of force, threat, or fear acts on his own fixed standards, intensifies differences, and rides roughshod over the individuality of the other person who must either resist or submit. In compromise, each party calls a truce and maintains the status quo while retaining his own point of view and worldly goods. Integrative behavior maximizes individuality and cooperativeness in seeking a common purpose with another, yielding present values and setting aside status differences in order to reach a mutually satisfying solution. Integrative behavior is the most advanced reaction, since it requires awareness of another's needs, foresight, and planning—all of which increase as the child develops. Anderson found that both domination and integrative behavior are reciprocated in kind. Even in childhood, power plays provoke power struggles, while a willingness to cooperate is the prelude to negotiation.

In his empirical investigation Anderson found that there is little relation between domination and integrative behavior; instead, each represents a different way of managing interpersonal confrontations. Domination is characterized by forceful appropriation of or defense of a desired object and forceful directing, forbidding, and criticizing. Integrative behavior is characterized by requests and suggestions in order to secure material, compliance with the suggestions and requests of others, and "sharing a common purpose." Anderson found crude and tentative forms of integrative behavior in children two-and-a-half to six-and-a-half years old, but his leads have not been followed into middle childhood and adolescence.

Competition and Cooperation

Competition, and its counterpart, cooperation, have been investigated more extensively than compromise or integrative behavior. Many of the studies have been concerned with the relative motivating power of the two—does the child do more or better under one condition or the other? Occasionally investigators have shared Adler's and Sullivan's concern with the effects of the two kinds of behavior on the child's social relations and on his self-acceptance.

Cooperation is defined as two or more individuals joining their efforts to reach a goal which is mutually desirable; in competition, the individuals vie with each other for the same goal. Although conceptually distinct, the two may overlap; e.g., members of a team cooperate to defeat a rival team while competing with one another to be better players. Competition is regarded as more primitive, both behaviorally and developmentally. One of its precursors may be the sibling rivalry for parental affection which is clearly present in the two-year-old. Although this involves one-sided rather than mutual competitive interactions, it may well serve as the foundation for the more complex manifestations to come. Competition in regard to a task, such as making a "bigger" or "prettier" construction, does not appear until around four years of age. By six years of age, task competition is well developed, due to an increased understanding of the rules of the game and an increased ability to compare the self with others. The more regressive nature of competition is reflected in the finding (Stendler et al., 1951) that second graders dramatically increased their negative social interactions when they were thrown into competition for a prize instead of being rewarded for group effort. However, competition and aggression are not uniformly related in the early years. This independence should be kept in mind because of the tendency of psychoanalysts and objective investigators to regard competitiveness or even assertiveness as closely related to or as derivatives of hostility (see Chapter 7).

Cooperation is considered more advanced than competition because it is dependent upon the growth of social awareness, social skills, adaptability, and altruism (see Wolfle and Wolfle, 1939). A study of the changes which must occur before cooperation is possible was done by Smith (1960), who asked children of different ages to make up a story as a group endeavor. The significant development was not in task orientation, but rather in the change from egocentricity to cooperation. Preschoolers introduced new ideas with no attempt to coordinate them with those of other children; they did not elaborate upon their ideas or ask others to do so. When they did react to someone else, it was usually to reject his contribution. The discussion, therefore, was more of a proclaiming of individual thought and no common theme was developed. In the nine- and twelve-year-old groups, collaboration and cooperation increased, autonomy was moderated by compromise, ideas were tentatively accepted and explored, and clarification was sought. Under these circumstances a common theme could be evolved.

Objective studies show that children in middle childhood prefer to work for themselves rather than for the group and will work harder under such conditions. This may well reflect the competitive orientation both at home and in the school where individual achievement is prized above that of the group. The findings themselves are clearly culture bound, since other societies teach children to be self-effacing about personal achievement or to work primarily for the prestige and glory of the group.[8]

Objective investigators share the clinician's concern over the effects of competition and cooperation on personality development. Competition can stimulate high achievement but risks alienating the child from his peers. Cooperation maximizes social awareness, but potentially can sacrifice individuality to group goals. Unfortunately, objective data on this crucial issue are meager. Stendler and his colleagues found that, with second graders, positive social behavior was greater than negative behavior when the children worked for group prizes, but the reverse was true when they worked for individual prizes. There was suggestive evidence of increased boasting and depreciation under competitive conditions because of the increased demands on each child to prove himself. The study suggests that, while competition may be more highly motivating, it has more undesirable social consequences. The authors caution against broad generalizations on the basis of limited evidence, and a more definitive exploration of the issue remains to be done.

Social Power and Leadership

Developmental Overview

The topic of social power and leadership has stimulated extensive research (see Thompson, 1962 and Hurlock, 1964). As early as ten months of age there is evidence of a dominance-submission pattern in the triumphant smile of the dominant infant who forcibly takes a toy away from his unhappy, tearful companion. From the second year on, the aggressively dominant child can be distinguished from the poised diplomat who leads by initiating and demonstrating activities (Buhler, 1933) and who

[8] For a revealing account of the Russian system of motivating the individual by stressing the importance of the group, see Bronfenbrenner (1962).

controls a larger number of children than the more brutal boss (Parten, 1933). By five years of age children exhibit a variety of behaviors designed to dominate or lead their peers. Hanfmann (1935) observed two kinds of forceful domination. One child used destructive methods and dominated by continually destroying his partner's work. When he was not being destructive, however, he was under the control of children with more constructive leadership abilities. Another child's behavior was marked by expansion. Because he forcefully appropriated anything and everything he wanted with total disregard for the rights and feelings of others, he was characterized as a gangster. He ran roughshod over everyone but the destructive child. At the other extreme were social isolates who were easily dominated by more assertive children. They preferred to play by themselves, wanted to be left alone, and resisted attempts to draw them into social interaction. One might question Hanfmann's labeling such behavior as "negative domination." Yet the isolate is controlling the social situation as much as the gangster by saying, in effect, "I insist on doing things my way." Just as silence can be a means of communication as well as speech, isolation as well as active engagement can be a technique for forcing others to accomodate to oneself.

Between the extremes of destructiveness and isolation Hanfmann found two kinds of leadership. The objective leader was task oriented, planned activities, obtained play material necessary to achieving them, and resolved conflicts by compromise. He was adept at cooperating and sharing in order to insure the successful pursuit of current interests. The social leader was more concerned with initiating and maintaining harmonious social interaction than he was in getting things done. He was unusually sensitive to the attitudes of his companions and tactfully catered to them in order to win cooperation. He explained, he joked, he overlooked, he gave in—all in the service of maintaining a harmonious relationship.

One should caution against adultizing such social behavior: a four-year-old is not a miniature adult. Parten's study showed independent play to be more characteristic of the children than social interaction, and a slightly different phrasing of a communication—"You sit there" versus "Do you want to sit down?"—could make the difference between a child's being classified as a bully or a diplomat. Though we are dealing with tentative beginnings, the versatility of responses is striking. Equally impressive is the fact that the same patterns, although more stable and complex, appear in middle childhood and adolescence: the authoritarian leader, the bully and the despot who control by power, the task oriented leader who is adept at getting things done, and the socially oriented leader who is skillful in maintaining pleasant interpersonal relations. Hanfmann's finding that four-year-olds prefer the leaders to the destructive child or the gangster is borne out in the trend for children increasingly to prefer democratic leadership as they develop. The adolescent leader must be especially subtle and avoid direct domination. True to the paradoxical nature of adolescence itself, he cannot deviate as widely from group norms and values as he did in middle childhood; rather, he must be more adroit in getting the group to do what they prefer doing in the first place (Horrocks, 1969).

Leadership

Leadership has proven to be one of those concepts which is difficult to define and to measure.[9] It is related to, but should not be equated with popularity, since a child can be high on the former and not the latter. Because leadership may involve different psychological factors at various age levels, it would be a mistake to assume that the preschool "gangster" is in any way related to the destructive gang leader in middle childhood or adolescence. Moreover, leadership itself has many faces: it can be democratic or dictatorial, task oriented or interpersonally oriented; there can be a single leader or multiple leaders; a leader can be chosen by the group or chosen for the group; he can be a constructive, high-status, high-minded child, or he can be a hot-headed, low-status child who galvanizes the mounting rebelliousness in the group. (We will also note in passing that the psychology of following may be as complex as the psychology of leadership, but the topic has been almost unexplored.)

In a general way, one might expect leaders to have above average intelligence, self-confidence, initiative, cooperativeness, social sensitivity, visibility, and popularity. But there is great variability of characteristics and many exceptions to this listing because leadership results from an interaction of traits variable with situational ones. Just as groups vary widely in their needs and goals, so will the kinds of traits necessary to realize them; in fact, leaders usually are regarded as deriving their power from the group rather than bending the group to their will. DeHaan (1958) offers the best rule of thumb in regard to leadership: while leaders may have different traits from followers, the situation will determine what traits will set which leaders apart.

In addition to trait factors, there is a skill factor: one learns to be a leader by leading. Some children may have greater potential than others for meshing traits with group needs, but all must practice. The group adds a reputation factor to leadership by regarding behavior as more consistent than it actually is. Psychologists sometimes refer to a "halo" effect: the individual comes to be evaluated not in terms of his behavior but in terms of his reputation for leading. The group also extends the leader's influence beyond his manifest behavior through a phenomenon labeled contagion, i.e., "picking up" or imitating his behavior even when he is making no attempt to influence others (see Polansky, 1952).

While leadership and social power have been studied extensively, comparatively little is understood about the factors responsible for their appearance and development. We do not know, for example, what psychological progress must occur before a four-year-old can show the tact and sensitivity of a social leader or the task oriented skill of an objective leader. Nor have the observations and objective studies been placed in a broader context by relating them to other significant personality developments. As we have seen, both Sullivan and Piaget regard peer relations in middle childhood as correctives for the egocentricism of the preschool period, while Adler's concept of social feelings seems to be exemplified by objective and social leaders.

[9] See Browne and Cohn (1958), Gardner (1956), and DeHann (1958) for a discussion of the multiple meanings of leadership and examples of the practical problems involved in its measurement. Horrocks (1969) offers a particularly apt resume of the literature.

The effect of leadership patterns on the emotional and cognitive developments which these authors discuss is only one of a host of developmental questions which remain unanswered.

The Group

The child's participation in informal and organized peer groups will have unique effects on his personality development. As groups progress from their insubstantial beginnings in the preschool years, through the gangs of middle childhood, to the adolescent cliques and crowds, they play an increasingly vital role as an extrafamilial influence preparing the child for participation in the adult community. Because we focus on the individual, our concern is the special contribution group membership makes to his development; however, the rich literature (see, for example, Hare, Borgatta, and Bales, 1955) tends to focus on the group as the unit of study rather than the individual, and the mainstream of thinking has not been concerned with child development.

The Power of Early Groups

The social expansiveness of the preschool years witnesses the emergence of the group, which, even in this period, can be remarkably potent.

One of the most dramatic examples of group identification is reported by Freud and Dann (1951) who observed six children between three and four years of age who had spent most of their lives in a Nazi concentration camp. Four had lost their mothers immediately after birth, and the rest within the first year of life. Having no family, they were passed around from one place to another, receiving the bare minimum of attention from the depleted and overworked adults assigned to the Ward for Motherless Children. There was no affection from adults, and no play. Only one child became attached to an adult. When transferred to a therapeutic setting in England, the children were restless, hypersensitive, difficult to handle, and treated adults with primitive but strangely impersonal aggression–biting, spitting, defiant urination on toys or clothes. Yet, group cohesiveness was remarkable. The children cared for one another with great warmth and looked after each other's needs. They were frantic when one member was separated from the group, regardless of the reason. There was almost none of the jealousy, rivalry, and competition which is so striking among siblings and peers. They spontaneously would take turns, eager for each member to have his share, and did not begrudge each other any special possession. There was no single leader, but one child would lead during mealtime, another during play, and so on. Thus, in the face of extreme deprivation, the children provided emotional support for one another and evolved a mode of functioning which offered them security and preserved their initiative.

The power of the group to determine the behavior of the individual is illustrated by Merei's (1949) ingenious study of children between three and seven years of age.

After selecting children low on leadership qualities, he let them interact to the point of forming a cohesive social unit, and then he introduced a child high on leadership ability. In every instance except one, the group did not follow the leader; on the contrary, the leader was forced to change in order to accommodate the group; e.g., a leader who formerly had gained power by commanding others was not heeded unless he changed and conformed to group interests. One kind of leader, called the diplomat, was able to adapt to the group temporarily while ultimately maneuvering them into acceding to his wishes. In general, the group was able to control the leader, even though each member alone would have yielded to him.

Boll (1957) utilized adult's recall of unsupervised groups (i.e., groups which developed more or less spontaneously in the neighborhood rather than in a structured setting such as a nursery school) to describe different kinds of groups. In some, the members were congenial, roles were clearly defined and agreeable, and there was a general air of getting along well together. In other groups where there was constant friction and continual jockeying for status and roles, the individual learned how to stand up for himself. Finally, some groups derived their closeness less through inner congeniality than through the need to protect themselves against other closely knit neighborhood groups. Boll draws a parallel between these preschool groups and adolescent gangs which are unified by similar concerns over the power of other gangs.

Factors Influencing Group Formation

The developmental factors which make the formation of groups possible have not been satisfactorily isolated, but Murphy and his co-workers (1937) make a number of pertinent observations. The developmental status of the average two-year-old handicaps him in regard to group formation. He has little grasp of time and, since he lives primarily in the present, cannot plan ahead or think of sequences of events. When left to his own devices, he cannot sustain meaningful contact with more than one or two other children, nor can he accept or apply rules of the game. Finally, he is limited in his use of words and therefore in his ability to communicate. In sum, the two-year-old cannot conceive of, sustain interest in, or communicate the kind of activity which requires individuals to perform distinct functions and interact in a coordinated way in order to achieve a mutually desired goal.

By four years of age the child has developed to the point where group activity is possible. Murphy lists maturity, security, initiative, and imaginativeness as necessary personality characteristics, although he provides little information concerning the manner in which they function to produce group behavior. Mallay (1935) found that active engagement was essential for group formation. Regardless of the child's intelligence or adjustment, engaging in similar or interdependent activity was necessary for the successful maintainance of social contact. If the child were an onlooker, a commentator, or if he were engaged in unrelated activity, social contact would be unsuccessful.

A host of situational factors affect group formation (Swift, 1964). We learn, for instance, that physical equipment such as dolls, clay, blocks, swings, and wagons facilitate the formation of cooperative groups, while limited physical space increases conflict. Since the children are too young to be unsupervised, the adult plays an important part in facilitating group formation and determining its nature. An example of the importance of situational factors comes from Kitano's (1962) comparison of problem and nonproblem children. He found that, in situations which were highly structured so that expectancies and appropriate behavior were clearly defined–situations like lunch, dressing, going home–there was no difference in the behavior of the two groups. Differences appeared only in unstructured situations, such as play, in which the children were free to do whatever they wished. Recall also that the striking development of group identity reported by Freud and Dann was due in part to the unusual challenge to survival which the children faced.

The Middle Childhood Gang

Groups, although possible among four-year-olds, are typically insubstantial and can form and dissolve in a matter of weeks. Status within them shifts so that November's leader becomes January's follower. Under normal circumstances, solidification and a sense of being part of a superordinate organization does not develop until middle childhood when the child enters what some psychologists call the "gang age." At this time the gang becomes sufficiently potent to compete with the family for the child's interest, loyalty, and emotional involvement.

Because the transition from preschool group to gang is gradual, one can still see the shifting, utilitarian behavior characteristic of the earlier period when the children are between six and eight years of age. There are few formal rules for governing the gang, and a rapid turnover in membership occurs. While the groups are formed by the children themselves, participation is based more on expediency than on any set characteristic of the child. Children come together to carry out a specific function, such as playing house, but are apt to disband when this interest wanes. If the purpose is to play cowboys and Indians, a boy may be included primarily because he happens to have a toy rifle. The principal selective factor is sex, since seven or eight years of age marks the beginning of same-sex associations. Individuals cooperate in order to satisfy their own needs and the spirit is one of "my" rather than "our" activity. Adult supervision–another link to the preschool era–is still needed to manage conflicts.

Only when children are between eight and twelve years of age does the gang satisfy Sherif's (1961) definition of a true group: a social unit consisting of individuals who have interdependent status and role relationships, and which possesses a set of implicit or explicit norms or values which regulate the behavior of the individual members in matters of consequence to the group. At around nine years of age, the child begins to subordinate his own interests and demands to the goals of the group, tries to live up to group standards, and criticizes those who do not. Thus, the gang takes precedence over the self; "we" is more important than "I." Groups of eight- to ten-year-old children still need a definite plan in order to structure their activity. They must have rules to follow or repetitive and ritualized play. Groups of ten-year-olds are autonomous in the sense of being able to meet situations as they arise, cope with the unexpected and take on new ventures. To illustrate this transition: eight-to ten-year-olds either imitate older peers or prefer stereotyped games such as hop-scotch or jumping rope; after they are ten years old, they can respond to and implement the more general suggestions, "Let's make a clubhouse," or, "Let's give a party" (Buhler, 1933).

The factors which influence gangs are similar to those influencing preschool groups (see Campbell, 1964). The physical setting is important (since propinquity still insures interaction), as is the activity setting (since gangs behave differently during mealtime than during free play). Rewarding experiences increase group cohesiveness, unrewarding ones diminish it. Younger children are more susceptible to group influences than are older ones; girls are more susceptible than boys. The less clear the situation, the more susceptible the child to group influences; e.g., the newly arrived fourth-grader is more apt to accept the gang's views concerning the school than are the more experienced pupils. Small groups achieve more consensus than large ones; stable groups have more communication than unstable ones.

A vivid account of group formation in twelve-year-old boys at a summer camp is given by Sherif. He hypothesized that a definite group structure will evolve when there are goals which have a common appeal value and which require interdependent activities for their attainment. His hypothesis was amply demonstrated as two groups of boys faced the problems of providing food and shelter, activities, and recreation for themselves. The groups developed a strong feeling that places, objects, and even songs were "theirs," and a definite status hierarchy was evolved in the first week. Each group had its own norms; e.g., toughness and cursing were approved in one group but not in the other. Deviations were punished by "calling down" or ignoring the culprit. Rules for spontaneous games and a system for taking turns were standardized.

Objective studies provide information concerning the process by which the individual becomes a member of the gang. Phillips and others (1951), in a study of six-to seven-year-old girls, noted that the group rarely made an effort to include a newcomer but left it up to the individual to initiate contacts. One of the child's initial approaches was to imitate the leader, imitation often being the child's first approximation of desired behavior (see Chapters 4 and 5). Next, the newcomer would initiate activities. Usually she was ignored at first, but eventually she would be partially successful. Finally, she was included in the group by choice and, by the sixth play session, was successfully initiating and directing activities. Here, as in Mallay's study of preschoolers, active participation is the keynote, and the newcomer must establish her worth in pragmatic terms of what she can contribute to the group.

Phillips discusses assimilation from the child's point of view. It is essential that the child acquire the common frame of reference of the group, that is, the mutually agreed-upon activities and goals. He must know who the children are, how and what they want to play, what is pleasing and taboo, who the leader is, and what the roles of the members are. Only when he becomes familiar with the group's frame of reference can he communicate and participate. The point is probably valid at all levels of development. Even with preschoolers, a strong leader is ignored unless he adopts the interests and values of the group; the ten-year-old at summer camp for the first time and the college freshman typically go through a feverish period of gathering information concerning the new group.

Campbell adds a final factor which is involved in the child's assimilation into the gang. He states that the child not only observes and learns about the group, but also evaluates it in terms of his own likes and dislikes, his image of himself, and the congruence between this image and group values. In terms of our discussion, the child exercises judgment (see Chapter 4). As he acts upon group members and is reacted to, he appraises the role he is carving out for himself and the status he is achieving. Campbell's point is that the child is not merely taking on the characteristics of the group; rather, there is an active interplay between individual and group. The individual seeks a niche congenial to his particular personality, but he also affects the group in the process; e.g., if he becomes a willing follower, a group activity will have a somewhat different flavor than if he becomes a disgruntled follower.

The Significance of the Gang

The gang of the ten- to fourteen-year-old lacks the permanence and potency it will achieve in adolescence. Members change too frequently, the group itself can be short-lived, and an aura of improvisation and expediency persists. Yet the gang, because it is an autonomous, superordinate organization, constitutes a significant landmark in the developmental picture. Gangs have two other characteristics which are important for personality development. First, they are identifiable social units, perceived by the child as existing in their own right. They have names, insignia, meeting places, and often an air of secrecy, manifested by special passwords, languages, or codes. Initiation rites test a newcomer's skill, daring, or ability to endure humiliation, and give him a feeling of importance once accepted. Secondly, gangs are segregated by sex and acceptable behavior follows the dictates of sex typing. Boys and girls differ in the kinds of activities they enjoy doing together. Boys are action oriented, mechanically inclined, assertive, daring to the point of engaging at times in taboo activities such as gambling, smoking, and stealing. Girls are sociable, gossipy, domestic in their interests, with a tendency to snobbishness and minor defiance of taboos in order to appear grown-up and sophisticated. More important is the striking antagonism between sexes which reaches its peak around nine to eleven years of age when boys and girls continually disparage one another with bickering, quarreling, and namecalling. The lone girl who must wait for a school bus with a group of boys or the lone boy who must wait with a group of girls is mortified by the sheer physical proximity to the enemy.

What are the psychological consequences of being a member of the gang? One obvious innovation is that the child now can identify with a social unit composed entirely of peers. The gang comes to have the kind of palpable existence which formally characterized only adult-involved groups, such as the family, the school, the team, the scouts. Just as friendship advances mutuality, identification with the gang advances the sense of belongingness. The child is no longer only his parent's son, his teacher's pupil; he is also his club's member. Because it can devise and implement meaningful activities, the gang functions as an important adjunct to the individual's initiative (see Chapter 3); gang members who pool their ideas and enthusiasms can venture out further than the individual could or would on his own. The control the gang exerts buttresses the individual's self-control, since a cohesive group structure, like a cohesive family structure, reassures the child that his temper, his egotism, or his greed will not get out of hand. The gang offers continual training in interdependent behavior, each member both making his special contribution and adjusting his behavior to the achievement of mutually desired goals. Finally, the requirements of the group are "ours" (peer-derived) rather than "theirs" (adult-derived), which gives them a special quality of congruence. The members are on their own and responsible for their fate. The adult is no longer present to suggest what to do, to correct injustice, and to resolve conflicts. Whatever qualities the gang possesses are self-generated.[10]

Sullivan (1953) notes that gangs, like friendships, foster realistic self-evaluation and correct the illusion of specialness. The guilt an individual feels over being angry at his mother or masturbating is relieved by finding others who are also angry or who masturbate. The less an individual can test his feelings and self-evaluations against a norm of group experiences, the more apt he is to exaggerate them, both positively and negatively. Sullivan's point about self-evaluation is elaborated by Festinger (1954) who assumes that all human beings have a drive to evaluate their abilities and opinions. Unlike ideas concerning physical reality which can be objectively validated (Will a ball bounce? Will melting snow become water?), abilities and opinions often lack objective, nonsocial criteria for correctness and adequacy. While there are objective, nonsocial criteria for some abilities, (e.g., one can gauge running ability by Olympic records) it is difficult to judge how good a painter one is or even how intelligent or socially skillful. Opinions have almost no objective anchor in reality and, consequently, can be ephemeral and unstable. The individual usually compares himself with others in order to validate his opinion, typically seeking people who are not too different from himself; e.g., an individual with race prejudice does not seek out a liberal in order to validate his opinions. This drive for self-validation is a potent motive for associating with groups, since group participation gives the individual the satisfaction of knowing that his opinion is correct and his performance is adequate.

Note that Festinger stresses the validating function of groups, just as Byrne emphasized the validating function of friendships. Sullivan was more interested in the group's potential for correcting faulty opinions by confronting the child with evaluations different from his own. However, Sullivan would readily admit that individuals who are threatened by change often seek just those settings which perpetuate their faulty self-image. Adler also describes the versatility and craftiness with which an individual seeks to maintain his illusion of specialness (see Chapter 3). The wallflower wears an unbecoming dress, no makeup, and sits slouched on a chair. What other opinion can she elicit except that she is especially unattractive? From Anderson's research we learned that domination evokes domination in return. What other conclusion can the dominating child draw except that the world is a jungle in which he must overpower or be overpowered? Adler claims that this predilection to manufacture justifications for a faulty style of living is one of the principal difficulties in

[10] For a resume of the contributions of the gang to personality development, see Crane (1958).

helping neurotics. Thus, exclusive pursuit of self-validation can be an ominous sign, since the individual is cut off from the confrontation which would counteract his efforts to justify his maladaptive opinions and behavior.

The gang, with all its growth-promoting features, has its special dangers. It comes into its own during a time when children are still extreme and absolute in their values: in morality, right is right; rules are rules; exceptions or extenuating circumstances are not allowed (see Chapter 5). The strict separation of the sexes in gang behavior illustrates the same rigidity of thinking. Note that the early battle of the sexes caricatures the adult values on which it is based. No mature adult would set such firm boundaries on things male and female or exact such a penalty of ridicule and humiliation for crossing into the enemy camp. This slavish endorsement of a stereotype must be counteracted by subsequent developments before mature heterosexuality can be achieved.

It is also possible that, while the child is receiving valuable lessons in accommodating to the group, he is not being encouraged in the kinds of self-confidence which says, "I am what I am and I value myself regardless of my status." To illustrate: if the gang approves, a boy may curse, gamble, and do other things which are forbidden at home, thus broadening his range of experience. Yet, the gang will not allow him *not* to curse or gamble without paying the price of loss of status. We can speculate that there are two potential dangers in gangs. The first is that they will "run wild," as parents fear; or, in more formal terms, that the acting out of antisocial impulses will become contagious and overwhelm the forces making for control. But the other danger is that the gang will deal too harshly with its members and that the nonconformist will be made to suffer unduly. Thus the gang raises the same problem for the individual which he faced earlier when his initiative was first being curbed by parental prohibitions (see Chapter 3): the problem of balancing the advantages of conformity with those of autonomy. As we have seen, there is no agreement among psychologists, parents, and the American culture concerning the issue of individual freedom and responsible control. The gang represents a variation of the same theme: how much does participation in an organized social structure fulfill development of the individual, and how much does it block the realization of potentials? Perhaps the answer lies in the balance between the expansive, humanizing forces and the rigid, absolutistic forces, both of which are characteristic of gangs.

The Adolescent Clique and Crowd

The high point of group involvement is the adolescent period,[11] when the group comes into its own as an autonomous social organization with purpose, values, standards of behavior, and means of enforcing them. In its stability, cohesiveness, and differentiation, it is more like adult groups than those of middle childhood.

Adolescent groups vary in structure and nature. The small, close-knit clique, whose members are constantly together, sharing interests, problems, aspirations, and secrets, is bound together by a high degree of personal compatibility and mutual admiration.

[11] For a more comprehensive account, see Ausubel (1954) and Horrocks (1969).

Although a clique has no formal organization, the intense affective involvement of its members makes it invulnerable to adult attempts to change or disband it. The crowd is a larger aggregate, united by mutual interests and common goals. It is neither spontaneous nor influenced by propinquity as are childhood gangs. Though it encompasses various degrees of personal closeness, the crowd does not demand the high personal involvement of the clique. It is not homogeneous in regard to status; rather, certain crowds are particularly high in status and power, and being a member is one of the surest ways of gaining popularity. The gang still survives and requires more solidity and loyalty than the crowd. It is often more hostile to adult society and has a specific goal—sexual, athletic, delinquent. Like its preadolescent precursor, it retains an emphasis on adventure and excitement as well as the formal trappings of organization, such as name, dress, and initiation rituals. However, the gang is usually replaced by the clique or the crowd in middle class society.

Sex differences in group behavior persist. Boys, who use the group largely as support in their quest for autonomy, are strongly committed to its impersonal features: its function of setting and enforcing standards and its utility in increasing their power to deal with parents or peers. More personal and social, girls find the rewards of group life to be popularity, friendship, sharing confidences or, at a less intense level, chatting and gossiping. Once again, the male is more impersonal, task-, work-, and power-oriented, while the female is more personal, social, and feelingful.

The goals and structures of adolescent groups are so diverse that it is difficult to generalize about them (Coleman, 1961). The individual is confronted with a wide spectrum of possibilities. He can identify with the intimate clique, with the general population of "teenagers," or he can find groups devoted to athletic, recreational, vocational, religious, sexual, aggressive, nihilistic, or idealistic pursuits. On the surface, the chances for individual fulfillment in a congenial setting would seem to be high. However, the adolescent group, like the gang of middle childhood, poses certain threats to the individual which we shall discuss after describing the developmental function of the group.

It is not difficult to understand the importance of the group in this period. The future oriented adolescent is no longer limited to the here-and-now world of childhood but is capable of evolving long-range plans. Social expectations and sexual maturation are constant reminders of his impending adult status. In middle class society, the group serves as the adolescent's primary bridge to the future. It provides him with a variety of social encounters and an arena for finding his place in a social milieu which resembles that of adult life. It helps him make the transition from a same-sex to a heterosexual orientation, offering him both the necessary protection and provocation (see p. 277). While the final choice of a mate depends on added factors of intimacy and mutuality, the group is a supportive setting in which the choice can be made.

The group provides the adolescent with valuable emotional support by offering him status—not in the sense that he occupies a leadership role, but rather in that he feels he is somebody. The sense of belonging is never more important than in this transitional period when the adolescent is no longer a child and not yet an adult. If parents are understanding and give him the respect and responsibility which is his due, and if adult society provides him with meaningful vocational opportunities, the discrepancy between present and future will be felt less keenly. When the adult world is less sympathetic, the group of his contemporaries is a source of strength and solace. (Parenthetically, Coleman (1961) regards adolescent society as a distinct subculture or what we have called a world of its own. Coleman is concerned over the anti-intellectual values of high school adolescents and their growing tendency to be cut off from adult society. Other authors share Coleman's concern over the isolation of the adolescent, some emphasizing the adolescent's need to react against adult authority, others, like Coleman, emphasizing the adult's failure to challenge the adolescent and offer him a chance to function in the adult community.) The frustration of having maturity so near and yet so far is eased by the assurance that he is on a par with his companions. Recall that the word peer means equal.

The group also helps the adolescent to master his uncertainty by providing clear guides to behavior, down to prescribing the clothes he will wear and the language he will use. Until he can evolve more mature criteria, he will lean heavily upon the crutch of group stereotypes. Like the gang in middle childhood, the group enables the individual to extend himself to a greater degree than he could on his own, while fostering self-control through its demands for self-restraint and cooperation. The group values the individual according to his merit and encourages him to maximize his assets for the good of the group. Finally, the group supports the individual in his opposition to parental control. From preschool on, the child has sensed the strength which lies in unity. Now he makes full use of his knowledge. "Everybody else is going to stay out until two, so why can't I?" With "everybody else" using lipstick, driving a car, or going on a weekend ski-party, the parent is placed in the position of championing an unjust, outmoded, minority viewpoint.

The adolescent's allegiance to the group is now strong enough to compete with parental ties. However, some authorities claim that the group does not stand in ideological opposition to the home and that the "battle of the generations" is fought only fitfully rather than continually.[12] They do not deny the difficulties, uncertainty, and tensions of the adolescent transitions, but they do deny that the majority of adolescents are primarily rebellious or alienated. The delinquent adolescents who are a concern to society and the lost and lonely adolescents who are described by novelists and discussed by commentators on the current scene are special subgroups which attract by the drama of their lives. Most adolescents are rebellious or alienated at times, but their lives are not governed by such feelings. Research surveys reveal that adolescents value personality traits of cooperation, helpfulness, self-control, dependability, and social skills, such as dancing and good manners. Boys do not like girls who are too sophisticated, giggly, and aggressive; girls do not like boys who are conceited, self-centered, and sloppy. As in middle childhood, the congruence of peer values with middle class standards is striking. The change in ideal sex role is also no cause for alarm. The boy adds poise in heterosexual relations and neatness to his middle

[12] See Campbell (1964), Douvan and Gold (1966), and Grinder (1964) for accounts of the alienation versus apprenticeship issue.

childhood values of athletic ability, assertiveness, and daring. Girls change from the "little lady" ideal of middle childhood to the ideal of being either extraverted, a good sport and companionable with boys, or glamorous, sophisticated, and fascinating to boys. Between twelve and fifteen years of age, then, the ideal image shifts—but hardly in the direction of becoming wild and sexually promiscuous. The anti-intellectual orientation of adolescence seems to be an exception to the rule of adolescent adherence to middle class standards. Parents value intellectual achievement highly, while adolescents consistently rank it low as an ideal and as a basis for popularity, prestige, or heterosexual attractiveness; adolescent girls are particularly wary of being considered studious by boys. Yet Coleman remarks that parents probably have a double standard, valuing academic achievement while wanting the child to be popular and successful with his peers. Douvan and Gold wryly comment that even teachers, principals, and superintendents value athletics and extracurricular activities as highly as academic achievement.

What mars adolescent groups is their rigidity and their demands for conformity. Adolescence is the high-water mark for group prejudice, when caste and class lines are sharply drawn and inclusions and exclusions are absolute. The picture is in striking contrast to that of middle childhood when an individual was included merely because he was handy or had a desired skill—or toy.[13] Coleman notes that the powerful and prestigeful "leading crowd" is composed of adolescents who exemplify to an extreme degree the class background dominant in a particular high school setting. Adolescents with different class backgrounds have considerably less chance of becoming popular and being the pacesetters for peer society as a whole. Cliques encourage exclusiveness, snobbishness, and intolerance—discrimination which may partly reflect the adults' concern with going with the "right" kind of person but is also the result of the adolescent's lack of generosity of spirit. Only friendship allows the adolescent to choose his companion solely on the basis of personal attraction.

The adolescent is also more conforming to the group than he has been before or will be again. Though adolescents may be united in their defiance of adults, they are conventional in their own peer culture—quite literally so, since they vigorously uphold the conventions of their chosen group. They champion fads in style and taste as if they were a matter of principle, and punish slight deviations with ridicule. It would not be correct to say that the adolescent sacrifices his individuality to the group, because he is not fully individuated; but he does pay a price for the protection the group offers him.

Thus the adolescent group retains some of the rigidity, absolutism, and concern with appearances which marked and marred the middle childhood gang. Perspective and flexibility, a realization that people should be evaluated in terms of personal worth, loyalty to the group without chauvinism, social commitments which transcend immediate group interests—all these lie in the future.

[13] Prejudice does not first appear in adolescence, since racial prejudice has been found as early as the preschool period (see Thompson, 1962).

Sibling Relations and Ordinal Position in the Family

The world of children contains siblings as well as peers. Adler championed the importance of ordinal position in the family as a determinant of personality, and his ideas generated a host of objective studies. After twenty years of research, the evidence was so contradictory that interest in the topic all but disappeared. Authorities alleged that Adler's observations had been disproved and were willing to consign the study of sibling relations to the scientific graveyard. However, in psychology as well as in folk tales, long-buried bones suddenly begin to whistle and get up and dance. After another twenty years of disinterest, ordinal position is again considered a significant factor in personality development. The renewed interest is due partly to the stubbornness of a few investigators who refused to abandon the idea and partly to new data which, wholely by accident, pointed to its relevance.

It will be instructive to compare Adler's clinical observations, begun in 1918, with the review of objective evidence by Sampson (1965) almost fifty years and at least two hundred empirical studies later. Sampson does not set out to report clearly established facts, since the evidence is still conflicting; rather, he presents his own educated guess regarding the importance of ordinal position in light of his study of a vast quantity of literature.

Both Adler and Sampson agree that ordinal position per se is not significant; i.e., it makes no sense to talk of the "psychology of the youngest child" as if position fated the child's personality. Rather, the size of the family and the child's position within it create special psychological environments, and the child's personality is affected by his efforts to cope with such environments. To illustrate: emotional intensity is likely to decrease and centralization of parental leadership to increase if a family has eight children rather than one. Or, again, the only child must continually contend with the "vertical" parent-child relation, unmodulated by "horizontal" child-to-child experiences. Both Adler and Sampson mention the sex and the spacing of siblings as significant variables (e.g., a boy with two brothers is in a different psychological situation from a boy with two sisters), and each in his own way stresses the importance of parental attitudes and child-rearing practices. The spirit of the two authors is somewhat different; Sampson emphasizes environmental factors, while Adler gives the child a more active role in shaping his destiny.

Because no two children are exactly alike and no two families are exactly alike, one should expect variability in children occupying the same ordinal position. However, both Adler and Sampson maintain that there is sufficient similarity in the psychological situation of children in the same ordinal position to expect that certain common personality characteristics probably—although not inevitably—will appear.

Adler depicts the oldest child as a dethroned king, initially the center of parental affection and attention, then forced to contend with the reality of being just another child. None of his siblings will feel displacement so keenly, since none will know the pleasures of being an only child, at least for a time. However, the oldest child is granted the status accorded to the firstborn; the custom of primogeniture is preserved psychologically by giving the eldest special responsibilities and rewarding him with parental pride. Adler states that the firstborn is drawn to the father, since pride in meeting responsibilities is more characteristic of the man. Misunderstanding and bitterness over not being loved exclusively make the relation with the mother difficult. Growing up with a sense of paradise lost, the oldest child admires the past and is pessimistic about the future. Authority and power are crucial to him, and he is likely to exaggerate the importance of rules and regulations. Although he tends to be a leader, he has a strong conservative bent and a nagging concern about being overthrown by his followers.

Sampson's picture enlarges Adler's but is similar. The parents of the firstborn are ambitious for him but inexperienced, confused, and anxious in their handling. Parental influences and the child's tendency to identify with parents are strong. When a sibling arrives, the child is no longer the center of the stage, and he must learn to cope with a rival for love and power within a framework heavily weighted with adult values. As he develops, his internalized parental ambitiousness drives him on to intellectual achievement, while his relation to siblings makes him chary in regard to sociability, play, and companionability. Only in times of stress does he turn to his contemporaries, perhaps for emotional support, perhaps for that sense of assurance and perspective which comes from sharing one's uncertainties with another.

Objective studies reveal that the firstborn is more likely to experience a sudden shift in status within the family, especially in regard to attention and affection, and to be in conflict over dependence (i.e., having parental love) versus independence (i.e., living up to parental ambitions). He is apt to be intellectually eminent as an adult, especially in the scientific (detached and impersonal?) fields, but he is not likely to be sociable, outgoing, empathetic, and sympathetic. He expresses aggression less and seeks the company of others primarily when anxious. In terms of our previous discussion of leadership, he is more task oriented than socially oriented.

The secondborn child must face his own constellation of advantages and disadvantages. Adler notes that, though he may be displaced subsequently, he will be spared the bitterness of being dethroned. His parents will be less ambitious for him and more relaxed and certain in their handling. His interaction with an older sibling provides direct access to the world of childhood, with its special affects, values, and sources of status. In his older sibling he has an easier model and pacemaker than his parents are. However, his pride—or what Adler prefers to call his need for superiority and power—is bruised constantly by the fact that his sibling is more mature, more skillful, more capable than he. As other siblings appear, he may suffer parental neglect, being neither the big boy nor the baby of the family. The compensatory desire to be first becomes central. Life is a race and he must travel at full speed. Unlike the firstborn, he does not accept the authority of established rules and regulations; on the contrary, he chafes at strict leadership and believes that any power can be overthrown. Concomitantly, he is less strict with his feelings, especially hostile ones. Adler, always mindful of the complexities of personality, maintained that the second child did not necessarily have to be the hotheaded rebel; he could undermine subtly by praising an authority extravagantly and creating an image which is impossible to realize. On the positive side, Adler (wrongly, as we have seen) claimed that second children are often more successful than first. On the negative side, they suffer from envy and the temptation to set impossibly high goals for themselves, thus insuring their downfall.

Like Adler, Sampson stresses the advantage of having a sibling who is a "closer, (more) manipulable, less powerful" model than parents. The second born, who is less involved in parental affections and ambitions, suffers less from their threats and withholding of love. Generally, he is more peer oriented than adult oriented, more sociable, outgoing, and companionable than his inward, uncertain, relatively isolated sibling. He has greater self-confidence and self-reliance in his manipulations of others. In direct contrast to the elder sibling, he withdraws into himself when difficult problems and choices arise, rather than seeking out others. Thus, Sampson lists more assets and fewer vulnerabilities than does Adler and does not emphasize the driven quality of the secondborn, perhaps because studies indicate that the firstborn has a stronger need to achieve and actually does achieve eminence more frequently.

According to Adler, the youngest child has the supreme advantage of never being dethroned, although, as the baby of the family, he tends to be spoiled and faces the problems of the pampered child (see Chapter 3). Unlike the only child, he is surrounded by powerful siblings who can set a pace for him or overwhelm him. Typically, he rises to the challenge and determines to outdo all the rest. In fact, the theme of the disadvantaged youngest son who overcomes apparently hopeless odds is a universally popular one in fairy tales, according to Adler. The youngest cannot afford to be as directly challenging of authority and as overtly angry as the middle child–the odds against him are too great. Instead he becomes more subtle and manipulative, using his understanding of human nature to maneuver others into doing his bidding. Tom Sawyer's adroitness in getting other boys to whitewash the fence for him illustrates Adler's point perfectly. The danger is that he may come to use people selfishly with little regard for them as human beings.

Adler observes that the only child is in the most difficult position because his undiluted involvement with parents leaves little opportunity to develop social feelings through interacting with siblings. He tends to be pampered, although sometimes he is unwanted and therefore hated. If spoiled, he will have little chance to revise the idea that he is the center of the universe and will be handicapped when faced with the problems of social living. If he is angry with his father, he is directly opposing the most formidable foe of his young life in an unequal, frightening contest. Since objective studies (see Sampson, 1965) often lump only children and firstborn together for reasons more justified on statistical than on psychological grounds, we cannot tell whether Adler's rather pessimistic picture of the only child is justified. However, there is no clear evidence that only children have more problems or are more psychologically disturbed than others.

In general, Adler's clinical observations have fared reasonably well, with the exception of the relative ambitiousness and success of first and secondborn children. The choice of theoretical accounting for clinical and objective findings has remained a matter of personal preference. Adler's explanation is in terms of a universal and fundamental drive for power, superiority, and mastery in order to overcome the equally universal and basic fear of helplessness and inferiority. Other hypotheses offered by social learning theorists, cognitive theorists, and role theorists receive attention in Sampson's detailed review and need not concern us here. The data on ordinal position are still too inconclusive to constitute a critical test of any one approach.

Comment: Sibling Attachment. We have seen that siblings can serve in a variety of capacities—as rivals, pacemakers, models, socializers. Yet, we know little about the attachment of siblings. Most studies of early friendships, sympathy, and empathy have been done with peers. One kind of relationship described by Adler and others in which the older sibling assumes something of a parental role—helping, guiding, teaching, feeling responsible for the younger ones—might be explained in terms of the older child's identification with the parent or in terms of a technique for obtaining parental approval. There is a more intense attachment which is baffling because it defies our usual notions concerning human behavior. A mother remarks: "I don't know why Tim (a two-year-old) keeps on tagging after Lanny (his four-year-old brother) who is always pushing him down, grabbing his toys, and making him cry. But two minutes later, there he is, tagging along, imitating everything Lanny does and says, happy as can be." Such reports are perplexing because the behavior runs counter to theories and expectations. We expect attraction to develop on the basis of pleasure, not pain; and we expect it to be facilitated by sensitive responsiveness to the child's needs, not by rank indifference to them. Because the conditions for developing a strong attachment are so unfavorable, its appearance is all the more fascinating and worthy of investigation.

The Influence of Family and Groups

What is the relationship and relative potency of the family and of peers as the two principal influences in shaping personality development? The general picture is as follows: The family is central during the first six years of life. Intrafamilial relations are intense, sustained, and varied. Peer relations are an unstable foreshadowing of things to come. In middle childhood friendships and group relations become sufficiently stable and intense to serve as nuclei for the development of distinct personality characteristics. The adolescent's concern with emancipation from the family and with preparation for adulthood causes a shift in which friends, cliques, crowds, and gangs may become more important in determining his behavior than the family. Peers are the bridge to adult life which, in middle class society, requires that social relations be primary and family attachments secondary. The import of family and group, therefore, will depend on which personality characteristics are being studied at what stage in the child's development.

There is a knottier problem which concerns the relative potency of family and of group characteristics in determining peer relations. Strictly as a heuristic device, let us present two "nothing but" propositions—one, that peer relations are nothing but a reflection of family characteristics, the other, that peer relations are nothing but the result of group characteristics. In the first proposition, group characteristics have no influence in determining the individual's relations to his peers; in the second the family has no influence.

The Case for the Family

Many empirical studies[14] support the proposition that the family plays a decisive role in determining peer relations. The combination of early, emotionally charged experiences in a cultural setting which gives the family primary responsibility for socialization makes the family the logical arena for establishing basic social attitudes. The data show that parents who are child-centered, democratic but firm in their discipline and harmonious in their own relations, have socially active, assertive, and responsible children. As parents deviate in the direction of being cold, rejecting, authoritarian, neglectful, indulgent, intrusive, dominating, and disharmonious with one another, the child's social relations also deviate from the middle class ideal. Baldwin (1949) found that a warm, democratic home environment, in which emotionally supportive parents offered explanations, encouraged experimentation, and discouraged infantilization, produced socially outgoing children who displayed both friendly and hostile or dominating behavior, were successful in their undertakings, and enjoyed high status among peers. Indulgent parents produced the opposite effect—shy, apprehensive children who were physically unskillful in nursery school. Baldwin's subjects were three-to five-year-olds, but compatible results have been obtained in middle childhood and adolescence. In a study of third- to sixth-graders, Hoffman (1961) found that positive assertiveness and influence in the peer groups, as well as liking and being liked in return, were related to a family in which the father was more powerful than the mother, was firm in his discipline, and a warm companion. The results were more significant for boys than for girls. Anderson (1946) concluded that parents of adolescents who were leaders respected their children as individuals, did not protect them from the normal responsibilities and risks of life, were not restrictive in their control, expected the children to use judgment, and included them when making decisions. Just as Baldwin found that deviations from the democratic model produced comparable deviations in peer relations in nursery school, Clark and Van Sommers (1961), using nine- and ten-year-olds, found poor peer relations in children whose home life was marked by contradictory parental demands, usually due to a domineering maternal grandmother and the mother siding against the father. McCord and his co-workers (1962) found that adolescent boys who shied away from adults and who were very dependent on peers for guidance had rejecting or neglecting parents who encouraged them to leave home and failed to exert supervisory control over them. Peck and Havighurst (see Chapter 5) found a high correlation between parental behavior and different degrees of maturity of social behavior.

[14]For a general discussion of studies of parental and home influences which includes research on peer relations, see Thompson (1962).

The Case for the Group

Such consistent and impressive findings would seem to justify the position that the child who has a good foundation in the home will be successful in his peer relations. Before regarding the matter as closed, let us look at the proposition that group forces are so potent they determine peer relations, regardless of the child's family background.

We have already encountered examples of the power of the peer group in determining the behavior of the individual member. Even among preschoolers, Merei found that an erstwhile leader had to alter his behavior in the face of group pressures, while Freud and Dann's study revealed the children's ability to develop a high degree of self-regulation and humanistic concern for one another with almost no adult guidance. Hartshorne and May (see Chapter 5) found that moral behavior was determined by the group to a striking degree; honesty or cheating, for example, was more characteristic of the class the child was in than of some hypothetical trait of honesty he possessed. In adolescence, the bright girl often decides to hide her intelligence regardless of parental reaction, so that she will not alienate her friends or be unattractive to boys. This is also the period in which members of a clique continue to be together whether or not parents approve of the association.

The forces generated by relations between groups are also potent determiners of peer relations, especially if the groups are competitive or hostile. Sherif (1961) vividly describes the change from a "my" to a "we-they" psychology which occurred when two groups of campers were put in a situation of hostile competition. Each group became more tightly knit, more self-conscious, more vigilant, more prone to inflate its assets and disparage the other. In time, negative attitudes and derogatory stereotyping of the "villainous" outgroup were crystallized. The reaction is reminiscent of the civilian population in wartime, when personal dedication can be uncommonly high, while prejudice can be unusually blind and vicious. In fact, the development of prejudice is the most striking product of intergroup hostility.[15]

When an adult leader is added who consciously manipulates the group, he has similar power to affect peer relations regardless of the family background. The study by Lewin, Lippitt, and White (1939) is a classic in this field. One group of eleven-year-old boys had a democratic leader who actively participated with them on various projects, achieved a balance between group freedom and responsibility, and produced a friendly, cooperative relation among the members. The authoritarian leader, who remained aloof and dictated policies and procedures, stimulated irritable and aggressive interactions among the boys, two of whom were so unhappy over their role as scapegoat that they left the group. The laissez-faire leader, who was passive except when the boys approached him for information or guidance, also created an atmosphere of irritability.

In a similar vein, Thompson (1944) exposed two groups of nursery school children to two different atmospheres; in the first the teacher was friendly but detached, while

[15] For a review of the literature on intergroup attitudes with special emphasis on prejudice, see Proshansky (1966).

in the second the teacher was friendly and involved in the children's activities. The children in the second group became significantly more mature socially, more constructive in the face of failure, showed more initiative and leadership than the children in the first group who in turn evidenced greater social disharmony through threatening, attacking, ignoring, or rejecting their peers. Both studies imply that, regardless of background, the contemporary situation as established by an adult leader can affect peer relations significantly.

What can we conclude from all this evidence except that the group per se, or the group as manipulated by an adult leader, exerts irresistible pressures on the individual, regardless of how warm and democratic or cold and authoritarian his home life? And have we not thereby impaled ourselves on the horns of a dilemma by justifying both of two incompatible propositions? We have, but only in order to learn in the process of disengagement.

Resolution of the Dilemma

Investigators rarely champion a "nothing but" stance, nor do their data justify one. Typically, they have been concerned either with the family or with the group, and their research has reflected this interest; rarely have studies been designed to pit one influence against the other. Let us take an example. In order to have a stringent test of the primacy of the family, we must be sure that the school environment (which is the social setting of most of the research) is significantly different from the home. Concretely, children from democratic, authoritarian, and laissez-faire homes should be exposed systematically to democratic, authoritarian, and laissez-faire school settings. If the child from an authoritarian home continues to act in a docile manner in spite of a friendly, stimulating, democratic teacher, or if the assertive product of a democratic home remains undaunted by a strict, authoritarian teacher with little patience for such behavior, we can be sure the home is as potent as its defenders claim. This is just the kind of definitive information we do not have. Moreover, there is a real possibility that the child in school is treated in a manner roughly comparable to that in the home. Parents try to select schools which are congruent with family atmospheres, and teachers themselves share middle class values. It may be that the bright, assertive, creative preschooler is as much a delight to his teacher as to his parents; that the less interesting, conforming child is rewarded for being good or is ignored; and that the troublemaker is as exasperating to the teacher as he has been to the parents. In sum: we do not know whether the family is a potent determiner of peer relations in its own right or whether the child's subsequent social environment subtly conspires to perpetuate the patterns established within the family. (Investigators of peer relations have been equally uninterested in varying family background, but we do not want to belabor the point.)

We can learn two more lessons from our "nothing but" caricature. The first is a general point about personality. Personality is complex and can be studied legitimately from different vantage points. Obesity, for example, can be studied by the physiologists, by the psychoanalysts delving into inner conflicts, by the social learning theorist examining familial and peer attitudes toward obesity, and by the social psychologist exploring the role assigned to obese children in different classes or societies. The division of labor is arbitrary but essential, since each approach is complex; working out the interaction of all the variables at all levels lies beyond the present limits of our science. We can resolve our present dilemma by recognizing that both home and group can affect peer relations significantly. We have yet to learn the limits of each influence and the effects of having them in opposition, in harmony, and tangential to one another at different stages in the child's development.

The second point concerns deviant development and psychopathology. Far from being the key to normal development, each "nothing but" approach has ominous overtones. The Freudian position has been misinterpreted as claiming that all peer relations merely recapitulate relations in the nuclear family during the first six years of life. In school, for example, the teacher is nothing but a mother figure and peers are nothing but rivalrous siblings. The correct interpretation of the psychoanalytic position is that experiences within the family, because they are early and highly charged emotionally, are the most potent determiners of the child's personality. They are never outgrown in the sense that they gradually fade away, since they continue to exist in the unconscious. They also determine significantly the child's reaction to subsequent extrafamilial relations. The continuity of development requires this to be so, for the present can be assimilated only in terms of the past. Having established a special relation to the mother, the child initially interprets his relation to other women in similar terms. However, the Freudians also maintain that other women are not mother, the classroom is not the family. The healthy child has sufficient reality contact to grasp the differences and respond accordingly. He can learn from experience. If he can interpret a realistically different situation only in terms of familial experiences, his grasp of reality is dangerously weak. In an extreme form, the "nothing but" approach is evidence of neurosis. It is only the disturbed child who interprets a low grade only in terms of loss of love or views an athletic contest as a terrifying battle against an enemy determined to mutilate him.

The peer-dominated child would be equally unfortunate. His behavior would be totally determined by group pressures and group leaders. He would be kind and altruistic, prejudiced and destructive, submissive and conservative, as he moves from setting to setting. His only stability or continuity would be derived from the stability and continuity of his milieu. Such an image would be difficult to defend as an ideal of healthy development. When clinicians find a child who is exceptionally facile at making social adaptations regardless of their nature, they suspect that he is failing to develop an identity of his own (see Chapter 9). He has no core values, no convictions, no attitudes which are peculiarly his own regardless of setting, even regardless of his behavior in a setting; e.g., a child who has core values may cheat because of group pressures, but he condemns the behavior as incompatible with his basic beliefs. In other deviant conditions, the individual does not lack identity so much as affection and social feelings. Again, he is highly adaptable and even socially charming and could be mistaken easily for "a good guy," "a prince of a fellow," "an all-American type." Yet his shallowness of feeling for others makes him capable of "conning," manipulating, betraying, or even destroying others with a minimum of concern or guilt. Once again, sheer cognizance of and responsiveness to group-sanctioned behavior is no proof against the development of severe personality defects.

We would like to make a final point, leaving "nothing-but" propositions aside but keeping their lessons in mind. It is a well-established fact that the individual tends to conform to the behavior of the group. At some undetermined point he probably develops a need to conform. Such a need has been demonstrated in adults. Individuals constantly scan the social scene to determine what comprises socially acceptable behavior and try to act accordingly. Socially acceptable behavior is reassuring; wide deviation from the group causes uneasiness. The development of the need to behave in a socially desirable way is an important area of study in its own right.

Having said this, we can now distinguish behavior which is primarily a manifestation of a need to conform from conforming behavior which serves to advance the child's emotional and cognitive development. We have in mind Sullivan's concepts of mutuality, cooperation, and compromise, Adler's concept of social feelings, and Piaget's theory concerning the ability to understand another person's point of view (see Chapter 6). None of these authors interprets social behavior as a mere reflection of group norms or a seeking of safety through conformity. Instead, the process of accommodating to the group forces major revisions in or adds new dimensions to the child's personality. To take Piaget's study of children playing marbles (see Chapter 5) as an example: the child was not learning how to conform to the rules of the game; he was gaining insights into the nature of social regulation of behavior. We also must distinguish mere conforming from a personal commitment to group-engendered values. The psychoanalysts distinguish imitation, which is a surface reflection, from identification, which is a more abiding, self-perpetuating modification of personality. Other psychologists use the term "ego involvement" or "identity" to signify a deep, personal commitment to ideals, values, attitudes, and opinions. We must make a similar distinction in regard to the individual's response to the group. One child may conform to group life in a summer camp and all but forget the experience in a month; another may discover a feeling of belonging and camaraderie which will become a nucleus of social development for several years. In sum: conformity to the group can be "nothing but" conformity; it satisfies the need to be acceptable to peers and avoid the anxiety of being different. But accommodating to the group can also be the vehicle for effecting significant reorganizations of the child's intellectual and emotional life and for providing new sources for deeply personal commitments.[16]

Conclusions

The special contributions peer relations make to personality development grow out of the fact that peers are equals, that they relate on the basis of shared interests and esteemworthiness, and that their interaction is marked by give and take. Peer

[16] For a searching analysis of this distinction in terms of role taking and role playing, see Jackson (1970).

relations are especially suited to correcting the emotional and cognitive egocentricity of early childhood–the child's exaggerated ideas about his goodness or badness, and his inability to see others as independent individuals. His self-knowledge is enhanced by the continual feedback from peers, which ranges from a joking taunt concerning his dress to a biting critique of his personality. Gradually he arrives at a realistic perception of his assets and liabilities and learns to value himself in terms of his proven worth. Friendships teach him to value individuality, to be concerned with the welfare of, to accommodate to, and be loyal to another. In doing so, they prepare the child for the heterosexual relations of adult life. The child comes to identify himself with a group and learns to find his way around an organized social setting, thereby preparing himself for living in the adult world of social groups. He also learns the specific social techniques of cooperation, competition, and compromise, and develops his style of exercising power over others.

It would be a mistake to view peer relations solely as correcting the errors of the earlier, family-centered years. The child does not set aside family values as he becomes peer oriented; these values are championed in middle childhood and constitute the implicit context within which the adolescent structures his future. Many of the personality assets developed in the home continue to be serviceable in the group. For their part, peer relations have special liabilities. Except for friendship, they can be marked by an intolerance of individuality, stereotyped standards, cruel treatment of deviant behavior, snobbishness, prejudice–in short, a narrow group-centeredness which is the counterpart of the child's earlier egocentricity. Such rigidity and narrowness is understandable in terms of the developmental status of the child and adolescent; yet their undue prolongation runs the risk of blocking the achievement of social maturity. While it is essential that the child grow into the group, it is essential that he grow out of it. Since the adolescent group still lacks flexibility in its standards of behavior, an appreciation of personal worth, and a social vision broader than its immediate concerns, the more mature forms of group behavior must develop at some later date.

9 Play and Work

Play

The Myth of Play

Play often means to the parent that the toddler or preschooler is enjoying himself doing something trivial. This is a myth. The parent is thinking in adult terms: play is the opposite of work; work is earnest and essential; play is happy and dispensable; early childhood is golden because, before school spoils things, the child has nothing to do but play all the time. Parents are also realistically self-centered. The attentive mother may be relieved when the special demands of infancy are past and her self-sufficient toddler or preschooler gives her increasingly long periods to attend to household chores, to socialize, to do what she wants. The release is not total—she must be on hand to protect the child from physical harm, to protect vulnerable objects from the child, and to referee arguments and fights between children—but whatever freedom there is, she treasures.

The playing child generally means two things to the parent: he is in a state of physical well-being in that he does not need to be fed, toileted, or medicated; and his behavior does not need to be socialized. Play seems unimportant to parents because it is irrelevant to "*No-Don't*," "*Good boy-Bad boy*." It does not challenge or even engage social values. Expressed in another way—play alters the balance of choices available to mother and child. Even the most liberal mother cannot choose not to socialize her child. A toddler cannot poke a pencil at a baby's eye or urinate in the supermarket. And, ultimately, the most benevolently socialized child cannot choose not to conform. Socialization is a requirement. But a mother can choose not to play with her toddler or preschooler. She may want to play because she enjoys it herself or because she wants the experience to be constructive for the child, but she is not required to. As for the child, play is like initiative: it comes from within. He is in his own world doing whatever is most meaningful to him. Play is never something a child must do or must be made to do. He chooses to play because play is natural to him.

As we shall see, play is neither trivial nor necessarily fun. Parents who think so believe in a myth. But their misconception allows the growing child to live in his special world and deal with life on his own terms. When we understand play and fantasy better, we will consider whether the myth should be destroyed. Are parents missing a unique opportunity to make their children brighter, more socially responsible, more creative than they are? Or are they giving children an essential respite from coping with reality and an essential freedom to be themselves?

There is another reason for claiming that play may not exist. Play has been given such diverse functions by psychologists that it is difficult to discover a unifying thread. With no pretense at comprehensiveness, one could list a baker's dozen concepts of play—it is nonpurposive, it is practice, it is mastery, it is cognitive learning, it is social learning, it is catharsis, it is excess energy, it is creativity, it is exploration, it defines reality, it establishes sex role, it provides erotic gratification, it is the child's work. While some differences are complementary, there is enough disagreement to make a consistent conceptualization impossible. Perhaps play refers to a loosely associated family of concepts, no one of which is essential. Berlyne (1969) concludes his extensive review of play in animals, children, and adults with a recommendation that psychologists abandon the category altogether. It is possible that once we understand each component, play will vanish. Instead of saying the three-year-old is playing, for example, we would say he is exploring and mastering a novel object which periodically gives him certain erotic gratifications and cognitive pleasures of recognition. Which is a dour prospect. But if Freud could spend his professional lifetime writing about sex without once being "sexy," we have no reason to expect the analysis of play to be fun.[1]

Play and Exploration. We do not plan to abandon the concept of play; instead, we shall relate play (and, for that matter, work also) to initiative (see Chapter 3). In the first few months of life, the infant begins to take matters into his own hands. When he no longer is engulfed by the distress of physiological needs, his focus is outward, on the world of people, on the world of objects, on the actions of his body. By the time he is one year old, he is an avid explorer. The awesome task of learning about the world is matched by the intensity of his desire to know. His motivation seems intrinsic; the environment can facilitate or impede his progress, but the natural eagerness to learn is his. His characteristic affect is interest, if interest can be called an affect. In observing the exploring one-year-old, one is not struck by the hedonistic element. Intense pleasure can enter the picture as can intense frustration and rage, but these come and go. The child is not seeking pleasure or avoiding pain; he is doing what is most compelling, most interesting, most meaningful to him at his stage of development.

Play and exploration share important characteristics. Play is intrinsically interesting, absorbing, compelling. Pleasure weaves in and out but is a byproduct rather than a hallmark or goal. Some psychologists equate play with exploration by calling it,

[1] For a review of objective studies of play, see Hurlock (1964). For a review of theories of play, see Slobin (1964), Gilmore (1966), and Klinger (1969). The relation of play to general developmental trends is presented in Hartley and Goldenson (1957).

"the child's work." We prefer to keep the two concepts distinct because of certain qualitative differences. More specifically, the toddler and preschooler are not continuously exploring and assimilating the environment; periodically they take time out to indulge in a more inward kind of activity. Just as the busy executive or mother may drift off into daydreams which are a more intimate expression of their personalities, the child alternates between striving to cope with reality and an activity which is pursued for satisfactions other than those of mastering the situation at hand. We shall elaborate on this distinction subsequently.

We do not mean to imply that because play is more inward it is less important than exploration. Because we can daydream whatever we wish, it does not follow that daydreams do not matter; on the contrary, play and daydreams are important just because they can reflect private motives and longings. In play, the child is free to think and feel and act as he pleases, relatively unconcerned with whether parents will smile or frown, and relatively indifferent to the realistic properties of the physical world. In sum, play eludes the constraints of social acceptability and of mastering the physical environment. It lies beyond socialization and exploration. It is the most personalized statement a child can make about himself.

Descriptive Overview. Since many different kinds of behaviors have been called play, it is best to orient ourselves with a descriptive account of its development. In infancy, play is usually applied to sensory and functional pleasures, a delight in sheer perceiving and doing. In the preschool years, dramatic play takes over. The child proliferates characters and plots which he acts out with varying degrees of obliviousness to reality. In school such play wanes and, after a couple of years, is replaced by games with rules. Some, such as sports, are public and congruent with the adult world; others, such as hide-and-seek, Red Rover, Red Light, and May I, belong to the world of children. Although these three types of play mark a forward progression, successive stages often incorporate features of preceding ones, so that games with rules can also involve pleasure in functioning and/or dramatic play.[2]

Sensory Pleasure and Pleasure in Functioning

We shall first consider the earliest manifestations of play. In infancy, one form of play consists in what Stone and Church (1968) call the "sheer appreciation of sensory experience." If infancy is a time of distress and bafflement, it is also a time of great delight in simple, primitive experiences. The body is particularly rich in positive sensations: the baby is delighted when his skin is stroked or when he is tickled and by his own activities of sucking, feeling textures with his fingertips, and kicking his legs. Thus, play is viewed as the infant's seeking of these bodily sensations and indulging himself in them. Parents unwittingly confirm this concept when they label the child's masturbation "playing" with himself.

This form of early play seems only natural: since man is often regarded as a pleasure-seeking organism, there is no mystery about trying to prolong intensely

[2] For a brief, vivid account of the stages of play, see Stone and Church (1968).

positive sensations. But there is another form of play which is more puzzling. It involves functioning for the pleasure of functioning. Watch an infant closely and you will see a good deal of what has been called curiosity, exploratory behavior, mastery, competence, or what we have called initiative. Yet the progression of his behavior is not relentlessly onward and upward. Having mastered a given problem, he can be observed returning to the activity over and over for the sheer pleasure of repetition. A two-month-old, after learning to throw his head back in order to look at familiar objects from a new angle, repeated the movement with ever-increasing delight; a seven-month-old who learned to bat down a piece of cardboard to reach a desired toy, proceeded to bat it down over and over for pleasure. A three-year-old will build a tower and knock it down endlessly, delighting in the destruction. As many exasperated parents will testify, a child who initially asks Why and attends to the answer can become dazzled by the magic of repeating the word and getting a response, the content of which matters not in the least.[3] The learning process per se is over, and the infant engages in what Piaget calls a "happy display of known actions."

Now we do stand perplexed. Why should such behavior exist? We can understand exploration and pleasure-seeking, but why this strange combination of the two whereby behavior, having served its mastery function, becomes pleasurable in itself? Functional pleasure introduces a new element into our thinking about personality development. From our discussion of initiative and cognition (see Chapters 3 and 6), one might picture the child as systematically exploring the world of people and things and progressively achieving an understanding of them. Generally speaking, the human is constructed to seek new things to master; recent acquisitions are soon put in the service of new ones. Now we have an exception to this constant search for novelty and drive to master reality. Instead of being bored by the familiar and seeking novelty, the infant-toddler delights in repetition and temporarily suspends his venturing out.

One explanation of early play is in terms of surplus energy (see Lehman and Witty, 1927). After the serious business of physical survival is assured, the infant has energy left over to be expended in the trivial activities of play. Our contention, however, is that the hungers of the brain (i.e., its need for varied stimulation) are equally as important as the hungers of the stomach (see Chapter 3); assigning priority to the satisfaction of physiological needs is unwarranted. Other theorists regard play as preparation for adult life. Deprived of inbuilt guidance by instinctual patterns, the human infant must continually engage in activities which rehearse the complex behaviors necessary to adult living. Such theories do not recognize that play appears after mastery has been achieved. Even if tasks mastered in the past prepare for tasks to be mastered in the future, why should the infant savor victories already won?

We find Piaget's explanation of early play most satisfactory. He maintains that, while the human is constantly striving to adapt to its environment, its attempts at mastery are not uniformly successful. Especially in infancy, the processes involved in learning are in disequilibrium. We have seen one example of this disequilibrium when we discussed imitation (see Chapter 4): the child acquires the surface manifestations rather than a true understanding of what he does. Play, according to Piaget, results

[3] Our examples are taken from Piaget (1962).

from a disequilibrium which is the counterpart of imitation. Instead of taking on too much reality to be understood, the infant disregards reality and luxuriates in his acquisitions. The questions, What is real? and What is allowed? are not asked; the child is satisfied with reiteration.

This disequilibrium should be translated into the terminology of Piaget's theory: imitation and play represent extreme manifestations of the two basic processes of cognition–accommodation and assimilation (see Chapter 6). Accommodation is a taking-in process through which the organism actively adjusts itself to a new situation. The infant must accommodate to the breast before he can suck; the graduate student must have the proper mental set when he confronts a new idea. After the information has registered, it must be assimilated into the existing body of accumulated understanding. In the process, this accumulated understanding is enriched or even reorganized. In adaptation, assimilation and accommodation are nicely balanced. However, this harmony is a developmental achievement. Especially in infancy and the preschool years, the child tends to go from one extreme to another. In imitation, the child is focused on reality and accommodates without sufficient assimilation. His behavior literally conforms to an external event, but he does not fully understand what he is doing. In play he is unconcerned with reality and he deals with it arbitrarily, capriciously, cavalierly–in other words, playfully. His focus is inward, on his accumulated understanding with little concern for accommodating it to reality.

To what purpose? In play does the intellectual machinery temporarily slip a cog and spin meaninglessly around before getting back to its basic function of adaptation? As we shall see, the answer is No. Play performs a variety of functions important to personality development. However, these functions are more readily understood in the context of dramatic play, which we shall now discuss.

Dramatic Play

Sensory and functional pleasures may be of interest to the professional infant-watcher, but they do not embody what ordinarily is meant by play. At best, they seem to be precursors. However, from toddlerhood through early childhood play runs rampant and is easily recognized. Its beginnings are simple: a three-year-old will suddenly drop on his hands and knees and say, "Grr; I'm a lion." By the time he is six years of age, he will be elaborately dressed as a cowboy and acting out complicated plots of daring and heroism. The period of dramatic play or make-believe is in its heyday. The child is free to be anything he wishes and to change roles at will. In like manner, a physical object can freely change its nature. Such "games of illusion" are the antithesis of reality. Listen to a bright four-year-old talking to his friend. "I'm going to have two kites. One has a (picture of a) car on it and one has a jet. And the one with the jet goes up in the sky and it goes so fast it really is a jet and it goes zoop in the sky. And the kite with the car crashed. And his mother bought him another kite. She bought him 100 kites and they were all red, and he had 100 arms and 100 legs." Note the many transformations: a kite becomes a jet, "I" becomes "he," the body sprouts 100 arms and legs. Whether the boy actually believes kites could become jets is not certain. He had just flown a kite for the first time, and he may not yet be sure of their nature. His uncertainty may also have been increased by the fact that recently he was fantasying about a boat which could become a car and a plane which could be a boat and was told by his parents that both were possible. The kite-into-jet transformation may well occupy that twilight zone between belief and pretense which is so characteristic of the preschool period. However, the possibility of having 100 arms and legs was obviously ridiculous to the boy. The point is that he was unconcerned about all the transformation. Both solid facts and ambiguities were treated with the same indifference. He was not trying to understand reality; on the contrary, he was suspending the effort. And how rich ideation becomes when reality constraints slacken!

The Development of Dramatic Play. A descriptive overview of the development of dramatic play will be helpful to our discussion of its mechanisms and functions. Pretending can readily be observed in a two-year-old who pretends he is various animals and people he has encountered. He becomes a kitty, drives a car, cares for the baby, cleans house. The activity is highly motoric; he pretends with his whole body. He also can hold a pretended conversation: "Hello, Daddy; come home soon and bring some bread. Bye." The fleeting episodes arise from some inner prompting, are acted out, and are finished, with little concern that others participate or even notice them. The toddler is oblivious of reality; twigs, stones, and bits of grass can serve as props for a teaparty as easily as toy plates, cups, and saucers. The three-year-old can sustain the pretense longer, for example reliving fragments of a trip to the zoo or to the doctor. He becomes more social, pretending with another child and even exchanging roles so that a follower becomes a leader. The typical picture is still a fluid transition from one isolated theme to another, but there are some exceptions. The imaginary companion flourishes around three-and-a-half years and can become a richly elaborated character present for weeks or even months. Or the child himself can pretend for a few weeks to be another person or an animal, such as a kitty or a baby.

The play of the four-year-old is more sober. Lions and tigers give way to truck drivers and nurses, as the child plays out his version of adult behavior. He likes toys which mimic real life, such as cooking sets, electric shavers, medicine cabinets. He also is beginning to need realistic props, although a single item, such as a hat or a gun or a pair of shoes, can stand for an entire costume. In spite of the increased realism of his play, instability still exists, and a block can become a plane or a train or submarine in quick succession. The trend toward realism continues during age five and complete and realistic costumes are required; not only dress, hat, and shoes, but also lipstick and nail polish. A coherent, realistic sequence can be acted out requiring a number of children to assume and sustain special roles; e.g., a grocery store episode can be enacted, with shoppers, grocery man, and cashier. Girls enjoy sequences of realistic doll play—dressing, undressing, feeding, bathing, spanking, and doctoring their babies endlessly. Boys tend to games in which action and noise are more important than complexity of plot. Children also begin to act out stories they have heard.

For the six- and seven-year-old, play themes can be sustained over a period of days. More than two children can be involved, and roles are interdependent, with give and take rather than the former leader-follower arrangement. For example, a doctor in outer space may have little to do until the captain is wounded by a laser ray from the Mars creatures; but then he rushes in, sews the victim up, gives him a special transfusion, and saves his life. The play follows reality constraints: one person cannot become another, and the dead must remain dead, at least for that episode. For the eight- or nine-year-old, dramatic play becomes more a consciously assumed set, introduced by "Let's play... ." Although it is now more richly elaborated and coherent than ever before, it is also more circumscribed. Sports and games, as well as school work and social interests, take up an increasing amount of time. Soon they will supersede dramatic play and leave the child to act out his adventures in the realm of imagination and fantasy.

We also have information concerning the content of dramatic play. Domestic themes, such as playing house, are favorites, along with animal activities, playing nurse, store, policeman, fireman, and themes of killing, dying, and giving parties. However, in viewing children's play from two to eight years of age, one is not struck by the thematic development as much as by the change in the structure of the activity itself. Play proceeds along the characteristic path of becoming increasingly sustained, complex, and organized. In its interpersonal aspects it follows the familiar line from egocentricity to cooperativeness. As play becomes more verbal it can be at the same time more ideational (since words are more versatile than physical action) and more social (since words are the principle vehicle for communication). Play becomes progressively more realistic. At first the constraints of reality are more or less ignored; with time, magic transformations are replaced by realistic object identity, social reality, and logic. If magic enters, it is purposely introduced. And, finally, play becomes a consciously assumed set. Play flows in and out of the preschooler's activities. It is sufficiently discrete not to be identified with reality, but the line of demarcation is so thin and wavering that pretense and reality can become confused (see Chapter 6). By middle childhood play has become a drama which children agree to enact.

Before leaving our survey, we should note a disappearance. Pleasure has vanished from play. In the stage of sensory and functional play, illustrative examples typically include behavior suggesting that the infant is having a good time. With dramatic play, intense pleasure appears infrequently. Yet it has qualities of lightness and mobility which serve to distinguish it from exploration; e.g., a three-year-old examining a toy in order to learn how it works is different from one pushing a block across the floor and pretending it is a truck. The latter child is free from concern with physical and social reality, but his freedom is often unaccompanied by intense delight.

Piaget's Theory of Dramatic Play. There is a nice complementarity to the theories of dramatic play. Piaget has been concerned primarily with the cognitive development evidenced in play, while the Freudians have explored its functions in the child's emotional development. Each occasionally ventures into the other's realm with interesting results, but their primary focus is different.[4]

Piaget's thesis is that the development of play parallels cognitive development. Dramatic play appears when, around eighteen months, the child begins to symbolize. In fact, Piaget's term for dramatic play is "symbolic games," as distinguished from earlier "practice games," or what we have called the pleasure of functioning. The younger child can explore, anticipate, and solve problems, but his world is limited to his perception of the immediate situation and to his repertoire of motor behaviors designed to deal with it. Like an animal, he is a creature of perception and action. However, toward the end of the second year, intelligence shifts from action to representations of actions or symbols—a shift which opens the door to pretending and hence to dramatic play.

Let us cite some examples of early symbolizing. A little girl has a ritual of curling up and sucking her thumb when she goes to sleep. One day while playing in her room (but not at bedtime) she sees the pillow, curls up on it, and sucks her thumb. Later, the sight of her mother's fur piece or even the tail of a toy donkey elicits the same behavior. She is beginning to pretend to go to sleep. Another child, having learned to drink from a cup, can make the drinking gesture even when the cup is empty. Later, he can use a box to drink from or even pretend to drink by making the appropriate gesture with hand and arm and smacking his lips. Pretending is the ability to use an initially irrelevant object to evoke a familiar bit of behavior. Piaget's point is that the irrelevance of the object is essential, along with the child's unconcern about its irrelevance. When the child pretends to go to sleep, any object can serve his purpose. He assimilates in a playful manner or utilizes what Piaget calls ludic assimilation.

Ludic assimilation can be contrasted with adaptive assimilation. A two-year-old discovers that he can use a stick to rake in a toy which is out of reach. The nature of the object is essential in trying to retrieve a too-distant toy. Having learned that a stick can serve an adaptive function, the child will look for stick-like objects in the future. He has achieved an adaptive assimilation. The cardboard box he pretends to drink from, however, could be a napkin ring or a peanut shell. It is a temporary convenience which permits the drinking gesture; when the sequence is over, there is no effort to profit from experience, because there is nothing in the box or ring or shell which will make it particularly appropriate next time the toddler wants to pretend to drink. He does not study the object in order to understand its nature; his sole purpose is to satisfy a need to repeat a drinking gesture, and a variety of objects will do. Such ludic assimilation does not enrich the toddler's grasp of reality. We have here another instance of Piaget's thesis that play is assimilation without sufficient accommodation; acquired behavior sequences are utilized with little effort to adjust them to reality.

Piaget then proceeds to trace the development of symbolic games or what we are calling dramatic play. As we have seen, these games initially represent the repetition of an habitual action (e.g., drinking from a cup) with an arbitrarily chosen (or what Piaget calls an inappropriate) object (e.g., a box.) Between the ages of two and four,

[4] For a presentation of the classical Freudian theory of play, see Wälder (1933). Our discussion of the psychoanalytic theory will be based primarily on articles by Peller (1952, 1954, and 1964).

the child elaborates this basic symbolic process into exceedingly complex behaviors. One of the first advances is the ability to shift the locus of activity by pretending that other persons or objects are performing the activity. The child can pretend that his teddybear, not he, is going to sleep. Or he can put one shell on top of another and say, "It's going on the potty." His symbolic games also begin to include activities which have not been part of his repertoire but are borrowed from models: a child picks up a stick (an inappropriate object) and holds a telephone conversation (a behavior in his mother's repertoire but not in his own). Imitation itself can expand so that it includes not only specific behaviors, such as a phone conversation, but a complete identification, e.g., the child calls herself "Mommy" and says, "Come, kiss Mommy." In sum, in dramatic play, the child can shift the locus of causation and can utilize a vast array of behaviors outside his own repertoire. In a double sense, then, he is beginning to be freed from his self-centeredness.

By three years of age, the child engages in complex symbolic combinations. He becomes absorbed in fantastically long doll play, with innumerable real life and imaginary episodes strung together without logic or purpose. The four- to six-year-old expands this tendency to string ideas together as he extends his play geographically to include settings such as a farm scene or street scene. Piaget makes the interesting point that the child's play should not be called imaginative; what may seem like imagination is a lack of coherence. The child is merely reproducing episodes from his real life by means of symbolic representation, just as in infancy he reproduced a new skill in his pleasure in functioning. In essence, he is saying "This happened and this happened and this happened," as, one after another, ideas come to mind.

Between four and seven years of age, symbols tend to lose their ludic character. They are less arbitrary, less determined by inner needs, and more closely tied to reality. Games become increasingly orderly and organized, and involve more collective symbols, i.e., symbols with agreed-upon meaning. This last characteristic means that grocers and spacemen and soldiers and mothers will act according to commonly held ideas concerning their behavior rather than changing to suit the whim of the child. According to Piaget, play undergoes these changes because the child's thinking has become more coherent and social. Cognitive and social growth are complementary; idiosyncratic ideation is inimical to social relations. In order to communicate, the child must use symbols as everyone else uses them. In sum, the child's social need for communication and reciprocity forces his play to conform increasingly to collective symbols and logic.

Let us return to the symbolic games of the three- to four-year-old, since they are of special interest. Remember that this is a time before order and collective symbols have affected play. Because they are still subjective, early symbolic games are particularly suited to expressing personal feelings. The very commonality of meaning which makes collective symbols ideal for social communication makes them unsuited to expressing the idiosyncratic ideas of the child.

In this early phase, Piaget describes three kinds of symbolic combinations. The first is a compensatory combination in which the child does something in play which he has been forbidden to do in reality. Here reality is "corrected" rather than reproduced; i.e., the fantasy concerns being gratified rather than being forbidden. Piaget cites a typical example: after being forbidden to play in the water, a child takes his cup to the empty tub and pretends to play. Next, there is catharsis in which a child neutralizes a fear by playing at something he dare not do in reality. Here "correcting" reality purges the experience of its distressing features. A little girl, too frightened to play theatre with her peers in a neighboring barn, enjoys her own theatre in the safety of her room. Finally there are liquidating combinations which relive difficult experiences at the symbolic level. Here the child re-exposes himself to the risks and dangers of reality, but can assimilate and triumph over them because they have been translated into symbols. The classic example is of a child playing dentist after a stressful visit to the dentist's office. In all three types of play, reality is either corrected or its frightening features are relived and mastered symbolically.

The trio of symbolic games is interesting in that it forces Piaget to devise new explanations. Play is no longer sheer repetition of assimilated experiences for the pleasure it affords; rather, the child is seeking to compensate for or revenge himself upon reality or he is trying to diminish the distress of an experience in order to master it. These explanations appear unexpectedly in Piaget's system instead of being carefully derived from previous developmental stages. The trio itself comes as something of a surprise and rests uneasily in Piaget's framework. It is presented descriptively without Piaget's usual painstaking attention to precursors. However, the trio has flourished within the psychoanalytic theory, and it is interesting to note how closely Piaget and Freud agree concerning the nature and even the function of symbolic play. As we have seen (Chapter 6), Piaget is concerned primarily with cognition and only secondarily with emotions, while the opposite is true of Freud. Piaget observed normal children; Freud studied only disturbed adults. In light of the differences in approach, the concordance of the two explanations is impressive.

Psychoanalytic Theory of Dramatic Play. We shall turn to the psychoanalysts because they have explored extensively the motivational and functional aspects of dramatic play which Piaget only mentions in passing. One function of play is the mastery of an experience. Taking our cue from Freud, we will try to understand mastery through play by first examining failures to master (see Erikson, 1950 and White, 1964). Adults can be exposed to situations which are traumatic, that is, overwhelmingly frightening; e.g., being trapped in a theater which is on fire, being in an airplane crash. The effects of traumatic experiences have been studied most extensively with soldiers in battle. Traumatized soldiers typically show signs of extreme disturbance: they are restless and irritable, they cry easily, they cannot concentrate on work, their self-confidence is shattered, they have recurrent nightmares, they must go through senseless rituals; e.g., one man had to lie on his stomach with his face in the pillow and hold his breath if he wanted to go to sleep. Most important, they have no memory of the traumatic event itself, which is too terrifying to be recalled. Such men can be helped, particularly if therapy begins within a few days after the trauma. The essence of therapy is to induce sufficient relaxation so the soldiers can remember the experience and realize that it part of the past. Relaxation is induced by the reassurance of the therapist, by hypnosis, or by drugs: Whatever the technique, the soldier must re-experience the trauma in order that he may realize that the danger has passed.

A terrified soldier being given a drug may seem to have nothing in common with a four-year-old girl spanking her doll for being bad. Yet, Piaget claims that symbolic games liquidate realistic fears by transferring the experience to the symbolic realm. Too frightened to relive the past symbolically, the soldier is incapable of gaining power over it. Only as he is helped to do so will the fear be exorcised. One might say that he needs special therapeutic aid to do what the child does spontaneously. She has been frightened by a spanking, but is mastering her fear by reviving the experience symbolically and in a form which permits her to be the spanker rather than the spankee.

Child psychoanalysts have been impressed by the many traumas inherent in normal development (see Chapter 7). Birth itself is considered traumatic, and, in the next few years, the child will be subjected to separation anguish, fear of loss of parental love, extreme guilt, and fear of retaliation and mutilation. His helplessness is constantly underscored by the relative weakness of his ego—that aspect of his personality responsible for coping with inner impulses and external reality. Play is one of his principal mechanisms for mastery. By reliving an experience symbolically, it is possible for him to reduce its anxiety.

Peller (1952, 1964) discusses the mastery functions of play in terms of the stresses inherent in psychosexual development, but her illustrative examples are of general interest. Play in the first two to three years revolves around the attachment to the mother. Peller pays particular attention to the hide-and-seek game which is so characteristic. The child not only repeats it endlessly but, unlike his older counterpart, makes sure that the mother knows ahead of time where he is hiding! Any attempt on the part of the bored parent to introduce variation or to explain to the child that he has missed the entire point of hiding is strongly resisted. The motive underlying the game is mastery of separation anxiety by reversing the roles. Instead of the mother disappearing from the child, the child disappears from the mother. Yet the idea of a more serious disappearance, a hiding in which the mother really does not know where the child is, would mobilize too much anxiety and would disrupt play altogether.

In the preschool years, the simple play of infancy gives way to the elaborate dramatization we have already described. The child is primarily concerned with mastering the stress of being denied adult status and his fantasies are compensatory ones of being big and engaging in adult activities. He is supremely masculine or feminine in his choice of roles—astronaut, cowboy, doctor, fireman, mother, nurse, princess. He delights in miniature toys which reproduce the grown-up world. Instead of being the recipient of adult directives, he is now in command. Play also allows for the expression of intensely felt regressive needs. The child who is temporarily beset by the demands of being a "big boy" in real life may revert to the role of baby or baby animal, making special demands to be fed and cared for while protecting his self-esteem and avoiding the censure of parents and conscience alike. After all, he is only pretending.

However, development is not all trauma, and play is not motivated exclusively by distress. There is much in the child's life which he has difficulty in understanding.

Children and adults live in different worlds, and even the most sensitive parents cannot make adult reality or the socialization process meaningful and congruent. Arbitrariness and incomprehensibility are bound to exist. The child carries his perplexities into his play and, by reliving them symbolically, digests them. A routine visit to the doctor may be a strange experience. The doctor flashes a light in the child's eyes, presses his tongue down with a stick until he gags, and, after he partially undresses, places cold metal objects on his chest and back. Subsequently, the child may be observed playing doctor, bringing back bits of the experience, going over them again and again, to reduce their strangeness. On the next visit he will know better what to expect, and the chances of being overwhelmed by strangeness will be reduced. Waelder (1933), using a psychoanalytic framework, chooses the word "assimilation" to describe play, a word which is one of the cornerstones of Piaget's theory of cognitive development. An experience which is too large to be assimilated "at one swoop" is mastered piecemeal in play.

Psychoanalysts have been explicit in regard to the mechanisms underlying mastery through play. The first and most important is the shift from passivity to activity. An essential feature of trauma is helplessness (see Chapter 7). Not only is distress intense, but the individual feels that there is nothing he can do to relieve it. At best he must endure it; at worst, he anticipates ever-increasing terror. Separation anguish is a good example of helplessness in a distressing situation: the mother leaves and the infant cannot make her reappear (see Chapter 2). At a nontraumatic level, passivity, in the sense of "being-done-to" and having no control over what is happening, leads to milder forms of distress. Much of socialization and countless daily experiences contain this nontraumatic kind of passivity. Children inevitably are "done-to" by adults, often benevolently. They are fed and diapered, taken to doctors, to preschool, to relatives, taught self-control, manners, and morality—all quite literally for their own good, but not by their own choice. Life could hardly be otherwise. To put a child completely in charge of making decisions runs the risk of baffling and terrifying him (see Wenar, 1957).

The remedy for the uneasiness of "being-done-to" is to become an active doer. The child who is in control can decide for himself what will happen and when and how. The same nine-month-old who cries out in anguish when his mother leaves will blithely crawl away from her when she is present (see Chapter 2). It is not separation, but separation plus passivity, which produces the anguish. Or, let us return to the example of the child in the doctor's office. He does not decide whether or not he will get a shot and he cannot order the doctor to stop whenever the pain becomes too intense. This is where play enters. In Erikson's terms, the child transfers the problem from the macrosphere—i.e., the world shared by others—to the microsphere—i.e., the small world of manageable toys. With toys he is in control; he can decide; he is the doctor. The past is not only revived, as Piaget says, but is revived with the child in command. Sheer repetition for its own sake would not suffice; the reversal of roles holds the key. When play performs its function of mastery, the child can return to the macrosphere with a renewed sense of confidence.

The second reason why play aids mastery is that it provides catharsis by offering a safe context for releasing pent-up feelings. The three-year-old who is punished for hitting his baby brother will not be punished for knocking down a tower of blocks. As we have seen, play is important because most parents regard it as unimportant. They do not feel compelled to judge it as right and wrong. The child is therefore free to bring to play the affects he dare not express otherwise and find some relief from their pressure. Psychoanalysts traditionally have championed the idea that all human beings need periodic emotional release; lacking this, they must divert too much energy from the task of mastery and growth to the task of keeping their feelings in check.

Additional Functions of Dramatic Play. Dramatic play serves an important socializing function. In the preschool years, social interaction during play is typically haphazard or impromptu. No one child is essential, and no one part is stable. However, the basis for cooperation is laid and will come to fruition in the highly organized and well-differentiated dramatization of the eight-year-old. We have already discussed the importance of play in teaching the child to attune his actions to the behavior of others and in enabling him to see a situation from different points of view (see Chapter 5). Egocentrism is counteracted in the context of activity which is congruent and meaningful to the participants.

The roles the child plays help prepare him for adult roles. The heroic fireman is feeling his way into the masculine assertiveness which lies ahead, and the mother who cares tenderly for her doll is glimpsing the maternal tenderness to come. Cameron and Magaret (1951) stress this aspect of play. Their concept of role-taking skill includes two components: the acquisition of social perspective (described in the preceding paragraph) and practice in being one or another kind of grownup with special responsibilities and privileges. Play offers the child the option of adult roles and helps him to know which role he wants for himself.

Much of the writing about the socializing function of play has been speculative. However, Marshall (1961) furnishes objective evidence of its importance for children between the ages of two-and-a-half and six-and-a-half. She found that social interaction during dramatic play was different in many ways from realistic social interaction. The language development was different over time, and the friendliness and hostility in one realm was not related to that in the other. Most important, children who engaged in more dramatic play were preferred by peers to those who played less. In fact, participation in dramatic play held the key to social acceptance and friendly interactions; whether the child was winning or obstreperous in his realistic transactions did not matter. Finally, children who engaged in a good deal of dramatic play had parents who talked to them about topics which could be utilized in such play. Such parents were specifically oriented to the children's interests and, at least in this study, their stimulation was more effective than television.

Important from many points of view, this study suggests that the father who talks to his boy about trains and fire engines is helping him become popular at nursery school. Emotional security at home or even a warm democratic atmosphere may not be the only factors which facilitate positive social relations (see Chapter 8). The study also illustrates our point that dramatic play is the child's own language. He has strong emotional attachments to adults, but the adult world is essentially foreign. Left to himself, he gravitates to his special way of thinking and feeling and acting and is attracted to others who can share this world of dramatic play with him.

Finally, dramatic play has been regarded as one of the precursors of creative thinking. Recall that Kogan and Wallach found that part of the creative process involved a willful suspension of logic and the constraints of realistic thinking (see Chapter 6). Having shifted his critical faculties into neutral, the individual is free to go wherever his ideas lead him. Neo-Freudians have a similar concept which they call "regression in the service of the ego" (see Kris, 1952). Freud himself did not see the issue clearly; he tended to contrast rational and irrational thinking, the former being one of the basic criteria of mature reality contact, while the latter characterizes the thinking of early childhood and disturbed adults. Followers of Freud objected to the idea that rationality is the principal vehicle for making a mature adaptation to reality.[5] There is more to maturity than being reasonable, logical, and appropriate in one's thinking. Reason both clarifies and confines, and periodically one must break the mold and return to playful thinking. Here ideas which logic separates can be juxtaposed; fantastic consequences can flow from antecedents; impossible combinations and distinctions can be made freely. As a final step, the creative individual reintegrates such ideas into the context of reality and produces the kind of imaginative thinking valued by artists and scientists alike.

The Diagnostic and Therapeutic Uses of Dramatic Play. In the preschool years and in early childhood, speech is not the child's language for expressing deep feelings and relating troublesome experiences. Ask him, "How do you get along at home?" and you frequently will receive a meager answer. The child is not necessarily being defiant or uncooperative; it is just that many of his intense experiences have not been translated into words. A boy may know that every time his mother serves him food his stomach churns; a girl may know that when her father calls her "Daddy's girl" she wants to scream and kick; but only rarely is a child sufficiently facile with words to verbalize these reactions. Feelings are rarely the subject of conversation.

Conversing has another drawback. In order to be understood, the child must use words which are common symbols. This very aptness for conveying agreed-upon meaning makes language insensitive to idiosyncratic thoughts and feelings. When a child talks about a "car" he may be using the acceptable verbal symbol for a four-wheeled vehicle, but he may also be thinking of a powerful machine for destroying everything which stands in its way. Give him a toy car and he may express this personalized meaning. Because play is nearer to action than is talking, it is also nearer to the experiential level. In sum, play gives the child an opportunity to relive his concerns in a medium he finds most natural for expressing them.

[5] Interestingly, Piaget also has been criticized for regarding play as an infantile kind of activity, which must be replaced by more realistic thought. He is as much bound to the primacy of logic as Freud was to the primacy of rationality. This criticism was raised by Sutton-Smith (1966).

However, seasoned clinicians caution against a simplistic interpretation of play. While a child who has been punished by her mother may spank her doll, one cannot observe a child spanking her doll and infer that she has been punished by her mother. The same theme in dramatic play can have a multitude of meanings. It is only the amateur who readily decodes play by using a handbook on symbolic meaning. Both Freud and Piaget stressed the individualized nature of symbol formation and the importance of determining the meaning of a symbol to the particular individual. The experience of a psychiatrist with a severely disturbed adolescent girl illustrates this point. One symptom of her psychosis was a continual scratching of her face and arms until she produced large bleeding sores. An amateur Freudian might readily interpret her behavior as symbolizing a strong self-destructive tendency, and her history would make such an interpretation plausible. As she recovered and could talk about herself, she gave a different account. She said that, in the depths of her disturbance and despair, she felt that all life had been drained out of her. Often she literally questioned whether she were dead. Only when she could scratch herself and feel pain did she know that she was still alive. Because she could feel something, she could avoid total hopelessness. Far from being self-destructive, her mutilation represented her desperate clinging to life.

The equivocal nature of manifest play has also been a thorn in the side of objective researchers who have tried to use it to assess personality variables. In their review of research utilizing doll play, Levin and Wardwell (1962) discuss the difficulty of deciding whether a given bit of play represents the repetition of a reality situation or a wish fulfillment. If a child plays out a scene in which a doll is given many presents, is he reliving his own pleasure in receiving gifts, or is he trying to compensate in fantasy for the deprivation he feels? The clinician has the time to explore the question with the child; the typical researcher does not. To date there is no easy method for decoding play symbolism. Doll play is something like a hypothetical intelligence test in which a given response indicates that the child is either very bright or very dull, and the examiner has no guides to which is the correct alternative.

We shall briefly touch on the therapeutic use of dramatic play.[6] In diagnosis, the clinician is alert to deviations in the healthy flow of dramatic activity: the pseudo-maturity which produces a total disdain of play; the exaggerated need for self-control which makes play formal and stilted; the preoccupations which force the child to repeat one theme again and again; the sudden upsurge of fear which makes him abandon play in the middle of a sequence; the anxious excitement which drives him from theme to theme in rapid succession; the weak self-control which makes him scatter the toys violently in all directions. The therapist uses these diagnostic signs as a point of departure for exploring, through play, the unmastered anxiety which lies behind them. In some instances, therapy can remain entirely at the level of play. In reliving a distressing experience symbolically, certain children can master it. As their play returns to normal, their general adjustment improves. This kind of "release

[6] For a more detailed presentation of the role of play in psychotherapy, see Erikson (1950), Amster (1964), and Woltmann (1964).

therapy" capitalizes on the cathartic effects of play (see Levy, 1939). With other children the therapist must make the connection between symbol and reality, such as a painful conflict within the family. He concentrates now on play, now on feelings about members of the family. With his help, the child can assimilate and master the situations which formerly had overwhelmed him.

There is a less obvious aspect of play therapy as it affects the child-therapist relationship. Many children come to psychotherapy because the ordinary growth-promoting, human feelings–love, interest, respect, pleasure, trust, pride–have been insufficient to his needs. The therapist brings such positive feelings to his transactions and they act as powerful healing agents. However, the child often needs to feel that he is understood, which is not synonymous with being loved and valued. A loving parent, for example, can fail to understand the child's viewpoint. The feeling of being misunderstood can be an impediment regardless of where it appears, whether it be in psychotherapy, in the generation gap, in labor-management disputes, or in international conflicts. The therapist's interest in the child's play is one of the principle signs of his desire to see the situation through the child's eyes. In attending to play he is saying, "Tell me about yourself in your own way, and I will try to understand." "In your own way" shows that he understands what others have failed to grasp–that play is the child's language for expressing his personal concerns.

Summary and Comment. Piaget and Freud agree on many of the functions of play. In dramatic play, the child can have a satisfaction which was denied him or an experience which was too frightening for him to have in reality. Piaget regards the child as "getting revenge" on reality, while Freud prefers to characterize play as "wish fulfilling." Liquidating anxiety by reliving an experience in play is exactly the mastery function Freud hypothesized for all fantasy–daydreams and night dreams, as well as dramatic play. He attributed its efficaciousness to the mechanism of turning passivity into activity. Catharsis provides release in play for burdensome affects originating in the child's effort to cope with reality. The repetitiveness of play diminishes the strangeness of a new experience, as familiarity serves to facilitate assimilation. Play serves a socializing function by counteracting egocentricity and affording the child a preliminary glimpse into adult roles. Finally, dramatic play has been regarded as a precursor of creative and imaginative thinking.

Now we can return to the question posed in the introduction to our discussion: Should the myth of play be destroyed? Phrased more sensibly: Should parents and teachers become more cognizant of the importance of play so that they can utilize it to further the child's development? (We are not raising the question of whether they should use play diagnostically and therapeutically, which deserves discussing in its own right.) The Russians, for example, have been more consciously concerned with "early character education" than we in the United States. In certain of their preschools, the teacher utilizes play to underscore adult values, such as the importance of tenderness in caring for children, of finishing a job, of sharing, of living up to obligations. One cannot quarrel with these values or deny that they can be introduced tactfully and skillfully by a sensitive teacher. We also know that the traditional methods of teaching values in our country have little effect; parents do not make children responsible by giving them chores, nor does Sunday School make them moral. When character becomes a lesson to be learned, it does not take. Why not inject values into the children's world of play where he can assimilate them at his own level?

We do not know the answer. From a pragmatic standpoint we do not know if such a program would work. If it did work, we do not know whether the child would pay a price for his added strength of character. But we can see certain dangers. Because play bypasses reality and socialization, it can serve the mastery and cathartic functions attributed to it. The more play becomes involved with adult values, so that some kinds are "right" and others "wrong," the more the child is denied a refuge from socialization, a chance to release its attendant tensions, and an opportunity to assimilate it in his own terms and at his own pace.

These same considerations apply to the issue of preschool education. Teaching academic skills runs the same risks as teaching values. Undoubtedly, children can learn reading and other academic subjects before entering school. But knowing that a child can learn to read does not mean that he necessarily should learn to read or that he is being deprived if he does not learn to read. The question is, How will an emphasis on this aspect of the child's development affect other aspects? Recall that dramatic play, not reading, holds the key to social relations in preschool and may be an important ingredient in creativity. At the very least we should ask: During the time the child is learning to read, what are the things that he is not doing? In this way we can weigh what he is gaining against what he is losing. The question has no easy answer, but raising it serves as a reminder that there is more involved in preschool education than teaching a skill at an earlier age than is traditional.

Fantasy

Dramatic Play and Fantasy. Dramatic play declines in middle childhood. The psychological processes themselves are not outgrown, however. They go underground. Instead of acting out symbolic games, the child or adolescent carries on the drama in daydreams or fantasies. The mechanisms and meanings of dramatic play apply equally well to fantasy (see White, 1964).

The transition from dramatic play to fantasy reflects a significant development which we have discussed in detail elsewhere (see Chapter 6); namely, the disentanglement of subjective ideas from objective reality. As we have seen, the omnipotent infant erroneously but understandably believes that his wishes alone cause events in the external world. The preschooler is not quite so gross in his confounding of subjective with objective events. His grasp of the workaday world is good; he knows that chairs and tables do not come alive and that parents do not have to obey his wishes. But the line between reality and thought is still not hard and fast. He believes that his dreams really happen and that his imaginary companions are living beings; he is apt to think that the people on TV are really there—perhaps in the box somewhere. Even in his dramatic play, he becomes what he does—he is a cat, not a boy pretending to be a cat. If questioned, he admits that he really is not a cat, showing that he has more insight into his play than into his dreams. In middle childhood, objective reality has been firmly grasped. The child realizes that the thoughts inside his head are different from events going on outside. He recognizes pretending as a special kind of behavior which is distinct from realistic happenings in the environment.

Like play, fantasy flouts reality. It is unconcerned with the here-and-now. A boy in the classroom can be riding a rocket to the moon in his daydreams. Fantasy is heedless of social realities and the constraints of propriety and conscience. It uses people capriciously and amorally. Brutality, licentiousness, and monstrous egotism come and go, sometimes savored briefly, sometimes hardly noticed. But fantasies are not airy nothings. Like play, they are important precisely because they do not matter. They are what a person thinks when the need to adapt to reality, the pressures of communication, the danger of action, and the threat of guilt have all been bypassed. Fantasies are what a person is when no one is looking, not even his self.

Fantasy, like play, is motivated by the tensions, frustrations, and distresses the child encounters in reality. Theoretically, transient difficulties will result in fleeting fantasies, while more pervasive problems will produce recurrent themes. Symonds (1949), for example, found that the fantasy of normal adolescents reflected their conflicting needs for independence and security, their contradictory impulses to love and hate, their moral sensitivity and striving to be successful, their discouragement and efforts to bolster their spirits.

Fantasy is also like play in that it allows the child to master, in his inner drama, the situation which was too taxing in reality. White notes that the pleasure generated by daydreams helps offset the tensions which motivated them. In the most familiar examples of compensatory daydreaming, mastery is not enough—the child triumphs gloriously. The wallflower does not dream of becoming just another popular girl, she is a dazzler; the retort to an insult an adolescent thinks of while lying in bed is not just apt, it is devastating. Or let us listen to the fantasy of an eighteen-year-old girl. "I'm in this room and this boy is furious at me, see, but I'm very cool. I begin to undress slowly, real slowly, like I'm not even studying him. Then I'm all undressed except my stockings and I sit in this chair and just look at him in this slow way. He tries to be angry but he breaks out into a cold sweat. Then he can't help himself. He tears off his shirt and is hot all over and jumps on top of me." The girl is talking to her therapist. She has come because she is terrified of boys. But she is allowing herself to imagine things which are beyond her capabilities in reality. The fantasy is a rehearsal; the excitement a lure.

The "conquering hero"motif is frequently met in childhood, but so is the "suffering hero," especially in puberty. Here the child is the innocent martyr, misunderstood and mistreated by parents, peers, or teachers. In the end, the persecutors see the error of their ways, and the hero is both exonerated and enjoys that unique pleasure which comes from forgiving a penitent enemy. Such fantasies strongly suggest the special form of hostility in which the individual punishes others by making them feel guilty over his suffering. In the child's words: I'll die, and then you'll be sorry (see Chapter 7).

The themes of fantasy are not limited to those we have presented. Jersild (1933) collected information on daydreams from 400 children between the ages of five and twelve. Approximately 11 percent had daydreams concerning commonplace events. At times, the fantasies seemed no more than fleeting images with little imaginative or affective content; at times they carried a mild charge of positive or negative affect; e.g., fantasying about unfinished homework, about fooling somebody, about seeing skyrockets go off. Such fantasies typify the basic process of symbolically reliving an experience. Approximately 60 percent of the children had positive fantasies, the younger ones tending to dwell on physical possessions, amusements, and friendly companionship, the older ones on conquering hero themes. Some of these themes might have represented a recapitulation of pleasant experiences, though many involved what psychoanalysts term wish fulfillment or what Piaget described as the child "revenging himself" on reality. So far the results conform to our understanding of fantasy. However, about 8 percent of the children had fantasies with negative affect—of school failures, of being left alone, of robbers, ghosts, and giants who came to do them harm, of accidents to self or injury and death of parents. Do these represent an exception to the rule concerning mastery of distress? Not if we are correct in regarding fantasy as the inheritor of dramatic play. Preschoolers are attracted to frightening and catastrophic themes which they master by reversing roles from passive to active. In like manner, dysphoric fantasy may represent an attempt actively to control a distressing situation rather than being helplessly victimized by it. As in play, the attempt is not always successful. Anxiety may be too great and may produce a morbid preoccupation which robs fantasy of its characteristic mobility.

Eroticized Fantasies. Fantasies can become eroticized, which is further proof that they are more than airy nothings. Because daydreams often lack the quality of wholehearted involvement which characterizes dramatic play, it is difficult to take their pleasures seriously. However, the affective charge of eroticized fantasy, especially that which accompanies masturbation, is compelling. Understandably, our information concerning masturbatory fantasies is limited, since most of it has been obtained from children and adults in psychotherapy. It also comes primarily from psychoanalysts who have been particularly interested in studying erotic gratification.

The glimpses psychoanalytic data afford into masturbatory fantasies reveal a richness which rivals that of dramatic play. The content may be an ordinary event, such as skimming over the water in a sailboat with a pretty girl or watching a fast-moving basketball game. At times, it is more fanciful, such as watching a gypsy dancer whose wild, "pumping motions" make her breasts grow larger and larger. Or it may reflect that dark inner world which Freud explored so courageously, such as the fantasy of cutting off a man's head and watching the blood spurt out. Psychoanalysts interpret these themes as products of the various anxieties inherent in the psychosexual stages of development. While some themes have no obvious relation to psychosexual theory, others directly involve sucking and biting, elimination and urination, inflicting or receiving injury, sexual organs and sexual activities.

The psychoanalysts also claim that masturbatory anxiety is a response to the impulse underlying the fantasy. Parental threats and shaming may play a part, but the child is primarily frightened by his unconscious feelings and desires. An eight-year-old boy, for example, had the following fantasy during masturbation: he put chains around a hairy giant and squeezed him tighter and tighter. The giant burst the chains with a tremendous effort and fell dead. The fantasy symbolized the boy's competitive feelings toward his father and his desire to eliminate him as a rival. His subsequent guilt represented the punishment of his conscience. Parental shaming for masturbating may heighten his guilt, but it would be present even in a benevolent setting.

Although the psychoanlysts have fought hard for a more liberal attitude toward childhood sexuality, they do not regard all masturbation as natural. The fantasy is normal only if it reflects a tension appropriate to the child's stage of psychosexual development. In a like manner they recognize dangers in masturbation itself, especially when the pleasures of self-manipulation impede the child's progress in finding erotic gratification from attachments to others. Ironically, in a White House Conference on mental hygiene held in 1950, a psychoanalyst was the sole objector to the idea that masturbation was perfectly natural and healthy in adolescence (see Eissler, 1951 and Blos, 1962; for a more detailed account of masturbation fantasies, see Wermer and Levin, 1967).

The Value of Fantasy. While we have emphasized the functional similarity between fantasy and dramatic play, there are important differences between the two which arise from the child's stage of development. The gap between fantasy and reality in middle childhood is wider than that between dramatic play and reality in the preschool period. The self is more unformed in the preschooler, making it easier for him to assume a variety of roles and to identify himself with each. Because dramatic play comes at a time of incomplete psychological differentiation, reality tensions can be more readily reduced in symbolic games and the resolutions achieved in play can be more readily generalized to reality. Equally important, dramatic play serves a socializing function, while fantasy runs the risk of social isolation. The daydreamer does not capture the interest of his peers as did the preschoolers in Marshall's study.

Because of the greater gap between fantasy and reality, psychologists are less sanguine about the value of daydreaming than they are about the value of dramatic play. Fantasy is suspect because it can isolate the child from reality rather than rehearse him for coping with it and because its seductive pleasure may lure the child into a world of false triumphs where realistic problems are no longer faced. Psychologists do not claim that daydreaming is the forerunner of emotional disturbance, but they note that excessive daydreaming can indicate that too many of life's problems are generating too much distress and causing too profound a retreat. The concern is that the child has been overwhelmed by life itself rather than merely preoccupied with his daydreams.

Summary. Fantasy is emotionally gratifying and aids mastery. It provides compensatory pleasures to counteract the tensions, distresses, and boredom of reality. It also can furnish substitute gratification for needs which cannot be otherwise satisfied in reality. As an aid to mastery, it liquidates some of the distress of reality, it turns passivity into activity, and it can serve as a rehearsal for realistic coping. However, every fantasy does not serve all these functions, nor is every fantasy successful in providing pleasure and mastery. Moreover, objective evidence concerning all of fantasy's constructive functions is meager. We know little about the conditions under which an imagined gratification will satisfy a need, when it will temporarily hold the need in check, and when it will serve to increase its intensity. We also know little about the conditions under which fantasy will serve as a rehearsal for mastery and when it will seduce the child away from mastering his anxieties.

Games and Sports

Developmental Overview. Dramatic play declines in middle childhood as children become more realistic, more organized, and more involved in group activities. Dramatic play initially arose on the spur of the moment, with no prelude or preparation. Unpredictable in its course and brief in duration, it ignored both physical and social reality. As the child matured, play became less spontaneous, more social and communicative. Crude preparations were required such as a tacit or explicit agreement to "play something," and crude props were used. Children played together, although any child might play any role and individuals were free to come and go as they pleased. In middle childhood dramatic play became a distinct activity requiring a special set with elaborate props. Special roles assigned to individual children were dutifully executed in coherent and sustained plots. Now, in games, formal and social elements gain ascendance over the spontaneous, personal ones. Games have beginnings and ends and are governed by rules which all must obey. Regulation is essential; idiosyncratic symbolism and improvisations are minimal. A child may have strong personal feelings and vivid fantasies about participating in a game, but these are his own business; they have no place in the context of what he is required to do.

Hurlock (1964) provides an overview of the development of specific kinds of games. "Mother games," so called because they are played with the mother, begin in the first year of life; e.g., peek-a-boo, pat-a-cake, finger play. Their appeal seems universal and they are handed down from one generation to another. The infant is the recipient and pleasure runs high. Around four or five years of age, neighborhood games appear. They are simple, brief, with few rules. Some, like tag, hide-and-seek, advancing statues, have continuity over time; others are created on the spot. The five-year-old is also interested in activities requiring skill, such as jumping rope or bouncing a ball, but the social and competitive elements are relatively unimportant. By the gang age of eight to twelve years, boys organize competitive games with rules. Basketball and football replace "kid games," such as cops and robbers. In adolescence the child has the skill, the cooperativeness, the grasp of rules, to qualify as a team player. Paralleling the development of sports are indoor games in which physical skill is unimportant compared with understanding, strategy, and luck; e.g., card games, Monopoly, craps.

The Spirit of Games. In an inventory of games, one can distinguish those which will continue into adulthood, such as football, baseball, cards, and checkers, and those peculiar to childhood, such as hopscotch, jacks, hide and seek, and jump rope. And, among the latter, one can distinguish fads from games of abiding interest. Fads date a person as accurately as rings on a tree. The yo-yo, the hula hoop, the knock-knock craze made their marks and passed away. However, Red Light, Crack the Whip, May I, along with Igpay Atinlay endure forever, handed down from generation to generation. Adults have no part in teaching or perpetuating them; they are part of the children's own culture.

There is a special spirit of play which gives games their childlike aura (see Stone and Church, 1968). The spirit is expressed partly in the rituals and chants which surround games and are an intrinsic part of them. Which is more absorbing to the little girl, the skill of jump rope or the chant of "Ladybug, Ladybug, turn around, round, round?" And choosing sides by One potato, two potato, three potato, four, is as important as any rule of the game itself. Other rituals and rhymes weave in and out of play, independent of any specific game. There are the rituals of taunts—"I'm rubber, you're glue; everything you say bounces off me and sticks on you"; and the rituals of secrecy—"Needles, pins, triplets, twins; What goes up the chimney? Smoke. I wish our wish will never be broke." Some rhymes are for special occasions—"I scream, you scream, we all scream for ice cream"; others are magical incantations—"Rain, rain, go away, come again another day"; while still others express popular superstitions—"Step on a crack and you break your mother's back; step on a line and you break your mother's spine." Along with game-relevant skills go stunts and tricks the child does with his body—such as wiggling his ears, crossing his eyes and belching voluntarily—to win the admiration of his peers.

Such chants and rituals and taunts, such secret signs and body skills, are more evocative of childhood than games themselves. Stone and Church speculate about their role in development. During the preschool years and during adolescence, the child is intensely involved with parents, although for different reasons. The preschooler is concerned with them as sources of love, punishment, and direction, while the adolescent is striving to emancipate himself from the family and become an independent adult. In middle childhood children turn away from the family, although they are unprepared for such independence. They cannot care for their bodily needs, and they do not have the control, judgment, and skills essential for true self-regulation. In their uncertainty, they cling to one another for the strength to be found in unity. Thus, one function of children's games is to foster a feeling of belonging and of being "in the know." The secret nature of some of the children's activity serves the double function of insuring closeness and excluding undesirables, especially parents. Rituals give the child a sense of mastery of reality. They reassure him that if he does just the right thing in just the right way, he will be all right. When one is unsure, authoritative prescriptions are welcome.

But the rituals of middle childhood have a special quality. They are not the wasteful behavioral prisons one sees in the compulsive neurotic, such as the irresistible need to wash one's hands repeatedly; nor are they like earlier omnipotence in which the infant-toddler fully expects his wishes to be immediately gratified. Instead, rituals are done with a feeling of propriety and, for all their magic, they are part solemn, part playful.

In the spirit of games, one senses the pleasure of doing for the sake of doing, the exercise of power for the sake of feeling one's power, and the delight in the magical and fantastic rather than in the realistic and utilitarian. Games, for all their emphasis on formality and regulation of behavior, still partake of the qualities of earlier play.

The Functions of Games. Games, like play, are multifaceted. We will review two of their functions discussed in detail elsewhere and then present three new ones.

The cognitive element in games—namely, the comprehension of rules—has been studied by Piaget (1948; see Chapter 5). Using the game of marbles as a prototype, he traced the child's progressive insight into the meaning of rules and his ability to use rules to regulate behavior. The three-to-five-year-old can verbalize the rules of his elders but proceeds to play in a totally egocentric manner, doing whatever he sees fit, blissfully oblivious of the discrepancy between word and deed. The child of seven to eight years begins to regulate his play according to rules, but his comprehension is still hazy. Not until he is twelve years of age can he utilize rules appropriately as guides to behavior. His grasp of the nature of rules also shows a developmental trend. At first he regards them as sacred and unchangeable, handed down from above (even though he breaks them constantly in his play). Ultimately he comes to see rules for what they are—means for regulating play and therefore subject to revision as the situation requires. Piaget maintains that the child's developing understanding of the nature of rules is one facet of his progress toward achieving moral insight. Games are the counterpart in play of the concern with obligations to one's fellow man.

Next, there is the sex typing function of athletics. More than any other single activity, sports epitomizes masculinity in middle childhood and adolescence (see Chapter 7). A boy's athletic ability significantly determines his self-esteem and his prestige and power with peers. He evaluates his body in terms of its being an asset or a handicap in sports. Both the physical skill necessary to sports and the related attitudes of assertiveness and good sportsmanship are highly prized. For approximately ten years, or roughly half the span from birth to maturity, sports provide the main proving-ground for masculinity.

But there is much more to games than their cognitive element and sex typing function. Peller (1954), writing in the psychoanalytic tradition, discusses their structure and impersonality as aids in mastering emotional problems. Recall that dramatic play offered the younger child an opportunity to "play out" his personal problems, which often centered around stresses in the nuclear family. In middle childhood, an effort is being made to leave these stresses behind as the child wants to turn away from the jealousies and anxieties, the hopeless striving to be as powerful as adults. The impersonal and highly regulated nature of games offers him an opportunity to find the refuge he needs.

Games also satisfy a deeper need, according to Peller. Like Stone and Church, she views the child as fearful of being on his own. Participating in games with peers gives him a feeling of strength, while rules provide authoritative directions as to how he must behave. Finally, games provide an important vehicle for shifting emotional ties from the family to peers. Attachments to and identifications with playmates are the "guiding stars" of games, since children are together as equals as they never have been in the past. Peller's point complements Sullivan's concerning the importance of give and take, competition and compromise, which the child learns in the context of playing the game and being a member of the team (see Chapter 8).

Competition. The role of competition in social development has already been discussed (see Chapter 8). Sutton-Smith and Roberts (1964) address themselves more specifically to the nature and function of competition in games. Their thesis is that various forms of competition are related to socialization practices. Moreover, games both relieve anxiety about competition and rehearse the child in attitudes which will be useful in adjusting to the adult culture. Thus, competition both reflects the past and prepares for the future.

The authors describe three kinds of games: games of chance, such as coin matching, in which the individual trusts to luck; games of physical skill, such as bowling, in which motor skills must be used successfully; and games of strategy, such as chess, in which the individual must make sagacious choices among possibilities. Naturally, games can be a combination of these three types; e.g., poker involves both luck and strategy. Theoretically, reliance on chance is related to socialization which requires the child to discharge routine responsibilities and which provides little opportunity for initiative. Such a strict, demanding environment produces a child who regards the "lucky break" as the only means of improving his lot. Games involving physical skill reflect socialization which stresses achievement. In order to win, the individual must compare his performance with that of others and exert a more heroic effort. The key to success lies in "more." The child who prefers games of physical skill is preparing himself for a future of striving to outdo his fellow man. Games of strategy are at the highest cognitive level and probably do not flourish until adolescence when the child's thinking is sufficiently abstract for him to understand the principles of such games and to grasp his opponent's style and use it to his own advantage. He also is sufficiently subtle to obscure his own style. Preference for games of strategy reflects intense, demanding socialization, implemented by psychological techniques (e.g., reasoning) rather than physical ones. The aggression engendered by the demands of the socializer is diverted into symbolic competition. The child learns how to dissemble—to appear to be good and docile while waiting for a chance to usurp his socializer's power. In his games of strategy, he makes moves which appear harmless, so that his opponent will be deceived or thrown off guard. Although these hypotheses have not been satisfactorily verified, there is evidence that children can reliably classify one another as "fortunists," "potents," and "strategists," and that strategists are brighter, most subtle in social relations, and among physical games prefer those requiring skill but not physical contact.

Deviant Development. As often happens, psychologists working with emotionally disturbed children are sensitive to variables which elude investigators who study normal behavior. So it is with games. Redl (1959), who for many years has conducted a therapeutic program for violently aggressive boys, is particularly interested in the development of impulse control. Just as Erikson regards dramatic play as the child's inner life depicted in the microsphere, Redl calls games a "universe in miniature." "Playing the game" is an important aspect of self-control, as games reflect the children's own attempt to regulate their behavior. Games serve the dual function of arousing affect and keeping it in bounds. In certain "quiet" games, such as Monopoly and checkers, the affect may be at a moderate level. However, other games involve a high level of emotional arousal: the excitement of tag games, the suspense of hide and seek, the aggressiveness of hitting games, the sexuality of kissing games. If the game serves its proper functions, the child can express his feelings within the protective confines of rules and regulations.

Redl is fascinated by the many checks and balances built into the structure of games. Being cooperative ventures, games require a certain de-individualizing of the participants. An erstwhile bully must control his urge to push others around if he is to play a hitting game; a shy child can allow himself to hit without being guilty or anxious because it is part of the game. Along with the impersonality of rules goes a group code which serves as a kind of game morality. In part, the code regulates behavior in order that the game can proceed on its course. The individual cannot demand time or attention or special concessions without being censored by the group. Hogging the ball to keep a better player from getting it, breaking out of a waiting line in a relay race, would be regarded as unfair. In addition there is a code of fairness and sportsmanlike behavior; a child must not cheat in order to win, the victor must not gloat over his victory or unduly disparage the vanquished.

Games provide a host of controls other than rules. The geography of a game is important. In circle games, the players are always in plain sight of the entire group and therefore are under maximum group surveillance. The "in bounds–out of bounds" dimension which is found in football and checkers alike, serves as a geographical check against capricious movement. Random turn-taking prevents power plays and favoritism. Some games allow the child to decide how long he will be the center of excitement and at what point he will terminate this role. Certain tag games, for example, give the "chasee" the option of prolonging the chase or seeking the safety of "home"; thus the child can regulate for himself the amount of exertion, excitement, and suspense he can tolerate. Games control the time dimension, which is another aspect of self-control. Some games consist of repeating a simple sequence of activities with different participants and can be terminated at any point. Others require an elaborate working out of sequential events and cannot be terminated until they have run their course. The latter obviously demand more self-control than the former. Thus, by structuring space and time, by providing supervision, by regulating order and length of participation, and by a host of other control mechanisms, the game can enable the child to manage the excitement it has aroused.

One sign of disturbance in children is their inability to "play the game." For the kind of impulse-ridden children Redl treats, the excitement stimulated is more than they can tolerate. The usual childhood games end up in fights or in "wild" behavior.

In certain groups where the power structure is rigidly fixed and defended, the leveling effect of reciprocal activity is intolerable. Neither leader nor follower can relinquish his position in order to participate in the fluid interchange of roles, such as required in a tag game. Redl's impression is that his disturbed children have far less capacity than have normal children to alter rules in order to meet special needs or unforseen contingencies. Finally, Redl contends that games, because they share the same intrinsic appeal in middle childhood that dramatic play had for the preschooler, can be used therapeutically to enable impulse-ridden children achieve the self-control they need.

Neurotically inhibited children can be equally unable to play the game. Just as the dramatic play of the troubled preschooler loses its light, improvisational quality, games can lose their playfulness. Winning and losing, instead of being part of the game, bite deeply into the child's self-esteem and touch off reactions of excessive vanity or excessive humiliation. In Adler's terms, the pampered child cannot abide the leveling effects of a game which makes him no different than his peers, while the hated child cannot tolerate the idea that others are getting the best of him. Games lack the symbolic richness of dramatic play, but they still have a thematic content. Redl's list of frequent themes includes race, chase, attack, capture, harassment, search, rescue, and seduction, any one of which can be frightening to an anxious child. The simple gesture of jumping an opponent's man in checkers may symbolize a destructiveness which the child finds intolerable; a marble taken away can mean being helplessly at the mercy of the victor. The excessive personalizing of essentially impersonal activities betrays the child's psychological disturbance.

Work

Overview

Play is the child's world; work is the adult's. The decision to enter the labor market on a full-time basis marks the end of childhood as much as any single decision can.

There are many testimonies to the importance of work. The most succinct is Freud's statement that the normal man should be able "to love and to work"—although, ironically, psychoanalytic theory has contributed little to our understanding of the psychology of work.[7] Roe (1956) lists the many basic human needs work satisfies. It provides money for food, shelter, clothing, medical care, and a secure future. Work answers the social need to be part of a group, since congenial interpersonal relations are important to job satisfaction. Self-esteem is contingent upon having a job, especially for men, and the man's status or prestige depends primarily on his occupation. Work can answer the need for exploration, expansion, and variation. In many instances, it is also the vehicle for self-actualization, that special fulfillment which comes

[7] Hendrick (1943) provided an opening wedge for a psychoanalytic study of work over twenty-five years ago, but the topic has received only fitful attention. For a review of the psychoanalytic literature on work, see Neff (1965).

when an individual regards his work as the expression of his unique talents and values.

Super (1957) lists a number of ways work influences one's life. In our fluid, industrial society, "Who is X?" is typically answered in terms of what X does. Work has its special physical settings (compare a hospital with a body repair shop), its special social climates (compare the social climate of workmen eating lunch outside the Club with that of the faculty eating lunch inside), and its special daily schedules and routines. Work determines friendships and social activities, clothing, language, and conduct. It affects the amount and use of leisure time, and the kind of leadership role one assumes in the community. Work not only expresses personal values but also determines them; e.g., lawyers tend to see themselves as pillars of society, physicians as members of a guild rather than helpers, and chiropractors as an oppressed minority. Work is a way of life or, in Wilensky's (1964) phrase, employment is the symbol of "one's place among the living." The nonworker, whether he be the unemployed, the dropout, the aged, or the unemployable, is isolated from the mainstream of community life.

Note that Super's catalogue, while extensive, still does not touch upon the interrelation of work with the adult functions of spouse and parent. Such interrelations form the fabric of adult life. They range from the simple situation in which a father who fails to get a raise whips his child to the complex problems of achieving a balance between the demands of home and a career.

In spite of its acknowledged importance, work has received scant attention from child psychologists whose attitude toward it resembles the preFreudian attitude toward sex: work appears fullblown in adolescence after a period of vocational innocence. Fortunately, in the past two decades, counseling psychologists have shown increasing interest in career development and are now engaged in promising longitudinal studies which will trace the course of vocational development in childhood. Until they are completed, we must rely on a literature which is more theoretical and programmatic than well documented and specific.

Our discussion of work will concentrate on three areas. The first is the origins of work, broadly defined as the ability to exert sustained effort in order to accomplish a task. Next we will treat the issue of occupational choice, epitomized by the changing responses the child gives to the question, What are you going to be when you grow up? Our focus will be on the child's developing understanding of the adult world of work. Finally, we will selectively explore certain vocationally relevant variables, concentrating on interests, abilities, self-concept, and parental influences. We will make only passing reference to intelligence, skills, talents, and values, all of which are important but would make our presentation too unwieldy. Because we have limited our discussion to the middle class, we must neglect one of the most potent determiners of occupation, namely, socioeconomic status. We shall regard this factor as establishing a frame of reference, a hierarchy of occupational possibilities for the middle class child.[8]

[8] See Roe (1956) and Super (1957) for reviews of the complex roles which socioeconomic status, interests, abilities, values, and personality factors play in determining occupational behavior. (Cont'd.)

Our discussion would have been simpler if development were logical. One might hope that interest and talent would go hand in hand, reinforcing one another in a happy partnership, and that career choice would be merely a matter of finding an appropriate reality setting for the adolescent's particular gifts. But research with adults shows us that work-related behavior is rarely either simple or logical. Adults do not necessarily choose a job they are interested in or have talent for; and, if they do choose on the basis of interest, there is no guarantee they will be successful. Therefore we have no precedent for assuming that an adolescent will be interested in his work or strive to fulfill the promise of his talents. A child may be uninterested in an activity, not because he has tried it and been bored, but because it is for sissies or tomboys. Variables other than sex role appropriateness weave in and out of the developmental picture, now subordinating, now subservient to, now mingling with abilities and interests. However, it is just because personality development is so erratic in its course and so intertwined in its substantive variables that child psychology can justify its claim to be a specialty area within general psychology.

Finally, a word about nomenclature. Job, occupation, and career can be distinguished from vocation. The first three are work specific. A job is a specific paid activity in a specific setting. Occupation refers to tasks which are similar from setting to setting. A man can say, "My occupation is a salesman, and I have a job at the Super Novelty Company selling toys to retailers." A career is a general life pattern which revolves around one's chosen occupation. A vocation, as we shall use it, is even more general, and refers to a mission, a life purpose, a calling. However, we will not deal with the concept of vocation until we have completed our discussion of work.

Origins of Work

The remotest origins of work can be found in the task orientation, planfulness, and frustration tolerance which are an intrinsic part of initiative.[9] Even the one-year-old has a remarkable capacity to concentrate on the task at hand. Like any effective worker absorbed in what he is doing, he is not distracted by inner needs and passions or by extraneous external events, people, or objects. One can observe the toddler exploring objects (turning a plastic cup over and over, patting it, hitting it on the floor, mouthing it, turning it over again, regarding it, all with great seriousness and concentration), experimenting with them (tapping a block now on the floor, now on a chair leg to discover the different sounds he can produce), solving problems (standing on a chair to look out of a high window), constructing (making a crude enclosure with blocks and calling it a garage for his toy car). It would be a mistake to dismiss such activities because of the toddler's lack of skill and intellectual sophistication or to claim that the child begins to work only when he is capable of carefully planned and realistic constructions. All origins are crude and unstable compared to fruitions. Nor is it correct

[8](Cont'd.) For a review of developmental research, see Borow (1966). For an excellent presentation of the area of career development, see Osipow (1968).

[9] We agree with White (1960) who regards work as the culmination of "competence," a concept which closely resembles initiative.

to say that the toddler lacks the ability to sustain effort in the face of frustration, which is the hallmark of work. On the contrary, toddlers can be remarkably persistent, especially in light of their limited skill and understanding; e.g., they can be observed repeatedly trying to build a block tower in spite of developmentally inadequate motor coordination.

The one-year-old is an embryonic worker, not a worker in miniature. Because his attention span is short, he does not stay with a task more than thirty to sixty seconds on the average. His short memory span limits his ability to return to unfinished business. He cannot set long range goals and coordinate discrete activities. But there is another quality which prevents the toddler from being a true worker: he is free to follow his interests wherever they lead. He might happen upon a bookcase and concentrate on taking the books out one at a time, then go to a paper bag and explore the sounds it makes when he hits it, then become interested in the problem of crawling up and down stairs. The child does what he wants to do because he wants to do it; the adult works whether he wants to or not. Unlike the other work-relevant variables which are present in embryonic form, the ability to do what one is required to do is absent from initiative. We shall be interested in discovering when this "requiredness" appears on the developmental scene.

Work and Play. The literature on early childhood contains the phrase, "Play is the child's work." We disagree, but the phrase indicates that observers have been struck by the workmanlike quality of the toddler's behavior. We suspect that the activity observed was exploratory, rather than pleasure in functioning or dramatic play, and could be subsumed under our concept of initiative. We prefer to distinguish work from play, rather than to equate the two. Work, as an aspect of initiative, is reality oriented and constrained by reason and logic. The child is concerned with exploring, learning about, and mastering the real world. In play, as we have seen, the child is relatively unconcerned with mastering the real world and relatively free of the constraints of reality, reason, and logic.

In distinguishing work and play, we do not mean to dichotomize them. Reality influences the content of play; the dramas acted out by the twentieth century five-year-old have different characters and props from those of his twelfth century counterpart. Furthermore, creative work in adults requires the ability to return to the fluid world of playful ideation. Nor do we imply that play is more rewarding than work in early childhood. The toddler or preschooler appears to be expressing himself as much in exploration as in play. He concentrates on mastery of the environment because this is the most compelling activity at the time; then he slips into dramatic play for the special pleasures it holds. The precursors of work seem as intrinsically rewarding as is play; the rewards are merely different.

Work in Middle Childhood

Industry and Inferiority. In middle childhood, psychologists recognize the child as a worker. However, the literature is meager and we must deal either with generalities or fragmentary research.

Erikson (1959), who is concerned with the growth of the self, regards middle childhood as a time when industry or inferiority develops. On the positive side, the child needs a sense of usefulness; he must feel that he can make things well and that he can win recognition by producing things. No longer limited to mastering objects by using his body alone, as a tool-user he extends his body. In his identifications he seeks adults who know things and who know how to do things. He models himself on "teachers," using that noun in its broadest sense. By developing a feeling of industry and by experiencing pleasure and pride in a job well done, the child is preparing himself to be an adult worker and a provider.

A failure in this development will produce a sense of inferiority. The child may be still involved with family relationships which should have been resolved: he may be too concerned with separating from his mother or with envying his father's power to participate in the impersonal activity of producing things. He may encounter teachers who disparage his efforts or who make him a pet and lure him into being a good little helper. Or he may meet prejudices which define him as inferior rather than evaluating him at his real worth.

Because of his psychoanalytic orientation, Erikson claims that the particular ratio of industry to inferiority which characterizes middle childhood persists into adult life. Although it is possible to change subsequently, change is slow and difficult to achieve.

Fantasy Choice. The most detailed hypotheses concerning career development come from Ginzberg (1951). His ideas are speculative, were derived from studying a limited number of children, and have not uniformly fared well in terms of objective verification. However, he has been more daring and more explicit than others who have written about middle childhood.

Ginzberg calls the age span between six and eleven years the period of fantasy choice. The child has grasped the idea that he will work when he grows up. He is eager to be adult and imagines himself in adult occupational roles. His response to the question, What would you like to be when you grow up? may tend to be less adventuresome and grandiose, more in line with reality, than those of a preschooler. Why, then, label this as fantasy choice? Because of the close connection between wish and reality in the child's mind. The child selects an occupation on the basis of what is attractive or pleasurable at the moment, and he thinks that the choice itself is sufficient to decide the reality of his occupational future. Liking to be a doctor and wanting to be a doctor mean that he will be a doctor. He has little grasp of means-ends relations and of the complex weighing of factors involving himself, the occupational world, and the preparation necessary to realize his goal. Above all, his time perspective is distorted. He knows only "now" and "later on," which he equates with "not now." He does not understand the concept of progressive change over time. His fantasy, his wish to be what he would like to be, is the link between himself and the future. Interestingly, this description sounds like a variation on the familiar theme of egocentrism (see Chapter 6)—reality is constructed from the point of view of the child's wishes.

Toward the end of this period the child is in a state of transition, since many forces conspire to bring him out of the fantasy period. He is beginning to be able to view himself objectively. No longer totally absorbed in doing, he has some capacity to stand aside, as it were, and reflect on what kind of person he is. With self-awareness comes the idea that he must size himself up if he is to achieve occupational goals. He dimly understands that the answer to the question, What kind of person am I?, is essential to answering the question, What kind of work do I want to do? Next, he is increasingly knowledgeable about reality. His horizons have expanded beyond the home to school, neighborhood, and city, and he has been exposed to adults in a variety of occupations. He views occupations more realistically and in a more differentiated form. Mere attraction is not enough; he now begins to weigh advantages and disadvantages. To want to be a pilot because it is exciting is regarded as "kid's stuff"; the excitement must be balanced against the dangers and risks. Parents and teachers also begin to talk more seriously about preparing for the future. The end of grammar school is the end of occupational innocence; while the first year of junior high school is not a year of vocational choice, vocational choices are "in the air." Finally, the physiological changes of the prepubertal period serve as biological reminders that childhood is drawing to a close and that maturity will be a reality.

The child between eleven and thirteen years of age can no longer accept the fantasy stage, but he is uncertain as to what he wants to be. He might turn to his parents, not in a dependent desire for protection, but in a search for guides to the future. Often, however, he is not satisfied. He realizes that occupational choice must ultimately be his own responsibility. At this point he enters the new phase called the stage of tentative choice. We will return to this stage when we discuss adolescence.

Although many of Ginzberg's speculations have yet to be validated, one empirical investigation is supportive. Gunn (1964) noted that the adolescent's ranking of occupational prestige was essentially the same as adults' and studied its development in boys. She found that most first- and second-graders knew what they wanted to be and, in keeping with Ginzberg's ideas, equated wanting with being. The reasons for their preferences were egocentric—they wanted the occupation which would give them the most pleasure and excitement, such as being a policeman. Even when they ranked a teacher high, it was for egocentric reasons, such as "A teacher helps you if you get stuck when climbing a tree." One is reminded of Ginzberg's observation that youngsters are concerned primarily with their own feelings and have little realistic understanding of occupations. Their fathers' occupations were high on the list, which probably reflected their limited perspective. Finally, Gunn found that the children were incapable of true ranking of the eleven occupations. They would choose a few they liked best and lump the rest together. Even among their preferences, one occupation was as good as another.

The third grade was a transitional time when the children began to see jobs in terms of their importance to the community rather than as pleasurable or exciting. By the fourth to sixth grade, occupations were ranked rigidly in terms of their service to the community; self-interest or father's occupation were no longer central considerations. In one case, father's occupation of landscape architect was ranked below that of filling station attendant because it "just makes things look fancy and is not really needed." To help, to cure, to fill a real need were the keys to prestige. The finding is interesting in light of our discussion of moral development. It is congruent with Kolhberg's "good boy and authority maintaining" morality and to Jacobson's picture of the too pure, too unworldly conscience (see Chapter 5).

Interests and Abilities. Interests play an important but complicated role in occupational choice, even at this early stage. Definitions of interest usually include a preference element and a satisfaction element: an interest is what a child would choose to do when he is free to choose, and what satisfies a need or brings him pleasure. In general the development of interests parallels broad trends in physical and mental development; the toddler will have different interests from the preschooler, the bright child from the dull (see Hurlock, 1964).

One might assume that interests and specific abilities go hand in hand, but such is not the case. There are many reasons why a child is not necessarily interested in what he can do well. The child is expansive in this period and his enthusiasm is intense but prone to evaporate suddenly (Harris, 1950). Many a parent has rued the day he bought a musical instrument, only to see the child's interest languish before the last of twelve monthly installments was paid. A dislike for a teacher can spread to a dislike of the subject taught, even if the child is doing well (Hurlock, 1964). A child can develop an interest in a skill after he discovers it is highly valued by peers (Harris, 1950). Thus, the child's temperament, his likes and dislikes of teachers, his need for status, recognition, and approval, are potent determiners of interest, independent of ability. Tyler (1951, 1955) found a striking sex difference in interests as early as the first grade. By ten years of age, sex appropriateness is clearly more potent in determining interests than experience with activities. Activities which boys or girls group together as disliking have little similarity in content but are all sexually inappropriate. Boys dislike blocks, jacks, playing doctor or school, cleaning house, and sewing, while the girls dislike football, arithmetic, and wrestling. Clearly children are not free to follow their developing abilities wherever they might lead, nor are they impartially rewarded for doing well. An interest which runs counter to the dictates of sex typing has a difficult time surviving.

The Influence of the Home. The home is often mentioned as an important influence on occupational choice. Super (1957) lists some of the ways it prepares the child for work. He notes that parents, and especially the father, serve as models of workers, a point corroborated by Gunn who found that, especially in the first few grades, boys frequently wanted to do what their fathers did. Super makes the more subtle point that parents also serve as models for authority and loving support, the father typically epitomizing the former, the mother the latter. He thus touches upon a dimension of work which we will discuss subsequently—the individual's attitude toward authority and power, and his need for understanding and kindness in the work situation. Implicitly Super suggests a tie between love oriented and power assertive discipline (see Chapter 4) and the interpersonal conditions necessary for work productivity.

The home is a place of work—it is a hotel, a restaurant, a school, a carpentry shop. As he grows, the child is given increasing responsibilities for the work aspects of family life—for cooking meals, setting the table, cutting the grass, caring for younger sibs. According to Super, these work opportunities enable the child to explore himself, to find what he likes and dislikes, what he does well and poorly. Super is careful to state that this is self-exploration, not occupational exploration, thus echoing Ginzberg's idea concerning the increased self-understanding in middle childhood. One cannot assume a direct relation between home experiences and occupational adjustment. Remember that Gunn found that the idealized father of the first-grader can be demoted lower than a filling station attendant by the fourth grader; equally relevant is Harris's (1954) finding that there is little relation between the assigning of routine tasks around the house and an attitude of responsibility in ten- and fifteen-year-old children. As many parents will confirm, the child is most eager to mow the lawn at an age when he does not have strength to move the mower. By the time children have the skill to be useful, household chores are a bore and a bother. In sum, mere exposure to models and assigning of chores do not mean that the child is being prepared for his adult life as a responsible worker.

Another approach to the relation between family and occupation has been made by Roe (1956), the outstanding spokesman for the view that vocational interests have their origins in specific parent-child relationships. The child who is adequately but not intensely loved tends to concentrate on objects rather than on human relations and is drawn to technical and mechanical interests. Intense parent-child relationships, on the other hand, will be a source of conflict which often centers around domination-submission. An individual greatly distressed by domination seeks to avoid situations which involve power struggles and will gravitate either toward service occupations (such as social work) which are interpersonal, helpful, social, and "feminine" or science, which is impersonal and intellectual. If the individual identifies with the dominating parent, he will gravitate toward business, which is exploitative, interpersonal, power oriented, nonintellectual and nonaesthetic. Incorporation of submissiveness may lead to a subordinate role in any chosen occupation. If the child concentrates on himself, either because of overconcerned parents or because he possesses special abilities, he develops the narcissistic attitude characteristic of people in the arts and entertainment.

Research efforts to relate early parent-child relations to occupation has had a now-you-find-it, now-you-don't kind of history. An early study by Friend and Haggard (1948) contrasted adults with good and poor occupational adjustments. The former had a history of a closely knit family, affection for the father, and early signs of independence, while the latter reported much antagonism especially to the same-sex parent, deviant family patterns in the form of delinquency, alcoholism, or family disruptions through death or divorce. However, the poor adjusters were more often the favorites of the family than were the good adjusters, suggesting that Adler was right in assuming that the pampered child was as unprepared for adult life as the hated one. Nachmann's study (1960) is also confirmatory. Using a psychoanalytic model, she successfully predicted the parent-child relations of three different occupational groups. To take one example of her approach: a lawyer is characteristically aggressive verbally, assertive, and concerned for justice, while a social worker cannot express aggression, is concerned with the spirit rather than the letter of the law, is benign, and helping. The father of the lawyer, therefore, should be strong, dominant, and masculine, while the social worker should have a weak father and a more dominant, adequate mother. These predictions were confirmed.

Yet, such successful studies are matched by an equal number of failures, including Roe's own failure to confirm many of her hypotheses. One is left with the impression that the family may be important, but is only one of a host of significant determiners of occupational interest and choice. To date there is no convincing evidence that it should be given priority.

It is possible that specific facets of work (rather than occupational choice) bear the stamp of early familial experiences. Let us take two studies concerned with achievement, or setting high standards for oneself, a variable which is especially important at higher occupational levels. Moss and Kagan (1961) found that intellectual achievement is one of the most stable personality characteristics. Preschoolers who have a strong desire to master intellectual skills retain this motivation during adolescence and into early adulthood. In addition, achievement is related to early maternal reward. We might therefore assume that a clear relation exists between maternal behavior and an important determinant of occupational choice and success.

However, Elder (1963) suggests that Moss and Kagan have isolated only one of many factors influencing academic achievement in high school. Achievement can wax and wane according to the realistic opportunities the environment offers for being rewarded for effort; the apathy and indifference of a slum child, for example, may be realistic if his occupational future is bleak. The school itself can serve to encourage or discourage high achievement. The achievement orientation of peers or of a best friend are also potent determiners of goals, especially in adolescence.

Moss and Kagan studied middle class families; Elder's was a rural population. It is possible that school, peers, and society conspired to support the intellectually aspiring child in the former study. The strength of the mother's influence may never have been put to the test by contrary extrafamilial pressures. Elder studied a group in which forces both within and beyond the family conspired to discourage intellectual ambition. His study suggests that a boy whose mother rewards intellectual achievement but whose peers resent it, whose teachers are threatened by it, and whose community has no occupational outlet for it, might not show the same consistency over time as did the children in the Moss and Kagan investigation. We are forced to conclude that the relation between early familial experiences and specific work-relevant variables is not clear.

Children as Workers. Most of the research on work is concerned with precursors of adult occupation in middle childhood. One of the few psychological investigations of children who work has been conducted by Engel and her colleagues (1967) who studied working boys between eight and fourteen years of age. Work was defined as earning money from a person outside the family. Engel developed the concept of work style which, by utilizing five dimensions, integrates various aspects of work. (1) Work-mindedness refers to the extent of the boy's commitment to work. At one extreme, work is merely a means to an end and lacks any intrinsic pleasure or interest; at the other extreme, boys work with seriousness, purpose, and zest, and their activity is intrinsically meaningful. Work engages their ego and becomes part of their identity. (2) Activity-passivity refers to the degree to which a boy tries to fashion the conditions of his work. Some boys accept their jobs without protest or criticism; others are go-getters who continually try to improve their lot in terms of pay, working hours or working conditions. One enterprising twelve-year-old organized a newsboys' strike to protest what he considered discrimination against younger workers. (3) Occupational orientation taps the boy's ability to conceptualize his vocational future and to understand the continuity between present work and the future. This is the time dimension which Ginzberg regards as so important to occupational development. Many boys do not entertain the idea that the present is linked with the future, while others regard it as a preparation for adulthood. (4) Concept of life span is related to occupational orientation but concerns the extent of the boy's time span rather than the continuity of present and future. Some boys live for the moment, while others can project themselves into adulthood. (5) Work relationships with adults refers to the depth of emotional involvement with work-specific adults. For some boys there is no evidence of emotional investment, while others form deep attachments.

Engel speculates that these five facets of work style do not develop at a uniform rate. Occupational orientation and concept of life span are unstable because the ability to think about the future in a realistic and complex manner might not develop until around eleven or twelve years of age. Activity-passivity reflects different degrees of initiative, while work relationships tap reactions to authority figures. Engel speculates that both issues have been settled in earlier childhood and should be stable during the middle period. Work mindedness as such is still in a state of flux. Like other personality variables, work style does not develop at a uniform rate; some of its components become fixed in early childhood, while others may not stabilize until adolescence or adulthood.

Engel's research represents a rare attempt to study personality characteristics as they are revealed in work. Note that she is not concerned with occupational choice or with the question of why the boys were attracted to one job rather than another. Rather, she delineates the kind of workers the boys were—their ego involvement in what they did, their initiative, their attachment, their time perspective. As we shall see (p. 357), Drucker will repeat her theme: the kind of worker one is depends on the kind of person he is. What are you like as a person? is as important a question as, What are you interested in doing? or What are your skills?

Summary. Middle childhood is the period when industry comes to the fore. The child becomes a producer, a tool user. When he actually holds down a job, he can derive intrinsic satisfaction from his work, show initiative, and develop a deep attachment to the adult for whom he works. However, his time sense is weak, and he still does not grasp the idea of realistic preparation for the future. In the early years of this period, the child equates wanting and liking an occupation with actually doing it in the future, and his ideas about various occupations are undifferentiated. Later in the period, he evaluates occupations in terms of their service to others, his choices showing the same concern with absolute goodness as his moral reasoning; however, his understanding of occupations is more differentiated. Interests are strongly sex linked rather than contingent upon experience and skill. In general, interests are related more to interpersonal factors, such as an activity's prestige value with peers, than to ability. The role of the family is unclear. Responsibility is unrelated to doing chores around the house. The attempt to relate specific parent-child relations to occupational interests and choices in adulthood has been only fitfully successful. However, the family may play a role in influencing specific work-relevant variables, such as intellectual striving, reactions to authority figures, and initiative.

School and Work

Imagine an adult who is giving an orientation speech to a group of newcomers.

> You have never been in this building before and you have never done anything quite like the work you are about to do. So let me tell you some things you should know. You will be spending most of the day in this room. While you are here, what matters most is how well you produce what we think is important. Some of you will enjoy working and will work well; others will not. Our aim is to help everyone do his best. Those who do well will be highly rewarded; those who do poorly will not. I will let the poor worker know how he has done and I will encourage him to improve. However, I do not have the time nor is it my function to cater to individuals. I will not play favorites and reward you for anything except what you do. On the other hand, I will not excuse you because of a trivial personal problem, such as temporarily feeling sick or bored or worried about what is going on at home. If you have some serious problem which interferes with your work, I will see that you get special counseling. But I can help you most by showing you how to work.

Where are we? Is this an intelligent, sympathetic employer speaking to a group of new workers or an intelligent, sympathetic teacher speaking to a group of first graders? It might be either. We are not suggesting that the first job and the first grade are identical; there are many significant differences. We are suggesting that a special psychological situation is established in school which is different from that which has existed at home but is on the same continuum as work. We need to define this continuum more explicitly. In order to highlight points, we will assume that the transition from home to school is more abrupt than it often is, since many middle class children have had some experience with school-like settings before they enter the first grade.

Home is familiar, school is strange. Many of the customary anchoring points of the physical and social environment vanish. The adults, the children, and their relation to each other are strange. The structuring of time is strange. "School" itself is not a psychological reality. The child probably has some factual information concerning what to expect, and he undoubtedly has his own fantasies, perhaps too frightening, perhaps too hedonistic. But the reality of being in school cannot register until school has been encountered and assimilated. The strain of adjusting to the first grade can often be evidenced by over-excitement, increased irritability, or excessive sleeping at home. The first weeks of school may be as taxing as the first weeks on a job.

The Impersonality of School. The most significant feature of school is its impersonality. The child is no longer special as he is at home or even in the neighborhood. He is one of many children whom he has not chosen to be with, and he is with an adult on whom he has no special claim and who herself feels obligated to show no favoritism. His physiological needs cannot be gratified at will; neither the refrigerator nor the toilet is available on demand; rather, eating and going to the bathroom must be done on a schedule which all must observe. Anna Freud (1965) elaborates on this theme of impersonality. At home, socialization is accompanied by the rewards of being loved. Parents are alert to the individual needs of the developing toddler and preschooler and to the times when demands should be relaxed and tenderness and sympathy increased. This is as it should be, since the child's ability to cope with expanding demands waxes and wanes. In school, however, rules are for all. The child's individuality is recognized to the extent that he is placed with children of similar age, but, within his age group, he is expected to conform to the norm. Anna Freud concludes that home and school are sufficiently different that one should not expect good adjustment in one setting to insure good adjustment in another.

A job represents a similar step in the direction of impersonality. In work, rules are for all; individual consideration is the exception. The rights of early childhood to dependency gratification are particularly inappropriate. The worker who is clinging, who shirks responsibility, who continually excuses himself on the basis of minor ills or fatigue, or the one who, like Adler's spoiled child, believes that his mere presence entitles him to special treatment is introducing an inappropriate personal note into the working world. Workers are expected to function and to be reliably productive in the face of contrary needs to be cared for.

Requiredness. Another impersonal feature of school is more closely related to the development of work. Before school, exploration was more or less the result of spontaneous curiosity peaked by environmental stimulation. The preschooler was free to pick and choose what he would do. In school, however, learning is required. It is this element of requiredness which accounts for one of the most frequently mentioned characteristics of work: work is something one must do whether one wants to or not. Schooling, like socialization, often can go against the grain. And, as in socialization, the child is not given the option of not complying.

Educational philosophies differ as to how best to handle the requirement to learn. Progressive approaches value the spontaneous curiosity of the child and his ability to learn through discovering for himself. In terms of our discussion, the progressive school gives the child's initiative free rein. Traditional education emphasizes discipline in learning and utilizes the adult's ability to guide wisely while serving as a model of disciplined thinking. A discussion of these contrasting philosophies of education (see Minuchin et al., 1969) lies beyond our scope, but we are struck with the parallel between the educator's dilemma and the parent's dilemma in regard to socialization (see Chapter 3). The good parent wants to foster initiative while requiring the child to conform to the demands of social living. The good teacher wants to keep spontaneous curiosity alive while requiring the child to master the content of his intellectual heritage. The hotly defended viewpoints concerning the relation of freedom to conformity has its counterpart in hotly defended opinions concerning unfettered curiosity and sound work habits.

Product Orientation and Public Evaluation. In school the product of the child's effort is valued as never before. The preschooler could be precise or haphazard in what he produced. His behavior might encounter parental censure (if it were destructive to property, for example) or parental praise, but his activities were not constantly monitored and there was no adult whose principle function was to evaluate what he did. The evaluation itself revolves around "right-wrong," which is different from the parental "good boy-bad boy." Being right, rather than being good, now opens the door to adult approval and the child's products take precedence over his conduct.

School also exposes the child to public evaluation. The first grader discovers that he receives a report card, which means that a nonparent has the right to judge his behavior and to summarize weeks of effort in a single grade. Parents must contend with this evaluation which, among other things, raises the question of who has the power to judge the child, the teacher or his parents? This may be the first time parental authority has been seriously challenged by an extrafamilial adult, and certain children are quick to sense a power struggle in the making.

Report cards not only reflect the evaluation of an authority, they also proclaim it. Grades become topics of conversation among peers and, because they are standardized, any child has a means of comparing himself with any other. Both aspects of the school's public evaluation have occupational overtones: the teacher, in her power to determine the child's fate by her evaluations in the classroom, foreshadows the boss; while grades begin the transition from the private, casual reaction to productivity within the family, to the standardized evaluations of the world of work.

Divided Loyalties and Gradeworthiness. There is another similarity which emerges as one juxtaposes the literature on school and work. School can confront the child with the issue of divided loyalties—to himself, to his peers, and to the teacher. His own interests and abilities may not be congruent with peer values, and the demands of the teacher may run counter to both. A girl who likes arithmetic or a boy who prefers studying to sports may suffer social isolation. The teacher's pet can be torn between the psychological seduction of the teacher and the loss of prestige among peers. By the stage of high school, conflicts of loyalties can be verbalized. A bright college girl recalls the following incident:

> I remember that in high school, I was smart, and I really liked Lit and the teacher we had. Some of my friends wanted me to sit near them so they could copy my paper during an exam. I didn't know what to do. I didn't want them to think I was square, but I knew it wasn't right, especially since I liked the teacher. I was mad, too, come to think of it; I had to work and they were going to get something for nothing.

School can set the stage for an important issue in the world of work: What do I owe myself? What do I owe the people I work with? and, What do I owe the people I work for?

Finally, school links self-esteem with productivity and achievement to a greater degree than in the preschool period. As we have seen, self-esteem depends on how loveworthy the child is to his parents and how respectworthy he is to his peers. Now we add the concept of being gradeworthy in school. This entails meeting the school's special intellectual and personal demands for productivity to the best of the child's ability and being fairly rewarded for his efforts. The child is now identified as "student" and "student" becomes part of his image of himself. His self-valuing is contingent upon how gradeworthy he is. The feeling of being a good student, of having done a good job in school, is an asset to carry into the world of work. The feeling of having failed in school may make the child doubt his ability to make the grade on the job.

There is another aspect to gradeworthiness. Grades become status symbols and, like any status symbols, they can have positive or negative effects. They can serve to spur the child on and they can be a sign of a job well done. But they also can become an end in themselves when the child begins to work only for grades and when grades become the sole focus of parental concern. As is true of any strong extrinsic motivation, the balance between effort and reward can be upset. The child comes to hate work, to hate his parents, to hate himself. Child guidance clinics always have a goodly number of bright underachievers with pressuring parents. Note, however, that we are only pointing to a danger of extrinsic rewards; we are not saying that pressuring is uniformly bad while intrinsic gratification is uniformly good. Status striving can maximize as well as overwhelm, and we know little about the conditions under which it takes one course rather than the other.

Summary. School is a halfway house between the relatively free exercise of initiative in the preschool years and the constraints of adult occupation. The child functions in an organized extrafamilial setting with its own routine, with objective standards of achievement which are uniformly applied to all, and with public evaluations of functioning. Rewards are product oriented and evaluations are in the hands of nonparental authorities. Most important, the requirement to perform is essential. The child cannot choose not to learn. He now has activities which are congruent with his interests and others which he must perform. Since the latter are likely to be disagreeable, work takes on its adult connotation of being "hard" or something one would not choose to do. The fact that the child's general attitude toward school often changes from positive to negative between the first and sixth grades indicates that school plays an important role in equating work with doing something disagreeable.[10]

We end our discussion of middle childhood with a reminder that initiative continues to flourish. The school presents courses and activities with such high appeal that the child feels that he is doing what he wants to do and would not choose to do anything else. Outside of school, boys build tree houses, collect rocks, take clocks apart, while girls design wardrobes, write poetry, and plan parties. Such activities are not play, since the children are mastering reality, not suspending it by the magic of pretense. The activities are effortful and goal directed but are not disagreeable as is "hard" work. Hobbies is too special a word to describe them, recreation too adult. They are intrinsically rewarding constructive activites. We shall meet the phrase and the idea again when we consider the matter of identity in adolescence.

Work in Adolescence

The Complexities of Occupational Choice. Adjectives used to describe the adolescent period are revealing—tentative, exploratory, trial, and even floundering. The picture is one of searching. Commitment to an occupation is in the making for four to eight years; finding a specific kind of work which is fulfilling takes even longer. The length of time is not due to the educational requirements for middle class occupations, although this is one variable. Rather, it is due to the complexities of occupational choice itself and to the unknown and unknowable factors involved in job satisfaction. What are some of the complexities?

Starting with the least psychological of variables, the job market itself is a complicated and changing reality (see Borow, 1966). The occupational structure of the country has shifted radically from rural to urban, from goods-producing industries to services, from unskilled, sales, or clerical work to white collar or professional jobs. The formal educational requirements for entrance into many occupations is on the rise. Once on the job, the worker is increasingly likely to receive additional training as his job becomes more demanding or to be retrained as his job becomes obsolete. The world of work increasingly is dominated by large scale industrial organizations, which means that the distance between worker and management and between worker and product has widened. Although being a member of a large organization may have psychological repercussions, there is no clear evidence that chances for upward mobility decline. Another occupational reality is the increasing proportion of women workers. Today's high school girl can expect to spend twenty-five years of her life as a worker. One final fact of occupational life is that the working class is working less, while professionals work as much as or even more than they did in the past! Since the occupational picture is constantly changing, there is no simple answer to the adolescent's questions: What jobs are available? What preparation will I need? Which

[10]While we have stressed the element of requiredness, a host of other factors are involved in making school disagreeable; the nature of the teacher and the curriculum, the child's relation to parents, siblings, and peers, his intelligence and sex. Hurlock (1964) provides an overview of these factors.

are the jobs with a future? Such questions are for experts, and experts are rarely available. The adolescent must piece together the picture as best he can. Often he is as poorly informed concerning the facts of occupational life as he is concerning the facts of sexual life.

If occupational choice were only a matter of becoming informed, the adolescent would not be so baffled. Let us add one more fact of occupational life: more jobs are lost for interpersonal reasons than for lack of skill or inadequate preparation (see Super, 1957). The world of work is the world of people. The university professor and the garage mechanic are alike in their need for an interpersonally congenial environment. Work may be relatively impersonal when compared with the intimacy of marriage and family life, but intense affect is still involved. Many of the needs we have discussed in other contexts are present in the work situation—the need to dominate and to be protected, to destroy and seduce, to expand and conserve, to uphold values and to manipulate them for personal gain, to have security and to have status, to maintain self-esteem and to placate others, and so on. In a humorous vein, Gross (1967) suggests that the occupation "newspaper carrier" should not be described in terms of duties to be performed but in terms of people to be dealt with: the customer who demands that a paper be delivered on time, in a convenient place, unaffected by rain, wind, and snow; the superior who tries to maintain the fiction that the delivery boy is an independent businessman but who really pressures him to bring in more new subscribers; the schoolteacher whose power to keep him after school is a potential threat to prompt delivery. Who must be satisfied and how are central considerations in any job, as well as what kind of people one's co-workers are. This interpersonal dimension introduces an element of the unknowable. The adolescent can be told as much as is known about it, so that he will not be caught unawares; but work atmospheres vary even for identical jobs, and particular encounters will be as varied as interpersonal relations themselves. In work, as in marriage, the true knowing comes through the living of it.

An extensive study of ninth-grade boys by Super and Overstreet (1960) reveals the long road which the adolescent must travel before reaching occupational maturity. Of the many characteristics studied, only one turned out to be relevant to occupational maturity—a general planfulness and ability to look ahead. Planfulness entails acceptance of responsibility for occupational choice and the ability to gather specific occupational information. As such, it represents a significant advance over the unrealistic, wishful approach to occupations characteristic of middle childhood. However, it is the sole asset in a long list of inadequacies. The ninth-grader does not have the understanding of his abilities and interests necessary for realistic or even consistent occupational choice, nor are occupationally relevant traits crystallized. He is in the position of anticipating his occupational future without the experience, the self-understanding, or the inner stability for making a wise choice. Incidentally, occupational maturity was not found to be related to family aspirations, peer acceptance, or personal adjustment. It showed some relation to living in a stimulating environment and to being bright and achieving, but even these were rather low.

The adolescent's progress along the road to occupational maturity is disjointed. There is little evidence that he becomes progressively aware of and interested in the occupation he is particularly suited for. We shall present only three variables and their developmental timetables–interests, values, and self-evaluation.

Interests, which were so insubstantial in middle childhood, become increasingly stable. Carter (1944) claims that they are as firm in high school as in college, while Roe (1956) detects a moderate rate of change in high school and stability only in post-adolescence. The exception is boys who embark on scientific careers, whose interests crystallize between the age of ten and fourteen (see Tyler, 1964). The sex differences of middle childhood are maintained, girls tending to prefer social, sedentary, aesthetic, and service activities, boys tending to prefer mechanical, scientific, practical, and physically active ones. Boys are more varied, venturesome, and imaginative in their interests; girls gravitate toward the modest triumvirate of teaching, nursing, and secretarial work.

Values show both stability and change between the eighth and twelfth grades (see Gribbons and Lohnes, 1965). Throughout the period, both boys and girls value occupations because they like them or because they regard them as a potential source of happiness and personal fulfillment. A shift occurs from idealism to realism; the eighth-graders are drawn to social service and personal goals of self-understanding or self-improvement, while twelfth-graders emphasize advancement, marriage, and doing what they can do well. Predictably, boys value salary and prestige more than girls; girls value personal contact and social service more than boys do. In this same eighth to twelfth grade period there is a change in the conception of occupational prestige (see Gunn, 1964). Recall that the seventh-grader ranked occupations according to service; the ninth-grader, now aware of social class distinctions, begins to rank accordingly and to show contempt for lower status jobs ("A janitor is an old man who can't do anything else") which contrasts the equalitarianism of middle childhood. Adolescents in grades ten through twelve rank occupations essentially as adults rank them. Incidentally, the increasingly realistic approach to values and prestige ranking is reminiscent of the transition from the idealistic to the worldly or realistically oriented conscience between middle childhood and adolescence (see Chapter 5).

Accuracy of self-evaluation, as measured by comparing self-ratings with ratings from objective tests, gradually but not uniformly improves (see O'Hara and Tiedeman, 1959). Seniors evaluate their interests and work values more accurately than freshmen did, but they continue to do poorly in regard to their aptitudes. Admittedly, it is easier to know one's interests and values than one's capabilities, since the former are subjective while the latter must be tested in reality. Yet the fact of the matter is that the high school senior must make crucial decisions concerning education or entering the job market with little accurate information concerning his special aptitudes.

In light of these developmental timetables, an early concordance among interests, values, and self-understanding seems unlikely. In certain instances, such as choosing to become a scientist or a mathematician, a decision in early adolescence can settle the child's occupational future. More often, the adolescent can be expected to be caught up in shifting interests and values and handicapped by lack of self-understanding. An early, idealistic dedication to teaching, for example, may give way as values become more realistic; while a subsequent choice of a more lucrative occupation may be abandoned when the adolescent discovers he has little aptitude for it.

Harris (1961) has related occupational tentativeness to the general personality instability of adolescence. He characterizes the adolescent as healthy, vital, exitable, iconoclastic, and inconsistent—a constellation of traits which clash with the constancy of effort, the planning, the ability to tolerate immediate irritations in order to achieve long range goals which characterize work. Ginzberg (1951) adds two more factors. The imperious demands for immediate sexual gratification divert the adolescent from thinking of the future and interfere with his ability to take stock of his assets. Emancipation alienates him from parents and parental advice. The boy's conflict with his father can be particularly handicapping in this respect.

Fortunately, adolescent inner turmoil is counterbalanced by stabilizing environmental factors which narrow the choices he must make. His socioeconomic status provides a potent directing force (see Borow, 1966). Membership in the middle class means that the adolescent will be strongly biased in favor of business and professional occupations rather than service and trade. This transmission of vocational preferences is probably due to a number of factors: parents, especially the father, provide models for identification and reward occupational choices congruent with their class; peers are usually from the same class and exert their own pressure to choose appropriate educational and occupational goals; the adolescent is also better informed about and more familiar with occupations popular in his class. Another stabilizing factor is the ranking of occupational prestige, which is remarkably consistent from year to year and even from class to class. Interestingly, no one factor can account for this unusual stability. The relative prestige of occupations is not a simple matter of income or autonomy, or education, or "head work versus back work," or risk taking. The school also serves as a stabilizing force by structuring the content of what the adolescent will learn and, through grades and promotions, evaluating his performance objectively. In addition, it provides temporal structure; entering and graduating from high school and from college are nodal points for making educational choices critical to an ultimate occupational decision. In general, then, the adolescent's turmoil takes place within a context which provides structure and guides and which protects him from being overwhelmed by having to undertake singlehandedly his own preparation for adult life.

We are now in a better position to appreciate the complexity of the vocational situation. The adolescent has his special interests, values, talents, along with his special personality traits and capacity for self-evaluation—multiple variables whose development and interaction are not necessarily orderly or logical. The environment has its structure of class and sex-appropriate occupations, its hierarchy of prestige, its requirements for occupational preparation. The world of work itself has its own special economic and interpersonal realities which must be assimilated and mastered. Thus, a host of changing intrapersonal factors must be brought into a harmonious relation with multiple, shifting, and subtle environmental demands.

The Vocational Self. We will follow Super's (1963) conceptualization of the adolescent period, since it relates occupation to identity and vocation on the one hand and initiative and judgment on the other. (For other conceptualizations, see Osipow, 1968.) Super casts occupational development in terms of the self-concept. By self, he means the individual's picture of himself, his comprehensive and honest answer to the question, Who are you? An individual may see himself as friendly, helpful, easily hurt or as forceful, stubborn, athletic, and so on. How others describe him or what objective evidence shows him "really" to be are not significant factors; his self-perception is the mainspring of his actions. If a person regards himself as "stupid," this is what will determine his behavior. Such a self-evaluation can persist in spite of contradictory evidence from friends and intelligence tests alike.

In describing the development of the self-concept, Super emphasizes exploration, identification, and role playing.[11] Exploration is similar to our term initiative. It is venturing out and testing oneself in a variety of situations. Every venture furnishes the individual with additional information concerning the kind of person he is. At home, he tests his interest in and talent for a variety of domestic pursuits; at school he can judge his success and failure in academic subjects and in extracurricular activities such as sports, dramatics, and journalism. The individual also identifies with models, thereby taking over wholesale the behaviors, attitudes, values, and goals of an admired adult. Boys are fortunate in having a variety of occupationally relevant models; girls have relatively few. Identification leads to role playing, in which the adolescent tries out the model's behavior to determine how it suits his own personality. A boy may successively assume the role of a baseball player, a scientist, an entrepreneur, in each instance trying to determine whether this particular role is the one which is the most fulfilling for him.

Against this general background, the adolescent begins to evolve a specific vocational self; that is, a constellation of self-attributes which he considers to be vocationally relevant.[12] He engages in vocational exploratory behavior—mental or physical activity whose purpose is eliciting information pertinent to preparing for and choosing an occupation. Obviously relevant exploration would be getting a job or gathering information about a job. The adolescent might also review experiences which, while not undertaken with an occupation in mind, now become pertinent; e.g., in considering the possibility of becoming an engineer, he may think back over his grades in mathematics. Such vocational exploration results in increased self-knowledge, increased knowledge of occupational possibilities, and a gradual increase in realistic planning during adolescence.

[11] Note that Super neglects those aspects of the self which result from adopting the labels given by the significant people in the child's life; e.g., whether a child thinks he is good or bad, attractive or unattractive, bright or dull, depends, to an important degree, on whether his parents have labeled him as such. As we saw in Chapter 3, Rogers, in particular, stresses the importance of these labels in his theory of the self.

[12] Super does not follow our use of the term vocation since, in our nomenclature, he is dealing with occupation and career. However, we will retain his terminology while presenting his ideas.

One might think that taking part-time jobs would be an especially fruitful way of exploring the world of work. However, Harris (1961) points out that it can be a mixed blessing. On the positive side, the adolescent's self-esteem is enhanced by being paid by an employer, for payworthiness is a source of pride and a symbol of his imminent adult status. The adolescent also has a chance to relate to adult workers, talk with them, be informed by them, and even model himself on them. On the negative side, the adolescent's work often is trivial and isolated, and he is considered immature and expendable. Instead of constructive interpersonal relations, a job may become another variation of the battle of the generations, with the employer viewing the adolescent as irresponsible and the adolescent viewing him as irritable and miserly. The menial nature of most part-time jobs does little to test the adolescent's capacity for work or prepare him for the complexities of the adult world. Part-time jobs can become merely an opportunity to earn a little money and feel somewhat more independent.

Super next delineates the specific stages of development which precede the achievement of occupational stability. The timing of these stages varies, since preparation for different occupations varies. Between fourteen and eighteen years, the adolescent is in the Tentative substage, in which Crystallization of Preference is required. Occupational choice now is serious and central, since the adolescent, for the first time, can contemplate the future as a reality. He is expected to have sufficient cognizance of fields and levels of work to commit himself to appropriate education or training for an occupation. However, his preferences are still general and tentative; e.g., he may choose a commercial course which could eventuate in a number of different occupations, or he might try a specific job with the understanding that he will drop it for another if it proves unsatisfactory.

Super[13] pays considerable attention to crystallization itself, which is a cognitive process of formulating generalized vocational goals. It involves a complex of activities which center around assessment of the self and of reality. Self variables include one's interests and values together with their specificity and relative importance. A young girl who vaguely wants to "help people" is in a different decision making position than one who specifically knows that she prefers to help others first by teaching disadvantaged children, then by discussing family management with their mothers, and finally by working in a community recreation center. Reality factors are assessed through an information gathering process. The adolescent may turn to others such as peers, teachers, or a favored adult; he may read literature, or he may seek part-time work. He informs himself both about jobs and about important contingencies: the need for parental approval and financial support, the risks and obstacles involved in different choices. Another factor in crystallization is awareness of present-future relations, or the ability to foresee the steps necessary to achieve a distant goal. This is the time dimension which, as we have seen, was so unstable during middle childhood and significantly interfered with realistic planning.

[13] His ideas are relevant to our concern with judgment and the child's ability to become his own legislator (see Chapter 4). For further reading on the choice-making process, see Tyler and Sundberg (1964) and Tiedeman and O'Hara (1963).

Super contrasts crystallization with what he calls pseudocrystallization which is based on identification. Here the adolescent decides on the basis of his attraction to a model rather than on the basis of exploration and testing. Super claims that such premature choices soon evaporate under the impact of reality. Like Piaget he maintains that there are no shortcuts to higher developmental levels. There is a crucial difference between going through the motions and real understanding.

The stage of Specific Vocational Preference comes between eighteen and twenty-one years of age when a commitment is made to a specific occupational choice. The choice need not be final in the young adult's life, but the tentative, preparatory stage is past. He now enters the world of work and exposes himself to its demands.

The Young Adult as Worker

We shall discuss the young adult only long enough to indicate that he is far from finding a congenial niche in the world of work. Dissatisfaction with the first job runs high, and a number of adjustments must be made. The worker goes from being in the majority, as he was as a student, to being in the minority, from a world of peers to a world of adults. The impersonality and the requirement to produce, which marked the transition from home to school, are even more pronounced in the transition from school to work; the worker's employer is not required to care about him as a person, but is responsible for his productivity. The college graduate experiences special frustrations; within three years of graduation, half of them have changed jobs (Schein, 1968). Typically, his view of work is abstract and idealistic. He tends to believe that good ideas alone are all he needs, since they will both serve the company and insure his own advancement. He discovers that he must deal with human beings with vested interests and personal quirks. His "good ideas" are undermined, sidetracked, or even sabotaged. His superior may not have a college education and may resent the college graduate's starting at a higher salary or his greater potential for promotion. He can be viewed by management as overambitious, inexperienced, demanding too much responsibility and money too soon. In addition, the young worker is subject to conflicting loyalties. Because he is new, he may see many shortcomings or inefficiencies which should be corrected but which would alienate him from his fellow workers. Should he be loyal to them, to the company, or to his own advancement? The college graduate's reaction to all these interpersonal difficulties is often a wish that the human element would go away so that he could exist in a purely rational environment. His effectiveness as a worker, and especially as a manager or executive, depends on his ability to come to grips with the realities of the human environment and on his facility in solving the interpersonal problems inherent in getting a job done.

Personality factors relevant to being a worker are spelled out by Drucker (1968). Instead of aptitudes, skills, and values, he stresses understanding the kind of person one is. Do you like security or do you like to take chances? Do you enjoy being a small part of a large organization or do you want to run the whole show? Do you like to start at the bottom and work your way up or do you want to begin near the top? And, finally, do you like specialization or do you enjoy general problem solving regardless of the content of the problem itself? Drucker also maintains that there is no reason for regarding the first job as the final one. A job offers the individual a chance to learn about himself as a worker. (Here Drucker echoes Super's concept of developing a vocational self.) He owes his employer enough loyalty that he should work well, but his primary loyalty is to his own development. Quitting a job while young has the advantage of teaching the worker how to take setbacks in stride and gives him a priceless sense of independence. In a similar vein, Drucker stresses the importance of having a meaningful life outside of work, instead of becoming a twenty-four-hour company man. Only then will an individual be his own man rather than just an employee. The world of work is the world of people. Intelligence and skill, therefore, are not what ultimately matters; what matters most are character and integrity.

Work and Personality

Before leaving our survey, we would like to point out the relation of work to other aspects of personality development. At times the literature suggests familiar variables which are now operating in the context of work. In middle childhood we find egocentricism in regard to occupational choice and rigid, "too good" values, which are followed by a more realistic grasp of the facts of occupational life and of the self, while sex typing is a potent determiner of interests from the first grade to adulthood. The survey also enriches our understanding of judgment: the intricate and prolonged process of occupational choice provides an excellent opportunity for examining the child's efforts to assess himself, to assess reality, and to develop guides for arriving at decisions. The literature on work has also provided a more detailed account of the development of time perspective than any other literature we have encountered.

Finally, our survey of work has enhanced our understanding of initiative. Toddlerhood emerges as a time in which initiative exists in pure culture. The constraints of school establish a distinction between doing what one wants to do and doing what one has to do. In seeking a satisfying occupation, the adolescent and young adult must come to terms with this distinction. For many reasons he may turn his back on initiative and decide in favor of other values, such as security, prestige, continuing a family tradition. Drucker advises against such a decision, as does Erikson, as we shall soon see. Both men are concerned with continuing into adulthood that sense of congruent expansiveness which is the individual's heritage from earliest childhood.

Vocation and Identity

Our discussion of occupations and jobs has avoided the broader concept of vocation—a topic which is beginning to receive more attention from vocational guidance counselors. The occupational situation itself is largely responsible for this increase in scope (see Robinson, 1965). Technological advances have resulted in a shorter work week and a concomitant increase in leisure time. There is also a concern that work itself is becoming more routine, mechanized, and meaningless. The problem of finding a suitable occupation has expanded to one of finding meaningful activities both within and outside the confines of the work situation. According to Robinson, little is known about satisfactions available from other areas—recreation, community, political, and religious organizations, aesthetic expression and appreciation. However, he recommends that the simple dichotomy of work and recreation should be replaced by a concern for "what a person can live by and for."

It is just this concern which is the central theme of Erikson's concept of identity. Writing from a psychoanalytic framework, Erikson conceptualizes ego development in terms of a series of crises. The identity crisis comes when the adolescent is seeking a vocation, a life's work, a calling. He is asking the fundamental questions, Who am I? and, What can I do which will be fulfilling?

Identity embraces the dual concepts of inner continuity and interpersonal mutuality. The intrapsychic and the interpersonal aspects are strictly complementary. Identity is both a selfsameness and a sharing with others. It is a feeling of well-being—of being at home with one's body, of knowing where one is going, of confidence that one's values are valued by others. Coming to terms with oneself and finding one's place in society are the two facets of achieving an identity. There is an identity crisis in adolescence because this is a period of great inner turmoil and equally great bewilderment regarding one's vocation.

At one level, all we have learned about occupational choice can fit into Erikson's notion of achieving identity. By the time of adolescence, the individual has a unique constellation of aptitudes, interests, and values, a unique personality and temperament. His occupation must offer him an opportunity to continue and fulfill this special heritage from his own special past. If he is bright, assertive, sociable, and mechanically endowed, his occupation must utilize these characteristics; if he is thin-skinned, aesthetic, moody, and musically talented, his occupation must express these qualities. Ideally, what the individual values most in himself will be valued most by others. These "others" may be the nation, an industry, the neighborhood, or a handful of select outsiders.

However, Erikson goes much more deeply into the issue of self-sameness and interpersonal mutuality. The individual's past also includes his particular resolution of four prior crises which are universal to ego development. All these crises have the same characteristic of being simultaneously inner and interpersonal. How they are resolved determines the adolescent's strengths and vulnerabilities in his search for an identity. We will describe these crises briefly in order to become familiar with Erikson's terminology, since we have already discussed them under somewhat different headings.

In earliest infancy, the issue of trust versus basic mistrust is the critical one. We have met this concept when we discussed the psychoanalytic theory of mothering (see Chapter 2). The infant who is sensitively and tenderly cared for experiences a variety of pleasures. He feels good inside and comes to regard intimacy as a source of good feelings. In brief, he develops a deep, somatic belief in inner and interpersonal goodness. Since his caretaker is reliable in her ministrations, he also develops a trust in others. If he is unloved or ambivalently and erratically cared for, he is subject to intense distress. He develops a basic mistrust of others, a feeling that inner impulses and interpersonal intimacy are bad in the sense of being sources of pain.

In the toddler stage, the issues of autonomy versus shame and doubt are at stake. Erikson's concept of autonomy corresponds to our constellation of initiative, willfulness, and negativism. It refers to the child's standing on his own two feet psychologically and asserting his selfhood, at times constructively, at times arbitrarily and defiantly. Parental handling will determine whether he will continue to enjoy his sense of autonomy in spite of realistic demands to curb its excesses or whether he will become ashamed of himself and be hamstrung by self-doubt.

Initiative versus guilt is the crisis around the ages of four or five. The child is "on the make" and glories in his powers. The boy wants to be the ultimate "big boy" and conquer the world; the girl wants to be the ultimate female and the sole object of adoration. Each in his way wants exclusive possession of the opposite-sex parent. At this critical juncture, however, the child is made painfully mindful of his littleness in comparison with his same-sex parent, while parents themselves are making increasing demands for responsible, self-controlled behavior. If the parents are too intimidating and if their demands are too harsh, the child will internalize their accusing voice in the form of a primitive, implacable, accusing conscience. Initiative, power, and competition become sources of guilt.

Finally, in middle childhood, there is the crisis of industry versus inferiority. The child turns his energies to mastering the world of objects and ideas and to the development of skills. His work becomes less whimsical; he begins to develop inner standards for a job well done and holds himself to them. The danger here is that, through inadequate or faulty teaching, through prejudice or seductive demands for good behavior, he fails to develop his ability to master the inorganic world and instead feels inadequate and inferior.

Each of these historic struggles assumes a form appropriate to the period of adolescence. We will return to them after sketching the adolescent's general psychological status. Like many psychologists, Erikson views adolescence as a time of instability. The adolescent is faced with a physiological revolution and with a concomitant uncertainty about his ability to master his sexual impulses and achieve a sexual identity. He is also unsure of his general ability to make far-ranging decisions about his life just at a time when he keenly feels the pressure to decide his occupational future.

The instability of the period has the advantage of allowing the adolescent to experiment with different roles. Because he is less accountable to and involved with parents and teachers, he is freer to venture out in all directions. Much of his seemingly erratic behavior is really the trial and rejection of different modes of living which fail to provide the fulfillment he is seeking. (Erikson's concept of self-testing and self-discovery is very similar to Super's and Drucker's view that jobs should be tools for finding one's occupational identity.) Erikson adds that society implicitly recognizes that the adolescent's task cannot be accomplished quickly and grants what he calls a psychosocial moratorium between childhood and adulthood. Instead of requiring final decisions concerning choice of mate and occupation, society gives the adolescent and even the young adult time to experiment in order to discover where his true self lies and to find the setting which will provide him with a sense of community with others.

In order to counteract inner uncertainty, adolescents band together, seeking security from rigid adherence to the values and fads of their particular clique and becoming snobbish and contemptuous of all outsiders. Because he does not know where he belongs, his need to belong is desperate. His sense of urgency may lead him to premature decisions concerning mate and occupation. He may plunge into marriage with impossible expectations and demands, only to be defeated by his incapacity for mature intimacy and—more importantly—by his lack of identity. For it is Erikson's contention that an individual cannot fully abandon himself to the pleasures of sexuality before he is certain of his identity; he cannot let go of self-control until he is certain that the self can be regained. The adolescent is also vulnerable to premature occupational choices, in which he latches on to the stereotyped images of the "successful man" without questioning whether this is his proper vocation. (Note the similarity to Super's concept of pseudocrystallization.) In trying not to lose a moment in his own version of the American success story, he finds himself trapped in meaningless activities.

The adolescent's excessive conformity and premature decisions can be remedied as long as he does not lose touch with his basic quest for identity. More ominous consequences result from the faulty resolution of the four crises in ego development. We now return to these crises and their consequences. Erikson pays particular attention to the adolescent's time sense which he regards as a derivative of the infantile crisis of trust versus basic mistrust. The adolescent who trusts in life plans realistically for the future. However, basic mistrust engenders what Erikson terms time diffusion. The adolescent feels a terrible urgency to change, but fears that any change will be for the worse. He cannot wait or work because he believes that gratitification is unpredictable. He mistrusts delays because they expose him to deception from others, and he mistrusts his own ability to hope. Immobilized by time diffusion, he wants to give up. His despair is not like that of a depressed person but is described as "the death of the ego"—the abandonment of the desire to grow, expand, find new meanings in life.

Erikson also discusses in detail the adolescent manifestation of initiative versus guilt. The former lays the groundwork for experimenting with new roles. Lacking such freedom, the adolescent is apt to develop what Erikson calls a negative identity. He scornfully rejects the proper roles offered by society and his family and rebels against his national, ethnic, or familial origins. He perversely identifies with just those roles which were presented to him as undesirable during his development. For example, if a mother has feared that her son might become an alcoholic like his father, it is just this role he will seize upon. He may even turn on himself, regarding everything he does as false. Care must be taken not to label such adolescents as "drunks" or "prostitutes" or "delinquents." They are defending themselves against the panic of having no identity and their resolution may well be temporary. However, if they are prematurely labeled, they will defiantly adopt the role as a more permanent identity.

Just as the adolescent must avoid premature identity resolutions, parents, teachers, judges, and social workers must avoid premature judgment of a negative role choice.

The adolescent version of industry versus inferiority is achievement versus work paralysis. The labels speak for themselves. In the former, the adolescent can rely on a basic capacity for work even when he is experimenting with the kind of occupation which suits him best. In the latter his capacity for work fails him—he cannot concentrate, he immediately forgets what he has read, he is careless, he botches and bungles.

Erikson characterizes adolescence as a time when ego identity versus diffusion are the issues at stake. We have already discussed identity. Diffusion is the result of negative resolutions of the four preceeding crises of ego development. It is a feeling of not being able to "take hold of some kind of life." The result may be an aimless, drifting existence in which the person never amounts to anything in his own eyes or in the eyes of others. More typically, in the middle class, the result is activity without fulfillment. The individual can be busy and even highly successful but derives no intrinsic gratifications. His inner emptiness is repeated in his external relations. The diffuse individual never feels a part of a community. He is socially rootless. But, more important, he has no ideology which is both larger than himself and expressive of himself. This larger purpose may be grand (e.g., the betterment of mankind) or modest (e.g., seeing that one's children have a better education than the parent himself had), but it is sustaining, it gives scope to activity, and it provides an ideal which is shared by others. The absence of an ideology, then, is a particularly devastating consequence of diffuseness.

Clearly, Erikson regards vocational choice as engaging the total personality in a struggle whose outcome will determine the level of future maturity. His concept of identity has been widely regarded as an important addition to the general literature on personality development. He has also captured something of the spirit of the times in the middle of the twentieth century. The collapse of ideologies and the ensuing sense of meaninglessness underlying staggering technological advances and frenetic busyness is a popular theme among social commentators and artists. Since it has not been put to the test of objective verification, identity is best regarded as a seminal clinical concept. Like defense mechanisms, it has an immediate ring of authenticity which overrides the scattered attempts at objective validation.

10 Constitutional Factors and The Body

We end at the beginning—with the newborn infant who enters the world bringing with him a genetic past stretching back to the dawn of human evolution, his special fetal history in the womb of his particular mother, and the imprint of the process of birth itself. We will deal only briefly with prenatal influences, and we will not perplex ourselves with the heredity-environment controversy, since these topics are traditionally covered in books on general child development. Instead, we shall select a few studies of constitutional factors and discuss their implications for personality development. The human body of the infant, its external surfaces and inner workings, will also play a crucial role in the development of his personality. We will examine this role in detail.

Constitutional Factors[1]

Genetic Influences on Personality

The list of human characteristics having a proven or possible genetic component is impressive. It includes physical features (color of eyes, shape of nose), anatomical (height, weight) and physiological (blood pressure, pulse rate, age of menstruation) characteristics, motor abilities (age of sitting and walking), certain physical defects and diseases (certain types of visual and hearing defects, hemophilia), certain mental defects and disorders (Down's syndrome in which a chromosomal defect has been demonstrated, schizophrenia and manic-depressive psychosis, in which a genetic component is suspected but not established), and intelligence (see Mussen, Conger,

[1] We use the term "constitutional factors" to cover both genetic and intrauterine factors.

and Kagan, 1963). But the most obvious fact is also the most important: genes insure the continued reproduction of human beings and guarantee that we will not suddenly branch off into fox terriers or sloths. As we shall see, the fact that we have a body which is human in form and function will influence many aspects of personality development.[2]

There is general agreement that the old controversy over heredity versus environment was futile and wasteful. The controversy was spawned by behaviorism, epitomized by Watson's boast that he could produce whatever personality he wished by utilizing psychological techniques alone on healthy infants. Instead of psychology and biology cooperating (as they do in human development), the disciplines were forced into the role of enemies. The error was further compounded by the assumption that hereditary factors were immutable, while environmentally determined ones were modifiable. Thus behaviorism, and subsequently learning theory, had a special aura of optimism—by environmental manipulation one could produce or modify personality at will. In reality, there is little reason for equating heredity with fate or environment with infinite malleability.

In the current reformulation, both heredity and environment are regarded as exercising a significant influence on behavior. In certain instances one influence is major; in other instances it is minor. The choice to concentrate on genetic or environmental factors reflects the interest of the individual investigator, not a division of labor in nature. Since each realm is complex in its own right, their interaction is exceedingly so. Yet, it is just such interrelatedness which eventually must be understood.

Instead of staying at the level of general issues, we prefer to narrow our focus in order to see the current reformulation concerning heredity and environment in action and to think through some of its implications for personality development.

Constitutionally Determined Reactions to Parental Behavior

Constitutional Factors and Attachment. We will start with one of the simplest studies—Schaffer and Emerson's (1964) investigation of cuddling in infants. With the exception of suckling, cuddling epitomizes loving maternal care more than any other single activity. Some authorities have gone so far as to claim that being held, rocked, and fondled are essential to healthy psychological development, since cuddling satisfies a basic and universal need (see Ribble, 1943).

Schaffer and Emerson, while studying the development of attachment in the first eighteen months of life, noted that infants seem to differ in their reaction to cuddling. Moreover, this difference was present from the very beginning. Nineteen of the thirty-seven infants were true cuddlers who accepted, enjoyed, and later actively sought all forms of physical contact. But nine were noncuddlers who protested, resisted, and avoided the physical contact of cuddling. The response was consistent over an eighteen-month period, and it was consistent over situations, such as illness, fatigue, and pain.

[2] For a resume of genetic mechanisms, see McClearn (1968) which contains a valuable list of references to the literature on behavior genetics. For a more extensive coverage, see Fuller and Thompson (1960).

When in distress, for instance, the noncuddlers had to be diverted by food or activities rather than being held and comforted. They were as involved with their mothers as were the cuddlers and eventually became as attached; however, they used different means of establishing contact—while the mother was regarded as a haven of safety, a noncuddler would hold her skirt, for example, or look at her or hide his face against her knee rather than seeking close physical contact. They did not react differently to physical handling or to skin contact per se, such as stroking and tickling. Rather it was the physical restraint inherent in cuddling which was noxious to the noncuddlers.

Although their data were not definitive, Schaffer and Emerson could find no evidence that cuddlers and noncuddlers were uniformly handled differently by their mothers. What did differentiate the groups from the beginning was activity level, a characteristic of infants we shall discuss more fully. Cuddlers were motorically quiet and were generally placid and content. Noncuddlers were motorically active, restless, and resentful of the restraint involved in being diapered, dressed, and tucked into bed. The reaction of the noncuddlers was not specific to the mother but was a general displeasure at any interference with motoric activity.

One conclusion from the Schaffer and Emerson study is that there may not be a universal need for contact, but that the need is dependent upon the constitutional factor of activity level. Another conclusion is more pertinent to our present concerns: the course of attachment itself is affected by a constitutional factor (see Chapter 2). While noncuddlers did develop an attachment, it was at a slower rate and via a different route. Extrapolating from the data, we might speculate that they would not be less loving but that their way of manifesting love would differ. Cuddlers somehow seem more "passive" in their attachment, more gut oriented, more desirous of being enfolded and protected; noncuddlers seem more "active" in their love, more muscle oriented, more given to action and interchange in their relations with the caretaker. We are suggesting the possibility of a stylistic difference in attachment comparable to different styles of self-control, family patterns, and cognition (see Chapters 4 and 6).

While our ideas concerning styles of attachment are speculative, we have some evidence that contact seeking can affect social and intellectual development. Wender, Pedersen, and Waldrop (1967) found that children who were contact oriented at two-and-a-half years of age showed more attention-seeking behavior in school four years later and also had shorter episodes of directed activity and a lower intellectual functioning on nonverbal tasks than did more autonomous toddlers. Incidentally, these differences were unrelated to parental behavior, just as Schaffer and Emerson's were unrelated to maternal behavior.

So far we have seen that a constitutional factor may affect the rate and quality of attachment. However, Schaffer and Emerson's data have other implications. If there is no universal need for cuddling, readiness to cuddle cannot be regarded as a universal requirement of good mothering. The mother who tenderly enfolds her infant may infuriate him rather than endear herself to him. This in itself is important. Phrased technically, the same stimulus may produce different responses depending on the inherent nature of the organism. Good mothering cannot be conceptualized in terms of maternal feelings, attitudes, and behaviors alone; rather, it is a product of the interaction between mother and infant.

This interaction model means that there is no fixed relation between the parent's behavior and the infant's response; the infant's characteristics play a part in determining the effect parental behavior will have. Next, the interaction model means that parental behavior cannot be categorized in absolute terms, such as being "good" or "bad," constructive or pathogenic. Many traits, attitudes, and behaviors are not positive or negative in themselves but only in relation to the characteristics of the child.

Primary Reaction Patterns and Socialization. Thomas and his co-workers (1963) extend Schaffer and Emerson's study by investigating a variety of infant reactions and their patterning. Their longitudinal study followed children from two months to two years of age. The investigators empirically derived nine categories of behavior which were present in all infants to various degrees, and which could be used to differentiate the infants from one another: (1) Activity level, or the extent of the motor component in the child's responses; e.g., an infant could be motorically inactive or vigorous. (2) Rhythmicity, or the predictability of functions such as the sleep-wake cycle, hunger, elimination. (3) Approach or withdrawal, or the response to a new stimulus such as a new food, toy, or person. (4) Adaptability, or the ease with which the infant's responses are modified in a desired direction. (5) Intensity of reaction, or energy level of response. (6) Threshold of responsiveness, or the intensity of stimulation necessary to evoke a response. (7) Quality of mood, such as jolly, pleasant, somber, fretful. (8) Distractability, or the ease with which extraneous environmental stimuli interfere with ongoing behavior. (9) Attention span and persistence, or the length of time an activity is pursued.

The investigators were concerned with delineating what they called primary reaction patterns or patterns of responding which individualized an infant, and with discovering the stability of these patterns over time. The nine categories proved to be stable over the period of the study (intensity being particularly so, followed by rhythmicity, adaptability, and threshold of responsiveness) and the primary reaction patterns were also stable; (a six-month-old who was intensely reactive, predictable, and positive in his response to novelty tended to evidence the same behavior at the end of his second year).[3]

Primary reaction patterns are similar to what used to be called "temperament," which was a pattern of traits characterizing an individual over his life span. The concept of temperament fell into disrepute in American psychology, partly because it ran counter to the environmental emphasis of behaviorism and learning theory, partly because it played only a minor role in psychoanalytic theory. Now it is being revived as an important variable in development.

[3] Thomas et al. do not claim that primary reaction patterns are exclusively determined by constitutional factors; they make allowance for the effects of early learning. However, we know that infants differ in activity level, energy, responsiveness, and the amount of smiling and crying in the early weeks of life. It seems likely therefore that primary reaction patterns are weighted with constitutional factors.

Thomas's study contains many accounts of the ways in which primary reaction patterns affect socialization. Babies high on adaptability and distractability were readily toilet trained and obeyed prohibitions. These are what mothers describe as "easy" or "good" babies. However, they were in danger of becoming too submissive or inhibited if parents were pressuring or rigid in their demands. Infants who reacted negatively to new stimuli, who tended to be nonadaptive and nondistractable, were difficult to train. Parental rigidity in this instance was responded to with negativism and anger. Again we see that the same stimulus–parental rigidity–is reacted to differently by "easy" and "difficult" infants.

Child-Engendered Parental Behavior

Is there evidence that inherent characteristics of the infant actually determine parental behavior? Or, to stand a popular notion on its head, that good infants make good parents? We are now raising the issue of directionality of influence in the parent-child interaction. Schaffer and Emerson's study is only suggestive. They deal briefly with "nonmatching couples"–cuddly infants with mothers who prefer to handle children with a minimum of close physical contact, and vice versa. In the former case, the infant obtained close contact from other members of the family or was so demanding that the mother had to give in. Typically, the mothers were flexible and adapted their mode of relating to the infant's needs. However, the fact that they had to adapt suggests that the infant played a part in determining which behaviors in her repertoire the mother would use. Had the mother been rigid or had she begun to interpret the infant's resistance as due to her inadequacy or to his rejection of her, a less healthy kind of relationship might have ensued.

Thomas and his colleagues have more definitive data on parental behavior being altered by the infant's primary reaction pattern. One mother of twins initially had the same positive attitude toward both. She continued to feel positively toward the twin who, because of a basic adaptability, learned to comply with maternal demands. The other twin, however, was persistent, difficult to distract and, consequently, difficult to socialize. The mother resented the infant's antagonism and the relationship developed into a "battle for control." In a subsequent publication (1968) the same investigators isolated a cluster of reactions–intense negative mood, nonadaptability, and withdrawal from new stimuli–which characterized a special subgroup of difficult children. These infants were irregular in their biological functions, slept at different hours or varying periods of time, and were unpredictable in their hunger and in the duration of feeding; they cried in response to any new stimulus and their negative affect was intense regardless of the situation, e.g., they would shriek at a scratch as well as a major cut; they adapted only slowly to parental demands. Initially, parental attitudes toward these infants were no different from attitudes of other parents in the study. Gradually, however, the parents began to change and develop special attitudes such as self-blame. The constant strain of caring for their children was a source of irritation, frustration, and guilt. A relatively high proportion of these infants developed behavior problems subsequently, although the group as a whole did not become emotionally disturbed. Had one of them been taken to a typical child guidance clinic for his problem, chances are extremely high that parental irritation, frustration, and guilt would be regarded as causative rather than as reactive. The widespread belief that disturbed parents are the source of disturbances in their children often goes unchallenged.

Bell (1968) develops a specific thesis in regard to child-engendered parental behavior. He marshalls evidence to suggest a congenital determination of assertiveness (i.e., intense, goal-directed behavior in the face of obstacles) and person orientation (or sociability). Next, he delineates two types of parental control: upper limit control in which the parent attempts to reduce or tone down the intensity of the child's response (e.g., his crying, hyperactivity, excessive assertiveness); and lower limit control in which the parent attempts to stimulate or tone up the intensity of the child's response (e.g., his lethargy, inhibition, lack of competence).

Using these four variables, Bell reinterprets certain studies of socialization. Aggressive children have been found to have punitive parents (see Chapter 4); parental behavior is usually regarded as the causative factor. However, it is also possible that the children were congenitally assertive, that lower limit control was both inappropriate and ineffective, so that the parents were forced to use upper limit control, including techniques such as physical punishment. Put concretely, sweet reasonableness may have had no effect on the child, so the parent was compelled to use firm, absolute mandates forcefully applied and reinforced by physical punishment. A congenitally hyperactive child would have had similar effects on parental behavior. Expressed in terms of our previous discussion (see Chapter 4), high assertive children are apt to elicit power assertive disciplinary techniques because only these can effectively control their behavior. Bell also reinterprets the finding that love oriented discipline is correlated with an intropunitive reaction to transgressions, in which the child blames himself for wrongdoing. One possibility is that the mothers were initially more affectionate because the infant had a congenitally high person orientation. Had the infant been low in person orientation, he would not have stimulated or rewarded the mother's affectionate overtures, and she might not have been inclined to use love oriented discipline.

To generalize: parents are not fated to react in one way to their children or to utilize a single disciplinary technique. Various options are available. The technique they adopt depends on a variety of factors, an essential one being the constitutional characteristics of the infant himself. Bell does not advocate substituting one kind of unidirectionality (child to parent) for another (parent to child). He is demonstrating the plausibility of alternative explanations of socialization.

The unidirectional model of parent-child interaction, in which parental behavior is cause and child's behavior effect, can no longer be accepted uncritically. The majority of studies of parent-child relationships are correlational and show only that a high (or low) degree of X in the parent is associated with a high (or low) degree of Y in the child. Correlation says nothing about causation. If secure parents have well-adjusted children, we cannot say whether parental security produced the child's good adjustment or the child's basic adaptability, good humor, and resilience reassured parents that they were doing a good job. In a like manner, a child whose development is inherently disjointed and unpredictable can produce bafflement, helpless rage, and protective withdrawal in an otherwise attentive parent, as readily as parental perplexity, rage, and coldness can produce the child's deviant development (see Chapter 2). Instead of unidirectionality, then, we must entertain three possibilities: (1) the traditional one that the child's behavior is due to parental behavior, (2) that the child's behavior is independent of parental behavior (e.g., it is part of his temperament), and (3) that the parent's behavior has been determined by the child (see Wenar and Wenar, 1963).

The Interplay of Constitutional and Environmental Factors

We are not needlessly complicating the issue of directionality. On the contrary, the more intensively one examines parent-child interaction, the more complex the issue becomes. Escalona's (1963, 1965) research on infants with high and low activity level illustrates the intricate interplay between constitutional and environmental factors. Individual differences in mobility can be observed in the first few days of life, are stable over time, and therefore seem to be constitutionally determined to a large degree (also see Schaffer, 1966). Escalona's highly active babies kicked, moved, thrashed about, and engaged in gross body motions frequently and forcefully, while the inactive ones tended to use gentle body motions and only seldom activated the entire skeletal muscular system.

In her sensitive report of four twenty-eight-week-old infants, Escalona describes the influences of activity level in their lives. Motorically active infants had a more intense and varied emotional life and shifted readily from distress to delight, while their inactive counterpart kept on an even keel. In feeding, for example, the active infant would literally throw his whole body toward the breast or bottle with an excitement which would lead to random movements, frustrating interruptions, and brief crying. When partially satisfied, he would become playful, touching the mother, vocalizing to her, and scanning the surroundings. Inactive infants were peacefully absorbed in feeding and oblivious of surroundings. The highly active infants were just as vigorous in their social responses; the mere sight of an adult would excite them, and even minimal responsiveness from adults would elicit squealing, cooing, babbling, and intense gazing. Less active infants also enjoyed social contacts but were more reserved. Thus, activity level turned out to be part of a broader behavior pattern characterizing the infant as generally active, wholehearted, and passionate, or reactive, moderate, and even-tempered.

Escalona observed the mother-infant interaction when the infants were twelve weeks and thirty-two weeks of age. In the earlier period she noted that active babies became quieter and more relaxed in response to direct contact with the mother, while inactive ones became more animated. That the same maternal behavior can have opposite effects is a familiar finding. But Escalona adds a new dimension: these contrary alterations in the infant's state had the similar effect of producing his most mature and complex behavior. Both kinds of babies became more coordinated and focused, the active ones as they relaxed, the inactive ones as they lost their lethargy.

By thirty-two weeks of age the active babies, when alone, spent a good deal of time in relatively complex and advanced behaviors, as compared with the inactive ones whose behavior had little focus and who seemed less mature when alone. However, under the influence of external stimulation, such as the mother playing with them or offering toys, the complexity of their behavior increased dramatically to equal that of the active group. Escalona concludes that active babies generate their own stimulation and the mere sight and sound of ordinary environmental events are sufficient for them to engage in mastery behavior; inactive babies must be energized by special external events, such as maternal attention, before their behavior reaches a comparable intensity and complexity. Consequently, the mother's style of relating is important. A relentlessly playful mother can overstimulate an active infant to the point that the constructiveness of his behavior deteriorates, while a placid mother can be so gentle that she fails to rouse her inactive child to his optimal level of functioning.

Escalona found that social behavior showed a pattern similar to mastery. Active infants were socially responsive even with minimal stimulation—the mere sight of a person would set them to squealing and cooing. However, inactive babies needed the adult's focused attention and playful stimulation before engaging in responses of comparable intensity and persistence. Since the active infant needs less stimulation, one might expect him to be socially responsive to a mother who is matter-of-fact in her child care activities and willing to leave him alone as long as he is content. His ability to generate his own stimulation might even enable him to develop a strong attachment to a mother whose contacts do not go beyond those necessary for good physical care. This same mother could fail to arouse her inactive infant sufficiently and could cause a lag in his development if he thrives only when the mother is spontaneous, effusive, and maximizes contact. Escalona's research, like that of Schaffer and Emerson, implies a constitutional factor in the formation of an attachment to the mother.

The active infant may seem to have an advantage over the inactive one, but such a conclusion is not warranted. While he is constantly engaging in exploratory behavior, his slower-paced counterpart is assimilating and consolidating his advances to a greater degree. The difference is not between rapid and slow development, but between expansive and reflective styles.

In her detailed observations of four infants, Escalona describes other advantages of inactivity. Since the active infant is intense and wholehearted, special adult attention—tickling him and clapping his hands together—is irresistibly exciting. He cannot help but react intensely but, more important, he has little ability to brake his responsiveness. The adult must sense when he is in danger of getting out of control. The inactive infant, because he responds moderately, can tune out excessive stimulation more readily. He has the ability not to respond to potentially overexciting adults and can rely on himself to maintain a comfortable level of arousal. The inactive in-
infant's low intensity also enables him to respond to multiple aspects of a

stimulating situation. When being tested, for instance, the active infants were oblivious of the tester; it was the inactive infant who could respond both to the adult and to the material, who could function simultaneously at a task oriented and an interpersonal level.

Thus activity and inactivity are styles of responding. In normal infants they cannot be ordered on a continuum, one being healthier or better than the other. Each has its assets and vulnerabilities. With no special attention, the active infant can master the world of objects and maintain a high level of social responsiveness, but he tends to concentrate only on one aspect of a situation and cannot protect himself from becoming unduly excited. The inactive infant must be aroused before he can function up to capacity and is apt to languish when deprived of external energizers; yet, his moderate nature allows him to make solid progress, to dose social stimulation, and to respond to multiple aspects of the same situation.

Conclusions

We now have a better appreciation of the complexities inherent in the programmatic statement, Behavior is the result of both heredity and environment. At the simplest level, the statement involves a search for correlates of constitutionally determined individual differences in behavior. In the studies we examined, for example, activity level was correlated with a wide range of behaviors: social responsiveness, focusing of attention, exploration of the physical environment, control of stimulation, and response to restraint. It was also an important variable in the development of attachment, independence, and intelligence.

Constitutionally determined behaviors complicate the outcome of parent-child interaction. First, we saw that the same parental behavior, such as cuddling or stimulating the infant, can produce different effects depending upon the infant's inherent characteristics. Moreover, effects which are behaviorally different may be functionally equivalent, as in the case of the quieted infant becoming as efficient as the aroused one in Escalona's study. An important implication of such findings is that parental behavior as such cannot be defined as soothing or frustrating or over-exciting: it must be defined in terms of its effect on the infant. This change in criterion might result in a different ordering of parental behavior, so that a "moderately stimulating" mother by one criterion would be "overstimulating" by another. It may even be that the congruence between the infant's constitutionally determined responses and the parent's behavior, between his temperament and her style of caretaking, is more important to personality development than any list of attributes in parent or child alone.

Finally, constitutionally determined behavior complicates the determination of directionality of influence. The traditional interpretation has been in terms of the parent determining the child's behavior. However, parental behavior itself may be a reaction to inherent characteristics of the infant; e.g., hyperactivity or hyperassertiveness may force the parent to utilize power-assertive techniques of discipline, while an intense social responsiveness in the child may elicit the love oriented responses in the parent's repertoire. It is also possible that certain constitutionally determined behaviors have little effect on parental behavior, although the studies reviewed contained no instances of this sort. In general, directionality of influence must be established rather than being assumed to go from parent to child.

Note that our discussion has primarily been concerned with infancy. Only occasionally were subsequent developments related to this early period. It is possible that the matching or mismatching of parent and infant behavior initiates reactions which figure prominently in subsequent stages. The exasperation of an effusive mother whose inactive infant "tunes her out" may color her handling of later developmental crises; the overstimulated, highly reactive infant may become increasingly difficult to manage in the preschool period. In addition, development carries its own mandates for change. Parent-child compatibility at one level does not guarantee it at another, just as incompatibility may give way to harmony. The working out of these and other developmental complexities has hardly begun.

The Body

The neonate has a human form and his body functions in a human fashion. He will have two eyes instead of six or none at all; he will be encased in a soft skin instead of a hard shell; he will digest food in his stomach instead of his mouth; he will be sentient rather than impassive, reacting negatively to some stimuli, positively to others; he will have legs which eventually can be utilized for locomotion and hands admirably designed for manipulation. The inventory could go on.

But why is it necessary to make it at all? Partly because our body is so obviously ours that we assume we have always been cognizant of it, just as the physical world is so obviously external to ourselves that we assume we have always thought of it as "out there" (see Chapter 6). Actually, the concept of "body," like the concept of "object," must be constructed from undifferentiated experiences. The idea that we have a body is obvious only because we learned it so long ago and are so consistently and constantly reinforced in our belief.

Next, the inventory is necessary because, as adults, we more or less take our body for granted. It was not always so. For the first half dozen years of life the child lives close to his body. Its workings and uses fascinate him; its pleasures and pains comprise some of his most compelling experiences. In addition, parents will direct the child's attention to bodily functions such as eating, elimination, and walking, and might well place evaluative labels such as "pretty" or "tough" on his appearance and actions. Thus, both intrinsic interest and the attention given it by others combine to place the body in the center of the psychological stage.

Our concern in this section will be with understanding the development of the concept of "body" and with the many ways the human body affects the development of personality.

Origins of the Body Concept

It is difficult to grasp the fact that the neonate literally cannot differentiate his physical body from objects and people in his environment. When the infant sucks, he is probably aware of vague sensations in his "stomach" and "mouth," and feels the pressure of his lips on the nipple; if he looks up he has some hazy impression of perhaps two dark dots which eventually will turn out to be his mother's eyes; the smell of the mother's body and the sound of her voice may also register. In this conglomerate of sensations there is no inside and outside–the dark dots, the pressure on the lips, the pangs in the stomach, all fade in and out of the timeless, formless stream of experience.

We can only speculate how some of these sensations begin to coalesce to form the nucleus of the experience of a body. To begin with, there is a shared distinctiveness to inner sensations; the movement of one's own arm or leg is qualitatively different from the sight of a rattle or the smell of a body. In addition, high and low correlations among stimuli begin to register computerwise in consciousness; impressions from a moving arm are highly correlated with one another and have low or no correlation with sights and sounds and smells. Thus inner sensations from the body, with their special qualities and special consistencies, begin to cluster together as "legs" and "arms" and "stomach" to form a dim, initial idea of "body."

Piaget on the Body Concept. Piaget speculates that the differentiation of the body grows out of the continual and increasingly complex interaction between the infant and the objects in his environment. Initially, the infant cannot distinguish his response from the object responded to. When he grasps and waves a rattle, the sight and feel and sound of the rattle and the sensations derived from his hand and arm through clutching and waving it form a homogenous constellation in which there is no "me" and "not-me" (see Chapter 6). As he continues to explore, he comes into contact with an increasingly wide range of objects which he handles with an increasingly rich repertoire of responses. His world includes not only a rattle but beads, his blanket, the bar of his crib, his bottle, colored cubes; and he not only grasps and waves, but he bangs and mouths and looks and feels. However, the infant is not merely widening his range of experience; he is also assimilating the information which comes his way. Through manipulating a rattle, he builds up an image of a rattle as something which is smooth, hard, noisy, brightly colored, easily graspable, wavable, and mouthable. In fact, what a rattle literally means to him is this constellation of manipulative and investigatory experiences. At the same time as he is assimilating the experiences derived from his exploration of objects, he is assimilating experiences arising from his exploratory responses. As he waves different objects which come his way, for example, the sensations involved in waving begin to cluster together into a special pattern. "Waving" begins to exist as an idea in its own right, along with "rattle." Next, the infant makes a crucial transition; he begins to understand that "waving" is something which can be done to many different objects. At this point he is on his way to distinguishing a thing from an activity. At an even higher level, he can systematically vary his actions when exploring an object; in terms of our example, he can wave an object now rapidly, now slowly, now rapidly again. As he enriches and deepens his understanding of objects, on the one hand, and exploratory activities on the other, he is increasingly able to discriminate a thing from an action. Things are "out there"; activities are "in here." "Me" and "not-me" are no longer intermingled; the eighteen-month-old can perceive himself as acting, doing, and causing changes in his physical environment.

The Psychoanalytic Approach. Piaget is concerned with the development of the concept of body as an agent; his emphasis is on activity. The psychoanalytic school is concerned with the construction of a mental image or picture of the body; its emphasis is on sensation and affects. The body image evolves from two sources: proprioceptive and organic sensations from within, and skin sensations from the surface. Internal sensations are ones of tension and relaxation, epitomized by the vivid experiences of hunger and satiety. As the infant's hunger is consistently gratified in the context of caretaking, he begins to anticipate that "something" will happen to relieve his distress, namely, that he will be fed. The anticipation of being fed carries with it a dim recognition that something will happen "out there" which is different from the distressing sensations "in here." The infant thus begins to sort out inner experiences from external reality.

Skin sensations are especially important in defining the limits and form of the body. The psychoanalysts state that skin stimulation produces a special erotic charge which emotionally underscores the information the infant is receiving. For instance, when the mother bathes the infant or rubs him with powder or oil, when she kisses his finger tips or the soles of his feet, she is defining the contours of his body in an exquisitely pleasurable manner. The infant, in addition, stimulates himself. Thumb-sucking provides a unique feedback (the pun is unavoidable) involving the dual pleasures of sucking and skin sensations; physical objects which are mouthed provide pleasure from mouthing alone. The infant now has another bit of information to help differentiate body from object (see Hoffer, 1949). The finger play of the infant, in which he lies in his crib touching and exploring one finger with another, serves the same self-stimulating, self-defining function.

The Role of Labels and of Vision. If the eighteen-month-old becomes intrigued with naming objects, attentive parents may begin labeling features of the face and parts of the body: "This is baby's mouth," "This is baby's hand." The mother points to a feature of her own face and a corresponding feature of the baby's, and frequently a reciprocal game develops. The infant seems particularly adept at learning that ears come in pairs; after pointing to one of his mother's ears, he will turn her head so he can point to the other. In these games affect is reciprocated as well as information—the infant's delight in discovery is complemented by his mother's delight in his newly acquired skill. The conditions for learning are ideal.

Labeling serves the important functions of calling the infant's attention to features and body parts; on his own he has been attracted to eyes and mouth and fingers, but he may not have paid attention to ears and chin and nose before his mother began naming them. Incidentally, labels force adults to cognize in the same manner. An astronomer can turn "a starry sky" into The Milky Way, Orion, and Scorpio. The reciprocal game both defines what bodies have in common and distinguishes the infant's body from the mother's. He sees his mother's finger point to a feature which is called "Mommy's nose," and then the finger turns and approaches a felt, but unseen, something called "Baby's nose." In addition, the general label "baby's" helps him to understand that there are a number of body parts which belong together because they are his.

Labels help the infant build up an image of his face. Unless he frequently looks in the mirror, he knows his face only in terms of nonvisual sensations. Through the reciprocal naming game, he learns that he shares visible facial features with the rest of humanity. When he points to the mother's visually apparent nose and then to some unseen region on his face, the label "Baby's nose" carries with it the implication that he has a visible nose also. The baby's image of his own unseen face, then, must be a fantastic conglomeration of his mother's features and any picture he has been able to evolve of facial features in general!

In order to learn what his face actually looks like, the infant must look into a mirror. When he does so, he must grasp the idea that he is seeing a reflection and he must realize that a particular face he sees is his. Dixon (1957) reports the following developmental sequence: around four months of age the infant regards his image briefly but somberly. When he is around six months old he reacts to his image as if it were another infant. After this period he begins trying to relate his image to himself—he looks alternately from his hand to the image of his hand, he opens and closes his mouth deliberately while watching his reflection, he studies his image while bobbing up and down. This systematic experimenting enables him to recognize himself around eighteen months of age, although his recognition is not yet firmly established.

Vision plays an interesting role in the early disentanglement of body from external environment. Again, we follow Piaget's ideas but in a simplified form (see Flavell, 1963). Initially, manual and visual explorations of the environment are entirely independent. The infant does not understand that the exploring hand and the seeing eye are two parts of a single organism. When, by chance, his hands come into his field of vision, he regards them as fascinating objects rather than as personal possessions. Around four months of age there are faint signs that a connection is being made between manual and visual response. When he chances to see his hands, he seems to want to keep them in sight, and manual manipulation increases while he watches. Next, he begins to understand that the hand can be used to grasp an object that he sees. However, both object and hand must be in the visual field at the same time. If, by chance, he loses sight of his hand, he grasps any object he happens to touch rather than the object he sees and wants. Here he is in a transitional stage. He knows that the exploring hand which is so interesting to watch is the same as the exploring hand he kinesthetically and tactually feels. In the final stage, around eight months of age, the connection between eye and hand is complete; the infant tries to grasp whatever he sees and tries to look at whatever he grasps, regardless of the position of his hands at the outset. This amalgamation of the two separate functions of visual and manual exploration is a milestone in forming an integrated notion of the body.

Conclusion. What a strange image the eighteen-month-old must have of his body. Let us take one who is precocious verbally, but who has had little experience with looking in a mirror. His kinesthetic image might be reasonably intact, although still incomplete. By "kinesthetic image" we mean the mental representation which enables us to know what parts of our bodies are moving and the location of each part relative to the others. When we move our right arm, for example, we know that it *is* an arm and that it is on the right side and that it is above the legs but below the head. Sometimes this kinesthetic image is called a "schema"; it is the kind of feel we have for our bodies when we close our eyes and move around. By eighteen months, then, the infant has moved his arms, hands, and fingers, his head, trunk, legs, and toes sufficiently to grasp their gross mechanics. Intense pains and pleasures, some arising consistently from stomach and skin, some occurring fortuitously from bumps and cuts, are also part of the body. He knows how his body looks from the chest down, but has no idea of the appearance of his face and the entire back of his body. Because of special parental stimulation, he may have labels for half a dozen body parts. Thus, there are some clearly delineated parts of his body, such as fingers and toes, which he can feel and see and label; other parts, such as his mouth, he can feel and label but his visual image of them is fuzzy and faulty; still other parts are both unseen and unlabeled but vividly experienced, like stomach pangs, or vaguely felt, like the back. One is reminded of the early maps of North America; the explored eastern seaboard is drawn with reasonable detail and accuracy, although familiar areas are exaggerated in size; the midcontinent has only a few prominent features, such as the Mississippi River; and the entire western third is lost in nothingness.

Obviously, the infant does not construct the concept of his body—any more than he constructed his concept of external reality—in an orderly, logical manner (see Chapter 6). He does not initially grasp the overall appearance and functioning of the body and then proceed to fill in the details. What he understands depends on the vividness of an experience, on his needs, on his exploratory behavior, on his ability to assimilate what he experiences, and on the stimulation he receives from others. Ultimately he will arrive at a reasonably realistic concept of his body, but only after many distorted images have been corrected along the way.

The eighteen-month-old has achieved a working distinction between his body and the external world, but a working distinction is a far cry from a clear separation. As an example of the lingering fuzziness concerning body and object, we quote an observation made by Preyer in 1890 (see Sherif and Cantril, 1947):

> Nay, even in the nineteen month it is not yet clear how much belongs to one's own body. The child had lost a shoe. I said, 'Give the shoe.' He stooped, seized it, and gave it to me. Then, when I said to the child, as he was standing upright on the floor, 'Give the foot' . . . he grasped it with both hands, and labored hard to get it and hand it to me.

The Body and Personality: Theoretical Approaches

Throughout childhood—and throughout life—the image of the body and the psychological significance of the image are continually changing (see Fisher and Cleveland, 1958). The child is constantly discovering new things his body can do; just watch a three-year-old trying to master the problem of having both feet off the ground when he is learning to jump. Then comes skipping, hopping, tricycling, skating, throwing; and, during middle childhood and adolescence, the whole world of strength and speed and grace. The athlete and the sissy experience their bodies differently, as do the popular dancer and the wallflower. The vividness of hunger pangs in early infancy are later supplemented by awareness of bowel motility and, during puberty, the demands of sexual impulses. The six-month-old who does not recognize his reflection grows into the eight-year-old who delights in mugging and grimacing in front of a mirror and the adolescent who is devastated by the appearance of a pimple. There are special psychological states accompanying sickness and physical defects, as well as special uses of the body to express affect. The body is involved in prejudices, in profanity, and in humor. Finally, there are expressions which epitomize the intimate relation of psyche and soma: "He's got guts"; "She's a brain"; "I'm broken-hearted."

The Psychoanalytic Theory. Of all theorists, Freud assigned to the body the most important role in personality development (see Chapter 7). The sexual instinct, libido, owes its existence to the biological constitution of the body. The stages of psychosexual development, those pivotal events of the first six years of life, are defined by the movement of libido from one erogenous zone of the body to another. But there is a more subtle point. The workings and appearance of the human body form the basic prototypes of crucial psychological variables. Being fed is the infant's initial and primary experience of being cared for, and being psychologically and physically nourished are forever equated. Moreover, the act of sucking itself is the vehicle which engenders the idea that something "outside" can get "inside." As the infant incorporates the milk through sucking, he also incorporates the mental image of the nurturant mother. It is no accident that the same term—incorporation—is used for the bodily process and the psychological event. Self-control revolves around the child's control of his anal sphincter, the body again providing the model for the psychological variable. Subsequently, the penis manifestly proclaims the superiority of the little boy and is the source of his pride and assertiveness. The body is not a metaphor symbolizing psychological developments; it is the essence of these developments. Let us put the same idea another way: if our bodies were so constituted that food was inserted directly into the stomach, and if waste were eliminated only by perspiration, and if the penis and the clitoris did not differ until the growth spurt in puberty, the basic nature of human personality would be radically altered. For Freud, the structure of the body literally determines the child's psychological destiny.

Sheldon's Constitutional Theory. Sheldon's (1944) constitutional theory states that there is an intimate relation between body build and broadly defined personality traits or temperament. The three chief components of physique are endomorphy, mesomorphy, and ectomorphy. Each individual is a combination of three components. However, when a single component dominates the other two, a characteristic picture emerges. The endomorph's physique is round and soft; the ectomorph is slender, poorly muscled, and delicate; the mesomorph is muscular, large-boned, and firm. Each physique tends to be related to a special temperament. The endomorphs tend to be viscerotonic—they are slow-moving and placid, need social contact and approval, are fond of eating and physical comforts. The ectomorphs tend to be cerebrotonics—they are thin-skinned, sensitive, and shy, are inhibited in their actions, and have a strong need for privacy. The mesomorphs tend to be somatonic—they are energetic and action oriented, competitive, aggressive, and direct. Sheldon's central thesis is that both physique and temperament are determined by constitutional factors. The environment may play an enhancing or a suppressing role, but the main contours of our bodies and of our personalities are not shaped by the environment.

Objective studies of children place Sheldon's constitutional theory in a never-never land; the evidence is never strong enough to validate his original findings but is never weak enough to dismiss his ideas altogether. One of the most interesting studies was done by Walker (1962) who predicted from their somatotype the behavior of preschool children between two-and-a-half and five years of age. As often happens, boys were more predictable than girls. Relations between personality and mesomorphy were predicted best. Both boys and girls were assertive, open in expression, fearless, and had a high energy level. The girls combined their assertiveness with socialness, cheerfulness, and warmth, while the boys tended to be impulsive and hostile. Predictions from ectomorphy were only moderately successful. Both sexes were rather aloof and emotionally restrained. The boys tended to lack energy reserve, were cautious and quiet, slow to recover from upset, and looked to adults for approval. The girls tended to be somber and tense, unfriendly, and irritable. Endomorphy failed to predict any behavior better than chance.

Walker rejects Sheldon's extreme constitutional position and concludes that both constitutional and environmental factors are important in personality development. Certain physique-behavior relations are inherent in the physical organism itself. The mesomorph, with his high energy level, his bodily toughness, and relative insensitivity to pain, is physiologically endowed to be action oriented, assertive, and indifferent to retaliation. The ectomorph, with his low energy level and high sensitivity, has less capacity for forceful, sustained action and suffers from painful encounters with peers. That social influences are also important is seen in the finding that mesomorphic boys channel their energies into aggressive activities, while mesomorphic girls are more sociable. A corroborating study (Glueck and Glueck, 1950) found that among delinquent boys there were three times as many mesomorphs as in a nondelinquent group, whereas in a private school setting, the short, stocky, well-coordinated child was highly sociable, self-assured, expressive, and affectionate. Whether the child becomes social or antisocial, then, is partly determined by the available outlets for his energies (see Sanford, 1943). (Somatopsychology is a third theory relating body to personality which we shall present when we discuss physical disabilities.)

The Visible Body: Normal Development

It is psychologically meaningful to separate the body from the skin out and the body from the skin in. The former is the visible body which is partly on display all of the time. The latter is the inner body, which, along with thoughts and fantasies, is the child's most private possession. The visible body and the inner body follow different developmental courses and have different psychological significances, so they will be discussed separately. However, in neither case will we be dealing with the physical body as such. We shall be dealing with the child's psychological image of his body which can either accurately represent or bizarrely distort physical reality.

Infancy and Preschool Period. Most of the literature concerns the reactions of others to the child's visible body and the effects of these reactions in determining his attitude toward his body and toward himself. The reactions begin at birth. Obstetrician and mother look to the body to answer the fundamental questions: dead or alive? intact or deformed? healthy or sickly? boy or girl? Then there are special maternal responses which suggest the dept of the mother's reaction to seeing her child for the first time:

> It was like a miracle. Everything so tiny and perfect. It was her hands, especially each little finger and those tiny, tiny fingernails.
>
> She was a mess! She had some kind of medical goo running out of her eyes, and her nose was all over her face, and that horrible black hair! She looked like a drowned rat. I cried and said, "On top of being a girl, you have to be so homely!"
>
> He was a real doll. Not at all wrinkled like I expected. He looked like Marlon Brando.
>
> I was disgusted at first, but then I felt so ashamed that I made up my mind that I was going to love that little piece of nothing if it killed me.

In the preschool years, every mother must concern herself with her child's body. She must pay at least a modicum of attention to the management of its secretions and excretions–urine, feces, and, to a minor degree, mucous. There are minimal rules for cleanliness which imply that a dirty body is bad and a clean body is good. The child must learn that certain bodily manipulations–rubbing his penis, picking his nose–cannot be done in public or, depending on the parent's tolerance, cannot be done even in private.

Aside from such realistic concerns, attitudes toward the child's body vary widely. Some parents are almost totally indifferent to physical appearance. They either do not think of their child in terms of appearance, or they use some general label such as "cute" or "pretty" in an offhand manner. Other parents have strong emotional investments in the way their children look. Face, body, clothing, and grooming are sources of pride or unhappiness. Clinicians, with some justification, describe this as a "narcissistic" emotional investment on the part of the parent—typically the mother. The preschool child has only an intellectual interest in his visible body. In the normal course of events, he would never regard himself as attractive or unattractive. In like manner, clothes are viewed with mild interest or resentment, but they carry no high emotional charge. Thus, the baby or preschooler who constantly looks pretty as a picture or is dressed like a perfect little gentleman is expressing the parent's need to have such a showpiece, and the compliments the child receives serve primarily to flatter the parent.

Because preschoolers are sensitive to the reactions of significant adults, they may learn to place a high or low value on their appearance. Little girls may be especially quick to see the advantage of looking pretty and showing off to an audience of admirers. However, psychologists have rarely studied this process of adopting parental values about the body; we have only occasional anecdotal evidence. A mother reports:

> I accidentally came across my (five-year-old) boy sitting in a chair and practicing smiling. He always had a special smile, like a little elf, but he never knew how cute he was before this. It made me sad to see him. It was like the end of innocence.

Not only had her boy grasped the source of adult admiration; his smile had also become a technique for conscious manipulation of others.

Middle Childhood. We know almost nothing of parental attitudes toward the visible body during middle childhood. Watson and Johnson (1958) argue that the child's understanding and realistic evaluation of his body are dependent on his general self-esteem which in turn reflects the love and approval of caretaking adults—particularly the mother. It is the well-loved child who knows and respects himself and his body. Some support for their speculation comes from Levy's (1929-30) study of children between five and fifteen years of age who were brought to a child guidance clinic. He found that the children who were excessively preoccupied with their bodies had overprotective mothers and a high degree of exposure to illness. Apparently the child picks up the mother's anxious overconcern that something might happen to him, while reality feeds his fears by providing many exposures to the consequences of illness.

Physical characteristics and appearance are important to the child's standing in the peer group and are correlated with popularity in the first, third, and fifth grades (see Tyler, 1951 and Tuddenham, 1951, 1952). However, there are qualifications. Personal attractiveness is consistently more important to girls than to boys, who are relatively indifferent to it in the third and fifth grades or who value height and body size probably because of their effect upon athletic ability and strength, both of which are highly prized (see Clarke and Clarke, 1961). Another qualification comes from studying children's wishes. Cobb (1954) found that, while girls wanted to be better looking and boys wanted to be bigger, these wishes were almost never as important as their concern about achievement, school, and personal-social relations. Generally speaking, then, physical attractiveness seems to be valued as a source of popularity for girls; for boys, height and size are extrinsically important as facilitators of athletic ability, and dissatisfaction with the visible body is rarely potent enough to be a primary source of unhappiness for either sex.

Puberty. The lukewarm interest in the visible body ceases at puberty, or between eleven and fifteen years of age for girls and twelve to sixteen years of age for boys. Even a resume of the facts concerning physical growth shows the dramatic alteration in the appearance of the body, especially after the relatively slow and even development of middle childhood. The entire body is in a state of transition. Around the eleventh or twelfth year, the pubescent growth spurt starts, and it is not unusual for the child to gain between ten and twenty pounds and to shoot up four to six inches in the year of maximal gain. Taking the entire growth spurt into account, there is a doubling of weight and a twenty-five percent increase in height (see Hurlock, 1964). Physical growth is not only explosive, it is disjointed. The nose matures before the mouth and jaws, causing the upper part of the face to protrude and the chin to recede. Hands and feet reach adult size before arms and legs, which gives the body a gawky appearance.

The secondary sex characteristics are more important to our discussion of the visible body than are the primary ones. They start to develop at the beginning of puberty but continue two to three years after the primary sexual changes have taken place. Thus, the girl is fifteen to sixteen years of age before she has the body of a young woman; the boy is sixteen to seventeen before he has the body of a young man.

> The sequence leading to the mature body structure is, in general, orderly. In the girl, breast development is one of the earliest signs of sexual maturation. This begins before the appearance of pubic hair. Axillary hair develops later and often forms only after the first menstrual period. Pubic hair is the first masculine secondary sex characteristic to appear, coming shortly after the primary sex organs, the penis and the testes, show evidence of increased size in comparison to the total body. Axillary hair and then facial hair follows. At the time of development of axillary hair, there is often some breast enlargement in the boy. This typically disappears after some months but may cause concern to his parents and the boy unless they are aware of how frequently it happens. Voice changes in the boy have traditionally been the criterion for determining the onset of adolescence, but actually this occurs relatively late in the period and usually indicates that the typical bodily changes are fairly well advanced. (Josselyn, 1955)

In addition, the arm, leg, and shoulder muscles develop, the skin becomes thicker and less transparent, the hair becomes coarser. Girls' hips are enlarged because of the broadening of the pelvic arch, and boys' shoulders tend to broaden. In summary, the entire body is transformed.

The adolescent's feeling of awkwardness and his self-consciousness is embodied in his physical appearance. His concern for his body is at the core of his concern about his rapidly changing self. The wish for a more attractive physique becomes primary. Jersild (1952) found that junior high school students were much more dissatisfied with their physical characteristics than with social or intellectual ones. However, girls continue to be more concerned than boys and to be much more particular about the changes they would like. "I would make myself thinner; I would make my ears lie back. I would make my forehead lower. I would take away my pimples and make my complexion clear and soft. I would make my eyes just a little bigger. I would make my feet smaller." Boys just want to be tall and strong (see Frazier and Lisonbee, 1950).

Complexion is the major concern of boys and girls; acne, pimples, moles, and blemishes are equally distressing to both sexes. The adolescents' concern for their body is not so much in terms of specific members, however; what worries them is any deviation from the cultural stereotype of a manly or a womanly physique (see Stolz and Stolz, 1944 and Frazier and Lisonbee, 1950). The boy's ideal is an athletic build: tall and muscular, with broad shoulders and deep chest. Facial and body hair are also essential. A boy who is fat around the hips, or has narrow shoulders, or small genitals, little pubic hair, or breast enlargement, is apt to be unhappy. If he has all these characteristics, he is vulnerable to greater distress. The girl is even more demanding; her facial features must conform to culturally accepted standards of attractiveness, and she must have a small-framed, hairless body with moderate sized breasts. To be tall or squat, to wear glasses or braces on her teeth, to have a massive jaw or a large nose, to have small breasts or large feet, to have a hairy birthmark or fatty buttocks are deviations from the feminine ideal which can trouble the girl.

Why this intense preoccupation? As we have seen, there are realistic factors. The growth spurt is just that, and its suddenness makes the adolescent's body unfamiliar. He trips over a rug, or steps on people's feet in the bus, or knocks the waterglass over at dinner, because the dimensions of his body schema are unfamiliar. Yet, awkwardness is neither a physiological necessity nor a fixed characteristic, since the gawky adolescent may also be an excellent swimmer or a graceful dancer. The discipline and directedness of a skill gives his body a smoothness of functioning which is lacking in the somewhat amorphous and unpredictable transactions of everyday life (Josselyn, 1955).

Adult society plays a role in accentuating adolescent awkwardness. On the one hand, the adolescent is teased and ridiculed when he is clumsy or when his voice cracks. On the other hand, he is exploited by movies, TV, and magazines which continually equate physical desirability with psychological attractiveness. It is the good, sweet teenager who has the smooth complexion; the funny one squints and wears glasses. However, these social factors play into existing uneasiness and serve to reinforce it; they are not causative. What are the deeper roots of self-consciousness?

To begin with the adolescent's self is, to an important degree, his sexual self and the body publicly announces the advent of sexual maturity. The budding breasts and the faint moustache are social indices of sexual maturity. The "coming, ready or not" quality of physiological maturation finds many children unprepared and, as we have seen, our society does not make the transition from childhood to adulthood an easy one. The adolescent is in the position of being on display when he is making his earliest fumbling attempts to master the sexual revolution inside him. Under such conditions, realistic self-concern can become painful self-consciousness. Pubescent boys can break into a cold sweat over the thought of being called on to recite in front of the class when they have an erection, and girls can have similar anxieties about displaying their breast development.

Other factors contribute to self-consciousness. The body of the opposite sex becomes a source of a new kind of delight. The boy looks at a girl's breasts or hips or buttocks with a special excitement, while the girl is similarly aroused by muscular arms or a deep chest or hairy legs. The body's potential for arousing erotic feelings gives it a new meaning, as if it were being seen for the first time. If an adolescent feels that he is on display, it is because—quite simply—he is! However, the feeling can also be exaggerated. The adolescent's heightened level of emotional arousal is often a state of hypersensitivity to the environment. No event is neutral. Just as the avid gossip is quick to sense a scandal in a chance meeting or a passionate liberal makes a moral issue of an offhand remark, the adolescent's sensitivity leads him to personalize the happenings around him. If she hears someone laugh as she goes onto the dance floor, the self-conscious adolescent girl is apt to think the laughter is directed at her.

Why are deviations from an ideal physique so unsettling? Partly because the ideal is shared with peers, and peer acceptance is vital to self-esteem during this period. In his emancipation from home, the adolescent turns to the group as a bulwark against loneliness and bewilderment (see Chapter 8). If his body conforms to criteria of acceptability, then he has one important source of security. The more he deviates, the more he is liable to ridicule, lower status, or rejection. His reaction is exaggerated because his investment in acceptance is so high. Cognitive factors are also involved. The adolescent is concerned about his sexual identity, but he is incapable of comprehending the nature of mature sexuality. It may be years before he experiences masculinity or femininity as an integral part of himself. In early adolescence, he knows only appearances; to be a man is to look manly, to be a woman is to look feminine. Since his criteria are external, departures from the ideal picture can make him question his future sexual identification. Thus, "feminine" body features on a boy may be interpreted as meaning he is really lacking in masculinity. Like the younger child who mistakes sensory experience for reality (see Chapter 6), the adolescent must first rely on appearances until he can experience and understand more deeply. In addition, physical deviations may come to symbolize psychological difficulties. In individual cases, worry over a pimply face may be rooted in guilt or disgust in regard to menstruation or masturbation. More generally, the concern that something is wrong with the body may reflect a feeling that something is wrong with being sexual.

Ironically, nature betrays the adolescent. Just when his need to look normal is keenest, his physical development is most out of joint. We have already noted the disproportionate aspects of physical development. Even more important are the individual differences in timing. Girls who dislike being tall have their growth spurt earlier than boys. Thus, the average thirteen-year-old girl is one-half to one inch taller than the average boy. Boys eventually become taller because their rate of growth is more rapid and lasts approximately a year longer, but this fact is cold comfort to the thirteen-year-old girl who towers over her partner on the dance floor.

Individual differences in the onset of maturity within each sex have a pronounced effect on personality (see Jones, 1957, Mussen and Jones, 1957, and Jones and Mussen, 1958). The early maturing boy not only looks more mature but also is accepted and treated as more mature by his peers and by adults. He tends to be a leader, and his status has the effortless quality of a "naturally born leader." The late maturing boy has a body which is short and childish during a time when a premium is placed on an athletic physique. He tends to feel inferior to and rejected by his peers. Sometimes he will capitalize on his little-boy status in order to gain special attention; other times he becomes rebellious or withdrawn. He is less popular with peers and tends to be restless and talkative. He also may be bossy and cocky rather than a natural leader, and he jokes and clowns instead of having a real sense of humor about himself. The picture in regard to girls is not clear. More (1953) found that the earlier maturing girls had a reputation for being self-assertive and forceful, at times to the point of bossiness, while the later maturing girls were tense, quarrelsome, and demanding of attention. Other studies found late maturing girls to be withdrawn. In a period when appearances mean so much, it would seem logical that the more mature girl would stand a better chance of being poised and self-assured than her immature age mate. However, studies (Macfarlane, Allen, and Honzik, 1954) show that the early maturers have more problems than girls who mature later.

In the adolescent's reaction to his visible body we have another example of behavior which is multi-determined. Because cultural, interpersonal, intrapersonal, and historical factors are involved in the adolescent's attitude, a tall, gawky, sixteen-year-old girl may be lively and popular and unself-conscious, while her physically attractive, feminine contemporary is humped or hobbled by some inner terror of sexuality. We rarely deal with inevitabilities in personality development; rather, we deal in successive probabilities. The more the adolescent's visible body departs from the culturally defined ideal, the more he is distressed by specific sexual developments and the general turmoil of his transition status, the more sensitive he is to the reactions of adults and peers and the more insensitively they react to him, the greater the likelihood that his body will become the focus of dissatisfaction and painful self-consciousness.

Adulthood. After adolescence, the visible body becomes less important. Its status is roughly comparable to what it was during middle childhood—more important to women than to men, but rarely of prime importance to either. The ordinary adult's image of his body is based on highly selective and biased samples of how he looks. Men glimpse it in the morning while shaving, women while putting on makeup, combing hair, trying on clothes, etc. They rarely know how they look while eating or driving a car or taking a bath or making love. With such limited information, they cannot see themselves as others see them.

When concern about physical appearance does become of prime importance, the clinicians begin to suspect an excess of body narcissism; the love which should be going out to others remains focused on the self. The woman who must be beautiful and glamorous at all times and the mirror athlete who spends hours watching his muscles develop seem to be caricatures of mature femininity and masculinity. They epitomize the adolescent's emphasis on physique while making its superficiality blatantly apparent.

The fate of the visible body in adulthood lies beyond the scope of our discussion. The preceding paragraphs were included in order to return to a general point. The image we have of our visible body is not fixed throughout life and only fitfully mirrors our actual physical appearance. In the early months it is fantastically distorted; in maturity it is intact but sketchy. At certain times the image fades almost completely, at other times it is central to self-esteem. The values we place on our bodies—pretty, ugly, exotic, powerful, cute, dumpy, even tall, short, fat, thin, young, and old—are a mixture of physical reality and psychological needs to evaluate ourselves in one way or another. An attractive girl can picture herself as being ugly because of her sexual fears; a mature woman can think of herself as a little girl because of her fear of growing up. Not infrequently, people try to make their bodies conform to their inner image. A person's appearance is the initial clue he gives to the image he has of himself, and it is invariably an important one.

Expressive Behavior

The visible body can express feelings, moods, and character traits. Personality is manifested not only in the content of what a person says but also in the expressiveness of his body. We form initial impressions of others before we know anything about them and before they utter a word. The face, in particular, is richly responded to and interpreted. When we say, "I don't like his looks," we are acknowledging the intimate relation of appearance and character. The mask-like expression of the Oriental is unsettling because it deprives the Occidental of one of his primary sources of information about another person; significantly, the cliché concerning Orientals is, "You never know what they are thinking."

For at least the first two years of life, the infant learns about people through their expressive behavior, since comprehension of speech is limited and fragmentary. He comes to know the mother's mood from her tone of voice, the expression on her face, the posture of the body, long before words are meaningful. Even after he understands verbal content, he continues to use the body as a source of information about personality. While the adult's use of the body to understand others has been studied under the heading of social intelligence or interpersonal perception or social sensitivity, almost no work has been done with children. We know more about the child's understanding of the world of things than we do about his understanding of the world of people; there is more data on how he learns to distinguish a circle from a square than on how he comes to distinguish the expression of love from the expression of anger. The process of learning to read behavior and the consequences of faulty reading have been largely unexplored.

A promising attempt to conceptualize and measure the area of social intelligence in children comes from J. P. Guilford's Psychological Laboratory (see O'Sullivan et al., 1965). The investigators use the term behavior cognition for the "ability to understand the thoughts, feelings, and intentions of people as manifested in discernible, expressional cues." The cues consist primarily of facial expressions, vocal inflections, postures, and gestures. The tests consist of photographs, realistic drawings, cartoons, silhouettes, stick figures and tape recordings, which were administered to 240 eleventh grade students. The investigators found that social intelligence was comprised of six factors, three of which are directly concerned with the ability to understand the language of the body. The first is cognition of behavioral units, or the ability to understand facial expressions, hand gestures, and postures. The second is the ability to group together different expressions having the same meaning, such as different ways of showing admiration or scolding. The third is the ability to comprehend the fact that the same expression can have different meanings, such as a hands-on-hips stance expressing either anger or provocative "come-hitherness."

While psychologists have not given the study of social intelligence in children the attention it deserves, they have not neglected the study of the child's body itself as a source of information about his personality. This approach to understanding is sometimes called the study of expressive behavior, since it focuses not on what the child does but on the manner in which he does it. For instance, the reaction of two fifteen-month-olds to the same new toy is described as follows:

> Beth sits back and sizes up the situation. There is no evidence of fear in this initial pause; it is not inhibition; it is just a stopping to ask, 'What do we have here?' After the brief sizing up, she approaches the (toy) fish with intense interest.
>
> Larry sees the fish while still in his mother's arms. He squirms and bucks vigorously trying to get down. He scoots across the floor on all fours as if his life depended on it and seizes hold of the fish. However, possession has a quieting effect and he examines the fish eagerly but with versatility—he devours it, but he also digests it.

Both children did the same thing; they showed interest in the toy, approached it, and examined it. But their expressive behavior stamps the sequence indelibly with their individuality.

Wolff (1946) explored expressive behavior in preschool children (and adults) and regarded it as a key to understanding personality. His analysis of the child's body is the most relevant for our present purposes. He used bodily posture as a clue to various psychological states and found that postures could be classified into categories expressing tension, balance, or indifference.

> Tension can easily be recognized. The limbs are strained, the position frequently is unsymmetrical. The posture is awkward. The bodily tension reflects a psychological tension which may also appear in nervousness, restlessness, aggressiveness, emotional instability, rigidity, inhibition, etc.

> The expression of balance is the opposite of the expression of tension. The limbs are relaxed, the position is symmetrical or balanced gracefully. The bodily balance reflects psychological balance which is characterized by adaptability, feeling for rhythm, calmness, free behavior, stability.
>
> Indifference is characterized by a lack of concern about the bodily posture. There is neither unconscious balance nor self-conscious expression of tension. The bodily posture of indifference may indicate indifference to the environment, shyness in social relations, a submissive attitude, and discouragement. A few individuals belong to one extreme, but most individuals show degrees of each type in a mixture. (p. 201)

The messages of the body are not ordinary communications through which the child intends to convey meaning to others. Rather, they are correlates of personality variables to be interpreted by those interested in understanding the child. Parents, teachers, and clinicians frequently use them. They are especially valuable because they reveal feelings, moods, and characteristics which the child is either unaware of or which he lacks the verbal sophistication to express.

The Visible Body: Deviant Development

Somatopsychology. Deviations in the visible body have been studied more intensively than have normal states, although rarely from a developmental point of view. Somatopsychology (see Meyerson, 1963; Wright, 1960), while basically a general theory, has addressed itself primarily to the psychological effects of extreme variations in physique—physical disabilities such as crippling, deafness, and blindness, and, occasionally, extremes in height and weight (see Mussen and Barker, 1944; Hanks and Hanks, 1948). The cornerstone of the theory is that behavior results from the interaction between individual and environment. With this in mind, the investigators set out to understand the disabled individual and the special environment in which he lives.

Somatopsychologists distinguish three aspects of the body. The body is a "tool for action" and certain deviations place realistic restrictions on what it can do; it is a stimulus to the individual who must manage the problems of adjustment arising from variations in physique; and it is a social stimulus through which the individual is evaluated. Somatopsychologists are particularly concerned with the labels "handicapped" and "disabled."

Being labeled handicapped or disabled varies from one society to another. It is not predicated on extreme deviations from the normal physique. The ancient Chinese used to bind the feet of a female infant so that in adolescence her foot would measure four inches. Although she could not run and could walk only with small, mincing steps, she was not labeled "deformed" but was considered attractively feminine. One culture's physical grotesqueness can be another culture's beauty. Many other instances of aribtrary evaluations can be cited: in some societies red-haired infants are considered evil and are killed; epileptics have been regarded as cursed or as especially blessed; if a blind man reads the Koran in Turkey his prayers are especially welcome to God, but a blind Catholic cannot become a priest. "Arbitrary" does not mean that the culture lacks practical or historical or economical justification for its evaluations; it means that there is nothing in the physical characteristic itself which defines it as a "handicap."

Somatopsychologists argue persuasively that the only valid approach to understanding the physically deviant individual is to view him as a person. All too often, the deviation is used to define the individual's status and even his personality.[4] For instance, a group of college students asked to rate cripples on twenty-four personality traits, such as self-confidence and conscientiousness, felt free to make such ratings without recognizing that there are many kinds of cripples. Inspirationalists make a similar error when they dwell on Beethoven's deafness or Teddy Roosevelt's puniness, as if the greatness of the man can be accounted for in terms of his effort to overcome a handicap. "The handicapped" as well as "the epileptic personality" or "the TB personality" are as much of a myth as "the foreigner." A handicap may cause an individual to have special problems, but it does not envelop his personality or determine the kind of adjustment he will make.

Another distinguishing feature of the somatopsychological approach is a dissatisfaction with descriptive accounts and inventories. Objective studies show that there are manifold reactions to a single handicap; e.g., a child crippled by polio may feel inferior, anxious, hostile, ashamed, indifferent, or he may accept his limitations realistically. Since different behaviors can have the same psychological meaning (e.g., an anxious person can withdraw or strike out or feign indifference) and the same behavior can have different psychological meanings (e.g., self-consciousness may be due to a sexual conflict, to delayed maturation, to excessive parental criticism), it is not behavior but the psychological meaning underlying behavior which must be studied. It is just this area of psychological meaning which the somatopsychological approach has investigated. We can present only a sampling of its many hypotheses.

The child who becomes disabled (see Richardson, Hastorf, and Dornbusch, 1964) is in a new situation psychologically. Because he does not know what to expect of himself and others, his behavior will be unpredictable, vacillating, and will have a trial-and-error quality. Even when he is trying to be cautious, he will make errors and false steps because he does not know the consequences of his actions. He will be alert to every cue concerning the success or failure of his actions; frustration and conflict will be high.

The disabled child may be subjected to "overlapping situations," i.e., situations in which contradictory demands are made upon him. Just as the adolescent is in the overlapping situation of feeling partly a child and partly an adult, the disabled child may feel overprotected and depreciated, or he may unrealistically be deprived of participating in family activities while being overvalued as a special cross the parents must bear. Such incongruities are a source of tension which may produce reactions of retreat or demandingness. Cataloguing these reactions is not important; understanding the child's psychological dilemma is.

[4] For a more detailed exposition of the somatopsychological approach and an example of their analysis of handicaps, see Barker et al. (1953).

Somatopsychologists also explore the reactions of the intact individual to the disabled one. A frequent response is pity which is rooted in the "requirement of mourning." In order to protect his status, the intact person needs to feel that the disabled individual is inferior and suffering. The handicapped, like the Negro in the old South, should "keep his place." If, as a socially labeled inferior, he acts as if he were as good as anyone else, the healthy individual loses his superior status and prerogative to pity or, in the case of children, to tease (see Wright, 1960). The "requirement of mourning" helps us understand why clinicians say that pity never should be the motive for undertaking psychotherapy with a child. Too often pity requires that the child remain pathetic, and it readily turns to anger if the child begins to be provocative and troublesome, even when such behavior is a step toward making a healthier adjustment.

The Clinical Approach. Clinicians interested in physical disability have explored its effects on the intrapsychic variables of self-control, conscience, anxiety, defense, hostility, and sex, as well as on the interpersonal variable of the parent-child relationship. They frequently try to understand present reactions in terms of the history of the child and his relations with his parents.

When clinicians regard physical disability as a stress which is essentially no different from other kinds of stress, their interpretations of its effects lack originality; e.g., a physical handicap accentuates castration anxiety or a malformation increases maternal guilt and subsequent overprotection. They make a more valuable contribution when they explore the symbolic meaning the body can have in individual cases. A child with only a minor visual defect may insist on wearing glasses because he feels they offer him psychological protection; a crippled arm may be interpreted by the child as justified punishment for an unacceptable impulse to hit a parent. Clinicians also make an important distinction between an expected reaction and a neurotic reaction to a disability. Both a slight and a severe deformity may be incorporated into an already existing and essentially unrelated problem. In such cases, the child seizes upon the disability as a means of solving his problems. The distinction has important implications for rehabilitation and therapy. An adolescent, whose feelings of inferiority and shame arise from many sources, but who attributes them to the fact that he has a large nose, and who in addition, uses his nose as an excuse for viewing people as unfriendly and secretively malicious, is not necessarily a good candidate for cosmetic surgery. Deprived of his excuse for being seclusive, he may be terrorized by the prospect of being close to others or assuming adult responsibilities. Likewise, an obese girl will not be happy when she loses weight, if obesity serves to protect her against the fear of being sexually attractive.

The General Effects of Disabilities on Personality. Research concerning the effects of disabilities on personality is incomplete and has been conducted from different points of view–the somatopsychological and the clinical. The picture which emerges, therefore, lacks cohesiveness. Barker (1953) gives a helpful resume of the studies. Though the physically disabled have a higher degree of maladjustment than the nondisabled, 35 to 45 percent of the disabled are adjusting as well as their physically intact counterparts. This overlap points up the error of equating disability with maladjustment. Since the kind of maladjustment is not peculiar to the kind of disability, one cannot expect to find a "polio personality" and an "amputee personality." There is some tendency for the disabled to be timid, withdrawn, and self-conscious, but they may also be just the opposite. The adjustive mechanisms which disabled people have at their command are the same as those of nondisabled people. When faced with a serious problem, they become frightened, or retreat, or fight back, or curse God, or face facts, or fool themselves into thinking that everything is all right. The chances of maladjustment are increased when the disability is severe rather than mild and when it is of long standing rather than of brief duration.

Unfortunately, Barker's summary applies primarily to adults. Objective studies of physical disabilities in children are meager. We have many leads concerning important variables—the age of the child, his general psychological resilience or vulnerability, the severity and duration of the disability, the attitude of parents and peers, the values of society—but information concerning these variables is incomplete. In addition, the conclusions drawn from studies often are a function of the way data are obtained, e.g., one can find that handicapped children are or are not different from nonhandicapped children, depending upon the assessment technique employed by the investigator. However, these are common shortcomings in a new field of inquiry and are mentioned here as a caution against accepting theorizing as established fact and existing evidence as definitive.

We will discuss three disabilities relevant to the visible body: obesity, crippling, and facial disfigurement.[5]

Obesity. Obesity is a mystery. It may be due to physiological and/or psychological and/or social factors, the etiological role of each being poorly understood. The physiological mechanisms for hunger and satiety and the essential conditions for accumulation of fat have not been definitely established. Therefore, we cannot tell whether obesity is due to a malfunctioning physiological process. There is some evidence that the obese child tends to be inactive, timid, submissive, and sensitive about his weight and that obesity is a socially defined handicap. Richardson (1961), for example, found that both adults and eleven- to twelve-year-old children ranked obesity as the least desirable physical handicap. Surprisingly, it was less desirable than facial disfigurement or crippling, regardless of the socioeconomic classification, race, and sex of the child, and whether he were handicapped or not. With the exception of Jewish children, the obese child must live in a society which looks upon him with great disfavor.

The most interesting speculations about psychogenic factors in obesity come from clinical studies. Bruch (1941, 1958), on the basis of interviews with parents of approximately 200 obese children, pictures the child as unhappy, socially isolated, immature, and studious. She attributes obesity to a faulty mother-child relationship in which the mother expresses her ambivalence toward the child in terms of feeding, the primary symbol of love (see Chapter 2). To deny her unconscious rejection and to

[5] For review of other disabilities, see Lowenfeld (1963), DiCarlo and Dolphin (1952), and Eisensen (1963).

reassure herself that she cares for him, the mother over-feeds and pressures her child to eat; the child, sensing rejection, constantly tests the sincerity of the mother's love by demanding more food. Thus, mother and child are locked in a vicious circle in which feeding and eating prevent the resolution of the basic problem of maternal rejection. There is another factor involved which will be relevant to our discussion of the inner body. Since the mother feeds the child in response to any distress, the child literally cannot distinguish the inner sensations of "hunger" from the inner sensations of "tension." He eats whenever he becomes tense, because in his experience, food has been used to reduce distress as well as to satisfy hunger. Once established, obesity isolates the child from his peers and from the rewards of group participation. Overeating then becomes a solace for loneliness. Instead of Sheldon's placid, sociable endomorph, we have a picture of a strong, depressive streak in a basically unloved, unwanted child. No objective evidence exists to indicate which picture of obesity is correct or whether both are valid, given certain unknown conditions.

Crippling. The literature on crippling in childhood contains a wealth of theory and a paucity of fact. With this warning in mind, it is possible to view the effects of crippling at different levels: the realistic limitations it imposes on activity, the effect on the child's personality, the reaction of parents, peers, and society in general.

Hollinshead (1959) gently cautions the clinically oriented investigator and the somatopsychologist not to lose sight of the realistic limitations on the crippled child. Important life goals may be unattainable because of such limitations, not because of psychological repercussions or social devaluation. Crippling may significantly interfere with mastering basic skills of eating, toileting, dressing, or self-help. It may block play outlets which require vigorous activity or fine motor coordination, such as using scissors or coloring. It may hinder the acquisition of basic academic skills which require copying or writing or drawing. Thus, Hollinshead warns against an uncritical acceptance of the credo that, if the handicapped child were accepted realistically, he would be as well adjusted as any other child. There may be extensive and unalterable frustrations which make the severely crippled child's life intrinsically more trying.

Psychoanalytically oriented investigation (see Barker, 1953) concentrates on the meaning of crippling to the child. Crippling focuses the child's attention on his body and himself. Overindulgent parents prolong his infantile, dependent attitudes; rejecting parents force him into premature and superficial independence. Babyishness or pseudomaturity are, therefore, common hazards. Crippling robs the child of normal play opportunities, one of the principal tools for encountering and assimilating reality. Since achievement requires special determination and effort, the handicapped child tends to be perfectionistic and to feel personally defeated by failure. On the positive side, the environment protects him by requiring less achievement of him than of his noncrippled counterpart. Crippling adversely affects the child's view of his parents and heightens ambivalence; unconsciously he blames them for letting the crippling happen and looks to them for a magical cure. Unfortunately, since most adults unconsciously equate crippling with wrongdoing, parents have difficulty in loving the child freely. Some crippled children never "outgrow" their body as do healthy adults;

instead they remain fixated on their bodies and consume their energy in self-concern, self-pity, or self-aggrandizement rather than in forming close interpersonal ties. This undesirable outcome is by no means inevitable, however.

Research concerning the effects of crippling on the child's personality presents a mixed picture. Bernabeu (1958), studying eight severly crippled children intensively, lists the most ominous reactions: frustration, anxiety, rage, depression, pathogenic defenses, deep body anxiety, mutiliation and castration fears. On the other hand, Barker, reviewing the childhood accounts of eight handicapped adults, concludes that some were severely disturbed, some moderately disturbed, and some fared well. As a group, then, they would be no different from any group of physically intact individuals. As often happens, the investigator who dives down deepest—Bernabeu—comes up with the richest harvest of pathology. Yet, it is difficult to know how widely her results can be generalized. Are these reactions common to all crippled children or are they specific to severe crippling? In her concentration on pathology, she also fails to delineate the assets and strengths which counteract the forces of devastation. Her report is best regarded as a description of possible rather than typical or inevitable consequences of crippling.

Mussen and Newman (1958) are concerned with discovering personality factors which may influence the child's adjustment. They follow the somatopsychologists in hypothesizing that the optimal attitude is a realistic acceptance of the handicap—"facing the disability without devaluing the self"—rather than acting as if it did not exist (denial) or retreating or trying to outdo the nonhandicapped. They found that the better adjusted crippled children in grammar school could acknowledge their greater need for help and did not set unrealistically high goals for themselves. They were generally more realistic about the range of friendships they could have and about their ability to participate in play. Like good, nonhandicapped middle class children, they inhibited overt expression of aggression for the sake of social harmony.

Parents of crippled children have been described as rejecting or overprotecting, as being disappointed in their child or pushing him to achieve. Marital disharmony also has been noted to increase when the child is crippled. Whether these reactions are due to the personality of the parents or to the realistic difficulties in dealing with the problems of a crippled child in our society is an unsettled issue. Boles's study (1959) is revealing, since it is one of the few to have a control group of mothers with nonhandicapped children. He found that the mothers of the handicapped children were more overprotective and had more marital discord than the control mothers; Catholic mothers were the most guilty, withdrawn, and unrealistic, while Protestant mothers were least anxious and withdrawn. However, there was no overall difference between the two groups in regard to anxiety, guilt, rejection, social withdrawal, or unrealistic attitudes. This does not mean that both groups were equally happy; rather, both were equally unhappy. Fifty-seven percent of the mothers of handicapped children said, "At times I feel that I am going to crack up,"—but forty-three percent of the control mothers said the same thing! Since so many studies lack control groups, we do not know how many attitudes have been wrongly attributed to having a handicapped child rather than having a child per se.

The meager evidence on peer relations suggests that crippled children in the elementary grades are chosen less frequently as friends and playmates than are their noncrippled contemporaries. As early as six years of age, the label "handicap" is used to indentify a child and to determine his inclusion and status in a group (Force, 1956). However, it is not known whether, at this early age, the crippled child is devalued on the basis of a label—i.e., whether true prejudice exists—or whether his realistic limitations prevent participation in the kinds of activities necessary for friendships and social status. Hollinshead warns of the danger in becoming prematurely "psychological" and in overlooking the implication of realistic limitations upon the child's functioning.

Barker and Wright (1954), using extensive data on three preschool children who were compared with nonhandicapped children, also found evidence of unsatisfactory peer relations. Conflicts with friends were more frequent, interaction was generally less harmonious, variability in affection and mood was greater. This pattern, together with the finding of a greater dependence upon the mother for nuturance, leads the authors to conclude, "In general, child associates were stress centers for the disabled children, while their mothers were centers for relief from stress (p. 470)."

Barker and Wright also studied the social interaction and overall environment of seven children who attended a generally pleasant special school. They found that the school provided greater security than a regular school setting but at the price of decreased variety of experiences. The children's behavior suffered a concomitant decrease in variety, and responsible participation in activities was low. Outside the school the children had lower status than nonhandicapped children and were never leaders. The interests of the two groups were strikingly different; the disabled were concerned with nutrition, physical health, and personal appearance; the nondisabled were concerned with sociability, play, work, and education. One is reminded of the previously mentioned psychoanalytic speculation that the disabled child continues to concentrate on his body rather than on interpersonal relations or activities.

In spite of the picture of a child who was more dependent upon his mother, more impoverished in terms of his experiences, less well accepted and integrated into peer groups, Barker and Wright found that the disabled child had the same amount of success and failure, gratification and frustration in his life as his physically intact counterpart. Life was generally good and satisfying for him. Is this because the authors chose to study a setting which was particularly congenial for all children? Is it because they studied only overt behavior rather than deeper, more personal reactions? We do not know. But the research suggests that a pattern of life which differs significantly from the average need not be distressing either (as the authors suggest) because of some protective mechanism in human nature or because of a different set of environmental gratifications and compensations.

In sum, crippling has a number of effects on personality. It can place realistic limits on the goals the child can achieve and make other goals more difficult to achieve. The child may live in a less varied environment than his nonhandicapped counterpart; he may have less status and more conflict with peers; and he may be more self-centered. His parents are apt to be overprotective and to have increased marital conflict. While he may suffer severe emotional disturbances, he may also be no better or no worse off than the nonhandicapped child in regard to general adjustment. Even when he has more frustrations and difficulties, he still may find a comparable amount of satisfaction as can the intact child. While psychological problems can be expected in parent and child, they are far from inevitable; "realistic acceptance" of the handicap seems to be the most desirable attitude for both.

Facial Deformities. The effects of facial deformities on personality is explored in an intensive clinical study of twenty-four children by Macgregor (1953). The general picture is a compendium of problems engendered by disabilities–the shocked and guilty reaction of the mother when the child was born, the subsequent overprotection or rejection, the prying curiosity and prejudice of neighbors, the teasing and isolation by peers in school, the sensitivity or attempt to deny on the part of the child, and the universal distress at adolescence. As is usually true, there was also a group of mothers and children who took all the stress in stride. However, the study makes two more novel points.

First, the authors maintain that the maternal reaction was a function of her personality, not of the objective severity of the deformity. Although there was a general guilt reaction in which the mother felt she was being punished for transgressions, such as sexual escapades, drinking, or not wanting the child, the guilt was managed differently. Some mothers attempted to avoid all situations which would expose the child to questions, or tried to hide the deformity through special clothes, hairstyles, postures, or denied that the deformity existed; others went to the opposite extreme of "undoing" or acting as if the child were particularly cute or special; and finally, some mothers faced realistically the existence of the defect and the problems it caused. The children of avoiding mothers were socially the most isolated; children of undoing mothers could utilize their false sense of superiority to become leaders of small groups of peers but were very upset by rejection; the children of realistic mothers had good peer relations.

Second, the authors present a valuable description of the specific ways in which maternal tension and denial affected the child's image of his face. The game of labeling facial features was so distressing to some mothers that they tried to avoid it completely. The child's natural curiosity about his face met with uneasiness and evasion. He began shying away from his deformity in an effort to avoid the disapproval he sensed in the maternal attitude. Later, when the child was teased by peers, the mother concentrated on handling the teasing and bypassed the role of the deformity. Consequently, some children had a generally negative idea of their facial features as being something they "should not have" rather than a realistic awareness of how they look. Other children literally excluded their face or its deformity from their image of the visible body. According to some children, it was only after corrective surgery that they could look in the mirror and discover what their face was like. Recall our discussion of cognitive development (see Chapter 6) in which we distinguished between an idea which a child could not think and one he dared not think. Here the child dared not perceive his own image, which is a more extreme avoidance of reality.

Psychopathological Disturbances. The intimate connection between the child's self and his body is illustrated by clinical descriptions of severely disturbed children.

Bettelheim (1950), in describing one group of severly disturbed children, refers to the physical "freezing" of bodily movements and compares their excessively inhibited, awkward movements to a bodily "armor." These children are terrified both of close emotional contact with an adult and of losing control of their hostile and sexual impulses. Their bodies mirror the protective psychological wall they have built around themselves and their desperate efforts to keep every action under strict control. Absence of extreme self-restraint would mean going to pieces, both psychologically and bodily. When they are about to relinquish conscious control, as they must before going to sleep, or when they are trying to regain it upon waking up, they can be observed making an inventory of their body to assure themselves that it is still all there.

Bender (1947) describes how six- to twelve-year-old schizophrenic children are oblivious to their body surface and indifferent to secretions; they act as if their tears, mucous, urine, and feces did not exist. They draw human figures with a multiplicity of heads and limbs, thereby expressing their confused notion as to the essential nature of the body. Their chaotic mental life has its counterpart in an equally chaotic image of their bodies.

The Inner Body: Normal Development

Significance of the Inner Body. Like the visible body, the inner body is a psychological construction. It is not an accurate reflection of physiological processes. Many of the vital workings of the body lie beyond the reach of consciousness, and experiences of the body can exist which have no physiological basis; e.g., a person may still feel his "arm" after it has been amputated.

While both the visible and the inner body are psychological constructions, they differ vastly in their significance for personality development. The visible body, particularly the face, is our most public self. Others know our visible body more accurately and comprehensively than we know it. The inner body is our most private possession. It is directly known only to ourselves. In telling others how we feel, we can only hope the information is accurately expressed and successfully transmitted.

The intrinsic privacy of the inner body raises unsolved problems for parents and psychologists alike. Parents must appropriately decode the infant's distress signals (usually crying) into hunger or colic or being stuck by an open safety pin. After the child has learned to label his inner states, parents must distinguish accurate reporting, say, of a stomachache, from minor malingering and attention-getting behavior. Their problem of evaluating verbal reports of inner states is shared by research psychologists. When two people say, "I am five feet, ten inches tall," psychologists can check their accuracy. But when two people say, "I feel very sick," there is no satisfactory way of knowing whether they are reporting the same inner event or of establishing greater or lesser degrees of that event. Individuals differ markedly in their sensitivity to internal stimuli: some register slight visceral quivers, while others respond only to explosions. Attempts to measure the physiological state directly have furnished only ambiguous indicies of introspective reports.

In spite of the methodological problems it poses, the inner body must be reckoned with, for it is important at all ages. The implicit or explicit messages, "I am hungry," "I have to go to the toilet," "I feel sexy," mark nodal points in development. The inner body is also important as a source of strong affect. Every felt emotion has its physiological concomitant. In fact, some psychologists maintain that emotions are basically one's awareness of altered physiological states.

Placing the psychophysiological nature of affect in a developmental context, we see that emotional reactions exist long before a child learns to speak. Love, hate, delight, distress, excitement, erotic arousal are keenly experienced in the first years of life when vocabulary is nonexistent or meager. A child's stock of labels may lack words to describe feelings, such as "worried," "frightened," "upset"; he has only the physiological concomitants of emotionality—a stomachache, sleeplessness, vomiting, diarrhea, restlessness. Sensitive parents, clinicians, and researchers are especially alert to such bodily signs of psychological disturbance. The child who senses parental concern can later use aches and pains manipulatively; e.g., he can manufacture a stomachache which miraculously vanishes when he is allowed to stay home from school. Such minor malingering is different from the original physiological reaction which was neither manipulative nor even communicative.

Finally, the inner body is intimately involved in the socialization of sexual and aggressive impulses (see Chapter 7). In successful socialization expression of impulses is controlled, but the child is still free to acknowledge erotic and angry feelings. In unsuccessful socialization, the child is alienated from his impulses. Normal reactions trigger off intense anxiety or guilt, and the child regards himself as bad or naughty. His plight becomes more extreme when he tries to deny his impulses by forcing them out of awareness. The impulses continue to exist and continue to trigger anxiety and guilt, but the child literally has no idea of the source of his distress. Let us make the same points in terms of the inner body. Successful socialization enables the child to be on good terms with his inner body. The physiological messages signaling the arousal of anger or erotic feelings are reacted to with a sense of congruence and a feeling of control. Once the child has become alienated from his impulses, however, his body becomes an inescapable source of anxiety and guilt. The child who regards his body as his enemy is less able to manage the normal difficulties of development than a child who is on good terms with his body, since his inner distress is constantly diverting him from constructive pursuits.

In light of this analysis, the ancient idea of a sound mind in a sound body takes on an added meaning. Soundness of body is more than a matter of exercise and diet; it is a matter of accepting one's deeply felt emotions. To be on good terms with one's body implies that the problems of intimacy, hostility, and sexuality have been mastered without undue self-alienation. Such harmony is a source of great security to the child—and perhaps to society as well.

The Construction of the Inner Body. One of the most extensive investigations of children's concepts of the inner body was conducted by Gellert (1962) who studied 96 hospitalized children between the ages of four and sixteen years. She did not find that the young child pictures the inside of his body as an undifferentiated dark tube for the storage of food, with perhaps a heart or another stray organ floating nebulously around. On the contrary, children populated the inner body with a number of appropriate items. The child's understanding of these items differed in accuracy, but generally became more realistic as he developed cognitively.

When asked, "What do you have inside you?" the mean number of responses increased from 3.3 to 13 items from the youngest to the oldest groups, while the number of different items progressed from 24 to 60. However, the rate of increase was not constant, there being a sharp rise in the items listed by the nine- to eleven-year-old group. This is the time when curiosity about nature in general and about the workings of the body in particular increases markedly. The younger children tended to conceive of the inner body as containing the items they observed going in and coming out of it—food, beverages, bowel movements, blood, and urine. However, even the youngest included one unseen and stationary element—the bones—and they also might have some knowledge of heart, brain, nerves, or stomach. Beyond the age of eleven years, the tendency to regard the body as a container disappeared, and the body was viewed as having structure, organs, and functions in its own right.

Children considered the heart to be the most important organ, followed by the brain, eyes, head, and lungs. Surprisingly, the stomach ranked a weak sixth; apparently, children are not as impressed as are psychologists with its importance. The reasons given for importance had to do with running or controlling bodily functions; e.g., "It's what makes us run" or "What makes us work" or "Gives us strength" or "What makes people breathe and move and makes blood circulate." Other organs were important because they were considered necessary for life. Perceiving, thinking, communicating, food intake, and elimination, all were less frequently mentioned reasons for importance.

The skin was often perceived as keeping the contents of the body from falling out and the bones as keeping the body from falling apart. As an eight-year-old expressed it, "Bones are to put ourselves together." In general, the function bones serve did not change with age, but the number of functions increased. Thus, both younger and older children related bones to growth, but the latter included additional functions. "Bones" typically included the extremities and skull; ribs were rarely mentioned and were incorrectly located by twenty-five percent of children at all ages. In light of the ease with which ribs can be investigated, this last finding is unexpected.

Most children correctly located the heart and were reasonably accurate in estimating its size. For children under seven years of age, the heart was described in terms of beating, being necessary to life, and as having religious or psychological significance, e.g., "That's where God lives." For older children, the heart was associated with being able to breathe and with blood and, by thirteen years of age, its role was explained in terms of supplying blood to all the parts of the body.

All but the oldest children exaggerated the size of the stomach, but most related the stomach to eating. At first it was seen as a container, but by eight years of age, the idea of digestion appeared; i.e., the child indicated that the food undergoes changes. The most prominent misconception in the early years was that the stomach instigates breathing. Most of the five- to eight-year-olds thought food could take only one path after going into the stomach; either it went out to various parts of the body, such as the head or the chest, or it was discharged as bowel movements. By contrast, most of the eleven-year-olds stated that some food stayed in the body while the rest was eliminated, thus grasping the fact that food can have at least two destinies.

For many five- to seven-year-olds, the purpose of bowel movements was "so that we can go to the toilet," or "so you don't do it in your pants." Even a nine-year-old, when asked what would happen if he never had a bowel movement, answered, "You'd never have to go to the bathroom; you wouldn't need a toilet bowl; we wouldn't have a bottom." The social significance of toilet training clearly outweighted its physiological import. However, half of the children in the five- to seven-year-old group and the majority of nine- to twelve-year-olds associated feces with food as well as with maintaining health or life. Only by thirteen to fourteen years of age did the majority of children explain feces in terms of eliminating waste matter. Surprisingly, only three of the children said feces were "bad" or noxious, in spite of parental attitudes of disgust during toilet training. As for the bladder—the idea of having one rarely occurred to the children!

Gellert's data are revealing for what they do not show as well as for the findings themselves. There is little evidence of a rich fantasy life accruing to the inner body, although its dark and mysterious nature would seem to be the ideal climate for breeding luxurious specimens of children's imagination. Ask a child a reasonable question about his inner body, and you will probably get a naturalistic answer. Eating and defecating are not weighted as heavily as they should be in light of their psychological significance; bones and heart, which are psychologically unimportant, receive even more attention and are assigned greater import. There is no evidence that the children's ideas are based on the kind and/or intensity of sensations or on the relative size of the different organs. Even the changes in ideas are not uniformly in the direction of greater clarity; the function of some organs is realistically grasped by a given number of the youngest children, while misconceptions are found in the oldest group.

Gellert's data suggest that understanding the inner body is an intellectual venture for the child. In the early stages he repeats explanations, interpretations, and misinterpretations he has heard, although there is always a question of how much he really understands. In his struggle to grasp the baffling and invisible phenomena of the inner body, he produces the literal and egocentric explanations characteristic of early childhood (see Chapter 6). First of all, he believes what he sees. What does the body have inside it? What goes in and what comes out of it, obviously. What could be more natural than to regard the body as an inert container like the familiar drawer into which some objects go and disappear and from which unseen objects are removed. Egocentric thinking is evidenced by his explanation of bowel movements. He believes that feces exist to provide material for toilet training, since toilet training is what he experiences most vividly. The concept of the inner body as an entity with characteristics and workings of its own requires a fund of information and, more important, the ability to reason and infer, which are beyond him. The surge of curiosity between nine and eleven years of age helps supply needed facts concerning the inner body, while his increased cognitive sophistication—especially the transition from literal, egocentric thinking to thinking based on reason and an understanding of the independent existence of objects—enables him to view the inner body as an independently functioning mechanism.

Gellert's study also suggests that the concept of the inner body might have a particularly interesting developmental history. The inner body has special characteristics which set it apart from the physical and social world, as well as from the visible body. It is invisible, many of its stimuli are vague and weak, and many of its functions cannot be experienced. The child cannot understand these stimuli and functions through the confrontation of faulty ideas with incompatible information, as he does when learning about physical and social reality (see Chapter 6). It may be that he must grasp the workings of the inner body in the same way as he grasps the workings of a machine; e.g., his understanding of the beating of the heart or the circulation of the blood might go through the same cognitive stages as his understanding of relevant mechanical principles. However, the inner body differs from a machine in two important respects: it is an intrinsic part of the child's self and it is the source of some of his most vivid positive and negative experiences. This combination of intensely experienced sensations and minimum information from reality might engender and perpetuate primitive concepts concerning the workings of the inner body or concepts based on defense mechanisms. A four-year-old believes that he has a stomachache because monsters are breathing fire inside him. If he does not communicate this idea, the belief may persist, since he may not encounter contradictory information. By contrast, his faulty beliefs about the physical and social worlds are continually being disconfirmed by a wealth of contradictory experiences. In general, then, we might expect that the concept of the inner body would include both the stages of cognitive development which Piaget delineated in regard to the workings of mechanical objects and the affectively determined, animistic, omnipotent ideation which Freud explored. Of all the realms we have discussed—the world of people, of objects, of the self, and of the body—the inner body has been the least investigated and presents some unique possibilities for illuminating cognitive development.

Puberty. Puberty deserves special attention, since it is a time of heightened sensitivity to the inner body. In part, this sensitivity is due to the profound physiological changes which are taking place, particularly in the year preceding sexual maturity. The homeostasis of the body is upset, as evidenced by a rise in blood pressure, basal metabolic rate, and pulse rate. The increased activity of the pituitary gland and the gonads change the endocrine balance. Poor eating habits during puberty may affect nutritional status and produce a subclinical malnutrition.

During this period of physiological change, the child is likely to complain of a number of rather ill-defined physical symptoms. He worries about constantly feeling tired, about palpitations of the heart, or shortness of breath, about headaches; girls frequently report abdominal pains a few months before the onset of menstruation. Such complaints can mean that the adolescent, like the infant, is in close psychological contact with his unstable physiological status. However, he also may be suffering a transient hypocondriasis. Even the normal adolescent is subject to sudden, brief feelings of panic during the day or to terrifying dreams at night. Less secure adolescents use the body to disguise more troublesome concerns over sexuality and emancipation. Worry over bodily integrity and resilience symbolizes his uncertainty about his ability to cope with the problems which beset him. Fortunately, the period of turmoil is temporary, for both the physiological and psychological upheavals subside during the later phases of puberty.

Menstruation and ejaculation have a special psychological role in this general picture of unrest. If breasts can be regarded as the body's public proclamation of sexual maturity, menstruation is its more intimate message. Ideally, the event should be greeted with pride or at least with realistic understanding. In these days of changing sexual attitudes, it is difficult to know how each new generation of girls is, in fact, reacting. The trend is toward realistic acceptance, but old prejudices die slowly. The stigma of menstruation as "the curse" remains. The feelings of apprehension and disgust are reinforced by headaches, backaches, cramps, and abdominal pains which can accompany early menstruation and by the association of bleeding with physical injury. Thus, cultural factors, physiological realities, and the girl's past associations conspire to make the onset of menstruation a stressful experience. The boy's first ejaculation apparently is less disturbing than a girl's menstruation. The cultural lore is neither as pervasive nor as negative. The attendant fears are due more to the boy's general concern about adult sexuality plus his anxiety and guilt over masturbation.

Once the storms of puberty subside, the inner body is no longer a central concern. Sexual impulses and needs continue, but they are more focused, more readily gratified, and arise in the context of a physiologically stabilized organism. Only under the special conditions of deprivation, illness, damage, or marked emotional upset will it become a cause for concern in its own right.

The Inner Body: Deviant Conditions

Physical Illness. The child who is on good terms with his inner body literally and figuratively feels secure "in his bones" and "in his guts." By the same token, physical illness represents a significant disruption of the harmony between body and self. Frequently, although not inevitably, the body becomes a source of discomfort and pain; the child's natural tempo is disrupted as his supply of energy diminishes; his time perspective collapses and he loses his tolerance for frustration and delay; the delights of the world become irritating intrusions, as he shifts his focus inward upon his hurt and aching body. Sickness has important interpersonal repercussions, for it touches off special attitudes in parents and special treatment by them. Some react to illness realistically; others resent the added burden and the babyishness of the child; still others feel most fulfilled by the child's dependency and try to prolong his helpless state. Illness is frequently complicated by medical procedures. Surgery is one of the most dramatic procedures, but prolonged bedrest, frequent injections, isolation, special diets not only compound the discomfort but may be more stressful than the illness itself.

Anna Freud (1952) sensitively analyzes several aspects of illness. The experience of being nursed is a mixed one. The child has sole possession of the mother's time and attention—a state so highly desired that he may be tempted to prolong it unduly. On the other hand, he is subject to a series of indignities centering around the fact that his body is no longer his own. He must passively submit to being on display, to being fed, dressed, washed, toileted, and dosed, to having injections and enemas. Just as adults rebel against being treated like a baby, so does the child. He has only recently been successful in his efforts at self-care and his concomitant struggle to establish an independent self; to force him back into a more infantile state is to deal a blow to his self-esteem. Being confined to bed when he does not actually feel depleted deprives the child of motor activity essential to releasing pent-up feelings. Dammed-up aggression, in particular, is likely to produce restlessness and heightened irritability.

One of Anna Freud's principal themes is that the child's reaction is always a function of the meaning the illness has for him. His response to pain does not depend so much upon the physical stimulus as upon his concern over being psychologically harmed. If he is worried over being punished or deceived or rejected, he will view pain in the light of his general psychological distress and will react to even minor hurts with anxiety. The secure child, on the other hand, will be able to tolerate intense pain well and forget it readily. Or again, four- to six-year-old boys who are making a good masculine identification are relatively indifferent to pain; those who are unsure of themselves and are fighting against strong tendencies toward passivity and femininity overreact. The child's attitude toward illness is also determined by the mother's domination of his body in the areas of health and hygiene. If she has insisted that he wash his hands, wear his rubbers, eat balanced meals, and dress warmly, the child is apt to adopt her concern. He becomes worried over his own body and devotes undue attention to it. Illness is regarded as added evidence that "mother was right," and he passively submits to his fate. The casual child of the casual mother does not have this self-concern. When he becomes sick, he will, at best, put up with the passivity of being cared for or, at worst, fight for his right to be sole master of his body.

Evidence that children react to illness in terms of their personality characteristics comes from a study by Prugh (1953). His subjects were 100 children between the ages of two and twelve who were hospitalized briefly for acute illnesses. He found that the greatest number of severe reactions were in children three years of age and under; the four- to six-year-olds had fewer severe reactions; and the six- to twelve-year-olds the least. This progression corresponds to the increased self-control, the stronger defenses against anxiety, and the greater variety of interests and emotional investments which characterized later personality development. The nature of the anxiety also was different. Separation anxiety was the principle cause of distress in the youngest children and was also seen in those older children who were having neurotic difficulties presumably with their parents. For the four- to six-year-olds, the psychological meaning of illness and treatment was a greater potential threat than separation. Since they were in the process of developing a strong conscience, they viewed illness in terms of punishment, and feared mutilation and death which, according to psychoanalytic theory, are common anxieties at this age. The oldest group presented a picture of good control. What anxieties they had were similar to those of the four- to six-year-olds.

Although the effects of illness upon a child are determined by his level of development and the attitude of significant adults, there are also some common reactions. Both dependency and rebelliousness are to be expected. In the former, the child indulges himself in the care and special attention required by his condition; in the latter, the child chafes against the restrictions his illness imposes and blames others for his incapacity. Regression also occurs. The younger the child and the more severe the illness, the more likelihood that he will revert to a more immature level of behavior, e.g., he may lose bowel and bladder control, go back to baby talk, become clinging and demanding. Recently acquired advances are usually the first to give way. Guilt is frequently mentioned as a common reaction if the child interprets sickness as punishment for wrongdoing. However, the evidence is not as convincing as one would like, since some children may merely be parroting parental attempts to relate sickness to disobedience. The effect of illness on self-esteem is still in doubt. It seems reasonable to assume that chronic illness would lower self-esteem, and even that low self-esteem would increase susceptibility to illness. There is some evidence to this effect, but it is by no means definitive. Finally, illness can be a constructive experience for those who are able to meet and master a stressful situation. Unfortunately, we know far too little about the children who are strengthened by stress rather than being blocked or overwhelmed by it (see Newman, 1963).

Psychosomatic Disorders. There are three possible relations between affect and physical illness. The first we have already considered; namely, that illness produces emotional reactions. The second is the possibility that emotional disturbances may increase susceptibility to illness and/or prolong its duration, and/or increase the probability of recurrence. Thus, a chronically fearful ten-year-old may have more frequent and prolonged sinus attacks than his emotionally secure counterpart. The final possibility is that emotional disturbances alone, if intense and prolonged, will place an intolerable physiological stress upon the organism and produce a structural change or a so-called psychosomatic disease, such as asthma, eczema, ulcers, ulcerative colitis, and rheumatoid arthritis (Selye, 1946). Since the concept of psychosomatic illness has been widely discussed and intensively researched, it deserves special attention.

There has been a tendency to dichotomize illness, on the one hand, as the traditional physical diseases due to germs, etc., and on the other, as diseases due to emotional problems. The former were the province of the medical man, the latter the province of the psychotherapist. Such an artificial staking out of territory could not last, for physical and mental health are too intimately intertwined to be categorized as either-or. For each disease and for each ill child, the problem is one of discovering the role of constitutional factors, infectious agents, nutritional difficulties, allergic sensitivities, emotional disturbance, familial disharmony, and cultural pressures. Some diseases fall on the "physical" end of the continuum, e.g., cerebral palsy, encephalitis, poliomyelitis; some bodily disturbances are located at the "psychological" end, e.g., in hysterical conversions in which the individual cannot move a limb or feel pain or see, even though his body is intact. The majority of illnesses lie in the vast middle region and result from a combination of physical and psychological factors.[6]

There has been little agreement among theorists concerning the etiology of psychosomatic disorders. Some champion a specificity hypothesis, claiming that each disease is either caused by or correlated with a specific psychological characteristic or conflict. Dunbar (1935) correlates different illnesses with personality traits; e.g., hypertensives are "top dogs" or "would-be top dogs," diabetics are "muddlers," accident-prone individuals are "hobos." Alexander (1950) states that for each psychosomatic disorder there is a typical conflict situation and a specific affect; if the conflict is unresolved, the physiological disturbance created by the affect will eventuate in physiological breakdown. The man who develops an ulcer, for instance, has an unconscious longing to be cared for, symbolized by a craving to be fed; the hypertensive has unconscious rage, and the asthmatic represses a cry for mother.

Grinker (1953) rejects the notion of specificity. He regards anxiety in infancy as the psychosomatic situation par excellence, since it is a state of intense physiological and psychological stress. In normal development, the individual has increasingly complex and differentiated mechanisms for mastery of anxiety. If these mechanisms for mastery fail to develop adequately, anxiety reappears in its primitive, global form and taxes the organism beyond its tolerance for healthy functioning. Physiological breakdown and disease result. In keeping with Grinker's developmental hypothesis, some theorists claim that the first year of life is the crucial period in deciding the child's vulnerability to subsequent psychosomatic disorders. During this period, physiological reactions are intense and unstable. If maternal care is adequate, physiological distress will be minimized and physiological integration will be fostered. If the infant receives inadequate maternal care, he will frequently be in a state of prolonged distress and physiological integration will be faulty. The stage is set for later physiological breakdown if the individual is under an emotional strain. The kind of stress, therefore, is not as important as its timing in the development process (see Garner and Wenar, 1959).

These theories, while plausible, share the fundamental weakness of being formulated at a time when little was known about physiological responses in early childhood and little direct observation had been made of early personality development. However, they served to renew interest in a badly neglected area of inquiry—the developmental study of psychophysiological responses in general and the developmental psychophysiology of affect in particular.

One of the most dramatic examples of the new interest in psychophysiological research in children is the case of Monica, a fifteen-month-old girl born without continuity between mouth and stomach (see Engel, Reichsman, and Segal, 1956). She

[6] For a somewhat different but compatible viewpoint, see Starr (1955).

was nourished by being fed through an opening in the abdominal wall. Through this opening, samples of gastric juice could be taken directly from the stomach and the rate of secretion of hydrochloric acid could then be correlated with her behavioral reactions. When Monica was emotionally withdrawn or depressed, or when her activity level was low, the production of hydrochloric acid was reduced. When she was responding vigorously to an adult, her production was high, regardless of whether the affect was pleasure or anger. The rise in production was highest during rage, when she was most intense, active, and least in control of herself. However, there was also a considerable rise when she was reunited with one of her favorite experimenters after a period of being with strangers. The authors interpret their findings as supporting Freud's concept of an oral period (see Chapter 7). In these vigorous, emotional involvements with people, the stomach behaves as if it were preparing for food. According to psychoanalytic theory, emotional attachments in infancy are mediated by the feeding experience, so there is good reason for linking digestion with strong positive and negative emotional responses to others.[7]

Surgery. Surgery, like illness, involves a radical departure from the child's usual relation with his inner body. The child interprets surgery at the level of his own cognitive and emotional development and is influenced by parental reactions. Unlike illness, surgery requires that he be hospitalized to face usually unknown and painful medical procedures administered by personnel with varying degrees of sympathy and understanding.

One of the most comprehensive clinical studies of surgery was done by Jessner and his colleagues (1952) who studied 143 children between the ages of two and fourteen during their two-day hospitalization for tonsillectomy and adenoidectomy. The investigators found that the experience of hospitalization was a constructive experience for a number of children: it improved their health; it was a challenge and increased their prestige; in certain instances, it represented atonement for guilt feelings. The operation was disturbing to children who had neurotic tendencies or who had experienced a frightening life situation, such as the death of a relative. In general, accurate information about the operation did not help children to cope with the frightening situation, if the mother was anxious, resentful, or ambivalent. The medical procedures also activated some disturbing or frightening fantasies. For some children, removal of the tonsils represented a threat to body intactness; for others, tonsils were the place where the "bad stuff" such as demons or dangerous enemies resided; for still others, "bad" had a moral connotation and the operation was viewed as an exorcism. The narcosis was perceived in terms of death or suffocation, as punishment, or as a murderous or sexual attack; older children were concerned with the loss of self-control. The operation was interpreted as mutilation by some children, while others

[7] Since psychophysiological research on children is a rapidly expanding area, articles soon become dated. Some idea of the intellectual orientation of the investigators and the complexity of their data can be gained from the Report of the Forty-fourth Ross Conference on Pediatric Research (1963) and from Richmond and Lipton (1959), Lipton, Steinschneider, and Richmond (1966), and Bridger (1962). For an exhaustive review of the psychosomatic area, see Prugh (1963).

feared that their sex would be changed; girls, and even a few younger boys, thought of the operation in terms of having a baby. Most children had mild postoperative reactions of irritability, demandingness, depression, or nightmares.

The authors noted that the children who could express their fears, either verbally or in play, assimilated the experience better than those who acted as if nothing were bothering them. The former could talk with parents or hospital personnel about being frightened and could cry. In their play, they symbolically rehearsed or relived the operation by taking the part of the surgeon or the comforting mother (see Chapter 9). The apparently unconcerned child who was inwardly or unconsciously fearful could become quite upset postoperatively.

Janis (1958) obtained similar results with adult patients and accounts for them in terms of "the work of worry." Reality inevitably confronts us with stressful experiences—a final examination, a crucial football game, or a job interview. Having been forewarned, the person begins to imagine the situation and the feelings it will provoke. This anticipatory anxiety or worry, while uncomfortable, does rehearse him for the greater stress to come. Those who refuse to imagine what the painful situation will be like, who pretend to be unconcerned or even happy, are not prepared; the inescapable blow hits them with full force and releases all the latent fear. Janis is not saying that all worry is adaptive, since in neurotics worry outstrips the stress involved in the event; he is saying that in a psychologically sound person worry serves a constructive function.

Phantom Limbs. One of the most elegant studies of the inner body is Simmel's (1966) research on phantom limbs. People continue to experience arms, legs, fingers, and toes after they are amputated; a child will report after an operation, "I looked and my leg was gone, but I still could feel it," or "I can feel my toes; I can wiggle them the same way you can wiggle yours."

Simmel accounts for such vivid and persistent phantoms in terms of the concept of a "body schema," which is roughly equivalent to a mental image of an animated stick figure of ourselves. The body schema is built up of kinesthetic sensations from postural changes. Every time the infant kicks his leg or moves his head or turns over, a special kinesthetic pattern is registered in the brain. Like a vast number of superimposed photographs, each new pattern is registered against the background of all previous patterns until an image of the broad outlines and the fine details of our body's position in space is built up. When the child raises his right arm, he knows that it is a right arm rather than a leg and that it is being raised instead of lowered because this one kinesthetic sensation occurs against a background of a myriad of sensations similar to it but distinct from sensations from other parts of the body. (The photographic analogy is not entirely correct because the body schema is fluid and constantly subject to modification by new experiences or by damage to the brain.) Because of its constant use, the body schema takes on a unity and a cohesiveness of its own. Even when a major component is lost through amputation, the total image remains until the schema can be slowly modified by new sensations (say, from a stump) or by continual failure to receive familiar sensations.

Simmel ingeniously tests the validity of the body schema hypothesis. Her finding that the phantom limb phenomenon occurs only in 33 percent of two- to four-year-olds and almost 100 percent of children six years of age indicates that the body schema takes time to develop cohesiveness. Sensory input is the essential condition for developing a phantom limb, because congenitally deformed children who were born without limbs do not have phantoms. When congenitally malformed limbs which have been devoid of sensation and motion are amputated, there is no phantom. When kinesthetic input is slowly diminished over time, as in the case of digital absorption due to leprosy, there is no phantom because the schema can be appropriately modified by the gradually changing kinesthetic messages. In sum, for the phantom limb to be experienced, the child must be old enough for a body schema to have developed and the limb must be suddenly lost so that he is prevented from gradually assimilating the information about the change in his body.

Psychopathological States. The final evidence that the inner body is a psychological construction comes from the clinical literature describing instances in which an essentially healthy body is experienced as malfunctioning. Hypocondriasis is a case in point. Hypocondriasis has roots in a variety of emotional difficulties. We have already mentioned the upset stomach or headache which appear before the child must face a difficult situation and which clear up when he is allowed to escape. Is the child really sick? Obviously not. But he is really unhappy and is using his body as the most likely resource for solving his problem. A more enduring tendency to interpret bodily sensations as noxious comes about through identification. Just as a child can model himself after a sweet, energetic, outgoing mother, he can model himself after a depleted, headachy, nauseated one. Children also can become helpless victims of their parent's hypochondriacal tendencies. The parent who is constantly worried about the child's health, who warns him of germs and infections, who talks about the nutritional value, the digestibility, or the cleanliness of food, who doses the child with laxatives and assaults him with enemas, is teaching the child to perceive his normally functioning body as vulnerable and sickly.

However, hypocondriasis can be of the child's own making. In some cases, the symptoms can take their cue from the physiological responses to stress. The anxious child experiences his heart beating rapidly, a shortness of breath, a strange and altered cluster of inner sensations. He can then displace the true cause of anxiety onto his body. An adolescent who is frightened of expressing anger toward a parent, for instance, becomes frightened of having a heart attack instead, and thereby saves himself from a painful confrontation.

In severely disturbed children the inner body can be bizarrely distorted. A child in a pathological state of euphoria brags that he is not only the biggest person in the world, but that he has more blood than anyone else. An adolescent complains that his brain is rotting and that his head is full of roaches which are feeding on it. A thirteen-year-old girl who had been mute for ten months finally explained that she was afraid to talk because feces would come out of her mouth if she opened it. These children, so often isolated from all positive human relations, have returned to the original, most primitive source of self to express their bewilderment, terror, grandiosity, and despair.

Concluding Remarks

In concluding our discussion of the body, we also complete our account of how the infant separates the stream of experience into self, people, and things. Adults think of self in terms of psychological characteristics; their bodies are important but not essential to defining who they are. In infancy the situation is reversed: quite literally the infant's self comes into being as the image of his body is formed. His clearest impressions of himself as an entity are in terms of stimuli from the body—arms and hands, legs and feet, skin and mouth and gut, all provide him with the most continuous as well as the most vivid experience of early infancy. Throughout the first six years of life, the body plays a significant part in determining the child's view of himself, as it does again in adolescence.

We have also seen that the body is a psychological construction. Accuracy in representing reality is achieved over a period of years. Even in adults, however, the psychological representation of the body is only a rough working model of reality. There are always aspects of his body about which the individual is uninformed, poorly informed, or misinformed. The representation can be distorted for motivational reasons, such as the need to perceive the self as more youthful or more sickly than is warranted. Under extreme stress, the representation can be dramatically distorted: manifest disfigurements can go unnoticed, animistic thinking can return, as in the case of the adolescent who believed his brain was being eaten away, or the integrity of the body itself can be shattered, as in the cases of severely psychopathological children.

11 Evaluation

This concluding chapter will return to certain ideas presented in the Overview (see Chapter 1) in the light of our subsequent discussions. We shall deal in generalities and impressions, since our purpose is to react and to provoke reactions, not to summarize and conclude.

We characterized the child as a learner, and we have been particularly interested in the process by which parental standards and values become the child's own. We have seen how socialization insures cultural continuity from one generation to another and provides a middle class frame of reference as a common denominator among children. We have also seen how individual families introduce their special variations on this common theme. However, the child does more than learn to think, feel, and behave in a prescribed manner. He evaluates, he integrates, he decides, he manipulates his environment. As parental socialization runs counter to his self-interests and pleasures, he weighs the consequences of compliance and defiance and decides which course to follow. Subsequently, certain prescriptions for acceptable behavior in the home run counter to those of the group, and the group's run counter to the identity he is evolving for himself. In addition to evaluating conflicting prescriptions, the child is capable of integrating them into higher-order guides and standards of behavior. Finally, the child is continually striving to manipulate the environment so that he may accomplish his own purposes. At first his techniques are crude—recall the willfulness and negativism of the toddler—but they rapidly become varied and subtle. We have been unable to find a single term which includes the functions of evaluation, decision making, integration, and manipulation. The characterization of child-as-agent, while not totally satisfactory, is a reasonable approximation if we remember the specific constellation of functions the term is designed to cover.

At this point we wish to evaluate how adequately psychologists understand the child-as-learner and the child-as-agent. Our criteria of adequacy are: (1) the existence of a general theory or of complementary theories which is (are) (2) developmentally oriented (3) relevant to observable behavior, and (4) supported by clinical and/or objective evidence.

The Child-As-Learner

Learning Theory

Whenever we have discussed the child-as-learner, we have felt ourselves to be on firm ground in regard to our criteria of adequacy. Learning theory and social learning theory in particular meet our standard unusually well. The central theme of the theories goes back to the birth of behaviorism: we learn our personality from our social (in the sense of interpersonal) environment. The tone was set by Watson's boast that he could make a healthy infant into any kind of person he wished by utilizing psychological techniques. In this approach the *how* of learning is generally more important than the *what*; anything can be done if one knows how to do it. The *how* includes all the well-established principles of learning, such as reward and punishment, imitation, stimulus and response generalization, and discrimination. The *what* can be broad social patterns (e.g., middle class and lower class sex roles), family patterns, peer group standards, etc. The challenge is to discover how the principles of learning can be utilized to account for a given aspect of the environment becoming part of the child's personality. The developmental dimension is introduced by determining which environmental factors change as the child goes from birth to maturity; e.g., what sex role behavior is expected of the four-year-old, the ten-year-old, the adolescent. In the learning theorist's program the emphasis is on rigor and precision; terms and issues are clearly defined, and there is an overall cohesiveness and a sense of direction. It is comparatively easy, then, to determine what the basic issues are, as well as the advantages, the questions, and the limitations of their approach. The emphasis on scientific rigor has served learning theorists well.

The learning theorists' program is worthwhile and reasonable. Certainly a significant amount of what we consider to be ourselves is what we share with members of our class and society. And certainly, we are what we have learned from parents and peers and other significant individuals. While social learning theory has tended to focus on broad social influences (such as class variables or patterns of child rearing), there is nothing which would preclude the study of the most individualized influences, say of a given parent on a given child.

When we delve into the kinds of explanations learning theorists offer, we encounter an interesting ferment. While we have seen that principles of learning work, it is not at all clear why they work. There are various explanations of imitation (see Chapter 5) and a lack of consensus concerning what is involved. There is even no general agreement as to why rewards and punishments have their predictable effect on behavior. Certain theorists recommend staying exclusively at the level of overt behavior and environmental manipulations; as long as we know that reinforcement and imitation work (in the sense of effecting behavior) there is no reason to question why (Bijou and Baer, 1961). Yet, social learning theorists do not usually take this stance, and it is instructive to examine the explanatory concepts they utilize. Aronfreed gives affect, especially anxiety, a prime role in determining internalization and refers to "cognitive templates" in accounting for imitation, while Bandura theorizes in terms of "images." In doing so they break with the traditional behaviorists who were suspicious of emotions and of cognition, both of which were too internal, too inferential, too mentalistic, for their taste. Having admitted affective and cognitive variables into their explanations, how far will learning theorists go in utilizing them? Will the inclusion of anxiety open the door to defenses against anxiety and to other intense affects as well? And will this in turn lead to Freud's thesis that defense mechanisms and intense affects can determine what a child perceives and thinks and how he behaves? Or, again, will acceptance of images and templates expand to the acceptance of schemata and ultimately of Piaget's thesis concerning the progressive stages of cognitive development? In short, are there things which the child dare not and/or cannot learn because intense affect, defense mechanisms, and his level of cognitive development systematically distort the information from his environment (see Chapter 6)? If so, the thesis, "We learn our personality from the environment" cannot be regarded as a simple matter of internalization of external events.[1]

Limitations of Learning Theory

Certain limitations to the research being done by learning theorists arise from their preference to study single variables in isolation. As we noted in the Overview, objective investigators prefer to proceed from simple to complex. While single variables are being explored with great ingenuity at present—e.g., Aronfreed's research on self-control, Bandura's on imitation—the simultaneous manipulation of a number of variables in order to observe the effects of their interaction is rare. In evaluating studies of parental discipline (see Chapter 4), for example, we noted how few multivariable studies are being conducted, in spite of their acknowledged importance and their feasibility. The same limitation appears in the learning theorist's approach to the developmental dimension. Separate groups of children are studied at discrete age levels, but there is little attempt to determine the effects of learning at one stage on learning at the next; e.g., an investigator may study the process by which an adolescent learns a new role of loving the opposite sex, but may ignore the fact that this involves a reversal of the previous role of antagonism toward the opposite sex. The final limitation of research in the learning theory tradition grows out of the emphasis on *how* and the relative neglect of *what*. Gesell, Piaget, and child psychoanalysts have rich descriptive data and have addressed themselves to accounting for it. In learning theory the living child is only rarely glimpsed. Instead, there is a tendency to be satisfied with general categories of behavior, such as aggression, as if there were no need to account for the difference between the furious striking out of the toddler and the adolescent's devastating insult. Changes in form within a general category of behavior are an integral part of development, and learning theorists' relative neglect of such changes makes their explanations seem oversimplified.

Finally, certain topics which we discussed seem to elude an explanation in terms of social learning theory. One example is play. If our conceptualization is correct, play and fantasy are not the result of the variables which lie at the heart of social learning theory. The motivation to play comes from the child and the activity is intrinsically absorbing. Reward, punishment, and imitation might account for the

[1] For a more detailed comparison of learning theories and Piaget's theory, see Kohlberg (1969).

frequency of play and its content, but not for its origin or development. In like manner, willfulness and negativism in the toddler do not fit the learning theory model. These behaviors appear on the developmental scene for some undetermined reasons, and parents must manage them as best they can. It would be difficult to make a case for their being caused or sustained by parental behavior.

Limitations of the Socialization Model

There are certain limitations which learning theory shares with other theories which subscribe to the socialization model, i.e., the model of personality as the product of influences from the socializing agents, particularly the parents. The first limitation arises from the fact that constitutional variables are once again being seriously considered as determinants of personality (see Chapter 10). These are not variables which are independent of the interpersonal relations and need only to be added to them like constants in an equation; e.g., they are not like differences in intelligence which might only influence the rate of learning and not the content of what is learned. Rather, they are forcing a revision of the simple image of child-as-learner doing little more than reacting to his environment. The child's constitutionally determined characteristics, or his temperament, may play a significant role in what his socializers will teach him, the techniques they will use, and the kind and intensity of affect they will display. An adaptable, socially responsive infant, for example, is apt to be labeled "good" and "sweet" by his parents and is apt to elicit love oriented techniques and warm feelings. In general, a unidirectional socialization model is no longer tenable; instead the parent-child interaction must be viewed as having three possible outcomes: the child's behavior is changed by the parent, the child's behavior is unaffected by the parent, and the child changes the parent's behavior.

There is no reason for learning theorists to limit themselves to a unidirectional socialization model as they have been inclined to do. Their theory is equally applicable to a more versatile model of parent-child interaction. All they need do is ask the same kinds of questions of the child as they have been asking of the parents: what are his vested interests (e.g., to be the center of attention? to have all the toys? to hurt his sibling?) and what techniques does he use to modify parental behavior in order to achieve his goals (e.g., temper tantrums? passive resistance? affection?) (Gewirtz, 1966, especially pp. 250-253). Instead of studying the parent's efforts to make the child more independent, for example, they could focus on the child's effort to remain dependent. By making this change in focus, learning theory may provide new insights into initiative, willfulness, negativism, judgment, and manipulation of the environment. Learning theory would then be in a position to make as many contributions to our understanding of child-as-agent as it has to our understanding of child-as-learner.

Comment: The Consequences of Being Socialized. Freud regarded society as opposed to the innate desire for immediate gratification of sexual and aggressive drives. The parent must bargain with the child, offering love in exchange for conformity to socially approved modes of behavior. The child accepts, but a part of him always resents the thwarting and postponement inherent in socialization; the socialized child is inevitably divided against himself. While Freud accurately depicted the situation of his tortured neurotic patients, there is a question of whether he was primarily describing the socialization process gone awry.

Looking back over our discussions, it seems that Freud's is not a generally valid conceptualization. If socialization involves *No, Don't, Bad Boy,* it also includes *Yes, Do,* and *Good Boy*. Phrased technically, positive reinforcement is a potent technique for strengthening desirable behavior. More important, there is an impressive amount of congruent behavior for the parent to reinforce. The child's eager exploration of the inanimate world, his ventures into realms which will lead to communication and to motor and academic skills, can readily be facilitated and expanded by the parent; e.g., the parent can help the child master the frustrations inherent in the goal-directed use of objects, can teach him to speak, to throw a ball well, to count, to recognize letters, to print. Certainly, becoming communicative, skillful, and industrious is part of becoming social.

In addition, Adler's point concerning man's basic social nature seems valid within limits. Infants have a compelling interest in human beings which may be buttressed by instinctual forces and innate perceptual preferences. The toddler and preschooler are strongly attracted to peers, in spite of the fact that peers do not satisfy the basic needs in Freud's motivational theory. If there is rivalry, hostility, and possessiveness, there also is sociability and friendship: the pre-schooler who will not share his toy spontaneously shares his fantasy play. Adler may be correct in stating that the parent who facilitates peer relations makes a basic contribution to the child's psychological growth. In sum, initiative and social feelings make socialization more than a confrontation between an asocial child and his parents. Becoming social involves both a countering and a facilitating, a blocking and an enriching of the child's preferred behavior. This conclusion does not solve the problem of achieving a balance between freedom and control; but it does allow for the observation that there are well-socialized preschoolers and adolescents who are living in harmony with themselves and with their environment.

The Child-As-Agent

Cognitive Theory

Our evaluation will first be in terms of Piaget's theory, since he has been our principle guide to understanding the child-as-cognizer. Piaget clearly depicts the child as an agent in cognitive development, constantly striving to understand his physical and social environment. In the early developmental stages, the child distorts reality to a fantastic degree, and his progress toward realistic thinking results from his efforts to accommodate to and assimilate new and disconfirming information. Piaget frequently does not address himself to the substantive problems which have concerned us, such as the conflict between child and parent or child and group. His image of the child-as-cognizer is closer to that of an intellectual problem solver than is our image of the child-as-agent who is concerned with the pleasures and pains which will result from a decision. However, cognitive integration is an essential element in our picture, and Piaget offers an unparalleled account of how integration takes place in a number of areas relevant to personality development. His approach amply meets our requirements for adequacy, since it contains a comprehensive theory, finely tuned to the reality of children's behavior and development and documented by a wealth of empirical evidence.

Our principle dissatisfaction is with the minor role affect plays in Piaget's program and his tendency to depict a dispassionate intellect. From earliest infancy, the need to know vies with the need for security (recall that the infant explores "from the secure base" of the mother), since novelty can frighten as well as lure. The child's desire to understand himself and his environment is constantly being diverted by other needs, such as the need to reduce anxiety, to bolster self-esteem, to justify delinquent behavior, to prove his helplessness. Our study of deviant behavior has revealed only a few of the innumerable irrationalities which arise from sources other than those of cognitive limitations. Hopefully, the interest in cognition, which Piaget has stimulated, will revitalize both the study of thinking which is motivated by needs other than the need to understand reality and the study of psychopathological thinking in children.

The Psychoanalytic Approach

The psychoanalytic approach meets our criteria of adequacy. Freud clearly regarded the child as an agent whose primary function is to manage the basic antagonisms between society and impulses. Our picture of the parent trying to socialize the child while the child is determined to pursue his own interests and pleasure is in the Freudian spirit. Freud was particularly insightful into the ways in which pleasure seeking and pain avoidance affect behavior: specifically, the ego evaluates an activity in terms of its eventuating in pleasure or in pain, and the ego's decisions are governed by the principle of gaining all the pleasure which society will allow while minimizing the pain inherent in social restrictions. As the ego develops, the toddler's primitive impulse to destroy his sibling becomes the athlete's determination to defeat his opponent; the preschooler's sexual curiosity becomes the adolescent's delight in tinkering with a jalopy. Neo-Freudians such as Erikson (see Chapter 9) and Jacobson (see Chapter 5) have addressed themselves to the growth promoting functions of the ego, especially in middle childhood and adolescence. They trace the process by which the child becomes integrated into society through achieving a mature conscience and an identity, and thus they enrich the classical Freudian picture of child-as-agent.

However, the psychoanalytic approach has its limitations. The first limitation results from the comparative weakness of Freud's structural theory as compared with his dynamic theory (see Chapter 6). Freud illuminated the purpose of a psychological activity more clearly than he illuminated its nature. One purpose of forgetting, for example, is to prevent anxiety-arousing information from becoming conscious, but

Freud had little to say concerning general factors which determine forgetting, such as retroactive and proactive inhibition. While forgetting may be used for defensive purposes, it is not a defense but a general psychological phenomenon. Thus, Freudian theory has not been helpful in clarifying the nature of the psychological activities which have concerned us—evaluating, deciding, and integrating.

The second limitation also involves structure but arises from the neo-Freudian's failure to realize the potential inherent in the structural theory of personality. The framework is comprehensive, since the id, ego, and superego include impulses, reason, and social values. Yet, psychoanalysts have contributed little to our understanding of the constellation of ego variables which includes curiosity, exploratory behavior, stages of cognitive development, and initiative. If Piaget has paid insufficient attention to affect, the psychoanalysts have paid insufficient attention to initiative and the inquiring intellect, and to the central place they occupy in personality development.

Other Approaches

When we leave cognitive and psychoanalytic theories, we are in troubled waters. There is a notable lack of satisfactory approaches to understanding the child-as-agent. Instead, we have had to rely on a fragment of Rogers's theory, a bit of description from Gesell, and a handful of objective studies which were not done with our special variables in mind. Our complaint is not that there are a diversity of approaches, since we favor a multifaceted examination of a given issue. Rather, the complaint is that one approach seems irrelevant to another or that the gulf between approaches is too wide to make possible a comparison of hypotheses and data.

To make the discontent more specific: the concept of a growth principle (such as found in Adler and Rogers) admirably captures the spirit of the vital, expanding child, but it is so general that it furnishes few leads concerning the child's actual development; on the other hand, the concept of child-centeredness admirably champions the child's uniqueness but is so highly individualized in its application that it does not help us understand the commonalities among children. In contrast to these general principles, there are concepts which are relevant to specific aspects of the child-as-agent, but which are difficult to fit into an overall scheme. We are referring to the concepts of internal locus of control (see Chapter 3), family themes (see Chapter 4), the vocational self (see Chapter 9), and the humanistic-flexible conscience (see Chapter 5). Each says something important in its own right, but they have little to say to one another.

The picture is not uniformly as bleak as we are painting it. In our discussion of the child-as-causer there was a dovetailing of Piaget, Freud, and Adler, each adding a new and relevant dimension—the development and dispelling of infantile omnipotence, the effects of omnipotence on the child's relations to his parents (e.g., in regard to toilet training in particular, or in regard to the power struggle inherent in the socialization of initiative in general), the conditions under which exaggerated notions of power are perpetuated. Relevant clinical and objective research either supported or failed to support the various positions. Although we still do not have an overall, unifying program here, the pieces fit together nicely.

Developmental Achievements

In the Overview we outlined the substantive issues in personality development we would discuss. It is now possible to select four developmental achievements which encompass many of the topics we examined and which are nodal points in personality as we have viewed it. The four achievements are: the ability to engage in a mature sexual act, the ability to perform a mature moral act, the ability to choose an appropriate vocation, and the ability to entertain a fantastic idea. (We use the term "mature" to mean characteristic of the young adult rather than to mean a final fruition or an ideal of behavior.)

The Mature Sexual Act

We agree with Freud that the mature sexual act represents the culmination of love (see Chapter 7) while bypassing his libido theory and certain specific assumptions concerning psychosexual development. Recall that we regarded the infant as passionate, and Freud has been our most serviceable guide to understanding man's passionate nature. His special province was the study of erotic pleasure from which develops the adult passion of love. We extract from psychoanalytic theory the idea that, in the first year of life, the infant comes to love his caretaker with a love so compelling that his greatest pleasures and his most terrifying fears and heartfelt anguish revolve around it. From the psychosexual stages we learn that there is a chronology to the development of love, an orderly unfolding such as has been postulated concerning cognitive development (see Chapter 7). The major gap in Freud's account is the development of mutuality in middle childhood, which Sullivan describes so well. We also believe that the management of intense affection, with its attendant anxieties, rages, and guilts, is crucial to the development of self-control and to the issue of normal or deviant development.

In the mature sexual act the young adult transcends immature (i.e., prior) aspects of intimacy: the demand to be given to while giving nothing in return, with its concomitant rage when not catered to; the irrational fear of loss of love, indicating a lack of trust in the continuity of love; the capricious alternation of love and hate; the ritualization of relationships when autonomy is sacrificed to maintain love; the desire for exclusive possession and the fear of competing with the same sex; the excessive guilt over the anger which inevitably accompanies closeness; the use of intimacy for the sole purpose of displaying one's power or attractiveness. All these residuals of childhood no longer dominate the act of love, although they do not vanish from the adult scene. Like the forepleasure (which is the erotic residual from childhood), certain of the residuals can be used to heighten the experience of intimacy; obviously, they also can block the achievement of maturity altogether.

We cannot go deeply into the nature of the mature sexual act itself. The one facet we discussed was mutuality—that is, each partner's desire and ability to accommodate to the other. The adult sexual act is both highly charged emotionally and highly idiosyncratic. To care about another's pleasure when one's own is at stake requires an ability to transcend the emotional egocentrisim of childhood. The final triumph of love is that as each partner accommodates to the other both are insured maximum pleasure and fulfillment.

The Mature Moral Act

The mature moral act results from utilizing realistic and humanistic guides to interpersonal transactions. It is the complement of the mature sexual act, since it involves relations to people-in-general rather than intimacy with a single individual. Adler comes closest to capturing its essence in his concept of social feelings, that characteristic which begins with the infant's interest in and accommodation to the mother and culminates in the adult's becoming a responsible member of the human community.

The mature moral act engages the total personality. It involves feeling, understanding, and behaving; moreover, the word "integration" often appears in discussing its development. The child is continually confronted with divergent standards and conflicting emotional ties; in coming to grips with them, he is stimulated to achieve the next level of development. If he is overwhelmed by them, his conscience will be primitive, fragmented, or ineffectual.

It is fascinating to watch the interplay of affective and cognitive factors in the home, with peers, and in the self, which fosters or impedes moral development. The forces are evident in the preschool years. In the home the child is confronted with mandates for socially acceptable behavior (*No–Don't; Good boy–Bad boy*) whose content ranges from idiosyncratic to religious and whose potency derives from the maintenance or jeopardizing of parental love. But the sensitive parent also capitalizes on the preschooler's empathy and on his inquisitiveness by asking, How would you feel if someone did that to you?, or by explaining rules whenever this can be done meaningfully. On his part, the child is internalizing parental standards, primarily on the basis of reward and punishment or imitation; but he is also testing the parents to determine how firm their directives are, how negotiable, and in what ways they can be bypassed or subverted. This testing often is motivated by a healthy self-assertiveness and a desire to pursue his vested interests. At the same time, he is confronted with peer group standards which differ from those of parents, and receives peer group emotional support when he dares to behave in ways unacceptable to his parents.

The child's emotional and cognitive development in middle childhood conspire to produce the "too good," absolute standards of goodness, which affect behavior at home and group acceptance, while engendering an exaggerated, perfectionistic ideal for the self. However, the standards are standards, and therefore are more general than the fragmented guides of the preschooler. In addition, counterforces are at work to correct their absolutism. In the home, reasoning parents now have a child who is more responsive to reason. Cooperation and compromise in the group and the personal interchanges in friendships foster mutuality and the ability to see a situation from different points of view. Peers and friends help humanize the child's self-image by requiring that he be esteemworthy rather than being loveworthy. The child also is becoming increasingly self-aware in the sense of learning where his talents, interests, and ego values lie. In a dim way he can grasp the distinction: home is home, peers are peers, and I am myself. He also can begin to think in terms of enlightened self-interest.

Between adolescence and early adulthood, the outlines of the mature conscience begin to emerge. The adolescent is self-concerned and future oriented. Intellectually, he is capable of dealing with abstract principles of behavior. In detaching himself from the family, he can gain perspective on family values; in preparing to face the world, he can adopt values which are more worldly. Even peer group standards, which were so supportive early in the period, begin to lose their appropriateness as guides to adult life. The young adult can retain what is valuable from the past (from parents and peers), but he is sufficiently independent to discard what is outmoded. To use the central word, he can integrate the past with his present needs to evolve a realistic, humane, self-satisfying guide to human relations.

We can make the same points in terms of deviations, calling on the men who understood such deviations best. Freud's is the classical account of the conscience which remains bound to the home; the child's needs for affection, his cognitive immaturity, and the behavior of the parents conspire to keep him tightly in the grip of an absolutistic, punitive superego. Sullivan depicted the danger of being enslaved by the group and of being unable to face the loneliness of turning away from group acceptance. Adler's genius lay in tracing the multiform manifestations and disguise of selfishness, in which a chronic terror over being helpless stifles social feelings and turns life into an unending struggle to gain power over others or to bolster a grandiose image of superiority.

Clearly, the mature moral act includes more of the topics we have discussed than any other developmental achievement. It involves an intricate interweaving of themes: venturing out and assimilating new standards (from parents, peers, "the world"), as the child's cognitive ability and self-reliance allow (since the adoption of each new standard raises the question, Do I dare jeopardize the support of my family or my peers?), all the while mindful of his own vested interests and the values engendered by the expanding self.

Comment: On Complexity and Contingencies. In the Overview we noted that self-control was determined by the interplay of a number of variables, but as we subsequently discussed other topics, the theme was repeated. Only rarely is personality development a matter of a simple relation between two variables which would allow us to say, If A, then B. More often it involves a series of contingency statements: e.g., if X and if Y and if Z, then the child will have a strong masculine identification or a humane conscience or a fulfilling vocation.

This awareness of complexities and contingencies makes one wary of simple statements about personality; e.g., it is bad to spank a child; realistic toys stifle creativity; masturbation is normal in adolescence; there is no such thing as a delinquent child, there are only delinquent parents. A host of provisos must be introduced before such statements have any validity. Understanding complexities and contingencies requires a high tolerance for uncertainty and runs the risk of being immobilized by alternatives. However, we need not conclude that, since development is complex, a definite statement is impossible; rather, we conclude that the choice of a statement should be a considered choice, the best choice possible in light of understanding all the variables and contingencies involved.

The Choice of an Appropriate Vocation

A vocation represents one of the developmental culminations of initiative. Ideally it involves congruent expansiveness, a sense of doing what one wants to do, which, even in the preschooler, is a source of healthy pride. Yet a vocation is more than a continuation of the unfettered expansiveness of the toddler. First, the child must come to grips with requiredness, which can turn expansiveness into "hard work," and he must adjust to a relatively impersonal, public, product oriented world. Next, the freedom to do what he pleases is replaced by judgment—that process of taking stock of oneself, of the environment, and then deciding on a course of action. The adolescent in the process of making a vocational choice or choosing a college which will lead to a vocation has provided us with our most detailed account of the exercise of judgment. On his part, the adolescent has his interests, skills, and talents, his work habits, his ambitions and values, his interpersonal sensitivities, assets, and commitments; the world of work has its requirements and rewards, its levels of prestige and status, its general work atmospheres, and special interpersonal relations. Judgment involves an assessment of areas of congruence and incongruence between self and an occupation, ascertaining the advantages and disadvantages of different occupational choices, the compromises which can be made, and the price to be paid for disharmony. Finally, the adolescent must weigh the relative importance of each consideration in order to arrive at a decision.

The child's understanding of his vocational self and of the world of work has its own developmental history which involves the familiar variables of initiative, egocentricism, internalization of parental values and sex role identification, as well as new variables, such as the achievement of time perspective, understanding the social status of occupations, and the psychology of being a worker. If vocational choice is not as rich an area as sexual and moral development, it is primarily because it has not been studied as intensively. In its own way, work complements love and conscience; finding one's vocation is essential to finding one's identity and one's place in adult society.

Entertaining a Fantastic Idea

Fantastic ideas abound. They are part of normal development and they are part of psychopathology. Realistic thinking is one of the major triumphs of cognitive evolution during childhood. The child comes to distinguish "me" from "not-me," the physical world from the interpersonal world, and to grasp the nature and workings of each. The process of disentanglement and construction defines the progress from bizarre distortions of reality to realistic thinking. However, while thinking can be studied in pure culture, it does not exist in pure culture. It can serve a host of needs other than the need to understand. Under such circumstances the child's progress may be blocked, and he may have a distorted view of the physical and interpersonal world, of himself, and of his body.

Our discussion did not end with a realistic grasp of the nature and functioning of people, things, and self, which has been achieved in middle childhood. Rather, it was extended to include the adolescent's ability to transcend reality by engaging in hypothetical thinking and assuming an "as if" attitude. His luxuriating in philosophies of life and solutions to universal problems is evidence of his new stage of cognitive development. The content of these abstract creations must be brought into line with reality, but the capacity for hypothetical thinking continues to be one of the hallmarks of mature ideation.

We could regard the ability to entertain a hypothetical idea as the culmination of cognitive development, but we prefer the ability to entertain a fantastic idea. By a "fantastic idea" we are not referring to the personalized daydreams which weave in and out of our thinking but which are primarily a private affair. Nor are we using the concept to lead up to the notion of "creativity," although the two may well be related. Rather, we are still in the realm of hypothetical, reality-relevant thinking but are merely adding an element of playfulness. The fantastic idea about reality is entertained for the pleasure it affords. The distinction between the possible and the impossible is consciously suspended. The idea is implicitly introduced by, "Wouldn't it be funny if..."

Why do we prefer this developmental achievement to the ability to entertain a hypothetical idea? Only as a reminder that play—and especially fantasy play—has a special status in childhood. It performs its many functions while remaining free, effortless, and natural. The child is unconcerned with accuracy, so distorted ideas are not ominous signs of fixation. Play adds a special qualification to the notion that development is the child's earnest effort to adapt to reality. In continuing to have access to playful thinking, the young adult continues to enjoy the special release, freedom, and pleasure which play affords. As in sexuality, morality, and work, the individual retains what is of value from the past and uses it to enrich the present.

We began with the infant who is passionate, exploring, learning, appraising, deciding, integrating, and we followed his development until he is capable of engaging in a mature sexual act, of performing a mature moral act, of choosing an appropriate vocation, and of entertaining a fantastic idea. This is our answer to the initial question concerning what must be discussed if we are to understand the development of personality.

References

Adams, D. K. The development of social behavior. In Y. Brackbill (Ed.), *Infancy and early childhood.* New York: Free Press, 1967.

Aichhorn, A. *Wayward youth.* New York: Viking Press, 1935.

Ainsworth, M. D. S. The effects of maternal deprivation: A review of findings and controversy in the context of research strategy. *Deprivation of maternal care.* Public Health Papers, No. 14. Geneva: World Health Organization, 1962.

Ainsworth, M. D. S. The development of infant-mother interaction among the Ganda. In B. M. Foss (Ed.), *Determinants of infant behavior.* Vol. 2. New York: Wiley, 1963.

Ainsworth, M. D. S. Patterns of attachment behavior shown by the infant in interaction with his mother. *Merrill-Palmer Quarterly*, 1964, *10*, 51-58.

Ainsworth, M. D. S. *Infancy in Uganda: Infant care and the growth of love.* Baltimore: Johns Hopkins Press, 1967.

Ainsworth, M. D. S. Object relations, dependency, and attachment: A theoretical review of the infant-mother relationship. *Child Development,* 1969, *40*, 969-1025.

Alexander, F. *Psychosomatic medicine*. New York: W. W. Norton, 1950.

Allport, F. H. *Social psychology*. Boston: Houghton Mifflin, 1924.

Amster, F. Differential uses of play in treatment of young children. In M. R. Haworth (Ed.), *Child psychotherapy*. New York: Basic Books, 1964.

Anderson, H. H. Domination and integration in the social behavior of young children in an experimental play situation. *Genetic Psychology Monographs*, 1937, *19*, 341-408. (a)

Anderson, H. H. An experimental study of dominative and integrative behavior in children of preschool age. *Journal of Social Psychology*, 1937, *8*, 335-345. (b)

Anderson, J. E. The development of social behavior. *American Journal of Sociology*, 1938-39, *44*, 839-857.

Anderson, J. E. Parents' attitudes on child behavior: A report of three studies. *Child Development*, 1946, *17*, 91-97.

Anonymous. Ambivalence in first reactions to a sibling. *Journal of Abnormal and Social Psychology*, 1949, *44*, 541-548.

Ansbacher, H. L., and Ansbacher, R. R. (Eds.) *The individual psychology of Alfred Adler*. New York: Basic Books, 1956.

Aronfreed, J. *Conduct and conscience: The socialization of internalized control over behavior.* New York: Academic Press, 1968.

Austin, M. C., and Thompson, G. G. Children's friendships: A study of the bases on which children select and reject their best friends. *Journal of Educational Psychology*, 1948, *39*, 101-116.

Ausubel, D. P. Negativism as a phase of ego development. *American Journal of Orthopsychiatry*, 1950, *20*, 796-805.

Ausubel, D. P. *Theory and problems of adolescent development.* New York: Grune and Stratton, 1954.

Ausubel, D. P. *Theory and problems of child development.* New York: Grune and Stratton, 1958.

Ausubel, D. P., et al., Perceived parent attitudes as determinants of children's ego structure. *Child Development*, 1954, *25*, 173-183.

Bakwin, H. Emotional deprivation in infants. *Journal of Pediatrics*, 1949, *35*, 512-521.

Baldwin, A. L. The effect of home environment on nursery school behavior. *Child Development*, 1949, *20*, 49-61.

Baldwin, A. L. The study of child behavior and development. In P. H. Mussen (Ed.), *Handbook of research methods in child development.* New York: Wiley, 1960.

Baldwin, A. L. *Theories of child development.* New York: Wiley, 1967.

Balint, A. *The psycho-analysis of the nursery.* London: Routledge and Kegan Paul, 1953.

Bandura, A. Vicarious processes: A case of no-trial learning. In L. Berkowitz (Ed.), *Advances in experimental social psychology.* Vol. 2. New York: Academic Press, 1965.

Bandura, A., Ross, D., and Ross, S. Imitation of film-mediated aggressive models. *Journal of Abnormal and Social Psychology*, 1963, *66*, 3-11.

Bandura, A., and Walters, R. H. *Adolescent aggression.* New York: Ronald Press, 1959.

Bandura, A., and Walters, R. H. *Social learning and personality development.* New York: Holt, Rinehart and Winston, 1963.

Barker, R. G., Dembo, T., and Lewin, K. Frustration and regression: An experiment with young children. *University of Iowa Studies in Child Welfare*, 18. Iowa City: University of Iowa Press, 1941.

Barker, R. G., and Wright, H. F. *The midwest and its children.* Evanston, Ill.: Row, Peterson, 1954.

Barker, R. G., et al. Adjustment to physical handicap and illness: A survey of the social psychology of physique and disability. *Social Science Research Council Bulletin*, 55, 1953.

Battle, E. S., and Rotter, J. B. Children's feelings of personal control as related to social class and ethnic groups. *Journal of Personality*, 1963, *31*, 482-490.

Baumrind, D. Child care practices anteceding three patterns of preschool behavior. *Genetic Psychology Monographs*, 1967, *75*, 43-88.

Bayley, N. Consistency of maternal and child behaviors in the Berkeley Growth Study. *Vita Humana*, 1964, *7*, 73-95.

Becker, W. C. Consequences of different kinds of parental discipline. In M. L. Hoffman and L. W. Hoffman (Eds.), *Review of child development research.* Vol. 1. New York: Russell Sage Foundation, 1964.

Behrens, M. L. Child rearing and the character structure of the mother. *Child Development*, 1954, *25*, 225-238.

Bell, R. Q. Retrospective attitude studies of parent-child relations. *Child Development*, 1958, *29*, 323-338.

Bell, R. Q. A reinterpretation of the direction of effects in studies of socialization. *Psychological Review*, 1968, *75*, 81-95.

Bender, L. Childhood schizophrenia: Clinical study of one hundred schizophrenic children. *American Journal of Orthopsychiatry*, 1947, *17*, 40-56.

Benedek, T. The psychosomatic implications of the primary unit: Mother-child. *American Journal of Orthopsychiatry*, 1949, *19*, 642-654.

Benjamin, J. D. Some developmental observations relating to the theory of anxiety. *Journal of American Psychoanalytic Association*, 1961, *9*, 652-668.
Berenda, R. W. *The influence of the group on the judgments of children*. New York: King's Crown Press, 1950.
Bergman, P., and Escalona, S. K. Unusual sensitivities in very young children. *Psychoanalytic Study of the Child, 3/4*. New York: International Universities Press, 1949.
Berkowitz, L. *Aggression: A social psychological analysis*. New York: McGraw-Hill, 1962.
Berlyne, D. E. Laughter, humor, and play. In G. Lindzey and E. Aronson (Eds.), *Handbook of social psychology*. (2nd ed.) Vol. 3. Reading, Mass.: Addison-Wesley, 1969.
Bernabeu, E. P. The effects of severe crippling on the development of a group of children. *Psychiatry*, 1958, *21*, 169-194.
Bettelheim, B. *Love is not enough*. New York: Free Press, 1950.
Bettelheim, B. *The empty fortress*. New York: Free Press, 1967.
Bialer, I. Conceptualization of success and failure in mentally retarded and normal children. *Journal of Personality*, 1961, *29*, 303-320.
Biber, B., et al. *Child life in school*. New York: Dutton, 1942.
Bijou, S. W., and Baer, D. M. *Child development: A systematic and empirical theory*. New York: Appleton-Century-Crofts, 1961.
Biller, H. B., and Borstelmann, L. J. Masculine development: An integrative review. *Merrill-Palmer Quarterly*, 1967, *13*, 253-294.
Birch, H. G. (Ed.) *Brain damage in children*. Baltimore: Williams and Wilkins, 1964.
Bloomberg, M. An inquiry into the relationship between field independence-dependence and creativity. *Journal of Psychology*, 1967, *67*, 127-140.
Blos, P. *The adolescent personality*. New York: Appleton-Century-Crofts, 1941.
Blos, P. *On adolescence*. New York: Free Press, 1962.
Bobroff, A. The stages of maturation in socialized thinking and in the ego development of two groups of children. *Child Development*, 1960, *31*, 321-338.
Boles, G. Personality factors in mothers of cerebral palsied children. *Genetic Psychology Monographs*, 1959, *59*, 159-218.
Boll, E. S. The role of preschool playmates: A situational approach. *Child Development*, 1957, *28*, 327-342.
Borow, H. Development of occupational motives and roles. In M. L. Hoffman and L. W. Hoffman (Eds.), *Review of child development research*. Vol. 2. New York: Russell Sage Foundation, 1966.
Bossard, J. H. S., and Boll, E. S. Child behavior and the empathic complex. *Child Development*, 1957, *28*, 37-42.
Bowlby, J. *Maternal care and mental health*. Monograph Series No. 2. Geneva: World Health Organization, 1952.
Bowlby, J. Some pathological processes set in train by early mother-child separation. *Journal of Mental Science*, 1953, *99*, 265-272.
Bowlby, J. The nature of the child's tie to his mother. *International Journal of Psycho-Analysis*, 1958, *39*, 350-373.
Bowlby, J., et al. The effects of mother-child separation: A follow-up study. *British Journal of Medical Psychology*, 1956, *29*, 211-247.
Brenner, C. *An elementary textbook of psychoanalysis*. New York: International Universities Press, 1955.

Bridger, W. H. Sensory discrimination and autonomic function in the newborn. *Journal of the American Academy of Child Psychiatry*, 1962, *1*, 67-82.

Bridges, K. M. B. Emotional development in early infancy. *Child Development*, 1932, *3*, 324-341.

Brim, O. G., Jr. The sources of parent behavior. *Children*, 1958, *5*, 217-222.

Brody, S. *Patterns of mothering: Maternal influence during infancy*. New York: International Universities Press, 1956.

Bronfenbrenner, U. Socialization and social class through time and space. In E. E. Maccoby, T. M. Newcomb, and E. L. Hartley (Eds.), *Readings in social psychology*. (3rd ed.) New York: Holt, 1958.

Bronfenbrenner, U. Freudian theories of identification and their derivatives. *Child Development*, 1960, *31*, 15-40.

Bronfenbrenner, U. Some familial antecedents of responsibility and leadership in adolescents. In L. Petrullo and B. M. Bass (Eds.), *Leadership and interpersonal behavior*. New York: Holt, 1961.

Bronfenbrenner, U. Soviet methods of character education: Some implications for research. *American Psychologist*, 1962, *17*, 550-564.

Bronfenbrenner, U. *Two worlds of childhood: U. S. and U. S. S. R.* New York: Russell Sage Foundation, 1970.

Bronson, G. W. The development of fear in man and other animals. *Child Development*, 1968, *39*, 409-431.

Brown, A. W., Morrison, J., and Couch, G. B. Influence of affectional family relationships on character development. *Journal of Abnormal and Social Psychology*, 1947, *42*, 422-428.

Browne, C. G., and Cohn, T. S. *The study of leadership*. Danville, Ill.: Interstate Printers and Publishers, 1958.

Bruch, H. Obesity in childhood and personality development. *American Journal of Orthopsychiatry*, 1941, *11*, 467-474.

Bruch, H. Developmental obesity and schizophrenia. *Psychiatry*, 1958, *21*, 65-70.

Bühler, C. The social behavior of children. In C. Murchison (Ed.), *Handbook of child psychology*. (2nd ed.) Worcester, Mass.: Clark University Press, 1933.

Burton, R. V. Generality of honesty reconsidered. *Psychological Review*, 1963, *70*, 481-499.

Byrne, D., and Griffitt, W. A developmental investigation of the law of attraction. *Journal of Personality and Social Psychology*, 1966, *4*, 699-702.

Caldwell, B. M. The effects of infant care. In M. L. Hoffman and L. W. Hoffman, (Eds.), *Review of child development research*. Vol. 1. New York: Russell Sage Foundation, 1964.

Cameron, N., and Magaret, A. *Behavior pathology*. Boston: Houghton Mifflin, 1951.

Campbell, J. D. Peer relations in childhood. In M. L. Hoffman and L. W. Hoffman (Eds.), *Review of child development research*. Vol. 1. New York: Russell Sage Foundation, 1964.

Carter, H. D. The development of interests in vocations. In N. B. Henry (Ed.), *Adolescence*. 43rd Yearbook, Part 1, National Society for the Study of Education. Chicago: University of Chicago Press, 1944.

Challman, R. C. Factors influencing friendships among preschool children. *Child Development*, 1932, *3*, 146-158.

Chauncey, H. (Ed.) *Soviet preschool education*. Vol. 1,2. New York: Holt, Rinehart and Winston, 1969.

Clark, A. W., and Van Sommers, P. Contradictory demands in family relations and adjustment in school and home. *Human Relations*, 1961, *14*, 97-111.

Clarke, H. H., and Clarke, D. H. Social status and mental health of boys as related to their maturity, structural, and strength characteristics. *Research Quarterly of the American Association for Health, Physical Education, and Recreation*, 1961, *32*, 326-334.

Cobb, H. V. Role-wishes and general wishes of children and adolescents. *Child Development*, 1954, *25*, 161-171.

Coelho, G. V., Silber, E., and Hamburg, D. A. Use of the student TAT to assess coping behavior in hospitalized, formal, and exceptionally competent college freshmen. *Perceptual and Motor Skills*, 1962, *14*, 355-365.

Coleman, J. S. *The adolescent society: The social life of the teenager and its impact on education*. New York: Free Press, 1961.

Coleman, J. S., Campbell, E. Q., et al. *Equality of educational opportunity*. No. D.S. 5.238:380001. U. S. Govt. Printing Office, Washington, D.C., 1966.

Crandall, J. Achievement. In H. W. Stevenson (Ed.), *Child psychology*. 62nd Yearbook, National Society for the Study of Education. Chicago: University of Chicago Press, 1963.

Crandall, V. C., Katkovsky, W., and Crandall, V. J. Children's beliefs in their own control of reinforcements in intellectual-academic achievement situations. *Child Development*, 1965, *36*, 91-109.

Crandall, V. J., Katkovsky, W., and Preston, A. A conceptual formulation for some research on children's achievement development. *Child Development*, 1960, *31*, 787-797.

Crane, A. R. Preadolescent gangs and the moral development of children. *British Journal of Educational Psychology*, 1958, *28*, 201-208.

Davids, A., and Parenti, A. N. Time orientation and interpersonal relations of emotionally disturbed and normal children. *Journal of Abnormal and Social Psychology*, 1958, *57*, 299-305.

Davitz, J. R. The effects of previous training on postfrustration behavior. *Journal of Abnormal and Social Psychology*, 1952, *47*, 309-315.

Davitz, J. R. Social perception and sociometric choice of children. *Journal of Abnormal and Social Psychology*, 1955, *50*, 173-176.

Décarie, T. G. *Intelligence and affectivity in early childhood.* New York: International Universities Press, 1965.

DeHaan, R. F. Social leadership. In N. B. Henry (Ed.), *Education for the gifted*. 57th Yearbook, Part 2, National Society for the Study of Education. Chicago: University of Chicago Press, 1958.

Deutsch, H. *The psychology of women: A psychoanalytic interpretation*. Vol. 1. *Girlhood*. New York: Grune and Stratton, 1944.

DiCarlo, L. M., and Dolphin, J. E. Social adjustment and personality development of deaf children: A review of literature. *Exceptional Children*, 1952, *18*, 111-118.

Dixon, J. C. Development of self-recognition. *Journal of Genetic Psychology*, 1957, *91*, 251-256.

Dollard, J., and Miller, N. E. *Personality and psychotherapy*. New York: McGraw-Hill, 1950.

Dollard, J., et al. *Frustration and aggression*. New Haven: Yale University Press, 1939.

Douvan, E., and Adelson, J. *The adolescent experience*. New York: Wiley, 1966.

Douvan, E., and Gold, M. Modal patterns in American adolescence. In L. W. Hoffman and M. L. Hoffman (Eds.), *Review of child development research*. Vol. 2. New York: Russell Sage Foundation, 1966.

Drucker, P. F. How to be an employee. *Psychology Today*, 1968, *1*, 63-66.

Dukes, W. F. Psychological studies of values. *Psychological Bulletin*, 1955, *52*, 24-50.

Dunbar, H. F. *Emotions and bodily changes: A survey of literature on psychosomatic interrelationships, 1910-1945*. (3rd ed.) New York: Columbia University Press, 1946.

Dunnington, M. J. Behavioral differences of sociometric status groups in a nursery school. *Child Development*, 1957, *28*, 103-111.

DuPan, R. M., and Roth, S. The psychological development of a group of children brought up in a hospital type residential nursery. *Journal of Pediatrics*, 1955, *47*, 124-129.

Eisensen, J. The nature of defective speech. In W. M. Cruickshank (Ed.), *Psychology of exceptional children and youth*. (2nd ed.) Englewood Cliffs: Prentice-Hall, 1963.

Eissler, K. R. Ego-psychological implications of the psychoanalytic treatment of delinquents. *Psychoanalytic Study of the Child, 5*. New York: International Universities Press, 1950.

Eissler, R. S. (Ed.) Problems of masturbation. *Psychoanalytic Study of the Child, 6*. New York: International Universities Press, 1951.

Elder, G. H., Jr. Achievement orientations and career patterns of rural youth. *Sociology of Education*, 1963, *37*, 30-58.

Elkind, D. Cognition in infancy and early childhood. In Y. Brackbill (Ed.), *Infancy and early childhood.* New York: Free Press, 1967.

Ellesor, M. V. Children's reactions to novel visual stimuli. *Child Development*, 1933, *4*, 95-105.

Engel, G. L., Reichsman, F., and Segal, H. A study of an infant with a gastric fistula. *Psychosomatic Medicine*, 1956, *18*, 374-398.

Engel, M., Marsden, G., and Woodaman, S. Children who work and the concept of work style. *Psychiatry*, 1967, *30*, 392-404.

Erikson, E. H. *Childhood and society*. New York: W. W. Norton, 1950.

Erikson, E. H. Identity and the life cycle. *Psychological Issues*, Monograph 1, Vol. 1, No. 1. New York: International Universities Press, 1959.

Escalona, S. K. Play and substitute satisfaction. In R. G. Barker, J. S. Kounin, and H. F. Wright (Eds.), *Child behavior and development*. New York: McGraw-Hill, 1943.

Escalona, S. K. Some considerations regarding psychotherapy with psychotic children. *Bulletin of the Menninger Clinic*, 1948, *12*, 126-134.

Escalona, S. K. Patterns of infantile experience and the developmental process. *Psychoanalytic Study of the Child*, New York: International Universities Press, 1963, *18*, 197-244.

Escalona, S. K. Some determinants of individual differences. *Transactions of the New York Academy of Science*, Series II, *27*, No. 7, 1965, 802-816.

Fantz, R. L. Pattern vision in young infants. *Psychological Record*, 1958, *8*, 43-47.

Fenichel, O. *The psychoanalytic theory of neuroses.* New York: W. W. Norton, 1945.

Feshbach, N. D., and Roe, K. Empathy in six- and seven-year-olds. *Child Development*, 1968, *39*, 113-145.

Feshbach, S. The function of aggression and the regulation of aggressive drive. *Psychological Review*, 1964, *71*, 157-272.

Festinger, L. A theory of social comparison processes. *Human Relations*, 1954, *7*, 117-140.

Fisher, S., and Cleveland, S. E. *Body image and personality.* Princeton: Van Nostrand, 1958.

Fite, M. D. Aggressive behavior in young children and children's attitudes toward aggression. *Genetic Psychology Monographs*, 1940, *22*, 151-319.

Flanders, J. P. A review of research on imitative behavior. *Psychological Bulletin*, 1968, *69*, 316-337.

Flavell, J. H. *The developmental psychology of Jean Piaget.* Princeton: Van Nostrand, 1963.

Flavell, J. H., Botkin, P. T., and Fry, C. L., Jr. *The development of role-taking and communication skills in children.* New York: Wiley, 1968.

Force, D. G., Jr. Social status of physically handicapped children. *Exceptional Children*, 1956, *23*, 104-107; 132.

Frazier, A., and Lisonbee, L. K. Adolescent concerns with physique. *School Review*, 1950, *58*, 397-405.

Frederiksen, N. The effects of frustration on negativistic behavior of young children. *Journal of Genetic Psychology*, 1942, *61*, 203-226.

Freedman, D. G. Smiling in blind infants and the issue of innate versus acquired. *Journal of Child Psychology and Psychiatry*, 1964, *5*, 171-184.

Freedman, D. S., Freedman, R., and Whelpton, P. K. Size of family and preference for children of each sex. *American Journal of Sociology*, 1960, *66*, 141-146.

Freud, A. Aggression in relation to emotional development: Normal and pathological. *Psychoanalytic Study of the Child, 3/4.* New York: International Universities Press, 1949.

Freud, A. The role of bodily illness in the mental life of children. *Psychoanalytic Study of the Child, 7.* New York: International Universities Press, 1952.

Freud, A. Adolescence. *Psychoanalytic Study of the Child, 13.* New York: International Universities Press, 1958.

Freud, A. *Normality and pathology in childhood.* New York: International Universities Press, 1965.

Freud, A. *The ego and the mechanisms of defense.* New York: International Universities Press, 1966.

Freud, A., and Burlingham, D. T. *Infants without families: The case for and against residential nurseries.* New York: International Universities Press, 1944.

Freud, A., and Dann, S. An experiment in group upbringing. *Psychoanalytic Study of the child, 6.* New York: International Universities Press, 1951.

Freud, S. *The problem of anxiety.* New York: W. W. Norton, 1936.

Freud, S. The interpretation of dreams. In J. Strachey (Trans.), *The standard edition of the complete psychological works of Sigmund Freud.* Vol. 4 and 5. London: Hogarth Press, 1953.

Freud, S. Formulations regarding the two principles of mental functioning. In P. Rieff (Ed.), *Collected papers of Sigmund Freud*. Vol. 6. *General psychological theory*. New York: Collier Books, 1963.

Friedlander, D. Personality of children who later became psychotic. *Journal of Abnormal and Social Psychology*, 1945, *40*, 330-335.

Friend, J. G., and Haggard, E. A. *Work adjustment in relation to family background*. Applied Psychology Monographs, No. 16. Stanford, Calif.: Stanford University Press, 1948.

Fuller, J. L., and Thompson, W. R. *Behavior genetics*. New York: Wiley, 1960.

Gardner, D. B., Hawkes, G. R., and Burchinal, L. G. Noncontinuous mothering in infancy and development in later childhood. *Child Development*, 1961, *32*, 225-234.

Gardner, G. Functional leadership and popularity in small groups. *Human Relations*, 1956, *9*, 491-509.

Garner, A. M., and Wenar, C. *The mother-child interaction in psychosomatic disorders*. Urbana: University of Illinois Press, 1959.

Gellert, E. Children's conceptions of the content and functions of the human body. *Genetic Psychology Monographs*, 1962, *65*, 293-411.

Gerard, D. L., and Siegel, J. The family background of schizophrenia. *Psychiatric Quarterly*, 1950, *24*, 47-73.

Gesell, A., and Ilg, F. L. *Infant and child in the culture of today*. New York: Harper, 1943.

Gesell, A., and Ilg, F. L. *The child from five to ten*. New York: Harper, 1946.

Gesell, A., and Ilg, F. L. *Child development*. New York: Harper, 1949.

Gesell, A., Ilg, F. L., and Ames, L. B. *Youth: The years from ten to sixteen*. New York: Harper, 1956.

Gewirtz, J. L. A learning analysis of the effects of normal stimulation, privation, and deprivation on the acquisition of social motivation and attachment. In B. M. Foss (Ed.), *Determinants of infant behavior*. New York: Wiley, 1966.

Gilmore, J. B. Play: A special behavior. In R. N. Haber (Ed.), *Current research in motivation*. New York: Holt, Rinehart and Winston, 1966.

Ginzberg, E. et al. *Occupational choice*. New York: Columbia University Press, 1951.

Glueck, S., and Glueck, E. *Unraveling juvenile delinquency*. Cambridge: Harvard University Press, 1950.

Golann, S. E. Psychological study of creativity. *Psychological Bulletin*, 1963, *60*, 548-565.

Goldfarb, W. The effects of early institutional care on adolescent personality. *Journal of Experimental Education*, 1943, *12*, 106-129. (a)

Goldfarb, W. Infant rearing and problem behavior. *American Journal of Orthopsychiatry*, 1943, *13*, 249-265. (b)

Goldfarb, W. Psychological privation in infancy and subsequent adjustment. *American Journal of Orthopsychiatry*, 1945, *15*, 247-255.

Goodenough, F. L. *Anger in young children*. Institute for Child Welfare Monographs, No. 9. Minneapolis: University of Minnesota Press, 1931.

Green, E. H. Friendships and quarrels among preschool children. *Child Development*, 1933, *4*, 237-252.

Gribbons, W. D., and Lohnes, P. R. Shifts in adolescents' vocational values. *Personnel Guidance Journal*, 1965, *44*, 248-252.

Grinder, R. E. Fidelity or alienation in the youth culture? *Merrill-Palmer Quarterly*, 1964, *10*, 195-203.

Grinker, R. R. *Psychosomatic research*. New York: W. W. Norton, 1953.

Gross, E. A sociological approach to the analysis of preparation for work life. *Personnel Guidance Journal*, 1967, *45*, 416-423.

Gunn, B. Children's conceptions of occupational prestige. *Personnel Guidance Journal*, 1964, *42*, 558-563.

Hagman, E. The companionships of preschool children. *University of Iowa Studies in Child Welfare, 7*. Iowa City: University of Iowa Press, 1933.

Hanfmann, E. Social structure of a group of kindergarten children. *American Journal of Orthopsychiatry*, 1935, *5*, 407-410.

Hanks, J. R., and Hanks, L. M., Jr. The physically handicapped in certain non-occidental societies. *Journal of Social Issues*, 1948, *4*, 11-20.

Harding, J. et al. Prejudices and ethnic relations. In G. Lindzey (Ed.), *Handbook of social psychology*. Vol. 2. Reading, Mass.: Addison-Wesley, 1954.

Hare, A. P., Borgatta, E. F., and Bales, R. F. *Small groups: Studies in social interaction*. (Rev. ed.) New York: Knopf, 1965.

Harris, D. B. How children learn interests, motives, and attitudes. In N. B. Henry (Ed.), *Learning and instruction*. 49th Yearbook, Part 1, National Society for the Study of Education. Chicago: University of Chicago Press, 1950.

Harris, D. B. Work and the adolescent transition to maturity. *Teachers College Record*, 1961, *63*, 146-153.

Harris, D. B. et al. The relationship of children's home duties to an attitude of responsibility. *Child Development*, 1954, *25*, 29-33.

Hartley, R. E., and Goldenson, R. M. *The complete book of children's play*. (Rev. ed.) New York: Crowell, 1963.

Hartmann, H. Ego psychology and the problem of adaptation. In D. Rapaport (Ed.), *Organization and pathology of thought*. New York: Columbia University Press, 1951.

Hartmann, H., Kris, E., and Loewenstein, R. M. Notes on the theory of aggression. *Psychoanalytic Study of the Child, 3/4*. New York: International Universities Press, 1949.

Hartshorne, H., and May, M. A. *Studies in the nature of character*. Vol. 1 *Studies in deceit*; Vol. 2 *Studies in self-control*; Vol. 3 *Studies in the organization of character*. New York: Macmillan, 1928-30.

Havighurst, R. J., Robinson, M. Z., and Dorr, M. The development of the ideal self in childhood and adolescence. *Journal of Education Research*, 1946, *40*, 241-257.

Havighurst, R. J., and Taba, H. *Adolescent character and personality*. New York: Wiley, 1949.

Haworth, M. R. (Ed.) *Child psychotherapy*. New York: Basic Books, 1964.

Haynes, H., White, B. L., and Held, R. Visual accommodation in human infants. *Science*, 1965, *148*, 528-530.

Heathers, G. Acquiring dependence and independence: A theoretical orientation. *Journal of Genetic Psychology*, 1955, *87*, 277-291.

Hebb, D. O. On the nature of fear. *Psychological Review*, 1946, *53*, 259-276.
Hebb, D. O. *The organization of behavior*. New York: Wiley, 1949.
Heckhausen, H. *The anatomy of achievement motivation*. New York: Academic Press, 1967.
Heinicke, C. M. Some effects of separating two-year-old children from their parents: A comparative study. *Human Relations*, 1956, *9*, 105-176.
Heinstein, M. I. Behavioral correlates of breast-bottle regimes under varying parent-infant relationships. *Monographs of the Society for Research in Child Development*, 1963, 28, No. 4 (Serial No. 88).
Hemming, J. Symposium: The development of children's moral values. I. Some aspects of moral development in a changing society. *British Journal of Educational Psychology*, 1957, *27*, 77-88.
Hendrick, I. Instinct and the ego during infancy. *Psychoanalytic Quarterly*, 1942, *11*, 33-58.
Hendrick, I. Work and the pleasure principle. *Psychoanalytic Quarterly*, 1943, *12*, 311-329.
Hess, R. D., and Handel, G. *Family worlds*. Chicago: University of Chicago Press, 1959.
Hill, K. T., and Sarason, S. B. The relation of test anxiety and defensiveness to test and school performance over the elementary-school years. *Monographs of the Society for Research in Child Development*, 31 (2), 1966 (Serial No. 104).
Hill, L. B. *Psychotherapeutic intervention in schizophrenia*. Chicago: University of Chicago Press, 1955.
Hoffer, W. Mouth, hand, and ego-integration. *Psychoanalytic Study of the Child, 3/4*. New York: International Universities Press, 1949.
Hoffman, L. W. The father's role in the family and the child's peer-group adjustment. *Merrill-Palmer Quarterly*, 1961, *7*, 97-105.
Hoffman, L. W., Rosen, S., and Lippitt, R. Parental coerciveness, child autonomy, and child's role at school. *Sociometry*, 1960, *23*, 15-22.
Hoffman, M. L. Techniques and processes in moral development. Report of research sponsored by The National Institute of Mental Health. M-2333. 1964.
Hoffman, M. L., and Saltzstein, H. D. Parent discipline and the child's moral development. *Journal of Personality and Social Psychology*, 1967, *5*, 45-57.
Hollinshead, M. T. The social psychology of exceptional children: Part 1. *Exceptional Children*, 1959, *26*, 137-140.
Horrocks, J. E. *The psychology of adolescence*. (3rd ed.) Boston: Houghton Mifflin, 1969.
Horrocks, J. E., and Buker, M. E. A study of the friendship fluctuations of preadolescents. *Journal of Genetic Psychology*, 1951, *78*, 131-144.
Hunt, J. McV. Experience and the development of motivation: Some reinterpretations. *Child Development*, 1960, *31*, 489-504.
Hurlock, E. B. *Adolescent development*. (2nd ed.) New York: McGraw-Hill, 1955.
Hurlock, E. B. *Child development*. (4th ed.) New York: McGraw-Hill, 1964.
Isaacs, S. *Social development in young children*. New York: Harcourt, Brace, 1937.
Izard, C. E. Personality similarity and friendship. *Journal of Abnormal and Social Psychology*, 1960, *61*, 47-51.
Jack, L. M. An experimental study of ascendant behavior in preschool children. *University of Iowa Studies in Child Welfare*, 1934, *9*, No. 3, 7-65.

Jackson, D. D. et al. Psychiatrists' conceptions of the schizophrenogenic parent. *American Medical Association Archives of Neuropsychiatry*, 1958, *79*, 448-459.

Jackson, D. W. Self as process: Implications of role behavior. Doctoral Dissertation. The Ohio State University, 1970.

Jacobson, E. *The self and the object world*. New York: International Universities Press, 1964.

Janis, I. L. *Psychological stress*. New York: Wiley, 1958.

Jersild, A. T. *In search of self*. New York: Teachers College, Columbia University Press, 1952.

Jersild, A. T. Emotional development. In L. Carmichael (Ed.), *Manual of child psychology*. (2nd ed.) New York: Wiley, 1954.

Jersild, A. T., Markey, F. V., and Jersild, C. L. Children's fears, dreams, wishes, day-dreams, likes, dislikes, pleasant and unpleasant memories. *Child Development Monographs, No. 12*. New York: Teachers College, Columbia University Press, 1933.

Jessner, L., Blom, G. E., and Waldfogel, S. Emotional implications of tonsillectomy and adenoidectomy on children. *Psychoanalytic Study of the Child, 7*. New York: International Universities Press, 1952.

Johnson, R. C. A study of children's moral judgments. *Child Development*, 1962, *33*, 327-354.

Jones, M. C. The later careers of boys who were early or late maturing. *Child Development*, 1957, *28*, 113-128.

Jones, M. C., and Mussen, P. H. Self-conceptions, motivations, and interpersonal attitudes of early and late maturing girls. *Child Development*, 1958, *29*, 491-501.

Josselyn, I. M. *The happy child*. New York: Random House, 1955.

Kagan, J. Acquisition and significance of sex typing and sex role identity. In M. L. Hoffman and L. W. Hoffman (Eds.), *Review of child development research*. Vol. 1. New York: Russell Sage Foundation,1964.

Kagan J., and Moss, H. *Birth to maturity: A study in psychological development*. New York: Wiley, 1962.

Kagan, J., Moss, H., and Sigel, I. E. Psychological significance of styles of conceptualization. In J. C. Wright and J. Kagan (Eds.), *Basic cognitive processes in children*. *Monographs of the Society for Research in Child Development*, 28 (2), 1963 (Serial No. 86).

Kanner, L. Autistic disturbances of affective contact. *The Nervous Child*, 1943, *2*, 217-250.

Kanner, L. Problems of nosology and psychodynamics of early infantile autism. *American Journal of Orthopsychiatry*, 1949, *19*, 416-426.

Kanner, L. *Child psychiatry*. (3rd ed.) Springfield, Ill.: Charles C Thomas, 1957.

Kanner, L., and Eisenberg, L. Notes on the follow-up studies of autistic children. In P. H. Hoch and J. Zubin (Eds.), *Psychopathology of childhood*. New York: Grune and Stratton, 1955.

Keister, M. E. The behavior of young children in failure. In R. G. Barker, J. S. Kounin, and H. F. Wright (Eds.), *Child behavior and development*. New York: McGraw-Hill, 1943.

Kellmer Pringle, M. L., and Gooch, S. Chosen ideal person, personality development, and progress in school subjects: A longitudinal study. *Human Development*, 1965, *8*, 161-180.

Kessen, W., and Mandler, G. Anxiety, pain, and the inhibition of distress. *Psychological Review*, 1961, 68, 396-404.

Kistiakovskaia, M. I. Stimuli evoking positive emotions in infants in the first months of life. *Soviet Psychology and Psychiatry*, 1965, *3*, 39-48.

Kitano, H. H. L. Adjustment of problem and nonproblem children to specific situations: A study in role theory. *Child Development*, 1962, *33*, 229-233.

Klinger, E. Development of imaginative behavior: Implications of play for a theory of fantasy. *Psychological Bulletin*, 1969, *72*, 277-298.

Kohlberg, L. Moral development and identification. In H. W. Stevenson (Ed.), *Child psychology*. 62nd Yearbook, National Society for the Study of Education. Chicago: University of Chicago Press, 1963.

Kohlberg, L. Development of moral character and moral ideology. In M. L. Hoffman and L. W. Hoffman (Eds.), *Review of child development research*. Vol. 1. New York: Russell Sage Foundation, 1964.

Kohlberg, L. Relationships between the development of moral judgment and moral conduct. Paper presented at the Symposium on Behavioral and Cognitive Concepts in the Study of Internalization, The Society for Research in Child Development, Minneapolis, March 26, 1965.

Kohlberg, L. A cognitive-developmental analysis of children's sex-role concepts and attitudes. In E. E. Maccoby (Ed.), *The development of sex differences*. Stanford, Calif.: Stanford University Press, 1966.

Kohlberg, L. Early education: A cognitive-developmental view. *Child Development*, 1969, *39*, 1013-1062.

Kohlberg, L., Yaeger, J., and Hjertholm, E. Private speech: Four studies and a review of theories. *Child Development*, 1968, *39*, 691-736.

Kohn, M. L., and Carroll, E. E. Social class and the allocation of parental responsibilities. *Sociometry*, 1960, *23*, 372-392.

Kris, E. *Psychoanalytic explorations in art*. New York: International Universities Press, 1952.

Kugelmass, S., and Breznitz, S. The development of intentionality in moral judgment in city and kibbutz adolescents. *Journal of Genetic Psychology*, 1967, *111*, 103-111.

Kuhlen, R. G., and Arnold, M. Age differences in religious beliefs and problems during adolescence. *Journal of Genetic Psychology*, 1944, *65*, 291-300.

Latif, I. The physiological basis of linguistic development and of the ontogeny of meaning. Part 1. *Psychological Review*, 1934, *41*, 55-85.

Laurendeau, M., and Pinard, A. *Causal thinking in the child*. New York: International Universities Press, 1962.

Lawson, R. *Frustration: The development of a scientific concept*. New York: Macmillan, 1965.

Lefcourt, H. M. Internal versus external control of reinforcement: A review. *Psychological Bulletin*, 1966, *65*, 206-220.

Lehman, H. C., and Witty, P. A. *The psychology of play activities*. New York: Barnes, 1927.

Lerner, E. The problem of perspective in moral reasoning. *American Journal of Sociology*, 1937, *43*, 249-269.

Levin, H., and Wardwell, E. The research uses of doll play. *Psychological Bulletin*, 1962, *59*, 27-56.

Levy, D. M. A method of integrating physical and psychiatric examination. *American Journal of Psychiatry*, 1929-30, *9*, 121-194.

Levy, D. M. Hostility patterns in sibling rivalry experiments. *American Journal of Orthopsychiatry*, 1936, *6*, 183-257.

Levy, D. M. Release therapy. *American Journal of Orthopsychiatry*, 1939, *9*, 713-736.

Levy, D. M. *Maternal overprotection*. New York: Columbia University Press, 1943.

Levy, D. M. Oppositional syndromes and oppositional behavior. In P. H. Hoch and J. Zubin (Eds.), *Psychopathology of childhood*. New York: Grune and Stratton, 1955.

Lewin, K. Environmental forces in child behavior and development. In C. Murchison (Ed.), *A handbook of child psychology*. Worcester, Mass.: Clark University Press, 1931.

Lewin, K. *Dynamic theory of personality*. New York: McGraw-Hill, 1935.

Lewin, K. Behavior and development as a function of the total situation. In L. Carmichael (Ed.), *Manual of child psychology*. (2nd ed.) New York: Wiley, 1954.

Lewin, K., Lippitt, R., and White, R. K. Patterns of aggressive behavior in experimentally created "social climates." *Journal of Social Psychology*, 1939, *10*, 271-279.

Lewin, K. et al. Level of aspiration. In J. McV. Hunt (Ed.), *Personality and the behavior disorders*. Vol. 1. New York: Ronald Press, 1944.

Lipton, E. L., Steinschneider, A., and Richmond, J. B. Psychophysiological disorders in children. In M. L. Hoffman and L. W. Hoffman (Eds.), *Review of child development research*. Vol. 2. New York: Russell Sage Foundation, 1966.

Loevinger, J. Patterns of parenthood as theories of learning. *Journal of Abnormal and Social Psychology*, 1959, *59*, 148-150.

Lowenfeld, B. Psychological problems of children with impaired vision. In W. M. Cruickshank (Ed.), *Psychology of exceptional children and youth*. (2nd ed.) Englewood Cliffs: Prentice-Hall, 1963.

Lucina, Sister M. Sex differences in adolescent attitudes toward best friends. *School Review*, 1940, *48*, 512-516.

Luria, A. R. *The role of speech in the regulation of normal and abnormal behaviour*. London: Pergamon Press, 1961.

Maas, H. S. The young adult adjustment of twenty wartime residential nursery children. *Child Welfare*, 1963, *42*, 57-72.

Macfarlane, J. W. Perspectives on personal consistency and change: The guidance study. *Vita Humana*, 1964, *7*, 115-126.

Macfarlane, J. W., Allen, L., and Honzik, M. P. *A developmental study of the behavior problems of normal children between twenty-one months and fourteen years*. Berkeley and Los Angeles, University of California Press, 1954.

Macgregor, F. C. et al. *Facial deformities and plastic surgery: A psychosocial study*. Springfield, Ill.: Charles C Thomas, 1953.

Maddi, S. R. Exploratory behavior and variation-seeking in man. In D. W. Fiske and S. R. Maddi (Eds.), *Functions of varied experience*. Homewood, Ill.: Dorsey Press, 1961.

Mahrer, A. R. The role of expectancy in delayed reinforcement. *Journal of Experimental Psychology*, 1956, *52*, 101-106.

Mallay, H. A study of some of the techniques underlying the establishment of successful social contacts at the preschool level. *Journal of Genetic Psychology*, 1935, *47*, 431-457.

Marshall, H. R. Relations between home experience and children's use of language in play interactions with peers. *Psychological Monographs*, 1961, *75*, (5, Whole No. 509).

Maudry, M., and Nekula, M. Social relations between children of the same age during the first two years of life. *Journal of Genetic Psychology*, 1939, *54*, 193-215.

McCandless, B. R., and Marshall, H. R. A picture sociometric technique for preschool children and its relation to teacher judgments of friendship. *Child Development*, 1957, *28*, 139-147.

McClearn, G. E. Behavioral genetics: An overview. *Merrill-Palmer Quarterly*, 1968, *14*, 9-24.

McClure, G., and Tyler, F. Role of values in the study of values. *Journal of General Psychology*, 1967, *77*, 217-235.

McCord, W., McCord, J., and Howard, A. Familial correlates of aggression in nondelinquent male children. *Journal of Abnormal and Social Psychology*, 1961, *62*, 79-93.

McCord, W., McCord, J., and Verden, P. Familial and behavioral correlates of dependency in male children. *Child Development*, 1962, *33*, 313-326.

McKeown, J. E. The behavior of parents of schizophrenic, neurotic, and normal children. *American Journal of Sociology*, 1950-51, *56*, 175-179.

Mead, M. A cultural anthropologist's approach to maternal deprivation. *Deprivation of maternal care: A reassessment of its effects*. Public Health Papers No. 14. Geneva: World Health Organization, 1962.

Merei, F. Group leadership and institutionalization. *Human Relations*, 1949, *2*, 23-39.

Meyerson, L. Somatopsychology of physical disability. In W. M. Cruickshank (Ed.), *Psychology of exceptional children and youth*. (2nd ed.) Englewood Cliffs: Prentice-Hall, 1963.

Midlarsky, E. Aiding responses: An analysis and review. *Merrill-Palmer Quarterly*, 1968, *14*, 229-260.

Miller, N. E. Experimental studies of conflict. In J. McV. Hunt (Ed.), *Personality and the behavior disorders*. Vol. 1. New York: Ronald Press, 1944.

Miller, N. E. Learnable drives and rewards. In S. S. Stevens (Ed.), *Handbook of experimental psychology*. New York: Wiley, 1951.

Miller, N. E. Learning of visceral and glandular responses. *Science*, 1969, *163*, 434-445.

Minuchin, P. et al. *The psychological impact of the school experience*. New York: Basic Books, 1969.

Minuchin, S. et al. *Families of the slums*. New York: Basic Books, 1967.

Mischel, W. Preference for delayed reinforcement and social responsibility. *Journal of Abnormal and Social Psychology*, 1961, *62*, 1-7.

Mischel, W. Theory and research on the antecedents of self-imposed delay of reward. In B. H. Maher (Ed.), *Progress in experimental personality research, 3*. New York: Academic Press, 1966.

Mischel, W., and Metzner, R. Preference for delayed reward as a function of age, intelligence, and length of delay interval. *Journal of Abnormal and Social Psychology*, 1962, *64*, 425-431.

Mittleman, B. Motility in infants, children, and adults: Patterning and psychodynamics. *Psychoanalytical Studies of Children*, 1954, *9*, 142-177.

Mitton, B. L., and Harris, D. B. The development of responsibility in children. *Elementary School Journal*, 1954, *54*, 268-277.
More, D. M. Developmental concordance and discordance during puberty and early adolescence. *Monographs of the Society for Research in Child Development*, 18 (1), 1953 (Serial No. 56).
Moriarty, A. Coping patterns of preschool children in response to intelligence test demands. *Genetic Psychology Monographs*, 1961, *64*, 3-127.
Moss, H. Standards of conduct for students, teachers, and parents. *Journal of Counseling Psychology*, 1955, *2*, 39-42.
Moss, H., and Kagan, J. Stability of achievement and recognition-seeking behaviors from early childhood through adulthood. *Journal of Abnormal and Social Psychology*, 1961, *62*, 504-513.
Mowrer, O. H. A stimulus-response analysis of anxiety and its role as a reinforcing agent. *Psychological Review*, 1939, *46*, 553-565.
Mowrer, O. H. *Learning theory and personality dynamics.* New York: Ronald Press, 1950.
Mowrer, O. H. Neurosis and psychotherapy as interpersonal processes: A synopsis. In O. H. Mowrer et al. (Eds.), *Psychotherapy: Theory and research.* New York: Ronald Press, 1953.
Munroe, R. L. *Schools of psychoanalytic thought.* New York: Dryden, 1955.
Murphy, G. *Personality: A biosocial approach to origins and structure.* New York: Harper, 1947.
Murphy, G., Murphy, L. B., and Newcomb, T. M. *Experimental social psychology.* (Rev. ed.) New York: Harper, 1937.
Murphy, L. B. *Social behavior and child personality.* New York: Columbia University Press, 1937.
Murphy, L. B. *The widening world of childhood.* New York: Basic Books, 1962.
Mussen, P. H., and Barker, R. G. Attitudes toward cripples. *Journal of Abnormal and Social Psychology*, 1944, *39*, 351-355.
Mussen, P. H., Conger, J. J., and Kagan, J. *Child development and personality.* (3rd ed.) New York: Harper and Row, 1969.
Mussen, P. H., and Jones, M. C. Self-conceptions, motivations, and interpersonal attitudes of late and early maturing boys. *Child Development*, 1957, *28*, 243-256.
Mussen, P. H., and Newman, D. K. Acceptance of handicap, motivation, and adjustment in physically disabled children. *Exceptional Children*, 1958, *24*, 255-260.
Muste, M. J., and Sharpe, D. F. Some influential factors in the determination of aggressive behavior in preschool children. *Child Development*, 1947, *18*, 11-28.

Nachmann, B. Childhood experience and vocational choice in law, dentistry, and social work. *Journal of Counseling Psychology*, 1960, *7*, 243-250.
Neff, W. S. Psychoanalytic conceptions of the meaning of work. *Psychiatry*, 1965, *28*, 324-333.
Newcomb, T. M. The prediction of interpersonal attraction. *American Psychologist*, 1956, *11*, 575-586.
Newman, J. Psychological problems of children and youth with chronic medical disorders. In W. M. Cruickshank (Ed.), *Psychology of exceptional children and youth.* (2nd ed.) Englewood Cliffs: Prentice-Hall, 1963.

Offer, D., and Sabshin, M. *Normality*. New York: Basic Books, 1966.
O'Hara, R. P., and Tiedeman, D. V. Vocational self-concept in adolescence. *Journal of Counseling Psychology*, 1959, *6*, 292-301.
Osipow, S. H. *Theories of career development*. New York: Appleton-Century-Crofts, 1968.
O'Sullivan, M., Guilford, J. P., and DeMille, R. *The measurement of social intelligence*. Cooperative Research Project No. 1976, 34, 1965. Office of Education, U. S. Department of Health, Education, and Welfare.

Parten, M. B. Social participation among preschool children. *Journal of Abnormal and Social Psychology*, 1932, *33*, 243-269.
Parten, M. B. Leadership among preschool children. *Journal of Abnormal and Social Psychology*, 1933, *27*, 430-440.
Patterson, G. R., Littman, R. A., and Bricker, W. Assertive behavior in children: A step toward a theory of aggression. *Monographs of the Society for Research in Child Development*, 32 (5), 1967. (Serial No. 113).
Peck, H. B., Rabinovitch, R. D., and Cramer, J. B. A treatment program for parents of schizophrenic children. *American Journal of Orthopsychiatry*, 1949, *19*, 592-598.
Peck, R. F., and Havighurst, R. J. *The psychology of character development*. New York: Wiley, 1960.
Peller, L. E. Models of children's play. *Mental Hygiene*, 1952, *36*, 66-83.
Peller, L. E. Libidinal phases, ego development, and play. *The Psychoanalytic Study of the Child, 9*. New York: International Universities Press, 1954.
Peller, L. E. Libidinal development as reflected in play. In M. R. Haworth (Ed.), *Child psychotherapy*. New York: Basic Books, 1964.
Perkins, H. V. Factors influencing change in children's self-concepts. *Child Development*, 1958, *29*, 221-230.
Phillips, E. L., Shenker, S., and Revitz, P. The assimilation of the new child into the group. *Psychiatry*, 1951, *14*, 319-325.
Piaget, J. *The language and thought of the child*. New York: Harcourt, Brace, 1926.
Piaget, J. *Judgment and reasoning in the child*. New York: Harcourt, Brace, 1928.
Piaget, J. *The child's conception of physical causality*. London: Routledge and Kegan Paul, 1930.
Piaget, J. *The moral judgment of the child*. London: Routledge and Kegan Paul, 1932.
Piaget, J. *The child's concept of the world*. London: Routledge and Kegan Paul, 1951.
Piaget, J. *The origins of intelligence in children*. New York: International Universities Press, 1952.
Piaget, J. *The construction of reality in the child*. New York: Basic Books, 1954.
Piaget, J. *Play, dreams, and imitation in childhood*. New York: W. W. Norton, 1962.
Piaget, J. *The psychology of intelligence*. Totowa, N. J.: Littlefield, Adams, 1966.
Piaget, J. *Six psychological studies*. New York: Random House, 1967.
Polansky, N. A. On the dynamics of behavioral contagion. *Group*, 1952, *14*, 3-8.
Proshansky, H. M. The development of inter-group attitudes. In M. L. Hoffman and L. W. Hoffman (Eds.), *Review of child development research*. Vol. 2. New York: Russell Sage Foundation, 1966.

Provence, S., and Lipton, R. C. *Infants in institutions.* New York: International Universities Press, 1962.
Prugh, D. G. Toward an understanding of psychosomatic concepts in relation to illness in children. In A. J. Solnit and S. A. Provence (Eds.), *Modern perspectives in child development*. New York: International Universities Press, 1963.
Prugh, D. G. et al. A study of the emotional reactions of children and families to hospitalization and illness. *American Journal of Orthopsychiatry*, 1953, *23*, 70-106.
Psychosomatic aspects of gastrointestinal illness in childhood. Report of the Forty-Fourth Ross Conference on Pediatric Research. Columbus, Ohio: Ross Laboratories, 1963.

Rabin, A. I. Some psychosexual differences between kibbutz and nonkibbutz Israeli boys. *Journal of Projective Techniques*, 1958, *22*, 328-332.
Rabin, A. I. *Growing up in the kibbutz*. New York: Springer, 1965.
Rank, B. Adaptation of the psychoanalytic technique for the treatment of young children with atypical development. *American Journal of Orthopsychiatry*, 1949, *19*, 130-139.
Redl, F. The impact of game-ingredients on children's play behavior. In B. Schaffner (Ed.), *Group processes*. Transactions of the fourth conference. New York: Josiah Macy, Jr. Foundation, 1958.
Redl, F., and Wineman, D. *Children who hate*. New York: Free Press, 1951.
Reichard, S., and Tillman, C. Patterns of parent-child relationships in schizophrenia. *Psychiatry*, 1950, *13*, 247-257
Rexford, E. N. (Ed.) *A developmental approach to problems of acting out: A symposium. Monograph of the Journal of the American Academy of Child Psychiatry*, No. 1. New York: International Universities Press, 1966.
Reynolds, M. M. Negativism of preschool children. *Teachers College Contributions to Education*. New York: Columbia University Press, 1928, No. 288.
Rheingold, H. L. The beginnings of social responsiveness in the human infant. In B. M. Foss (Ed.), *Determinants of infant behavior*. Vol. 1. New York: Wiley, 1961.
Rheingold, H. L. The effect of environmental stimulation upon social and exploratory behavior in the human infant. In B. M. Foss (Ed.), *Determinants of infant behavior*. Vol. 1. New York: Wiley, 1961.
Ribble, M. A. *The rights of infants*. New York: Columbia University Press, 1943.
Richardson, S. A., Hastorf, A. H., and Dornbusch, S. M. Effects of physical disability on a child's description of himself. *Child Development*, 1964, *35*, 893-907.
Richardson, S. A. et al. Cultural uniformity in reaction to physical disabilities. *American Sociological Review*, 1961, *26*, 241-247.
Richmond, J. B., and Lipton, E. L. Some aspects of the neurophysiology of the newborn and their implications for child development. In L. Jessner and E. Pavenstedt (Eds.), *Dynamic psychopathology in childhood*. New York: Grune and Stratton, 1959.
Robertson, J. Mothering as an influence on early development. *Psychoanalytic Study of the Child, 17*. New York: International Universities Press, 1962.

Robinson, F. P. Beyond vocational development. *Journal of Counseling Psychology*, 1965, *12*, 114.

Roe, A. *The psychology of occupations*. New York: Wiley, 1956.

Rogers, C. R. A theory of therapy, personality, and interpersonal relationships as developed in the client-centered framework. In S. Koch (Ed.), *Psychology: A study of a science*. Vol. 3. New York: McGraw-Hill, 1959.

Rogers, C. R. Toward a modern approach to values: The valuing process in the mature person. *Journal of Abnormal and Social Psychology*, 1964, *68*, 160-167.

Ross, D. C. A classification of child psychiatry. 4951 McKean Ave., Philadelphia, Pa., 1964.

Rotter, J. B. An analysis of Adlerian psychology from a research orientation. *Journal of Individual Psychology*, 1962, *18*, 3-11.

Rotter, J. B. Generalized expectancies for internal versus external control of reinforcement. *Psychological Monographs*, 1966, *80*, (Whole No. 609).

Ruttenberg, B. A., Dratman, M. L., Fraknoi, J., and Wenar, C. An instrument for evaluating autistic children. *Journal of the American Academy of Child Psychiatry*, 1966, *5*, 453-478.

Rutter, M. Concepts of autism: A review of research. *Journal of Child Psychology and Psychiatry*, 1968, *9*, 1-25.

Salzen, E. A. Visual stimuli eliciting the smiling response in the human infant. *Journal of Genetic Psychology*, 1963, *102*, 51-54.

Sampson, E. E. The study of ordinal position: Antecedents and outcomes. In B. A. Maher (Ed.), *Progress in experimental personality research*. Vol. 2. New York: Academic Press, 1965.

Sander, L. W. Issues in early mother-child interaction. *Journal of the American Academy of Child Psychiatry*, 1962, *1*, 141-166.

Sander, L. W. Adaptive relationships in early mother-child interaction. *Journal of the American Academy of Child Psychiatry*, 1964, *3*, 231-264.

Sanford, R. N. et al. Physique, personality, and scholarship: A cooperative study of school children. *Monographs of the Society for Research in Child Development*, 8 (1), 1943 (Serial No. 34).

Sarason, S. B. et al. *Anxiety in elementary school children*. New York: Wiley, 1960.

Sayegh, Y., and Dennis, W. The effect of supplementary experiences upon the behavioral development of infants in institutions. *Child Development*, 1965, *36*, 81-90.

Schaefer, E. S. A circumplex model for maternal behavior. *Journal of Abnormal and Social Psychology*, 1959, *59*, 226-235.

Schaefer, E. S., and Bayley, N. Maternal behavior, child behavior, and their intercorrelation from infancy through adolescence. *Monographs of the Society for Research in Child Development*, 1963, *28*, No. 3 (Serial No. 87).

Schaffer, H. R. Activity level as a constitutional determinant of infantile reaction to deprivation. *Child Development*, 1966, *37*, 595-602.

Schaffer, H. R., and Callander, W. M. Psychological effects of hospitalization in infancy. *Pediatrics*, 1959, *24*, 528-539.

Schaffer, H. R., and Emerson, P. E. The development of social attachments in infancy. *Monographs of the Society for Research in Child Development*, 1964, *29*, No. 3 (Serial No. 94).

Schaffer, H. R., and Emerson, P. E. Patterns of response to physical contact in early human development. *Journal of Child Psychology and Psychiatry*, 1964, *5*, 1-13.

Schein, E. H. The first job dilemma. *Psychology Today*, 1968, *1*, 26-37.
Sears, P. S. Levels of aspiration in academically successful and unsuccessful children. *Journal of Abnormal and Social Psychology*, 1940, *35*, 498-536.
Sears, R. R., Maccoby, E. E., and Levin, H. *Patterns of child rearing*. Evanston, Ill.: Row, Peterson, 1957.
Sears, R. R. et al. Some child-rearing antecedents of aggression and dependency in young children. *Genetic Psychology Monographs*, 1953, *47*, 135-234.
Selye, H. General adaptation syndrome and diseases of adaptation. *Journal of Clinical Endocrinology*, 1946, *6*, 117-128.
Sewell, M. Some causes of jealousy in young children. *Smith College Studies in Social Work*, 1930, *1*, 6-22.
Shakow, D. Experimental psychology. In R. R. Grinker (Ed.), *Midcentury psychiatry*. Springfield, Ill.: Charles C Thomas, 1953.
Sheldon, W. H. Constitutional factors in personality. In J. McV. Hunt (Ed.), *Personality and the behavior disorders*. Vol. 1. New York: Ronald Press, 1944.
Sherif, M., and Cantril, H. *The psychology of ego-involvements, social attitudes, and identifications*. New York: Wiley, 1947.
Sherif, M. et al. *Intergroup conflict and cooperation: The Robbers Cave experiment*. Norman, Okla.: University Book Exchange, 1961.
Sigel, I. E. The attainment of concepts. In M. L. Hoffman and L. W. Hoffman (Eds.), *Review of child development research*. Vol. 1. New York: Russell Sage Foundation, 1964.
Silber, E., et al., Adaptive behavior in competent adolescents. *Archives of General Psychiatry*, 1961, *5*, 354-364.
Silber, E. et al. Competent adolescents coping with college decisions. *Archives of General Psychiatry*, 1961, *5*, 517-527.
Simmel, M. L. Developmental aspects of the body schema. *Child Development*, 1966, *37*, 83-95.
Singer, J. L. Delayed gratification and ego development: Implications for clinical and experimental research. *Journal of Consulting Psychology*, 1955, *19*, 259-266.
Skeels, H. M., and Dye, H. A. A study of the effects of differential stimulation on mentally retarded children. *Proceedings of the American Association for the Mentally Deficient*, 1939, *44*, 114-136.
Slobin, D. I. The fruits of the first season: A discussion of the role of play in childhood. *Journal of Humanistic Psychology*, 1964, *4*, 59-79.
Smith, A. J. A developmental study of group processes. *Journal of Genetic Psychology*, 1960, *97*, 29-39.
Spitz, R. A. Hospitalism: An inquiry into the genesis of psychiatric conditions in early childhood. *Psychoanalytic Study of the Child, 1*. New York: International Universities Press, 1945.
Spitz, R. A. Hospitalism: A follow-up report on the investigation described in Volume 1, 1945. *Psychoanalytic Study of the Child, 2*. New York: International Universities Press, 1946.
Spitz, R. A. The psychogenic diseases in infancy: An attempt at their etiologic classification. *Psychoanalytic Study of the Child, 6*. New York: International Universities Press, 1951.
Spitz, R. A., and Wolf, K. M. The smiling response: A contribution to the ontogenesis of social relations. *Genetic Psychology Monographs*, 1946, *34*, 57-125.

Spitz, R. A., and Wolf, K. M. Anaclitic depression. *Psychoanalytic Study of the Child, 2*. New York: International Universities Press, 1946.
Spock, B. The striving for autonomy and regressive object relationships. *Psychoanalytic Study of the Child, 18*. New York: International Universities Press, 1963.
Starr, P. H. Psychosomatic considerations of diabetes in childhood. *Journal of Nervous and Mental Disease*, 1955, *121*, 493-504.
Stendler, C. B., Damrin, D., and Haines, A. C. Studies in cooperation and competition. I, The effects of working for group and individual rewards on the social climate of children's groups. *Journal of Genetic Psychology*, 1951, *79*, 173-197.
Stolz, H. R., and Stolz, L. M. Adolescent problems related to somatic variation. *Yearbook of the National Society for the Study of Education*, 1944, *1*, 80-99.
Stolz, L. M. Old and new directions in child development. *Merrill-Palmer Quarterly*, 1966, *12*, 221-232.
Stone, C. P., and Barker, R. G. The attitudes and interests of premenarcheal and postmenarcheal girls. *Journal of Genetic Psychology*, 1939, *54*, 27-71.
Stone, L. J., and Church, J. *Childhood and adolescence*. (2nd ed.) New York: Random House, 1968.
Stott, D. H. An empirical approach to motivation based on the behavior of a young child. *Journal of Child Psychology and Psychiatry*, 1961, *2*, 97-117.
Strodtbeck, F. L. Family interaction, values, and achievement. In D. C. McClelland et al. (Eds.), *Talent and society*. Princeton: Van Nostrand, 1958.
Sullivan, H. S. *The interpersonal theory of psychiatry*. New York: W. W. Norton, 1953.
Super, D. E. *The psychology of careers*. New York: Harper, 1957.
Super, D. E. et al. *Career development: Self-concept theory*. New York: College Entrance Examination Board, 1963.
Super, D. E., and Overstreet, P. L. *The vocational maturity of ninth-grade boys*. New York: Teachers College, Columbia University Press, 1960.
Sutton-Smith, B. Piaget on play: A critique. *Psychological Review*, 1966, *73*, 104-110.
Sutton-Smith, B., and Roberts, J. M. Rubrics of competitive behavior. *Journal of Genetic Psychology*, 1964, *105*, 13-37.
Swanson, G. E. Determinants of the individual's defenses against inner conflict: Review and reformulation. In J. C. Glidwell (Ed.), *Parental attitudes and child behavior.* Springfield, Ill.: Charles C Thomas, 1961.
Swift, J. W. Effects of early group experience: The nursery school and day nursery. In M. L. Hoffman and L. W. Hoffman (Eds.), *Review of child development research*. Vol. 1. New York: Russell Sage Foundation, 1964.
Symonds, P. M. *Adolescent fantasy: An investigation of the picture-story method of personality study.* New York: Columbia University Press, 1949.

Tasch, R. J. Interpersonal perceptions of fathers and mothers. *Journal of Genetic Psychology*, 1955, *87*, 59-65.
Thomas, A., Chess, S., and Birch, H. G. *Temperament and behavior disorders in children*. New York: New York University Press, 1968.
Thomas, A. et al. *Behavioral individuality in early childhood*. New York: New York University Press, 1963.
Thomas, N. *Life*, (Jan 14, 1966) Vol. 60, No. 2.
Thompson, G. G. The social and emotional development of preschool children under two types of educational program. *Psychological Monographs*, 1944, *56*, (5, Whole No. 5).

Thompson, G. G. Children's groups. In P. H. Mussen (Ed.), *Handbook of research methods in child development*. New York: Wiley, 1960.

Thompson, G. G. *Child psychology*. (2nd ed.) Boston: Houghton Mifflin, 1962.

Thompson, G. G., and Horrocks, J. E. A study of the friendship fluctuations of urban boys and girls. *Journal of Genetic Psychology*, 1947, *70*, 53-63.

Thompson, G. G. and Hunnicutt, C. W. The effects of repeated praise or blame on the work achievement of "introverts" and "extroverts." *Journal of Educational Psychology*, 1944, *35*, 257-266.

Tiedeman, D. V., and O'Hara, R. P. *Career development: Choice and adjustment*. New York: College Entrance Examination Board, 1963.

Tryon, C. M. Evaluations of adolescent personality by adolescents. *Monographs of the Society for Research in Child Development*, 1939, *4*, No. 4 (Serial No. 23).

Tuddenham, R. D. Studies in reputation: III. Correlates of popularity among elementary school children. *Journal of Educational Psychology*, 1951, *42*, 257-276.

Tuddenham, R. D. Studies in reputation: I. Sex and grade differences in school children's evaluations of peers, II. The diagnosis of social adjustment. *Psychological Monographs*, 1952, *66*, (1, Whole No. 333).

Tyler, L. E. The relationship of interests to abilities and reputation among first-grade children. *Educational and Psychological Measurement*, 1951, *11*, 255-264.

Tyler, L. E. The development of "vocational interests." I. The organization of likes and dislikes in ten-year-old children. *Journal of Genetic Psychology*, 1955, *86*, 33-44.

Tyler, L. E. The antecedents of two varieties of vocational interests. *Genetic Psychology Monographs*, 1964, *70*, 177-227.

Tyler, L. E., and Sundberg, N. D. *Factors affecting career choice of adolescents*. Cooperative Research Project No. 2455. Eugene, Oregon: University of Oregon Press, 1964.

Valentine, C. W. The innate basis of fear. *Journal of Genetic Psychology*, 1930, *37*, 394-419.

Wälder, R. The psychoanalytic theory of play. *Psychoanalytic Quarterly*, 1933, *2*, 208-224.

Walker, R. N. Body build and behavior in young children: I. Body build and nursery school teachers' ratings. *Monographs of the Society for Research in Child Development*, 1962, *27*, No. 3 (Serial No. 84).

Wallach, M. A., and Kogan, N. *Modes of thinking in young children*. New York: Holt, Rinehart and Winston, 1965.

Walters, R. H., and Parke, R. D. The role of the distance receptors in the development of social responsiveness. In L. P. Lipsitt and C. C. Spiker (Eds.), *Advances in child development and behavior*. Vol. 2. New York: Academic Press, 1965.

Watson, E. J., and Johnson, A. M. The emotional significance of acquired physical disfigurement in children. *American Journal of Orthopsychiatry*, 1958, *28*, 85-97.

Wenar, C. The therapeutic value of setting limits with inhibited children. *Journal of Nervous and Mental Disease*, 1957, *125*, 390-395.

Wenar, C., Handlon, M. W., and Garner, A. M. *Origins of psychosomatic and emotional disturbances*. A psychosomatic medicine monograph. New York: P. B. Hoeber, 1962.

Wenar C., and Wenar, S. C. The short term prospective model, the illusion of time, and the tabula rasa child. *Child Development*, 1963, *34*, 697-708.
Wender, P. H., Pedersen, F. A., and Waldrop, M. F. A longitudinal study of early social behavior and cognitive development. *American Journal of Orthopsychiatry*, 1967, *37*, 691-696.
Wermer, H., and Levin, S. Masturbation fantasies: Their change with growth and development. *The Psychoanalytic Study of the Child, 22*. New York: International Universities Press, 1967.
Werner, H. *Comparative psychology of mental development*. (Rev. ed.) New York: International Universities Press, 1957.
White, R. W. Motivation reconsidered: The concept of competence. *Psychological Review*, 1959, *66*, 297-333.
White R. W. Competence and the psychosexual stages of development. In M. R. Jones (Ed.), *Nebraska symposium on motivation*. Lincoln, Nebraska: University of Nebraska Press, 1960.
White, R. W. *The abnormal personality*. (3rd ed.) New York: Ronald Press, 1964.
White, R. W. The experience of efficacy in schizophrenia. *Psychiatry*, 1965, *28*, 199-211.
Wilensky, H. L. Varieties of work experience. In H. Borow (Ed.), *Man in a world at work*. Boston: Houghton Mifflin, 1964.
Winnicott, D. W. *Mother and child: A primer of first relationships*. New York: Basic Books, 1957.
Witkin, H. A., et al. *Psychological differentiation*. New York: Wiley, 1962.
Wolff, P. H. The developmental psychologies of Jean Piaget and psychoanalysis. *Psychological Issues*, Monograph 5, Vol. 2, No. 1. New York: International Universities Press, 1960.
Wolff, P. H. Observations on the early development of smiling. In B. M. Foss (Ed.), *Determinants of infant behavior*. Vol. 2. New York: Wiley, 1963.
Wolff, P. H. The development of attention in young infants. *Annals of the New York Academy of Science*, 1965, *118*, 815-830.
Wolff, P. H., and White, B. L. Visual pursuit and attention in young infants. *Journal of the American Academy of Child Psychiatry*, 1965, *4*, 473-484.
Wolff, W. *The personality of the preschool child*. New York: Grune and Stratton, 1946.
Wolfle, D. L., and Wolfle, H. M. The development of cooperative behavior in monkeys and young children. *Journal of Genetic Psychology*, 1939, *55*, 137-175.
Woltmann, A. G. Concepts of play therapy techniques. In M. R. Haworth (Ed.), *Child psychotherapy*. New York: Basic Books, 1964.
Woodcock, L. P. *Life and ways of the two-year-old*. New York: Basic Books, 1941.
Woodworth, R. S. *Dynamics of behavior*. New York: Holt, 1958.
World Health Organization. *Deprivation of maternal care*. Public Health Paper No. 14. Geneva, 1962.
Wright, B. A. *Physical disability: A psychological approach*. New York: Harper and Row, 1960.
Wright, M. E. The influence of frustration upon the social relations of young children. *Character and Personality*, 1943, *12*, 111-122.

Yarrow, L. J. Maternal deprivation: Toward an empirical and conceptual re-evaluation. *Psychological Bulletin*, 1961, *58*, 459-490.

Yarrow, L. J. Research in dimensions of early maternal care. *Merrill-Palmer Quarterly*, 1963, *9*, 101-114.

Yarrow, L. J. Separation from parents during early childhood. In M. L. Hoffman and L. W. Hoffman (Eds.), *Review of child development research*. Vol. 1. New York: Russell Sage Foundation, 1964.

Yerbury, E. C., and Newell, N. Genetic and environmental factors in psychoses of children. *American Journal of Psychiatry*, 1943-44, *100*, 599-605.

Author Index

Subject Index